Library of
Davidson College

Defects and Their Structure in Nonmetallic Solids

NATO ADVANCED STUDY INSTITUTES SERIES

A series of edited volumes comprising multifaceted studies of contemporary scientific issues by some of the best scientific minds in the world, assembled in cooperation with NATO Scientific Affairs Division.

Series B: Physics

RECENT VOLUMES IN THIS SERIES

Volume 10 – Progress in Electro-Optics
edited by Ezio Camatini

Volume 11 – Fluctuations, Instabilities, and Phase Transitions
edited by Tormod Riste

Volume 12 – Spectroscopy of the Excited State
edited by Baldassare Di Bartolo

Volume 13 – Weak and Electromagnetic Interactions at High Energies
(Parts A and B)
edited by Maurice Lévy, Jean-Louis Basdevant,
David Speiser, and Raymond Gastmans

Volume 14 – Physics of Nonmetallic Thin Films
edited by C.H.S. Dupuy and A. Cachard

Volume 15 – Nuclear and Particle Physics at Intermediate Energies
edited by J. B. Warren

Volume 16 – Electronic Structure and Reactivity of Metal Surfaces
edited by E. G. Derouane and A. A. Lucas

Volume 17 – Linear and Nonlinear Electron Transport in Solids
edited by J. T. Devreese and V. E. van Doren

Volume 18 – Photoionization and Other Probes of Many-Electron Interactions
edited by F. J. Wuilleumier

Volume 19 – Defects and Their Structure in Nonmetallic Solids
edited by B. Henderson and A. E. Hughes

Volume 20 – Physics of Structurally Disordered Solids
edited by Shashanka S. Mitra

The series is published by an international board of publishers in conjunction with NATO Scientific Affairs Division

A	Life Sciences	Plenum Publishing Corporation
B	Physics	New York and London
C	Mathematical and Physical Sciences	D. Reidel Publishing Company Dordrecht and Boston
D	Behavioral and Social Sciences	Sijthoff International Publishing Company Leiden
E	Applied Sciences	Noordhoff International Publishing Leiden

Defects and Their Structure in Nonmetallic Solids

Edited by
B. Henderson
Trinity College
Dublin, Ireland

and

A. E. Hughes
Atomic Energy Research Establishment
Harwell, England

PLENUM PRESS • NEW YORK AND LONDON
Published in cooperation with NATO Scientific Affairs Division

Library of Congress Cataloging in Publication Data

Nato Advanced Study Institute, University of Exeter, 1975.
  Defects and their structure in nonmetallic solids.

(NATO advanced study institutes series: Series B, Physics; v. 19)
"Published in cooperation with NATO Scientific Affairs Division."
Includes index.
  1. Semiconductors—Defects—Congresses. 2. Electric insulators and insulation—Defects—Congresses. 3. Point defects—Congresses. 4. Crystals—Defects—Congresses. I. Henderson, Brian. II. Hughes, Antony Elwyn. III. Series.
QC611.6.D4N37 1975   537.6'22   76-16280
ISBN 0-306-35719-4

Proceedings of a NATO Advanced Study Institute held at the
University of Exeter, August 24–September 6, 1975

© 1976 Plenum Press, New York
A Division of Plenum Publishing Corporation
227 West 17th Street, New York, N.Y. 10011

All rights reserved

No part of this book may be reproduced, stored in a retrieval system, or transmitted, in any form or by any means, electronic, mechanical, photocopying, microfilming, recording, or otherwise, without written permission from the Publisher

Printed in the United States of America

CONTRIBUTORS

K. H. G. ASHBEE — *H. H. Wills Physics Laboratory, University of Bristol, Bristol, England.*

R. DE BATIST — *S. C. K./C. E. N., Mol, Belgium.*

P. DOBSON — *Department of Physical Metallurgy, University of Birmingham, Birmingham, England.*

D. L. GRISCOM — *Naval Research Laboratories, Washington D.C., U.S.A.*

L. W. HOBBS — *Materials Development Division, A.E.R.E., Harwell, England.*

A. E. HUGHES — *Materials Development Division, A.E.R.E., Harwell, England.*

A. B. LIDIARD — *Theoretical Physics Division, A.E.R.E., Harwell, England.*

Y. MERLE D'AUBIGNÉ — *Laboratoire de Spectrométrie Physique, Université Scientifique et Médicale de Grenoble, France.*

R. C. NEWMAN — *Physics Department, University of Reading, Reading, England.*

H. PEISL — *Sektion Physik, Ludwig-Maximilians Universität, Munich, Germany.*

D. SCHOEMAKER — *Physics Department, University of Antwerp, Wilrijk, Belgium.*

W. A. SIBLEY — *Physics Department, Oklahoma State University, Stillwater, Oklahoma.*

J. M. SPAETH — *Experimentalphysik, Gesamthochschule Paderborn, Germany.*

G. SPINOLO — *Instituto di Scienze Fisiche, Università di Milano, Milan, Italy.*

A. M. STONEHAM — *Theoretical Physics Division, A.E.R.E., Harwell, England.*

W. VON DER OSTEN — *Experimentalphysik, Gesamthochschule Paderborn, Germany.*

G. D. WATKINS — *Physics Department, Lehigh University, Bethlehem, Pennsylvania.*

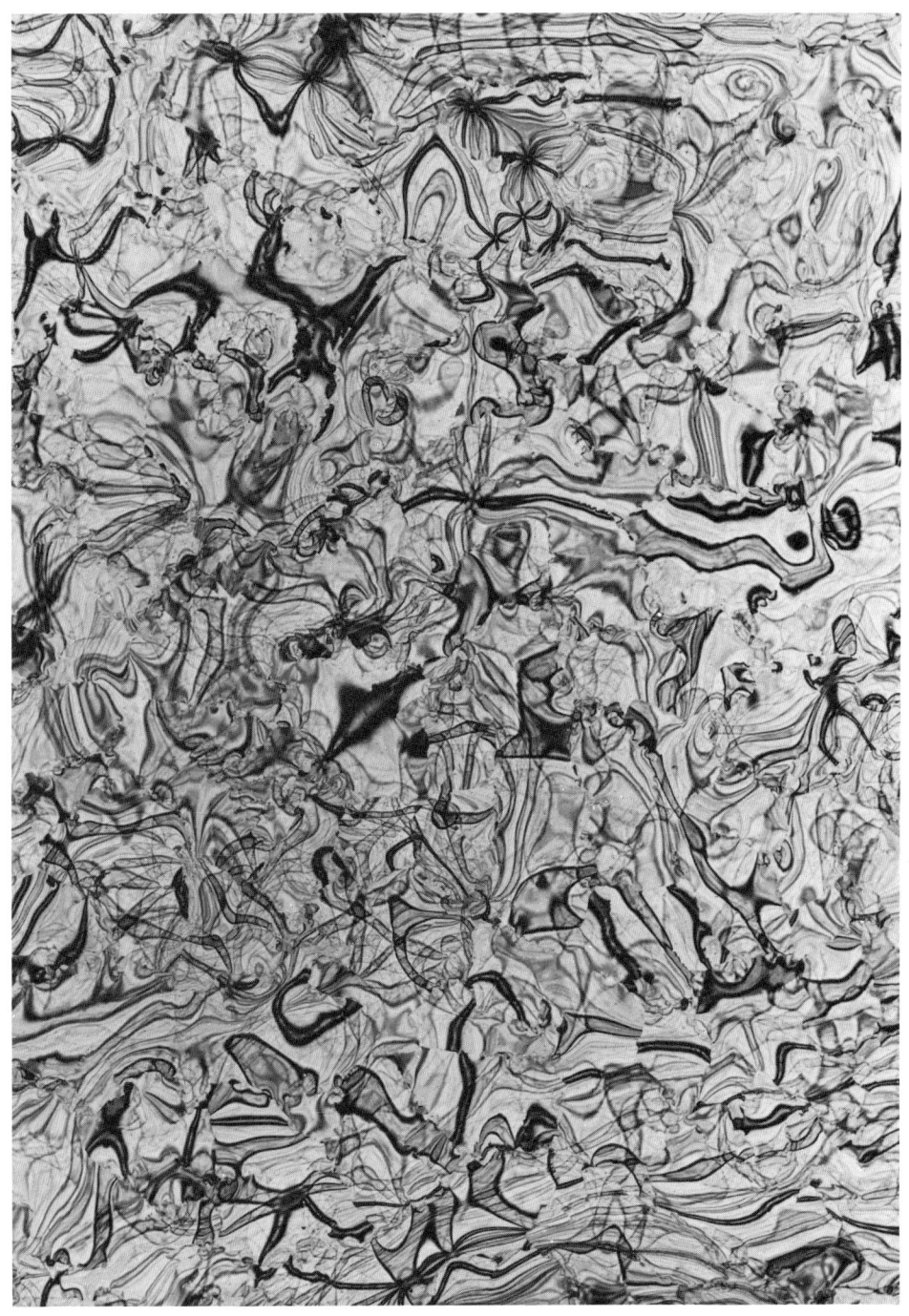

Bend Contours in an Anodic Film of Tantalum Pentoxide
(after R. E. Pawel)

Preface

The Advanced Study Institute of which this volume is the proceedings was held at the University of Exeter during 24 August to 6 September 1975. There were seventy participants of whom eighteen were lecturers and members of the advisory committee. All NATO countries except Holland, Iceland and Portugal were represented. In addition a small number of participants came from non-NATO countries Japan, Ireland and Switzerland.

An aim of the organising committee was to bring together scientists of wide interests and expertise in the defect structure of insulators and semiconductors. Thus major emphases in the programme concerned the use of spectroscopy and microscopy in revealing the structure of point defects and their aggregates, line defects as well as planar and volume defects. The lectures revealed that in general little is known of the fate of the interstitial in most irradiated solids. Nor are the dynamic properties of defects understood in sufficient detail that one can state how point defects cluster and eventually become macroscopic defects.

Although this book faithfully reproduces the material covered by the invited speakers, it does not really follow the flow of the lectures. This is because it seemed advisable for each lecturer to provide a single self-contained and authoritative manuscript, rather than a series of short articles corresponding to the lectures. These manuscripts are arranged in three parts. Part A contains material relating to the fundamental point defects in crystalline non-metallic solids, the structure of which are revealed largely by spectroscopy. In Part B are discussed defects in low symmetry systems such as are found in vitreous solids, on surfaces and in the neighbourhood of dislocation lines. Finally Part C reviews X-ray and electron diffraction studies of point defects and a wide variety of macroscopic defects.

In addition to the invited lectures a number of contributed papers were read during the numerous discussion periods: the titles

only of these are collected in Appendix I. No attempt is made to
report the content of the discussions of either invited or contributed papers. Appendix II is a list of the more general references
compiled by the invited speakers for the benefit of the other
participants.

        B. Henderson
               Co-directors
      A.E. Hughes

Contents

Part I
POINT DEFECTS IN INSULATORS AND SEMICONDUCTORS

Point Defects in Non-Metals: The Role of
       Structure 1
  A. B. Lidiard

Point Defects and Diffusion 95
  P. S. Dobson

From Spectroscopy to Microscopy - The Photon
       as a Probe 107
  W. A. Sibley

Optical Techniques and an Introduction to the
       Symmetry Properties of Point Defects 133
  A. E. Hughes

Magnetic Resonance Studies of Vacancy Centers
       in Ionic Crystals 155
  J. M. Spaeth

Interstitial Centres: Optical Absorption
       and Magnetic Resonance 173
  D. Shoemaker

EPR Studies of Lattice Defects in Semiconductors 203
  G. D. Watkins

Infrared Studies of Defects 221
  R. C. Newman

Vacancy Aggregate Centers in Ionic Crystals 237
  W. von der Osten

Perturbation Spectroscopy and Optical Detection
       of Paramagnetic Resonance 261
  Y. Merle d'Aubigne

Excited State Properties of Localized Centres 283
  G. M. Spinolo

Part II
DEFECTS IN LOW SYMMETRY ENVIRONMENTS

Internal Friction and Defects Near Dislocations 305
  R. De Batist

Defects in Non-Crystalline Oxides 323
  D. L. Griscom

Insulator and Semiconductor Surfaces 355
  A. M. Stoneham

Part III
DIFFRACTION TECHNIQUES FOR POINT AND EXTENDED DEFECTS

Structural Information and Defect Energies
       Studied by X-Ray Methods 381
  H. Peisl

X-Rays and Electron Microscopy 407
  K. H. G. Ashbee

Transmission Electron Microscopy of Defect
       Aggregates in Non-Metallic
       Crystalline Solids 431
  L. W. Hobbs

Appendix I. Contributed Papers 483

Appendix II. References – Defects in Non-Metallic
       Solids 485

Appendix III. Participants at NATO Advanced Study
       Institute, Exeter, 1975 491

Index 495

Part I

Point Defects in Insulators and Semiconductors

POINT DEFECTS IN NON-METALS: THE ROLE OF STRUCTURE

A. B. Lidiard

Theoretical Physics Division

A.E.R.E., Harwell, Oxon OX11 ORA

1. INTRODUCTION

The aim of this volume is to review the properties of defects in non-metals, particularly their structural properties. Apart from impurities, such defects are, of course, only the atoms of the solid arranged in a different way (and sometimes also in different states of ionisation). In general, therefore we might expect to see the differences between different classes of solids (metals, ionic solids, covalent semiconductors, etc.) reflected in the properties of the defects in these solids. Often we can, though it is the manifestations of the defects which frequently provide the most obvious differentiation. This is imply another way of saying that one often studies the properties of the defects through their effect on the most characteristic properties of the solid; e.g. in semiconductors through their effect on electrical properties, in wide-gap ionic materials through their effect on optical properties, and so on. Such differentiation is not particularly fundamental. Indeed when we look at lattice defects at the more fundamental level of atomic arrangement, electronic states, energy levels and so forth such differences between one class of substances and another may well disappear. Thus in the above example both the electrical effects in semiconductors and the optical effects in wide-gap ionic solids derive from the fact that defects may possess localised electronic states in the 'forbidden gap' between valence and conduction states, which thus allow the defect to act as donor and/or acceptor or as a colour centre according to the circumstances. At this level of description it becomes more difficult to see the characteristic differences between defects in semiconductors, in ionic solids, in molecular solids, etc. Indeed, it is the fact that these differences are small which allows us to define defect solid state physics as a distinct subject. We can define the same sort of entities - vacancies, interstitials, dislocations, etc. - in all classes of crystalline solid. We do not have one sort of defect in semiconductors, another sort in ionic crystals and so on. Such remarks may seem rather unnecessary today, although 25 years ago the unity of the subject was only just emerging (1).

This means that the differences between defects in the different types of solid often emerge at a secondary level of theory; this will be clear from many of the lectures. In the present part we have to deal with two general aspects of the theory of point defects and their aggregates namely (i) the numbers of defects arising in conditions of thermodynamic equilibrium or near to it, defect mobilities, their relation to macroscopic effects such as ionic conductivity, diffusion and certain relaxation processes, etc. (2) and (ii) the theory of the energies and other characteristic properties of individual defects. We shall do so in such a way as to bring out the dependence on the structure - both crystal structure and electronic structure - of the host crystal. We deal with the thermodynamic and kinetic aspects in Sections 2-5 and the atomic aspects in Sections 6 and 7. Useful comprehensive general references will be found in Refs. 2-7.

2. CONCENTRATIONS OF DEFECTS IN THERMAL EQUILIBRIUM

In this Section we shall review the concentrations of point defects and their aggregates as they arise thermally and from changes in chemical potential (including 'doping'). We deal first with the simplest case; in the present context of non-metallic solids this means molecular solids whose crystal structure has only one atom per unit cell, though the calculation is probably more familiar as an introduction to the study of point defects in metals. The important restriction is not the type of bonding but the fact that we are dealing with only one type of vacancy - or one type of interstitial - which can be present in only one charge state. We then consider the generalisation of the calculation to compounds and to other cases where more than one type of basic defect must be allowed for. We then go on to discuss briefly the formation of complex defects, non-stoichiometric oxides and solids which show high degrees of disorder in one sub-lattice. More extensive discussions of this material may be found in standard texts (2-5,7).

2.1 Concentrations of Defects in Thermal Equilibrium - a Simple Calculation

Though the elementary calculation of the thermal equilibrium concentration of vacancies and interstitials in a monatomic solid having one atom per unit cell is well known, we shall first repeat it here in order to bring out more clearly the differences which arise with ionic and covalent compounds and in other more complex situations. Consider then the thermal production of vacancies in a single neutral charge state either in a monatomic substance (such as a metal or a solid rare-gas) or in a molecular solid in which every molecular lattice site is equivalent*.

Let there be N atoms and n vacancies, i.e. N + n lattice sites. For equilibrium at constant pressure, P, and temperature, T, we want the Gibbs function $G(P,T)$ which is minimal. If interactions among the vacancies can be ignored then we can write $G(P,T)$ as

$$G(P,T) = G_o(P,T) + ng_v - kT \ln \Omega , \quad (2.1)$$

in which $G_o(P,T)$ is the Gibbs function for a perfect crystal of N atoms, g_v is the change in the Gibbs function on creating one vacancy at a particular but arbitrary site, while Ω is the number of accessible configurations of the defect solid containing n vacancies, i.e. it is the number of ways of arranging n vacancies on N + n sites. Thus

$$G(P,T) = G_o(P,T) + ng_v - kT \ln \frac{(N+n)!}{N!n!} . \quad (2.2)$$

The thermal equilibrium concentration of vacancies is always that which minimises G at each (P,T), i.e.

$$\delta G = 0 \quad (2.3)$$

or, in the present case,

$$\mu_v \equiv \left(\frac{\partial G}{\partial n}\right)_{P,T} = 0 . \quad (2.4)$$

By analogy to material particles we may call μ_v the chemical potential of the vacancies. By (2.2)

$$\mu_v = g_v + kT \ln (n/N+n)$$

$$\equiv g_v + kT \ln [v] , \quad (2.5)$$

---

*For the convenience of avoiding writing 'atoms or molecules', etc. we shall simply refer always to 'atoms' and to the substance as 'monatomic'. The calculation applies equally well to molecular solids, however, as long as the restriction to one molecule per unit cell is retained. For reviews of defects in molecular solids see, e.g. Flynn (2) or Chadwick and Sherwood (8).

where we use square brackets to denote atomic or mole fraction. Thus the equilibrium fraction of vacancies in this system is

$$[v] = e^{-g_v/kT} . \quad (2.6)$$

The free energy of vacancy formation, g_v may be written in terms of the entropy and enthalpy of vacancy formation in the usual way,

$$g_v = h_v - Ts_v$$
$$= u_v + Pu_v - Ts_v . \quad (2.7)$$

At normal atmospheric pressure the difference between h_v and u_v is almost always quite insignificant and we shall therefore use the terms defect enthalpies and defect energies (meaning internal energies) synonymously. In high pressure experiments by contrast the difference is significant, but we shall have little to say about these here. The entropy is normally significant, and the distinction between g and h must therefore be preserved.

We list a number of observations on this basic calculation.

(i) A very similar calculation may be made for interstitial defects, though in this case it is useful to go to the more general expression for the chemical potential of a defect namely

$$\mu_i = g_i + kT \ln (\gamma_i [i]) , \quad (2.8)$$

where γ_i is the activity coefficient. For the simple interstitials γ_i is a number deriving from (a) the ratio of possible interstitial sites to normal lattice sites and (b) the number of equivalent interstitial configurations associated with one site (e.g. the 3 equivalent <100> orientations of a dumbell interstitial in a f.c.c. lattice). It can easily be worked out for each case as appropriate by enumerating Ω and hence $G(P,T)$. The equilibrium concentration of interstitials is obtained by setting

$$\mu_i = 0 , \quad (2.9)$$

whence an expression for $[i]$ is obtained very similar to (2.6)

$$[i] = \gamma_i^{-1} e^{-g_i/kT} . \quad (2.10)$$

(ii) Since $g_i, g_v \gg kT$ the differences between g_i and g_v (effectively the differences between h_i and h_v) are overwhelmingly important in determining which is the 'majority defect'. In practice, it is often quite difficult to obtain accurate information even about the majority defects; minority defects can mostly only be studied when introduced in excess, non-equilibrium concentrations, e.g. by irradiation. Alternatively we can often obtain reasonable guidance from theoretical evaluations of the energies of formation and other properties. Though some uncertainties still exist in such calculations, they consistently predict $h_i \sim 3 h_v$ for f.c.c. rare-gas solids and for many f.c.c. and b.c.c. metals (9). In these materials then vacancies will be the dominant intrinsic defects and the concentration of interstitials will be completely negligible in thermal equilibrium. This seems likely to be the case in many molecular solids too (8).

(iii) Direct experimental information about defect type and equilibrium concentration is generally rather difficult to come by. An important method, but equally one of the most challenging experimentally, is the Simmons-Balluffi experiment, i.e. the simultaneous measurement of macroscopic expansion ($\Delta L/L$) and X-ray lattice parameter expansion ($\Delta a/a$). The difference yields the excess of vacancy defects (including small clusters) over interstitial defects

$$3 \left(\frac{\Delta L}{L} - \frac{\Delta a}{a} \right) = [v] - [i] . \qquad (2.11)$$

This method has been successfully used for Ne and Kr and a number of metals and alloys and confirms the dominance of vacancy defects in these systems. Other valuable direct methods, which have proved particularly useful for metals, include (a) the quenching-in of the high temperature defect population by very rapid cooling of thin wires and foils and (b) the measurement of the rates of annihilation of thermalised positrons (which become trapped at vacancies). These two techniques confirm the dominance of vacancies and give reliable values for vacancy formation energies.

(iv) By these and a variety of other detailed studies we find that h_v in the rare gas solids is about 2/3 of the cohesive energy (though in Ne it is closely equal to the cohesive energy). In metals, the relative ease with which the valence electrons can redistribute themselves around the defect and thereby lower the energy results in values of h_v of only about 1/3rd of the cohesive energy. Though diffusion has been measured in many molecular solids, the values of the vacancy formation energies remain rather uncertain. There seems no doubt however that vacancy disorder dominates in most cases (8).

2.2 Defects in Compound Solids - General

In the preceeding calculation we only considered one type of defect at a time. In compounds this is no longer sensible. For example, in a compound MX it is not possible at constant composition and structure to create vacancies in the M sub-lattice without simultaneously creating either M interstitials or X vacancies. Also for 'reactions' among defects we must obviously consider the simultaneous presence of different defects. In this section we indicate the formal changes which are required for these cases.

The fundamental thermodynamic principle by which the thermal equilibrium concentrations are determined is the same as before, namely that for equilibrium at constant (P,T) the Gibbs function $G(P,T; \ldots n_j \ldots)$ must be minimal with respect to variations in the defect concentrations, n_j. The first step therefore must be the specification of the defect types which are thought to be important. The next step in principle is to evaluate G using the methods of statistical mechanics. This must generally be done for each system as required, though many systems differ only in minor details. The task is usually straightforward for dilute systems without interactions among the defects or with only short-range interactions. For the more difficult cases of concentrated systems or of systems with long-range defect interactions (e.g. Coulombic) one may use some of the well-known approximations of statistical mechanics (e.g. mean field approximation, Debye-Hückel theory, etc.) though to go beyond these can be quite difficult[*]. Although it would be a major task to review the whole body of relevant statistical calculations, we can nevertheless see the essential thermodynamic structure of the equilibrium equations they lead to without difficulty. The reason is the close analogy between defects and reacting chemical species.

Minimisation of G at constant (P,T) thus means setting

$$\delta G = \sum_j \left(\frac{\partial G}{\partial n_j}\right)_{T,P,n_\ell (\ell \neq j)} \delta n_j$$

$$\equiv \sum_j \mu_j \delta n_j = 0 \, , \qquad (2.12)$$

where we define

$$\mu_j = \left(\frac{\partial G}{\partial n_j}\right)_{T,P,n_\ell (\ell \neq j)} \, , \qquad (2.13)$$

as the chemical potential of defect j. This minimum of G must generally be found subject to certain constraints namely those of (a) the given crystal lattice structure (b) conservation of chemical species and (c) bulk electro-neutrality. These constraints can all be expressed by equations, linear in the n_j; thus

---

[*]For examples of discussions in higher approximations see e.g. Schapink (10) for metal alloys and Allnatt and Loftus (11) for ionic crystals.

POINT DEFECTS IN NON-METALS

$$\sum_j \alpha_j n_j = 0$$

$$\sum_j \beta_j n_j = 0 \qquad (2.14)$$

$$\sum_j \gamma_j n_j = 0 \ .$$

By the method of Lagrange multipliers the equilibrium equation (2.12) thus yields

$$\mu_j + \alpha_j \xi + \beta_j \eta + \gamma_j \zeta = 0 \ , \quad \text{(all j)} \qquad (2.15)$$

where the three Lagrange multipliers ξ, η, ζ, are to be found by insertion into (2.14). This set of equations is the generalisation of (2.4).

However, we observe that for any <u>defect reaction</u> expressible as

$$\sum_j (\nu_j \text{ of defect } j) = 0 \ , \qquad (2.16)$$

(e.g. $A + B \rightleftharpoons AB$ is of this form when $\nu_A = \nu_B = 1$ and $\nu_{AB} = -1$) which satisfies all three constraints (2.14) we obtain

$$\sum_j \nu_j \mu_j = 0 \ , \qquad (2.17)$$

from (2.15) by multiplying by ν_j and summing over all j. Eqn. (2.17), in fact, is just the condition of quasi-chemical equilibrium for the reaction (2.16). Lastly we express the chemical potentials in the form

$$\mu_j = g_j(P,T) + kT \ln f_j$$

$$\equiv g_j(P,T) + kT \ln(\gamma_j [j]) \ , \qquad (2.18)$$

in which

$g_j(P,T)$ = Gibbs free energy of formation of a single defect j (relative to some standard state),

f_j = the activity of species j,

γ_j = the activity coefficient of species j,

$[j]$ = the mole fraction of species, j.

When we do this eqn. (2.17) yields the corresponding mass action equation. We make the following observations on these results.

(i) The explicit form of (2.18) in principle is to be obtained by differentiating the expression for G, according to (2.13). The activity coefficients are then functions of defect structure and of the mutual interactions among the defects (and thus of their concentration).

(ii) Though the form (2.18) is general, the similarity to (2.5) and (2.8) is plain. Indeed in very many circumstances of interest (e.g. low defect concentrations) the activity coefficients γ_j are either independent of defect concentration or only slowly varying functions of it, with the result that the mass-action laws which follow (2.17) and (2.18) are dominated by the concentration products. This fact is extremely useful and has been widely exploited, though its limitations must be recognised (see e.g. Kroger, 3). One of the more obvious exceptions is provided by the carriers (electrons or holes regarded as 'defects') in a degenerate semiconductor, where the chemical potential (Fermi energy) has a quite different dependence upon carrier concentration than is suggested by (2.18).

(iii) The same formalism allows us to include additional phases, as is necessary when we wish to discuss departures from stoichiometry. For a gaseous phase in the ideal limit it is convenient to write

$$\mu(P,T) = \mu_o(P,T) + kT \ln P \, , \qquad (2.19)$$

as the chemical potential of the gaseous molecules. The effects of non-ideality are generally quite negligible for the pressures and temperatures of interest here.

(iv) One of the most obvious physical consequences of interactions among defects is the formation of complex defects. We treat these as distinct entities in the above formalism, which is then a very convenient way to describe the dependence of their concentration upon P,T and the concentrations of other defects. Common examples of such complex defects are: (a) the pairing of divalent cations with cation vacancies in alkali and silver halides (11,12), (b) the pairing of Group V elements with vacancies in Si and Ge (E-centres; Dobson and Watkins, this volume), (c) the pairing of interstitial anions with trivalent rare-earth ions in alkaline earth fluorides (7,13). From the point of view purely of statistical thermodynamics, i.e. of statistical averaging, the division between complex defects and 'uncomplexed' or 'unassociated' defects is somewhat arbitrary, though, as in the theory of ion pairing in weak electrolytes, one may define an optimal division. The common-sense approach of regarding as distinct species those

POINT DEFECTS IN NON-METALS

defects whose individual properties one may expect to study by other means (see e.g. §§4,5) however carries us a long way and is also reflected in the better rate of convergence of statistical expansions which separate out the complex defects in the first place (11).

We now consider several particular systems in greater detail.

2.3 Group IV Semiconductors (Diamond, Si, Ge)

Although these substances are monatomic they do not fall within the class of substances considered in §2.1, because vacancies and interstitials (and their simple clusters) can exist in several states of ionisation depending upon the position of the Fermi level (Fig. 2.1). Each of these should be considered as a separate defect since their properties differ. However, it will be immediately seen that vacancies and interstitials in the neutral charge state do not enter into any of the constraining eqns. (2.14) i.e. all of α, β and γ are zero for these defects in these monatomic substances. Thus the equilibrium fraction of such vacancies V^o is given by

$$\mu(V^o) = 0 , \qquad (2.20)$$

or

$$[V^o] = \gamma^{-1}(V^o) \ e^{-g(V^o)/kT} \qquad (2.21)$$

In this case the activity coefficient is just the inverse of the

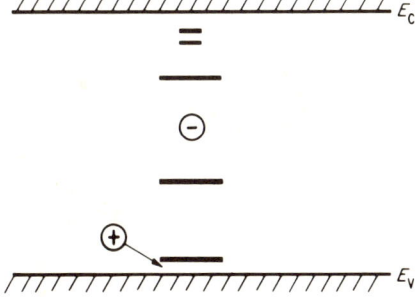

Fig. 2.1 Experimentally determined electrical level structure of the vacancy in Si; the charge state depends on the position of the Fermi level as indicated. (Watkins, this Volume).

partition sum over the various electronic states and orientations (cf. Jahn-Teller distortions) of the neutral vacancy*, while $g(V^o)$ is the free energy of formation of a single vacancy (in a particular but arbitrary site and orientation) in its ground state. Otherwise expressed,

$$[V^o] = \sum_\alpha \omega_\alpha \, e^{-g_\alpha(V^o)/kT} , \qquad (2.22)$$

where the sum is over all states and orientations, α, of V^o; and similarly for interstitials. In diamond thermally accessible excited states of V^o are known to exist (14). In Si, the ground state appears to have spin zero with the possibility of three equivalent <100> Jahn-Teller distorted states; no excited states have yet been identified (Watkins, this volume).

The important consequence of (2.20) and (2.21) is that $[V^o]$ is a function only of T and P. It does not depend upon the presence of donors or acceptors. However the concentration of vacancies in their electrically charged states is strongly influenced by doping. Consider the reaction

$$V^o + e^- \rightleftharpoons V^- , \qquad (2.23)$$

where e^- denotes an electron in the conduction band. Assuming that the electron distribution is non-degenerate, we then obtain from (2.18)

$$\frac{[V^-]}{[V^o]} = [e^-] \left(\frac{\gamma(V^o) \, \gamma(e^-)}{\gamma(V^-)} \right) e^{\Delta g/kT} , \qquad (2.24)$$

where Δg is the free energy gain when the reaction (2.23) goes to the right. The concentration of V^- is thus (approximately) proportional to the concentration of <u>free</u> electrons, i.e. it is enhanced in n-type material and depressed in p-type material. In the intrinsic region it depends on only temperature and pressure. Corresponding statements hold for V^+ and the free hole concentration.

2.4 Ionic Crystals : Frenkel and Schottky Disorder

Ionic crystals, such as alkali and silver halides and many oxides, have a wide separation between the valence and conduction

---

*This statement assumes that interactions among defects (including electrons and holes) are unimportant. The effects of Coulombic interactions can often be adequately described by means of the Debye-Hückel approximation when necessary.

POINT DEFECTS IN NON-METALS

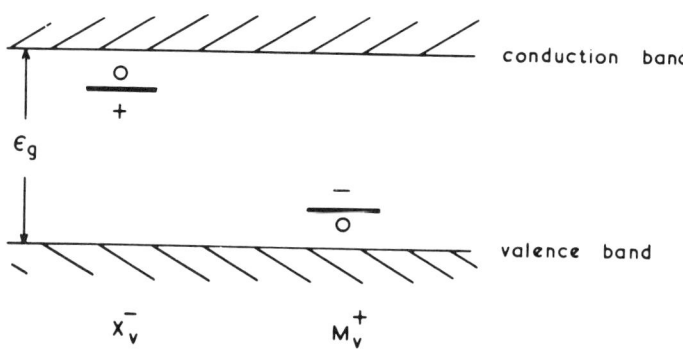

Fig. 2.2 Schematic electrical level structure of anion and cation vacancies in an ionic semiconductor, MX.

bands. As the energy levels separating the different charge states of the vacancies and interstitials lie as shown schematically in Fig. 2.2, there is thus a wide range of Fermi energy which leaves the dominant defects as those bearing effective charges appropriate to the simple idea of <u>ionic</u> defects, i.e. ion vacancies and interstitial ions. Furthermore as these materials are not easily made non-stoichiometric this range of Fermi level corresponds to a wide range of practical conditions of experimentation and specimen preparation. The appropriate analysis for these conditions can thus be made by concentrating on ionic defects and ignoring electrons and holes. Our model will thus be a perfect ionic conductor. However, many of the results we obtain will continue to be valid in the presence of electrons and/or holes since the defect reactions we consider will be a sub-set of all those which occur in the more general case.

Consider first the formation of vacancy defects (Schottky defects) in a substance MX. Corresponding to the reaction*

$$X_V^{\bullet} + M_V^{'} \rightleftharpoons \text{perfect lattice} , \qquad (2.25)$$

we have the equilibrium condition

$$\mu(X_V^{\bullet}) + \mu(M_V^{'}) = 0 . \qquad (2.26)$$

We note that we cannot equate $\mu(X_V^{\bullet})$ and $\mu(M_V^{'})$ separately to zero

---

*We use the notation $^{\bullet}$ and $^{'}$, common in solid state chemistry, when the effective or net charges on the defects are $+e$ and $-e$ respectively. In the present example it is only necessary for the charges on the complementary defects to be equal and opposite.

since the corresponding reactions would not satisfy the structural or the electroneutrality constraints. On the other hand it is only necessary that the charges on X_v and M_v be equal and opposite. They need not have any particular magnitude though, in fact, they will generally be $\pm e$ in the alkali and silver halides, $\pm 2e$ in the alkaline earth oxides, etc.

By (2.18) and (2.26) we then obtain

$$[X_v^\bullet][M_v'] = \gamma^{-1}(X_v^\bullet)\,\gamma^{-1}(M_v')\,\exp\!\left(-\frac{g(X_v^\bullet) + g(M_v')}{kT}\right)$$

$$\equiv \gamma^{-1}(X_v^\bullet)\,\gamma^{-1}(M_v')\,\exp(-g_S/kT)\;, \qquad (2.27)$$

where g_S is the Schottky formation energy. In the dilute limit the activity coefficients can be set equal to 1 (corresponding to (a) a simple lattice, all X sites equivalent and all M sites equivalent and (b) neglect of Coulomb interactions among the defects). We thus obtain

$$[X_v^\bullet][M_v'] = \exp(-g_S/kT)\;. \qquad (2.28)$$

This product relation always holds, independent of the relative concentrations of anion and cation vacancies – which can be altered for example by impurities, by irradiation or by changes in stoichiometry.

In place of the reaction (2.25) we could also have considered the formation of complementary interstitials and vacancies in either sub-lattice. We should then obtain the corresponding Frenkel product relations

$$[X_i'][X_v^\bullet] = z(X_i')\,\exp(-g_{FX}/kT)\;, \qquad (2.29)$$

$$[M_i^\bullet][M_v'] = z(M_i^\bullet)\,\exp(-g_{FM}/kT)\;, \qquad (2.30)$$

where we have again omitted the effects of Coulomb interactions but retained the appropriate site and/or orientation factors, z, for the interstitials.

The dominant defect type in any given material MX is thus determined by the lowest of g_S, g_{FX} and g_{FM}, which in practice means the lowest of the corresponding enthalpies.

In compounds of different chemical composition, e.g. $M_a X_b$, the Schottky reaction would become

$$b\,X_v + a\,M_v \rightleftharpoons \text{perfect lattice}\;, \qquad (2.31)$$

Table 2.1

Occurrence of Schottky and Frenkel Defects in Strongly Ionic Solids

Substance	Structure	Type of Disorder
Alkali halides	NaCl	Schottky
MgO etc.	NaCl	Schottky
AgCl, AgBr	NaCl	Cation Frenkel
CsCl, TlCl etc.	CsCl	Schottky
BeO	Wurtzite	Schottky
CaF_2, ThO_2 etc.	CaF_2	Anion Frenkel

Note that these are the <u>dominant</u> defects; there will always be a small proportion of the other minority defects. The separate product relations, (2.28) et seq. hold simultaneously.

with the result that in place of the product (2.28) we obtain

$$[X_v]^b [M_v]^a = \exp(-g_S/kT) \ . \qquad (2.32)$$

Schottky defects in CaF_2 are thus one Ca^{2+} vacancy and two F^- vacancies, in Al_2O_3 are two Al^{3+} vacancies, three O^{2-} vacancies and so on.

The dominant defects found in practice as the result of a great deal of experimental and theoretical study are as indicated in Table 2.1. As in other materials the study of minority defects can be quite difficult; in general, one must study non-equilibrium systems to gain this information, (e.g. irradiated crystals studied at temperatures low enough to slow down defect movements). However, when the intrinsic defects occur primarily in one sub-lattice (as in AgCl, AgBr, fluorite compounds) measurements of the diffusion rates of the other component necessarily involve one of the minority defects. In the practical quantitative application of the above Schottky and Frenkel relation it is now usual to include the activity coefficients in a form which allows for Coulomb interactions among all the charged defects (generally in the Debye-Hückel approximation).

In concluding this section we note that the models of ionic conductors which we have analysed are special cases of the more general model of a compound semiconductor (e.g. III-V or II-VI compounds). The particular Frenkel and Schottky product relations

which we have derived for these special cases also hold more
generally. Indeed, as eqns. (2.16) and (2.17) show, any number
of defect reactions may be in equilibrium simultaneously. The
experimental analysis of these more general cases, in which both
vacancy and interstitial defects can be present in several states
of ionisation can therefore be very challenging. The dominance
of the $kT \ln c_j$ term in the chemical potentials μ_j is then of
great help, a fact exploited very effectively in the approach of
Kröger and Vink (3). For up-to-date illustrations in the cases
of III-V and II-VI compounds see Casey and Pearson in Ref. 5.

2.5 Complex Defects

The attractions between some defects can be sufficiently
strong (e.g. one-tenth of an eV upwards) that pairs or larger
clusters form and remain together for times long enough for their
specific properties to be studied. For example, a divalent
cation in an alkali halide may bind a neighbouring cation vacancy
to it sufficiently strongly ($\sim \frac{1}{2}$ eV) that its lifetime as a pair
is not only very many lattice vibrational periods but also many
jumps of the vacancy around the impurity from one neighbouring
site to another. In these circumstances a number of properties
characteristic of the pair can be studied (optical, vibrational,
e.p.r., dielectric and anelastic relaxation, etc.).

The statistical thermodynamics of these pairing and aggregation reactions conforms to the framework we have already described.
We simply introduce the pair or cluster as a further distinct
defect species. All the previous equilibrium relations continue
to hold among the concentrations of the 'free' or 'uncomplexed'
defects, since the corresponding reactions remain unchanged. At
least, this is true as long as we can omit the dependence of the
activity coefficients upon defect concentrations. Such dependence
may arise from the introduction of other charged defects into the
total population (e.g. charged impurities or charged clusters).
Debye-Hückel theory shows that this will be the effect of lowest
order (corrections $O(c^{\frac{1}{2}})$ on the r.h.s. of eqns. 2.28-2.32). In
all cases there will also be some hindrance to the possible
arrangements of free defects caused by the presence of clusters
and other cluster components (which the free defects cannot
approach closely because they would then not be counted as free).
These hindrances reduce the configurational entropy associated
with the free defects and lead to correction terms $O(c)$ on the
r.h.s. of eqns. (2.28-2.32). The magnitudes of these effects must
be worked out for the various particular systems as required, but
for many systems of interest they are quite small.

If for qualitative purposes we omit these corrections then,
to lowest order, we may describe these pairing and aggregation

effects via simple mass-action-law equations. For example, corresponding to

$$\text{Impurity} + \text{Vacancy} \rightleftharpoons \text{Pair} , \qquad (2.33)$$

we obtain

$$\frac{[P]}{[I][V]} = z_p e^{\Delta g/kT} , \qquad (2.34)$$

where z is the number of distinct orientations of the pair and Δg is the free energy gain on bringing I and V together. If several paired configurations are of closely equal energy (as e.g. in the case for n.n and n.n.n. pairs formed from Mn^{2+} ions and cation vacancies in alkali halides or from trivalent $R.E.^{3+}$ ions and F^- interstitials in alkaline earth fluorides) then to obtain the total fraction of pairs the r.h.s. of (2.34) should be replaced by a partition sum over all relevant separations i.e. by

$$\sum_\alpha z_{p,\alpha} \exp(\Delta g_\alpha/kT) .$$

The magnitudes of the binding energies which occur are quite small in metals (~ 0.1 eV for vacancy pairs and impurity-vacancy pairs (15,16) but substantially larger in ionic crystals ($\sim\frac{1}{2}$ eV; Refs. 12,17) and even larger still in covalent semiconductors (Watkins and Corbett (18)).

An example of the dependence upon temperature of the degree of pairing in an ionic crystal is shown in Fig. 2.3

2.6 Non-Stoichiometric Oxides

Many oxides can depart substantially from stoichiometric composition. This is especially true of the oxides of the transition metals and other elements showing more than one distinct valency, though the occurrence of non-stoichiometry is by no means limited to such cases (cf. 'additively coloured' alkali halides, ZnO and other examples). The formation of point defects has long been discussed as a mechanism by which the excess or deficiency of oxygen is accommodated (19,20). Point defects would appear to be generally important at the higher temperatures and the lower degrees of non-stoichiometry. There is no doubt, however, that as the temperature is lowered and/or the degree of non-stoichiometry is increased (by changing the partial pressures over the compound), the clustering of point defects can become overwhelmingly important. The form that the clustering takes varies very much from one system to another. It ranges from planar aggregates in the so-called 'shear

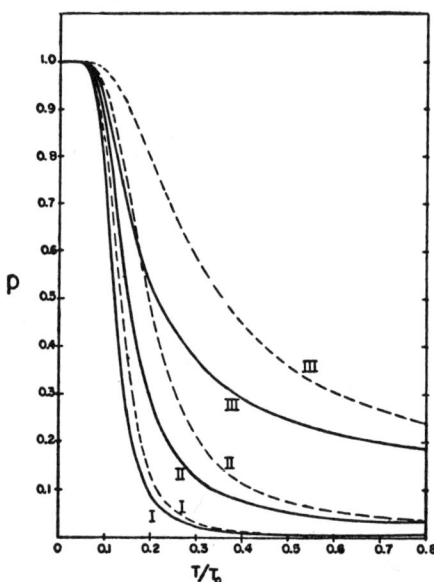

Fig. 2.3 Curves showing the fraction of impurity ions paired with vacancies as a function of reduced temperature in an alkali halide doped with divalent cations to three different concentrations; I, mole fraction $c = 10^{-4}$; II, $c = 10^{-3}$; III, $c = 10^{-2}$. The reference temperature is the impurity-vacancy binding energy/k. The dashed lines were obtained without consideration of long-range interactions while for the full lines these were included in the Debye-Hückel approximation. Only charge-compensation vacancies are included; thermal defects have been ignored.

plane structures' to three-dimensional clusters which can be regarded as nuclei of another phase corresponding to a different cation valency (e.g. Fe_3O_4 in FeO). At present it is not possible to predict in any very fundamental way which defect structures will occur in any class of system though progress in understanding the stability of experimentally identified structures is now being made quite rapidly (21,22).

Two extensively discussed examples are $Fe_{1-x}O$, where there is a deficiency in the cation sub-lattice, and UO_{2+x} where there is an excess of oxygen which is incorporated interstitially. The simplest reactions leading to the formation of simple point defects would be

$$\tfrac{1}{2} O_{2,gas} \rightleftharpoons Fe_v'' + 2 Fe_s^\bullet , \qquad (2.35)$$

POINT DEFECTS IN NON-METALS

and

$$\tfrac{1}{2} O_{2,\text{gas}} \rightleftharpoons O_i'' + 2U_s^\bullet , \qquad (2.36)$$

where we have indicated the electron holes by $Fe_s^\bullet$ and $U_s^\bullet$ in the expectation that they will be self-localised as higher valence states of the cations. By (2.17)-(2.19) and (2.35) and (2.36) we thus expect the density of Fe_v'' and O_i'' defects, i.e. x, to be proportional to $P_{O_2}^{1/6}$ at degrees of non-stoichiometry such that intrinsic thermal disorder is negligible by comparison. (When this condition is not satisfied the dependence of x and other thermodynamic properties upon the oxygen partial pressure, P_{O_2}, depends strongly upon the nature of the intrinsic disorder in the stoichiometric compound, i.e. upon whether Schottky defects, Frenkel defects or electronic disorder are dominant). However it has been known for a long time that the observed dependence of the degree of non-stoichiometry upon pressure corresponds to various higher powers (depending on the exact range of temperature and oxygen partial pressure), typically $P_{O_2}^{1/4}$. The simple mass-action relations therefore imply that there should be only about two defects formed on average for each extra oxygen atom incorporated in the lattice. Indeed various particular defect models consistent with such an inference have been analysed; though the agreement with thermodynamic data is sometimes good (e.g. Liebowitz, 23, for $Fe_{1-x}O$) the assumptions may not be consistent with more direct observations or calculations of defect structure (Catlow and Fender, 22). The precise relation of this more recent information about defect structure and clustering to the thermodynamic data still remains to be determined, even in these already well-studied systems.

2.7 Highly Disordered Ionic Solids

We have already indicated that, although the simple mass action relations are very adequate for analysing the qualitative features of systems of point defects, for accurate quantitative analyses it is necessary to allow for the effects of defect interactions (for example, through the activity coefficients). The effects of the long-range Coulombic interactions among defects bearing net electrical charges can often be adequately represented by means of the Debye-Hückel approximation (see e.g. 12,24). Not only does this appear to work well in practice but theoretical support for it is provided by more rigorous statistical mechanical theories (Allnatt and Loftus, (11)). Applied to the equations for intrinsic Schottky and Frenkel disorder, it leads, in fact, to products like (2.28) - (2.30) but in which the free energy of formation, g, is reduced by a mean interaction term

$$\Delta g_{D.H.} = \frac{-e^2 \kappa}{\varepsilon_o (1+\kappa R)} \qquad (2.37)$$

Here the defects are assumed to have charges $\pm e$, ε_o is the static dielectric constant and R is the distance of closest approach of two defects; κ^{-1} is the usual Debye-Hückel screening length such that

$$\kappa^2 = \frac{8\pi \Sigma n_i q_i^2}{\varepsilon_o kT} \qquad (2.38)$$

where n_i is the number of defects of type i per unit volume, having charge q_i ($= \pm e$ in this example). As (2.37) shows this term $\Delta g_{D.H.}$ is closely proportional to the square-root of the defect concentration. Indeed it is the long-range nature of Coulomb interactions which gives such a dependence. Nevertheless there have been indications for some time that in AgCl and AgBr this does not provide a complete description. Some clear evidence has recently been provided by Friauf and Aboagye (25) who showed that in addition to (2.37) there is another temperature-dependent term which at the melting point is several times larger. The temperature dependence of this term suggests that it is <u>linear</u> in the density of defects. Such a term could arise as the first-order contribution from the effect of defects in changing the lattice parameter, the dielectric constant and other properties of the lattice which enter into the energy needed to create (additional) defects. In any case, such a linear term (which corresponds to the addition of a quadratic term in the Gibbs free energy function) is assumed in the phenomenological models of disordering transitions in ionic solids (see e.g. Refs. 26-31). Both first- and second-order transitions may occur, depending principally upon the structure of the configurational entropy term, though models leading to first-order transitions have been mainly considered.

Thus, to return to the Ag-halides, we might suppose from the evidence of Friauf and Aboagye (25) that if AgCl and AgBr had not melted first they would have undergone a transition to a state of high disorder in the Ag sub-lattice. This tendency is stronger in AgBr than in AgCl and thus might be expected to be stronger still in AgI (with its still more polarisable anions). Unfortunately AgI is only stable in the same (rocksalt) structure as AgCl and AgBr at high pressure, though the high temperature α-phase has been known for many years to be highly disordered in the cation sub-lattice. In recent years several closely related silver iodides ot the same general formula MAg_4I_5 (M = alkali ion or NH_4^+) have also attracted attention since they too are disordered in the Ag sub-lattice and like α-AgI have very high ionic conductivities in consequence. The silver chalcogenides and some cuprous compounds are very similar (32). However, in none of these is a transition from a state with a small degree of disorder (Frenkel defects) to one with a high degree of disorder seen in the same crystal structure. Possibly, experiments on the high pressure phase of AgI or on mixed AgI/AgBr crystals might disclose such a transition.

Another class of crystals currently attracting interest as possible media for disordering transitions are the fluorite compounds. In this case there are indications from specific heats derived from enthalpy measurements that second-order transitions occur (Ref. 7, Chap. 3). Less direct indications are also obtained from the occurrence of high ionic conductivities at high temperatures (33). Since anion Frenkel defects are dominant at lower temperatures we would expect any large degree of disorder to arise in the anion sub-lattice, as indicated by various experiments.

We have discussed two classes of solids in this section, the silver halides and the fluorite compounds, since it seems to us that these offer the best chance for understanding the origin of these highly disordered states. A number of other materials showing high ionic conductivity are also of current interest but owing to their generally rather complex crystal structures they would seem less amenable to fundamental study. Of these other materials probably the so-called β-alumina is the most favourable from this point of view (34). In all cases, however, the following conditions would seem to be met. Firstly the mobile ions (e.g. Ag^+ ions in αAgI) are small by comparison with the lattice parameter of the sub-lattice defined by the immobile ions (e.g. by the I^- sub-lattice in α-AgI). Secondly, this rigid sub-lattice should define more sites for the mobile ions than there are ions to occupy them. Thirdly, the energy barriers separating these various sites should be small. To some extent this third condition follows from the first, though it will be more easily attained if the rigid sub-lattice is highly polarisable.

3. ELEMENTARY MOVEMENTS OF POINT DEFECTS

Atomic transport in crystalline solids is one of those important areas of solid state physics which rely heavily upon the theory of point defects. Properties such as solid state diffusion, ionic conductivity, annealing of radiation damage, oxidation rates, etc. may generally only be understood in terms of the nature and concentration of point defects and of their mobilities under the influence of thermal activation. There are, of course, exceptions. We have already mentioned in §2.7 the so-called 'fast' or 'super' ionic conductors which show high degrees of disorder in one sub-lattice and associated high ionic conductivities (up to ~ 1 $(ohm.cm)^{-1}$); in these cases the ideas of a small concentration of defects distributed throughout an otherwise perfect lattice structure do not apply. Even when point defects represent the only disorder present there may be circumstances in which they can move from one position to another without thermal activation. Thus in a semiconductor under irradiation it is possible for the state of ionisation of a defect to be continuously changing and if the stable structure of the defect is different in these different states it is possible for defect migration to occur as a result of successive electron and hole trapping. Such a mechanism has been proposed by Bourgoin and is discussed along with a number of other aspects of radiation damage and defects in semiconductors by Corbett and Bourgoin (35). Owing to the limited space available here we shall not consider these and other special situations, significant though they are. Instead we shall concentrate on the more general situation in which mass transport is describable only in terms of thermally-activated point-defect movements. Even so, our discussion will be rather brief and superficial since for the purposes of this Volume only rather general characteristics of defect jump rates are required and more fundamental discussions would be out of place (36).

We have discussed defect concentrations in the previous Section and have shown that they depend strongly upon temperature and chemical potentials. Here we comment upon the factors influencing the rate of movement of defects. We shall see that temperature is very important but chemical potential as such rather unimportant. Although the characteristic activation energies for particular movements of a defect may be significantly altered by the presence of other defects in the immediate neighbourhood, this can be understood in purely mechanical terms (§6). After this consideration of the movements of individual defects we shall then, in the next two sections, go on to consider two ways in which these movements manifest themselves macroscopically - in diffusion and in certain relaxation phenomena.

3.1 Defect Displacement Rates

Underlying a great deal of the theory of the defect solid state is the idea that defects - vacancies and interstitials - make a sequence of individual thermally activated jumps from one lattice site to another. These jumps are relatively infrequent by the standards of lattice vibration periods and generally involve sufficiently large local fluctuations in energy that successive jumps are dynamically uncorrelated with one another. In kinetic treatments it is generally assumed that vacancies move directly only between neighbouring lattice sites and that interstitials correspondingly jump between neighbouring interstitial sites - though this case may be more complicated for two reasons. One is the occurrence of certain less obvious structures (e.g. dumb-bell interstitials; Fig. 3.1). The second is the operation of 'indirect' or

POINT DEFECTS IN NON-METALS

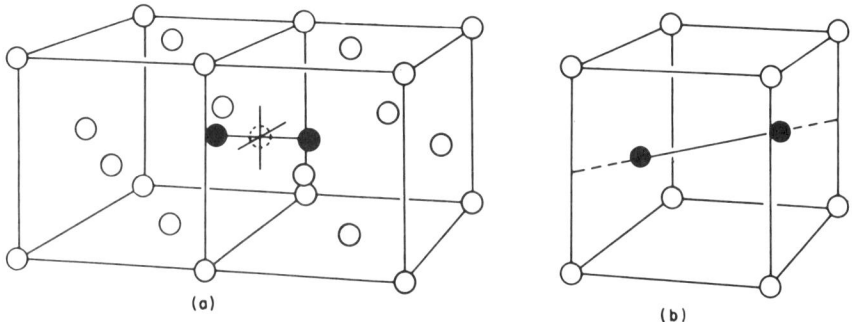

Fig. 3.1 Schematic diagram showing the stable structure of self-interstitials in (a) f.c.c. and (b) b.c.c. metals.

'interstitialcy' modes of displacement in which the interstitial moves by pushing a neighbouring atom off its normal lattice site into another interstitial position and itself taking the site so vacated.

We observe that by these assumptions we have reduced the potentially many-particle problem to one in which we focus on the movement primarily of one particle (i.e. the defect) or a very few (e.g. the interstitialcy) with associated relaxations and following movements of its immediate neighbours. Which defect movements are likely is then a question of the form of the energy surfaces in the space spanned by the displacement vectors of these atoms involved in the movements. Clearly, movements corresponding to passage over regions of exceptionally high energy will be very infrequent since the necessary fluctuations in energy will be very rare. The dependence of jump rate on energy barrier and temperature as deduced from the application of classical statistical mechanics has the well known Arrhenius form (37)

$$w = \nu \, e^{-g_m/kT} \qquad (3.1)$$

Here g_m is the Gibbs free energy of activation (for conditions of constant T and P), which can be decomposed into entropy and enthalpy terms in the usual way. We shall list a number of observations on this result, its implications and generalisations.

(i) The derivation of (3.1) assumes that once the system has passed over the energy barrier dividing the initial atomic configuration from the final configuration it relaxes to the final configuration and does not return to the initial one until a new fluctuation in energy occurs. Successive displacements of a defect are thus uncorrelated with one another; the system loses its memory of where the defect was previously by the time the defect makes another jump. (See McCombie and Sachdev, Ref. 36).

(ii) Since g_m is generally $\gg$ kT even for the most probable transitions, the exponential form of (3.1) guarantees that other paths and transitions can be neglected except when they have activation energies fortuitously close to that of the most probable transition, i.e. not more than a few times kT greater. Such near 'degeneracies' are most likely to occur with interstitial movements (§6.3.1).

(iii) Though (3.1) represents the jump frequency of a single isolated defect, jump frequencies are in fact, very little affected by the presence of other defects statistically distributed throughout the system. There may be small corrections (e.g. coming from the change of mean lattice parameter caused by the presence of other defects in the lattice and the 'drag' coming from the existence of a Debye-Hückel atmosphere of defects of opposite change) but these are not generally qualitatively important. This fact is of very considerable practical importance since it means that by studies at constant T and P we can (a) observe directly the effects of changing defect concentration (e.g. by doping, changing partial pressures, etc.) and (b) thereby often infer directly both defect jump frequencies and defect concentrations from these measurements. Of course, the binding of one defect to another in a complex will not usually leave the activation energy nor the jump frequency unaltered, but this is really another matter since we have already seen that such complexes are to be regarded as distinct defects.

(iv) When we can be confident that complications such as these (i.e. multiple paths, influence of other defects, etc.) are absent then empirically we find that the enthalpies and entropies of activation are very largely independent of temperature. In other words a plot of $\ln w$ vs. T^{-1} is generally linear.

(v) When one can make a quasi-harmonic approximation for the lattice vibrations at both the intermediate and the initial configurations (different force constants and frequencies in the two cases of course!) then the enthalpy of activation, h_m, is equal to the potential energy difference between these two configurations of the static lattice. This is a very useful result since it means that calculations of h_m - which can be used to select the modes of migration most appropriate for statistical treatment - can be carried out in a static lattice, which is a very valuable simplification (cf. §6). The same quasi-harmonic analysis also allows the calculation of the factor ν and the entropy of activation s_m but in practice most effort has gone into calculating h_m, simply because this determines the temperature dependence of w.

(vi) Though there have now been many calculations of h_m (based on the above result) most of these have assumed the 'obvious', generally the shortest, displacement path for the defect. There have been few comparisons of the activation energies of alternative

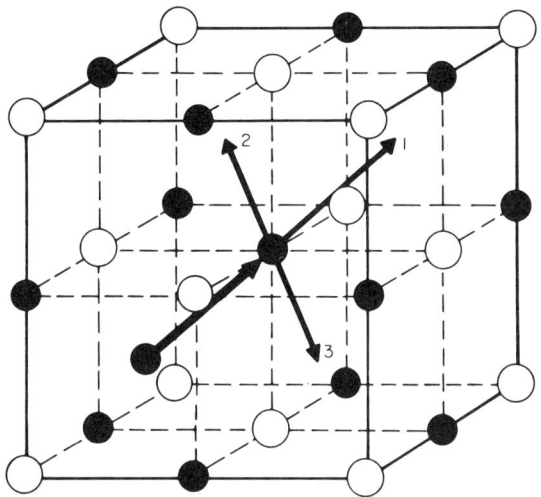

Fig. 3.2 Schematic diagram showing interstitialcy modes of movement in the NaCl lattice. The interstitial shown in the $(\frac{1}{4},\frac{1}{4},\frac{1}{4})$ position moves to the lattice position $(\frac{1}{2},\frac{1}{2},\frac{1}{2})$ and the ion which was at that position moves to one of the three interstitial positions shown. Displacement of both ions in the same direction generally requires the least energy of activation (alkali halides, Diller 83; AgCl and AgBr, 12).

paths. Such indications as there are (38) give general support to the assumption that for <u>vacancies</u> only jumps between neighbouring lattice sites are significant - as generally assumed in kinetic theories (39). <u>Interstitials</u> generally give more complex energy surfaces and a wider range of reaction paths. Those of lowest activation energy generally correspond to 'replacement' or 'interstitialcy' displacements (Fig. 3.2) or else to the motion of 'dumb-bell' or 'split' interstitials (Fig. 3.1). Such details may be detected in a few sufficiently refined experiments (comparison of ionic mobility and tracer diffusion coefficient and isotope effect measurements).

(vii) Isotopic effects in diffusion can provide information useful in the elucidation of defect mechanisms (40) and it is therefore important to know the dependence of the defect jump rate upon the isotopic mass of any of the atoms involved. If the atom which jumps into a vacancy could be regarded as simply moving in a given potential field then w would depend on the isotopic mass of this atom, m, through the frequency factor ν as $m^{-\frac{1}{2}}$. However, the atoms

of the lattice which determine the energy barrier also take part in
the dynamics of the jump process and do not simply define a (mean)
potential for the atom which jumps; their masses thus also enter
in. The quasi-harmonic approximation to Vineyard's formulation
allows the dependence to be calculated explicitly and shows that w
will vary more slowly than $m^{-\frac{1}{2}}$. How much more slowly will depend
on the extent to which the barrier atoms are also moving during the
jump. The detailed calculations which have been made for atoms
jumping into vacancies yield corrections to the $m^{-\frac{1}{2}}$ proportionality
which are smaller than those found experimentally, for reasons which
are still not completely clear. In close-packed metals (f.c.c.
and h.c.p.) the observed dependence is about 90% of that predicted
from a proportionality to $m^{-\frac{1}{2}}$, so that in these cases this proportionality would not be a bad first guess. In more open structures
(b.c.c. and diamond) the dependence is substantially slower,
typically only 50%. For an extensive review see Peterson (40).

(viii) So far we have relied entirely on classical ideas. We
should naturally expect quantum effects to become apparent at those
low temperatures where they become evident in other lattice dynamical
properties. Rather surprisingly, defect movements at quite low
temperatures are often found to follow the Arrhenius law (3.1)
closely. While we should not assume that (3.1) will necessarily
govern low-temperature defect migration, theory is not at all confident at present about what to put in its place. Qualitatively
though, we should expect the known quantum effects such as the
existence of zero-point motion, quantum mechanical tunnelling and
Bose-Einstein statistics all to lead to higher mobility at low
temperatures than would be expected by extrapolation of the high-
temperature classical expression (3.1), i.e. we expect a plot of
$\ln w$ vs. T^{-1} to be concave upwards rather than linear. Such curvature may not be easily detected by studies made by a single
technique, since the range of measured values of w may not be large
enough to disclose such a systematic curvature.

4. STATISTICAL THEORY OF DEFECT MOVEMENTS - I. DIFFUSION

In the previous Section we have summarised some of the important
features of the jump frequency of individual point defects. The
dependence of these jump frequencies upon lattice structure and type
of solid can be determined experimentally by appropriate measurements
on systems containing large numbers of such defects, but this requires
statistical analysis. In this section we review the sort of
analyses required for macroscopic diffusion experiments. In the
next section we review the theory of certain relaxation experiments
and we shall see how the information they yield can complement that
obtained from diffusion studies.

4.1 Random Walk Theory of Diffusion

We shall suppose we are dealing with situations where Fick's law

$$\underline{J} = - D \, \underline{\nabla} n \tag{4.1}$$

($\underline{J}$ = flux of atoms whose concentration is n) provides an adequate phenomenological description. This would generally be the case for measurements of the diffusion of radiotracers (both self- and impurity diffusion) for example. A more elaborate phenomenology is needed in situations where there are fluxes of more than one species (which may include defects), though in some of these the net result after appropriate analysis still has the form (4.1). For the treatment of these more complex situations we refer the reader to more comprehensive discussions (41).

If the atoms whose diffusion we are following execute a series of unbiased thermally activated displacements from one lattice site to another - a random walk in other words - we can use the Einstein equation to express the diffusion coefficient in terms of the mean square displacement of an atom in time t. For cubic crystals (whose diffusion behaviour is necessarily isotropic) the Einstein equation takes the form (see e.g. Chandrasekhar (42), Le Claire (43))

$$D = \lim_{n \to \infty} \frac{\langle R^2 \rangle}{6t}, \tag{4.2}$$

in which $\underline{R}$ is the radial displacement of an atom after n jumps requiring a total time t; the angular brackets denote a mean taken over an ensemble of equivalent atoms. If the individual displacements are denoted by $\underline{r}_1, \underline{r}_2, \ldots \underline{r}_n$ we have

$$\underline{R} = \sum_{i=1}^{n} \underline{r}_i, \tag{4.3}$$

and thus

$$R^2 = \sum_{i=1}^{n} r_i^2 + 2 \sum_{i=1}^{n-1} \sum_{j=1}^{n-i} \underline{r}_i \cdot \underline{r}_{i+j}. \tag{4.4}$$

The average value $\langle R^2 \rangle$ is thus

$$\langle R^2 \rangle = \sum_{i=1}^{n} \langle r_i^2 \rangle + 2 \sum_{i=1}^{n-1} \sum_{j=1}^{n-i} \langle \underline{r}_i \cdot \underline{r}_{i+j} \rangle. \tag{4.5}$$

In the elementary treatments of Brownian motion and random walks one usually drops the $\langle \underline{r}_i \cdot \underline{r}_{i+j} \rangle$ terms on the grounds that the direction of one jump is unrelated to that of preceding jumps. However, it is now clearly recognised that for diffusion in solids taking place via the agency of defects this is not generally true so that these terms must be retained. The convergence with increasing j is however such that for large n we can write

$$\langle R^2 \rangle = \sum_{i=1}^{n} \langle r_i^2 \rangle f \, , \tag{4.6}$$

where

$$f = \left\{ 1 + 2 \lim_{n \to \infty} \frac{\sum_{i=1}^{n} \sum_{j=1}^{\infty} \langle \underline{r}_i \cdot \underline{r}_{i+j} \rangle}{\sum_{i=1}^{n} \langle r_i^2 \rangle} \right\} \tag{4.7}$$

is called the correlation factor since it derives from the correlation between successive jumps. If all jumps are of the same length, s, (e.g. from one lattice position to its nearest neighbours only) then by (4.2) and (4.6) we have

$$D = \frac{1}{6}\left(\frac{n}{t}\right) s^2 f$$

$$\equiv \frac{1}{6} \Gamma s^2 f \, , \tag{4.8}$$

where Γ is the mean jump frequency of the diffusing species. We can distinguish two important classes of mechanism. In the first, successive jumps are uncorrelated with one another, $\langle \underline{r}_i \cdot \underline{r}_{i+j} \rangle = 0$ (all i,j) and f = 1. Examples are provided by purely interstitial solutes (e.g. C or N atoms in α-Fe and other b.c.c. metals, Li donors in Si and Ge) and by intrinsic defects such as vacancies and interstitials themselves. In these cases the surroundings of the 'particle' which has jumped (interstitial solute or defect) when viewed from the sites occupied by the particle appear the same after the jump as before it (translational invariance). By the assumptions of §3 successive jumps are therefore uncorrelated with one another.

The second class of mechanism is distinguished by the existence of correlations between the directions of successive jumps so that $\langle \underline{r}_i \cdot \underline{r}_{i+j} \rangle \neq 0$ and f < 1. This includes all cases where the atom, whose diffusion is studied, is only able to move through the involvement of a lattice defect, i.e. vacancy mechanisms and interstitialcy mechanisms (Fig. 4.1). These correlations are essentially geometrical; their occurrance is not in conflict with the assumption of rate theory that successive displacements are dynamically independent.

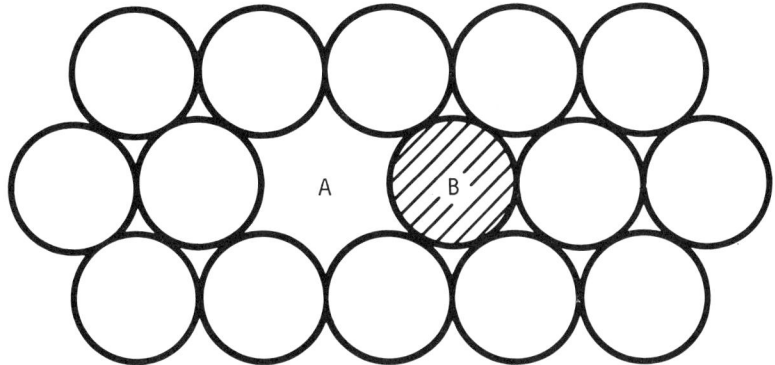

Fig. 4.1 A section of a (111) plane of a close-packed crystal, showing an impurity or tracer atom which has just jumped into a vacancy (initially on its right at B now on its left at A). Various further jumps of the vacancy will now occur, but the next jump of the shaded atom will clearly be more likely to be to the left, i.e. back to its original position, than in any other direction. From Fig. 3.2 it can be seen that a similar correlation between pairs of successive atom jumps will occur for those pairs where the impurity or tracer atom moves initially from the interstitial position to the substitutional position.

They arise because an atom which has just jumped into a vacancy is more likely at its next jump (not at that of the vacancy) to move back to its original position – where the vacancy now is – than to go to any of its other new neighbours. This correlation factor thus depends upon the details of the mechanism (vacancy, interstitialcy, vacancy pair, etc.). Broadly speaking, f is smaller, i.e. correlation effects are more significant, the less the freedom of movement of the defect. This is because the greater this freedom the greater is the probability that a defect which has just effected an atomic displacement will jump away without moving the displaced atom back to its original position. Thus f is smaller in lattices with small co-ordination number (e.g. the diamond structure) and it is generally smaller for complex defects (e.g. vacancy pairs, vacancies bound to impurities, etc.). We now consider the significance of the factors entering into D (eqn. 4.8) in more detail.

4.2 Significance of the Factors in D

We deal separately with the factor Γ, i.e. the mean number of jumps made by the diffusing particle per unit time, and with the correlation factor, f.

4.2.1 **The jump frequency, Γ.** We make the following observations.

(i) If we are following the motion of defects themselves, as e.g. in the annealing out of an excess defect population by migration to sinks, then Γ is just the defect jump frequency, i.e. w of §3, multiplied by the number of distinct but equivalent jumps (equal to the lattice co-ordination number, z, for vacancy motion between nearest neighbour sites). Similarly for interstitial solute atoms. In these cases Γ depends only on the thermodynamic variables T and P, at least as long as there are no significant interactions with other defects or with their own kind.

(ii) However if we are observing the motion of an isotope of the crystal lattice (say A^* in A) via <u>vacancies</u> then

Γ = (probability that A^* has a vacancy as neighbour) ×
(frequency that A^* exchanges with a neighbouring vacancy)

= z × (vacancy fraction) × w. (4.9)

A similar result will apply to self-diffusion of A^* in A via an <u>interstitialcy</u> mechanism though in this case the factor corresponding to z will depend on the interstitial structure (e.g. dumb-bell or cube-centre structure) and on the most probable displacements.

(iii) When considering substitutional impurity diffusion (via vacancies) we have to recognise that there will almost always be a non-zero interaction, frequently, though by no means always, attractive in sign. Then

Γ = (fraction of impurities paired with vacancies, p) ×
(frequency that the impurity exchanges with a neighbouring vacancy).

(4.10)

The first factor thus directly measures the number of impurity-vacancy pairs (cf. §2).

(iv) In both self- and impurity diffusion therefore, Γ directly reflects the defect population and its dependence on thermodynamic variables, including the coupling of defect populations in different sub-lattices of compounds, (cf. §2.4; for experimental illustrations see Refs. 44). A contrasting situation is provided by rare gas impurities, which often diffuse more rapidly as interstitial atoms than when trapped into vacancy defects; in these systems Γ is then <u>inversely</u> proportional to the vacancy population (since the more traps the smaller is the proportion of rare gas atoms in mobile interstitial states, (45)).

Table 4.1

Some Correlation Factors for Self-Diffusion via Vacancies and Interstitials

Lattice	Defect Mechanism	f
Diamond Lattice	Vacancy	$\frac{1}{2}$
Simple cubic (αCsCl)	Vacancy	0.6531
Body-centred cubic	Vacancy	0.7272
Face-centred cubic (NaCl and zincblende)	Vacancy	0.7815
Fluorite Lattice – Anion	Interstitialcy	0.986
Rocksalt Lattice	Interstitialcy* – collinear, 1 – non-collinear, 2 – non-collinear, 3	 2/3 32/33 0.9643

*Designations as in Fig. 3.2.

4.2.2 <u>The correlation factor, f.</u> We make the following observations.

(i) For self-diffusion via vacancies (A^* in A) we may assume that all possible vacancy jumps occur with the same frequency since A^* is chemically indistinguishable from the atoms of the host (we neglect the very small effects due to the changed mass, §3.1, for the moment). The correlation factor (4.7), which is dimensionless, is then a pure number. Some examples are given in Table 4.1. These have been obtained by mathematical methods which have been reviewed in detail by Le Claire (46). From the values in Table 4.1 it will be seen that f is largest in the close-packed lattices and smallest in the diamond lattice.

(ii) For impurity diffusion via vacancies there is, in general, necessarily more than one distinct jump frequency to be considered since the presence of the impurity is expected to alter the vacancy jump frequencies locally. One model which has been extensively used to discuss diffusion in f.c.c. lattices including rock-salt and zinc-blende structures) is illustrated in Fig. 4.2. For this model

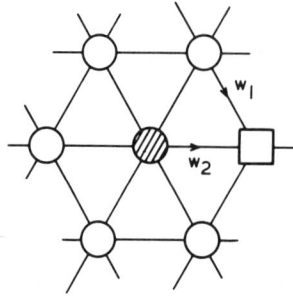

Fig. 4.2 A section of a (111) plane in a close-packed crystal showing an impurity atom (shaded) adjacent to a vacancy. Atomic jump frequencies w_1 and w_2 as indicated. In the '5-frequency' model, w_3 and w_4 additionally denote the frequencies with which vacancies jump respectively away from the nearest neighbour position of the impurity and back to it from more distant sites, while w_o is the corresponding jump frequency for isolated vacancies.

$$f = \frac{w_1 + w_3 F}{w_1 + w_2 + w_3 F} \qquad (4.11)$$

where the coefficient F is a function of $(w_4/w_o$ such that $1 \leq F \leq 7/2$ (39,46). Corresponding expressions have been derived for other lattices but it should be noted that in more open structures (e.g. b.c.c. and diamond lattices) there is no analogue of the w_1 term since none of the nearest neighbours of an impurity are also nearest neighbours of one another. The important point to notice in this case is that f is not a simple numerical coefficient but depends strongly upon the ratio of w_2/w_1. When $w_2 \ll w_1$ then $f \sim 1$ and the rate of diffusion is indeed determined by the impurity-vacancy exchange rate (eqn. 4.10). But when $w_2 \gg w_1$ then the product Γf and thus D becomes largely independent of w_2 and diffusion is limited by the magnitudes of w_1 and w_3 ie. by the speed with which vacancies can move around the impurity, rather than by the speed with which they exchange with the impurity. In lattices which are not close-packed, where none of the nearest neighbours of an impurity are neighbours of one another, this motion of the vacancy around the impurity necessarily involves at least partial dissociation of the impurity-vacancy pair - which can be a significant limitation when the pair is strongly bound (e.g. E-centres in Si). We see therefore that in impurity diffusion correlation effects can be of qualitative importance.

(iii) Isotope effects. It was first pointed out by Schoen (47) that the correlation factor can enter very directly into isotopic effects in diffusion. This can be seen immediately from (4.10) and (4.11) if we suppose that changing the mass of the impurity (without changing its chemical nature) affects only w_2 and none of the other jump frequencies. Then for isotopes α and β

$$\frac{\Delta D}{D_\alpha} \equiv \frac{D_\alpha - D_\beta}{D_\alpha} = f_\beta \frac{\Delta w_2}{w_{2\alpha}} \quad . \tag{4.12}$$

Since w_2 may be expected to be approximately proportional to $m^{-\frac{1}{2}}$ in many systems (cf. §3.1) such a direct relation is clearly very useful. Though (4.12) is by no means universally true it is far more widely applicable than our limited derivation via (4.10) and (4.11) indicates. In particular, it applies to diffusion via single vacancies in cubic lattices generally (48) and to certain situations where diffusion is via vacancy pairs and vacancies bound to impurities (49). The anlysis has also been developed for interstitials though in this case the relations are not quite so simple (46).

(iv) Relation of electrical mobility to diffusion coefficient. It is well known that the electrical mobility, λ, and the diffusion coefficient, D, of independent charged particles obeying classical Maxwell-Boltzmann statistics are related through the Nernst-Einstein relation (50)

$$\frac{\lambda}{D} = \frac{q}{kT} \tag{4.13}$$

where q is the (net) charge on the particle. Such a relation evidently holds for <u>ionic defects</u>, at least to the approximation that their movements are uncoupled. In the same way it holds for interstitial solutes (e.g. Li^+ in Si and Ge) moving directly from one interstitial site to another. However, it does not hold when the flow of the particles under consideration is strongly coupled to the flow of other species. For example in an ionic crystal the motion of substitutional ions via a vacancy mechanism necessarily implies movements of the host ions (otherwise as we have seen $f = 0$ and thus $D = 0$) and in this case λ/D depends upon the various vacancy jump frequencies and the charges carried by the impurity and the host ions (39,41). Even for isotopes of the host lattice ('tracers') there is a residual correction to (4.13) which, since all jump frequencies are identical in this case, is purely a numerical factor. This we now briefly consider; the discussion applies to both vacancy and interstitialcy migration and is very simple.

Firstly, we observe that the mobility of an ion (tracer or host – there can be no difference) is equal to that of the defect, λ_d, multiplied by the fraction of defects (we suppose for simplicity that only one type of defect is effective). By (4.8), with f = 1 for the defect and (4.13) we have

$$\lambda_T = \lambda_{ion} = \left(\frac{q}{kT}\right)\left(\frac{1}{6}\Gamma_d s_d^2\right)[d] \quad , \tag{4.14}$$

in which Γ_d is the defect jump frequency s_d is the defect jump distance and [d] is the fraction of defects. Secondly we can write

$$D_T = \frac{1}{6}\left(\nu \Gamma_d [d]\right)s_T^2 f \quad , \tag{4.15}$$

where ν is the number of ions moved in each defect jump (ν = 1 for a vacancy, ν = 2 for the interstitialcy jumps shown in Fig. 3.2) and s_T is the distance which the tracer moves in each jump (= s_d for a vacancy mechanism but different from it for interstitialcy jumps). By (4.14) and (4.15)

$$\frac{\lambda_T}{D_T} = \left(\frac{q}{kT}\right)\cdot\left(\frac{s_d}{s_T}\right)^2 \frac{1}{\nu f} \quad . \tag{4.16}$$

The generalisation when two or more types of defect movement occur simultaneously is straightforward. It is therefore clear that accurate measures of the ratio λ_T/D_T can be very useful in providing information about fundamental details of defect movements. In fact, the first experimental demonstration of the interstitialcy mechanism was obtained by Compton and Manner (51) in just this way (actually for Ag^+ ions in AgCl); some very refined analyses of AgBr and AgCl have now been made (52). Similar ideas are now sometimes employed in the analysis of Ag^+ ion movements in fast ionic conductors such as α-AgI, α-Ag_2S, etc. (32).

4.3 Conclusion

The subject of solid state diffusion is now very large and in this section we have not done more than touch upon a few basic features which disclose the relation of some measured quantities to the defect structure of a solid. The generalisation of the particular results we have given to more general situations is often straightforward in principle though in practice the resulting complexity can appear rather forbidding. Perhaps this is most evident in the III-V semiconductors where in addition to the features which arise in the ionic compounds (coupled defects in the anion and cation sub-lattices, departures from stoichiometry, etc) one may also have defects in several states of ionisation, the same impurity in both interstitial and substitutional configurations, exchange disorder, etc.*.

We should however emphasise that in our discussion we have had very much in mind the 'almost uniform' system. In particular, we have not considered examples where the diffusing elements themselves cause an appreciable change in the defect population. As long as the defect population is still locally in equilibrium it is possible to establish a fairly general framework for the analysis of these cases (see e.g. Refs. 41). One important class is provided by divalent cations diffusing in the alkali halides, where a series of exemplary experiments by Fredericks et al. (53) have fully confirmed the predictions of the defect theory based on these principles and those outlined in §§2.4 and 2.5. When the defects are not in local equilibrium the theoretical task may be complicated owing to the flows of defects to a distribution of sinks; even here however it is possible to do controlled experiments and analysis which permit the determination of fundamental information about solute-defect interactions (54).

5. STATISTICAL THEORY OF DEFECT MOVEMENTS – II. RELAXATION PROCESSES

In the previous section we discussed some aspects of the movements of atoms and defects over macroscopic distances. The thermally activated reorientation of defects having an electric or elastic dipole moment (generally, though not always, complex defects) can also give rise to readily detectable macroscopic effects in the response to time-varying electric or elastic fields. These appear as relaxation effects in the response of the system to the applied fields. The general phenomenology of such dielectric and anelastic relaxation effects and the analysis of relaxation modes in terms of defect structure and symmetry is discussed in a number of sources (55) so that we shall here confine ourselves to indicating salient points with one or two examples.

---

*This complexity (affecting the Γ's) may be one reason why discussions and reviews of diffusion in semiconductors often give little attention to correlation effects (f's); for instance, Casey and Pearson (44) do not even mention their occurrence. There is no reason for them to be any less important in semiconductors than in other materials, rather the opposite in fact.

5.1 Phenomenology

For simplicity we shall confine ourselves to crystals with cubic lattices. It is well known that by symmetry the dielectric constant of such materials is a scalar (56); this applies in the presence of time-dependent effects as well as to the static case. If we write the applied electric field as $\underline{E}(t)$ and the corresponding dielectric displacement as $\underline{D}(t)$ then by making a Fourier analysis we have

$$\underline{E}(t) = \int_{-\infty}^{\infty} \underline{E}(\omega) \, e^{-i\omega t} \, d\omega \, , \tag{5.1}$$

and

$$\underline{D}(t) = \int_{-\infty}^{\infty} \varepsilon(\omega) \, \underline{E}(\omega) \, e^{-i\omega t} \, d\omega \, , \tag{5.2}$$

with $\varepsilon(\omega)$ the frequency dependent dielectric constant. This is a complex quantity

$$\varepsilon(\omega) = \varepsilon_1(\omega) + i\varepsilon_2(\omega) \, . \tag{5.3}$$

Although elastic stress and strain are second-rank tensors — which would prevent us from using a single equation such as (5.2) generally — for a cubic solid subject to uniaxial stress along one of the three cubic symmetry directions (<100>, <110> and <111>) we can in fact use exactly the same formal relations. In these cases E is to be identified with the magnitude of the uniaxial stress while D is that of the strain in the same direction. We shall therefore use (5.2) for both dielectric and anelastic relaxation but we must keep the restrictions in mind.

For anelastic relaxation and for dielectric relaxation in an insulator we may write the coefficient of the response as ('Debye relaxation')

$$\varepsilon(\omega) = \varepsilon_\infty + \sum_j \frac{\delta\varepsilon_j}{(1 - i\omega\tau_j)} \, . \tag{5.4}$$

Here j specifies the various modes of relaxation (determined by the nature of the defects) while τ_j and $\delta\varepsilon_j$ are respectively the relaxation time and strength of mode j. Often one mode will dominate the response, which may make it difficult to measure the strengths and relaxation times of the others. The quantity ε_∞ is formally the response coefficient for the crystal without defects; for the frequency range usually employed ($\omega \sim \tau_j^{-1}$) it may frequently be

POINT DEFECTS IN NON-METALS

taken as a constant. However, if we are dealing with a conductor (ionic or electronic) instead of an insulator then we can usefully separate out the conductivity term in $\varepsilon(\omega)$. In Gaussian units we would then write

$$\varepsilon(\omega) = \frac{4\pi i}{\omega} \sigma(\omega) + \varepsilon_\infty + \sum_j \frac{\delta\varepsilon_j}{(1 - i\omega\tau_j)} \quad . \tag{5.5}$$

The frequency dependence of $\sigma(\omega)$ depends upon the nature of the conductivity (e.g. closely constant for ionic conductivity, Drude-like for free electrons, etc.). In connection with ionic conductors it is important to note that electrode processes (e.g. blocking) can give rise to very large effects having a frequency dependence like the Debye relaxation terms in (5.5). With dielectric relaxation experiments care is needed to ensure that these are not confused with defect relaxation processes; such electrode effects of course do not occur with anelastic experiments. We now note one or two consequences of (5.4) and (5.5).

5.1.1 <u>Single applied frequency, ω</u>. This is the commonest experimental situation; one measures the dielectric or anelastic loss, the latter either in a driven specimen or one excited into a vibration which is then allowed to decay freely. We have

$$D(t) = \varepsilon(\omega) E_o e^{-i\omega t}$$

$$= (\varepsilon_1 + i\varepsilon_2) E_o e^{-i\omega t}$$

$$\equiv (\varepsilon_1^2 + \varepsilon_2^2)^{\frac{1}{2}} E_o e^{-i\omega t} e^{i\delta} \tag{5.6}$$

where the loss angle δ is given by

$$\tan \delta = \varepsilon_2/\varepsilon_1 \quad . \tag{5.7}$$

By (5.4)

$$\tan \delta = \frac{\sum_j \frac{\omega\tau_j \delta\varepsilon_j}{(1 + \omega^2\tau_j^2)}}{\varepsilon_\infty + \sum_j \frac{\delta\varepsilon_j}{(1 + \omega^2\tau_j^2)}} \quad . \tag{5.8}$$

For dielectric loss in a conductor there will be an additional term $4\pi\sigma/\omega$ in the numerator.

In the special case where there is only a single mode of relaxation (5.8) gives

$$\tan \delta = \frac{(\varepsilon_s - \varepsilon_\infty) \omega \tau}{\varepsilon_s + \varepsilon_\infty \omega^2 \tau^2} \tag{5.9}$$

where we have written $\delta\varepsilon = \varepsilon_s - \varepsilon_\infty$ in which ε_s specifies the static response coefficient (i.e. $\omega \to 0$). Under the same conditions it can easily be shown that a plot of ε_2 vs. ε_1 is a semicircle whose centre lies on the axis of ε_1 midway between ε_∞ and ε_s (known as a Cole plot). When there is more than one relaxation mode such a plot may still be approximately semi circular but the origin will generally lie below the ε_1-axis.

5.1.2 <u>Sudden application of an electric field.</u> For an applied field E(t) equal to zero for $t < 0$ and to a constant value, E_o, for $t > 0$ the response at $t \geq 0$ is

$$D(t) = D_o + \sum_j E_o \, \delta\varepsilon_j \, (1 - e^{-t/\tau_j}), \tag{5.10}$$

when $\varepsilon(\omega)$ is given by (5.4). We observe that $D = D_o$ at $t = 0$ while $D \to D_s$ as $t \to \infty$ with

$$D_s = \varepsilon_s E_o \tag{5.11}$$

and

$$\varepsilon_s = \varepsilon_\infty + \sum_j \delta\varepsilon_j \tag{5.12}$$

the static dielectric constant as before. This method of determining τ_j has been used successfully by Dreyfus (57) but has not been taken up to any notable extent otherwise.

5.1.3 <u>Method of ionic thermocurrents (I.T.C.).</u> A related method in which the decay of electric polarisation following the removal of a static field is followed not isothermally but isochronally is due to Bucci et al. (58) and is known as the method of ionic thermocurrents. It relies upon the fact that the relaxation times τ_j are inversely proportional to the defect jump frequencies and therefore become very long at low temperatures where defect jumps are infrequent. The technique is as follows:

(a) Polarise the specimen in a static field at a temperature where the relaxation times are conveniently short.

(b) Quench the specimen (in the field) rapidly to a low temperature where the relaxation times are very long.

(c) Holding the specimen at the low temperature, remove the field; we now have a specimen with the high-temperature, polarised defect-distribution frozen in.

(d) Now warm the specimen at a controlled rate (generally so that dT/dt is constant, though dT^{-1}/dt has also been used) and measure the depolarisation current, $J = dP/dt$, as a function of temperature. This current will generally appear as a series of peaks corresponding to the different relaxation times.

Analysis of such ionic I.T.C. curves is often very convenient; one evident advantage over A.C. dielectric loss is the absence of the background conductivity loss which can make the analysis of tan δ into separate Debye terms difficult. The important features of an I.T.C. peak can easily be seen by examination of the case of a well isolated peak to which only one relaxation mode contributes. Then by (5.10) the polarisation P decays according to

$$\frac{dP}{dt} = \frac{(P_o - P)}{\tau} = -\frac{P}{\tau}, \qquad (5.11)$$

since $P_o = 0$ (cf. (c) above). This leads to the following useful results.

(i) Initially, i.e. on the low temperature side of the peak,

$$J = \frac{dP}{dt} = \frac{E_o}{4\pi} \frac{\delta\varepsilon}{\tau}, \qquad (5.12)$$

by (5.10). The initial growth of J vs. T thus allows the activation energy, h, governing τ to be determined*.

(ii) At a constant heating rate, i.e $dT/dt \equiv b$, it is easily shown that the maximum thermocurrent occurs at a temperature T_m, such that

$$T_m = \left(\frac{b \, h \, \tau \, (T_m)}{k} \right)^{\frac{1}{2}}. \qquad (5.13)$$

This allows the pre-exponential factor in τ to be determined, since E is known from (i).

---

*Though we are only dealing with one mode of relaxation it is not generally true that τ is determined by only one atomic jump frequency. Nevertheless one jump frequency may often predominate in the expression for τ so that it has the appearance of being a singly activated quantity.

(iii) The area under the I.T.C. peak gives the strength of the corresponding relaxation mode. Thus,

$$\int J dT = b \int J dt$$

$$= b \Delta P = \frac{b}{4\pi} E_o \delta\varepsilon \quad . \tag{5.14}$$

This is a very useful feature*.

These then are three principal ways in which information about defect relaxation modes can be obtained. The same or related relaxation times can also be obtained by other means, most notably (a) electron paramagnetic resonance used to study the rates of redistribution of defects among different configurations (e.g. Symmons (59)) and (b) measurements of the lifetime broadening of e.p.r. lines (60,61).

Information on the motion of defect complexes is also obtainable from measurements of nuclear magnetic relaxation times, though these are generally accompanied by contributions from other defects as well. The theory of the effect of thermally activated atomic movements and diffusion upon the various nuclear relaxation times is an important topic but it is one which has not engaged very full attention until recently. Measurements of N.M.R. relaxation times can provide important information which complements that obtainable from diffusion, ionic conductivity, dielectric and mechanical relaxation studies, etc. Illustrations of their value can be seen in Refs. 62.

5.2 Relation to Defect Properties

We shall now look at the relation of the observable relaxation times τ_j to the nature of defect movements which are responsible. Let us denote the possible configurations and orientations of a given type of defect by $u = 1, 2 \ldots n$. Also let n_u be the numbers of defects in configuration u; in general, this will be a function of time. If the rate of jumping from configuration u to r is w_{uv} then the equation governing the time-dependence of n_u is

$$\frac{dn_u}{dt} = \sum_{v \neq u} n_v w_{vu} - \sum_{v \neq u} n_v w_{uv} \quad . \tag{5.15}$$

---

*It may be noted that eqn. (3.69) of Ref. 7 should have a factor 4π inserted on the r.h.s. to be correct.

POINT DEFECTS IN NON-METALS

Since we are here only interested in linear responses to the applied fields we can expand the terms n_u and w_{uv} to first order in the perturbation and retain only first order terms in (5.15). Firstly, we set

$$n_u = n_u^{(o)} + \delta n_u \qquad (5.16)$$

where $n_u^{(o)}$ is the value of n_u in thermodynamic equalibrium. Also, by the results of rate theory

$$w_{uv} = w_{uv}^{(o)} \left(1 - \frac{\delta E_{uv}^s - \delta E_u}{kT}\right) \qquad (5.17)$$

in which δE_{uv}^s is the change in energy of the saddle point configuration for the transition $u \to v$ while δE_u is the change in energy of configuration u itself both resulting from the impressed field. Eqn. (5.17) comes from expanding the exponential in the rate process Arrhenius formula (3.1) in powers of $\delta E/kT \ll 1$. We now substitute from (5.16) and (5.17) into (5.15) and retain only first-order terms. Furthermore we also use the principle of detailed balance, i.e.

$$n_u^{(o)} w_{uv}^{(o)} = n_v^{(o)} w_{vu}^{(o)} . \qquad (5.18)$$

Thus

$$\frac{d}{dt}(\delta n_u) = \sum_{v \neq u} \delta n_v w_{vu}^{(o)} - \sum_{v \neq u} \delta n_u w_{uv}^{(o)}$$

$$+ \sum_{v \neq u} n_v^{(o)} w_{vu}^{(o)} \frac{(\delta E_v - \delta E_u)}{kT} . \qquad (5.19)$$

Observe that the changes in energy in the saddle point configurations have all dropped out.

The relaxation times can be found from the homogeneous terms alone. They are the roots of the determinantal equation

$$\det||w_{vu}^{(o)} + \tau^{-1} \delta_{vu}|| = 0 , \qquad (5.20)$$

in which we set

$$-\sum_{v \neq u} w_{uv}^{(o)} \equiv w_{uv}^{(o)} , \qquad (5.21)$$

for the diagonal elements. (This is easily verified by inserting the trial solution $\delta n_u = \delta n_u(o) e^{-t/\tau}$ into (5.19) and writing down the condition for a non-trivial solution, i.e. for $\delta n_u(o) \neq 0$). Eqn. (5.20) is the general equation from which the relaxation times can be calculated when the $w_{uv}^{(o)}$ are known. The values and the characteristics of the corresponding solutions ($\delta n_u^{(o)}$) depend upon the details of defect structure and modes of movement. However, as with other secular equations of the type (5.20), e.g. those arising in the analysis of molecular vibrations (where we would have $i\omega$ in place of τ^{-1}), we can simplify the task of solving (5.20) by using the methods of symmetry analysis. The techniques are general and are described in many texts (63; see also Hughes this Volume). We shall therefore only summarise one or two of the most commonly used results (for more detailed analyses see Refs. 55 and, more particularly, for the present results Franklin et al. (64)).

We refer only to cubic crystals, though symmetry analyses have been made for most other classes as well. Some of the simplest cases are those where the defects are either tetragonal (<100>) or trigonal (<111>) or orthorhombic (<110>) and reorient between equivalent positions such that the set of these physically equivalent orientations has the cubic 'point' symmetry, O_h. (The symmetry of this set of orientations may in fact be lower than O_h in cubic compounds, or more generally in cubic crystals with a 'basis'; an obvious example is provided by nearest neighbour pairs in a zinc-blende structure, i.e. trigonal defects whose set of equivalent orientations has only tetrahedral, T_d, symmetry). Examples are provided by simple molecular substituents (e.g. OH^- on anion sites in alkali halides), and by pairs of defects (e.g. M-centres, foreign ions paired with either vacancy or interstitial defects, etc.). In all cases the defects may give rise to several relaxation modes, each of which belongs to one of the irreducible representations of the symmetry group of the defect orientations O_h. The particular irreducible representation specifies the symmetry of the relaxation mode, and in turn this determines the nature of the field to which the mode can couple. In particular, only T_{1u} modes can couple to a uniform electric field while only A_{1g}, E_g and T_{2g} can couple to a uniform stress field. Which irreducible representations actually occur in the solutions to (5.19) and (5.20) is determined by the symmetry of the defect in itself. We shall now illustrate these general remarks by summarising several particular results. We do so first for the three simplest cases of defects which are either tetragonal, trigonal or orthorhombic (<110>) in all orientations and configurations included in (5.19) and (5.20).

Tetragonal defects. There are solutions of (5.20) corresponding to the A_{1g} (non-degenerate), E_g (doubly degenerate) and T_{1u} (triply degenerate) representations. The degeneracy specifies the number of distinct modes having the same eigenvalue, τ^{-1}. Of these, only T_{1u} (which is odd) can couple to a uniform electric field, i.e. only T_{1u} relaxation modes can be studied in dielectric loss or I.T.C. experiments. The doubly degenerate E_g modes can couple to uniaxial stresses applied along <100> or <110> directions but not along <111>. In principle, the totally symmetric A_{1g} mode can couple to a hydrostatic stress though for this to give rise to an observable relaxation it would be necessary for the defect to be able to 'extend' and not only to reorient.

Trigonal defects. There are solutions of (5.20) corresponding to the A_{1g} (non-degenerate), T_{2g} (triply degenerate), A_{2u} (non-degenerate) and T_{1u} (triply degenerate) representations. As before only T_{1u} modes can be seen in dielectric relaxation and I.T.C. experiments. The T_{2g} modes can be studied in mechanical relaxation experiments when the uniaxial stresses are applied in <110> or <111>, but not <100>, directions.

Orthorhombic <110> defects. Here there are solutions belonging to the A_{1g}, E_g, T_{2g}, T_{1u} and T_{2u} representations. The coupling of these to uniaxial stress fields and to uniform electron fields is the same as before. In particular, defects of this symmetry should give rise to mechanical relaxation effects for uniaxial fields along all three principal cubic directions.

These differences in response to uniaxial stress between the three types of defect are useful in determining the underlying symmetry of the defects giving rise to the relaxation. Dielectric relaxation by contrast tells us only that the defect has an electric dipole moment but cannot tell us anything about the axial symmetry of the defect. When the nature of the defect is reasonably well known however, a combination of dielectric and mechanical relaxation measurements, possibly supplemented by other kinetic information (e.g. lifetime broadening of e.p.r. lines, diffusion data, etc.), may allow the accurate determination of several jump frequencies and corresponding activation energies. We then need the particular solutions of (5.20) in terms of the various jump frequencies which characterise the defect. These must be worked out for each case as appropriate. A number of important cases of defect pairs have been worked out by Franklin et al. (64); for an example of a more complicated case see the study of the complex anion interstitial structures in the fluorite lattice by Socino et al. (65). Some of these practically important cases show that we need to include defect jumps which change the symmetry of the complex; for example, in the case of impurity-vacancy pairs in alkali halides there may be only small differences in energy between nearest and next-nearest neighbour separations, i.e. between orthorhombic and tetragonal configurations. In these cases the range of symmetries of the allowed solutions of (5.19) and (5.10) is correspondingly enlarged; in the preceeding example we obtain solutions belonging to all of A_{1g}, E_g, T_{2g}, A_{2u} and T_{1u} (2 distinct eigensolutions with this same symmetry but having different eigenvalues τ^{-1}).

So far we have concentrated on the nature of the solutions and the corresponding relaxation times, τ. In practice, however whether a given, active relaxation mode is observable depends upon its strength. This is determined by the particular eigensolution ($\delta \underline{n}$ in (5.19)) and upon the magnitude of its coupling to the applied field (i.e. upon the δE_u in (5.19)). It is only convenient to make general statements for defects possessing only one active mode of the type considered.

Thus if there is only one electrically active mode (T_{1u}) this is entirely responsible for

$$\delta \varepsilon = \varepsilon_s - \varepsilon_\infty$$

which therefore equals

$$\delta \varepsilon = \frac{4\pi c \mu^2}{3 v_m kT} \tag{5.22}$$

by the usual statistical calculation for the suceptibility of a system of non-interacting permanent dipoles (66). Here c is the molefraction of defects and μ is the molecular volume (so that n the number of defects per unit volume is c/v_m). Eqn. (5.23) does not include any internal field effects. Though there may be corrections of this kind they depend upon details of the defect and its structure. There seem to be rather few studies of these corrections but such as have been made (67) would imply that they are relatively unimportant for extended defects (such as, for example, impurity-vacancy pairs. On the other hand, we should recognise that it would be quite wrong to employ the Lorentz internal field correction which is valid for <u>point</u> dipoles.

There are expressions similar to (5.22) for the change in elastic compliance $\delta s = s_s - s_\infty$. In this case the place of the electric dipole moment, μ, is taken by the 'elastic dipole tensor' of the defect, viz.

$$G_{ij} = \sum_\ell F_i^{(\ell)} R_j^{(\ell)}, \tag{5.23}$$

where suffices i,j denote Cortesian components, $\underline{F}$ is the force exerted by the defect upon ion ℓ of the lattice at position $\underline{R}(\ell)$ (see e.g. Peisl, this volume). It is sometimes convenient instead of $\underline{\underline{G}}$ to use the dimensionless quantity

$$\underline{\underline{\lambda}} = \frac{1}{3Bv_m} \underline{\underline{G}} \tag{5.24}$$

where B is the bulk modulus ($3B = c_{11} + 2c_{12}$). For trigonal and tetragonal defects $\underline{\underline{G}}$, and thus $\underline{\underline{\lambda}}$, has only two principal components (G_1 and G_2, along and perpendicular to the defect axis) and the expression for δs takes the simple form

$$\delta s = \frac{\beta v_m c (\lambda_1 - \lambda_2)^2}{kT}$$
$$= \frac{\beta n (G_1 - G_2)^2}{9B^2 kT} \qquad (5.25)$$

in which $\beta = 2/3$ for tetragonal defects and $\beta = 4/9$ for trigonal defects*. For other more complex cases see Nowick and Heller (55).

When more than one mode contributes to either the dielectric or the mechanical relaxation then one must divide up the $\delta \varepsilon$ and the δs according to the relative strengths of the modes as given by the solution of (5.19). The results for particular defects (especially impurity-vacancy pairs) show that when there is more than one mode of a given type (e.g. T_{1u}) then that with the longest τ is often dominant (64). This may make the detection of the others of the same type difficult, though not impossible.

5.3 Some Examples

We conclude this chapter with a few brief comments on several well studied examples.

(i) Fluorite compounds doped with lower valency cations. As long as the foreign cations go into solution substitutionally we would expect their electrostatic attraction for anion vacancies to lead to the formation of stable nearest neighbour pairs in such systems; the co-ordination of the fluorite lattice then dictates that this defect would have trigonal (<111>) symmetry. If the only vacancy jumps to occur are the <100> jumps among the 8 nearest neighbour positions of the foreign cation with frequency w then the eigenvalues of (5.20) are

---
*It may be noted that for eqns. (3.63) and 3.64) of Ref. 7 to be correct as they stand n_D there should be interpreted as the mole fraction of defects while $\underline{\underline{G}}$ there should be interpreted as $\underline{\underline{\lambda}}$. Alternatively the additional factor $9B^2 v_m^2$ can be inserted into the denominator on the R.H.S. of eq. (3.64).

$$\tau^{-1}(A_{1g}) = 0$$

$$\tau^{-1}(T_{2g}) = 4w$$

$$\tau^{-1}(A_{2u}) = 6w \qquad (5.26)$$

$$\tau^{-1}(T_{1u}) = 2w .$$

That the ratio of τ^{-1} for the mechanically active mode to that for the electrically active mode, $\tau^{-1}(T_{2g})/\tau^{-1}(T_{1u})$, is 2 has been verified experimentally in ThO_2:Ca, CeO_2:Ca and CaF_2:Na. The fact that the T_{2g} mode is active for uniaxial stresses along <110> and <111> but not along <100> has been verified for CaF_2:Na. Such results are by no means a foregone conclusion; they would be altered if any appreciable fraction of the vacancies were in next-neighbour or more distant positions and they would also be altered if there was any significant probability of vacancies jumping along <110> from one nearest neighbour position to another.

(ii) Alkali halides doped with divalent cations, e.g. NaCl:Mn^{2+}. In this case pairs will form by the attraction of the divalent cations for the cation vacancies. To begin with let us again assume that only nearest neighbour sites are populated (Fig. 4.2). This case differs from the one before because the component defects are on the same (cation) sublattice; two distinct jumps are possible without extending the defect, namely jumps of the vacancy from one nearest neighbour site of the impurity to another (w_1) and vacancy-impurity exchange (w_2). The pair is orthorhombic (<110>). We then find

$$\tau^{-1}(A_{1g}) = 0 ,$$

$$\tau^{-1}(E_g) = 6w_1 ,$$

$$\tau^{-1}(T_{2g}) = 4w_1 , \qquad (5.27)$$

$$\tau^{-1}(T_{1u}) = 2(w_1+w_2) ,$$

$$\tau^{-1}(T_{2u}) = 6w_1 + 2w_2 .$$

Thus mechanical relaxation experiments (E_g and T_{2g} modes) should give information about w_1 while dielectric relaxation experiments (T_{1u} mode) should yield information primarily about w_1 or w_2, depending on which is bigger. Also we saw in the previous section (§4.2.2) that the diffusion coefficient of the foreign ion will be largely determined by the <u>smaller</u> of w_1 and w_2. Combined measurements can thus not only allow the inference of jump frequencies but also provide a check on the consistency of the model.

Broadly speaking, the activation energies of impurity-vacancy reorientation rates obtained from dielectric relaxation studies are significantly lower than those characterising the diffusion of the corresponding impurity ions*. For example, in NaCl a variety of foreign cations give activation energies for dielectric relaxation in the range 0.65-0.72 eV (which is much the same range as is spanned by different measurements of the activation energy for free vacancy migration) while reliable values of the corresponding activation energies for diffusion* range from 0.9 to 1.3 eV (with the exception Zn^{2+} which appears to be as low as 0.52 eV; see Fredericks, (53). We thus infer that in many cases $w_2 \ll w_1$. While this conclusion may still be valid, combined dielectric and mechanical relaxation measurements have shown that in the two cases studied, namely NaCl:Ca^{2+} and NaCl:Mn^{2+}, the nearest neighbour model is too simple and that we must allow the vacancy to jump at least to next-nearest neighbour sites as well. In the case of a paramagnetic ion such as Mn^{2+} it is also possible to measure the proportions of vacancies in nearest neighbour and next nearest neighbour sites through the different crystal field effects on the e.p.r. spectra (orthorhombic and tetragonal respectively). In these cases very detailed analyses into the various jump frequencies have been made (see e.g. Symmons, 59). That the conclusions for Mn^{2+} in NaCl are inconsistent with the diffusion measurements (particularly in the value of w_2) may simply indicate that if w_2 is indeed small then it will be very difficult to determine it accurately from relaxation experiments.

(iii) Alkaline earth fluorides doped with higher valency cations e.g. CaF_2:Gd^{3+}. These form another fairly extensively studied class, in some ways analogous to the last example. In this case the trivalent cations form pairs with interstitial F^- ions, but these may occupy nearest and next nearest neighbour (and possible even more distant) sites - as shown by many magnetic resonance experiments (7). Relaxation studies include dielectric and anelastic loss measurements, I.T.C. and e.p.r. line broadening on several closely related systems. The activation energy for jumps of the interstitial from one nearest neighbour position to another in CaF_2 appears to be $\simeq 0.4$ eV largely independent of which trivalent ion is involved.

---

*We are speaking here of the activation energy for diffusion in the limit when all impurity ions are paired with vacancies. There is then no contribution from the energy of association, only from w_2 and from the correlation factor. As we saw in §4.2.2 this means from the smaller of w_1 and w_2.

Assignments of various other characteristic activation energies
($\simeq$ 0.15 eV and $\sim$ 0.7 eV) can also be made, though no fully consistent analysis of all the experimental data seems yet to have been achieved. There is little doubt however that when all relevant factors are controlled the relaxation measurements will remain important in the final analysis.

5.4 Conclusion

In this section we have reviewed the relations between thermally activated defect reorientations and changes of configuration and observable dielectric and mechanical relaxation phenomena. We have seen that there is a close connection between these macroscopic effects and defect symmetries and jump frequencies - although observations of these relaxation rates alone may not be sufficient for a complete analysis of the underlying movements, particularly in the case of complex defects. Nevertheless supplemented with information from other experiments (e.g. e.p.r., diffusion) they are very valuable and enable us in many cases to determine quite intimate details of the characteristics of compound defects.

6. ATOMIC THEORY OF DEFECTS

In the previous sections we have reviewed the thermodynamics of point-defect concentrations and some aspects of the kinetics of defect movements. Broadly speaking, these theories yield general relations and leave the determination of specific features (e.g. dominant type of defects, mechanisms of movement, characteristic energies, etc.) to experiment. However few of these specific features can be determined in any direct way, mostly it is necessary to make assumptions or models which are carried along as hypotheses until it is necessary to modify them. We have already mentioned some of these in previous Sections (e.g. that defects in ionic crystals occur in fixed states of ionisation, that atomic jumps into vacancies only occur significantly from nearest neighbour positions and so on). It has therefore long seemed desirable to calculate these specific features from the general theory of solids. Such calculations constitute what we may call the 'atomic' theory of defects as distinct from the thermodynamic and kinetic theories. Early examples are provided by the calculations (i) of Mott and Littleton on Schottky and Frenkel defects in some alkali halides (68) (ii) of Tibbs (69) on the energy levels of the F-centre and (iii) of Huntington and Seitz on vacancies and interstitials in Cu (69). These calculations played an important part in establishing respectively (i) the dominance of Schottky disorder in alkali halides (ii) the vacancy nature of the F-centre and (iii) the relative magnitudes of the formation and migration energies of vacancies and interstitials in Cu and the responsibility of vacancies for diffusion in this metal. There has been a very considerable growth in this field of defect theory in the past few years (see e.g. the recent book by Stoneham, (6)). Important advances in the treatment of the electronic states of defects have been accompanied by greatly improved accuracy and reliability of methods of calculating the distortion of the lattice surrounding the defect. Unfortunately though, we still do not possess a practical general theory able to deal with problems of self-consistency simultaneously in the electronic states and in the atomic structure of the defect. In other words if the electronic structure of the defect is simple (e.g. 'closed-shell' defects) then the methods for determining defect structure and lattice relaxation can now handle complex defects and determine - not assume - the defect symmetry. Equally, if features of the atomic arrangement of the defect, such as symmetry, can be assumed then it is possible to determine the electronic states quite accurately. Unfortunately, it is not possible at present to deal confidently with cases involving both large atomic rearrangements and major changes in electronic states. In the present Section we shall therefore deal primarily with lattice relaxation and the atomic structure of defects while in the next we review the calculation of the electronic states of defects. Our aim is to show how these calculations can, within their limitations, provide greater insight into defect properties and aid the interpretation of experimental results.

6.1 General Formalism for the Static Lattice

We shall limit our discussion of the electronic and atomic structure of defects by neglecting all thermal motion. There are very many problems where it is quite permissable to do this. For example, we have already noted (§3.1) that, within the validity of the quasi-harmonic approximation, activation energies are equal to the corresponding potential energy changes in the equivalent static lattice. The same is also true of other thermodynamic defect energies (e.g. formation energies and interaction energies). In defect spectroscopy the Franck-Condon principle tells us that the peak of an optical absorption or emission band corresponds to an electronic transition but with the atoms fixed at the mean positions they have in the initial state. It is these mean positions which our calculations aim to determine. Of course, there are other problems where we need more than the average atomic position. For example, for defect entropies, among thermodynamic properties, and for Huang-Rhys factors and other consequences of electron-phonon coupling among spectroscopic properties, we need to determine the modes and frequencies of vibration of the defect and the surrounding lattice. However, we shall not go on to consider these motions here but will limit ourselves to the more basic problem of determining the mean structure and quantities which can be derived from it.

We therefore first write down the potential energy function of our defect plus the surrounding lattice, and then determine the atomic positions which minimise this function. All the methods which have been devised for this problem in principle divide the crystal into two regions - I and II. Region I contains the defect and an arbitrary number of its neighbouring atoms while Region II is the rest of the crystal. The object of this division is to separate out that part (II) where the distortions are small enough for, at least, a harmonic approximation to apply; analytic methods or other established results can then be applied. Conversely we must expect the distortions to be sufficiently large in Region I that numerical solutions are unavoidable. The boundary between Regions I and II is, of course, arbitrary though obviously it should be drawn sufficiently far from the defect that the results are independent of its exact position.

We therefore write the energy function as

$$E = E_I(\underline{\lambda},\underline{x}) + E_{I,II}(\underline{\lambda}:\underline{x},\underline{\xi}) + E_{II}(\underline{\xi}) \ . \tag{6.1}$$

Here λ denotes any variational parameters in the electronic wavefunction of the defect, while $\underline{x}$ and $\underline{\xi}$ denote the positional variables of the atoms in Regions I and II respectively. We do not limit ourselves to 'rigid' atoms so the $\underline{x}$ and $\underline{\xi}$ vectors also include internal co-ordinates, e.g. relative position of 'shell' and 'core' in a shell-model atom or ion. Given that Region II is harmonic, it is always possible so to arrange terms between $E_{I,II}$ and E_{II} that E_{II} is a pure quadratic function of $\underline{\xi}$, i.e.

$$E_{II} = \frac{1}{2} \underline{\xi}^T \underline{\underline{A}} \, \underline{\xi} \ . \tag{6.2}$$

This is then actually the energy of our Region II filled with perfect undistorted and unpolarised lattice in place of our defective I. The force constant matrix $\underline{\underline{A}}$ is then obtainable either from a model of the solid or, in favourable cases from a knowledge of the eigenvectors and eigenvalues (frequencies) characterising the vibrations of the perfect lattice. The energy of region I, E_I, and the interaction term, $E_{I,II}$, require either quantum mechanical calculations or else a potential model of the defect.

Let us deal first with the minimisation of E w.r.t. $\underline{\xi}$. By (6.1) and (6.2) the equation of equilibrium in $\underline{\xi}$ is

$$\underline{F} = -\frac{\partial E}{\partial \underline{\xi}}_{I,II} = \underline{\underline{A}} \, \underline{\xi} \ . \tag{6.3}$$

It is consistent with the use of the harmonic approximation to set

$$\underline{F} = \underline{F}^{(o)} + \underline{\underline{F}}^{(1)} \underline{\xi} \ , \tag{6.4}$$

whence

$$[\underline{\underline{A}} - \underline{\underline{F}}^{(1)}(\lambda,x)]\,\underline{\xi} = \underline{F}^{(o)} \quad . \tag{6.5}$$

The solution for $\underline{\xi}$ is therefore

$$\underline{\xi} = \underline{\underline{G}}\,\underline{F}^{(o)} \quad , \tag{6.6}$$

in which

$$\underline{\underline{G}} = [\underline{\underline{A}} - \underline{\underline{F}}^{(1)}]^{-1} \quad , \tag{6.7}$$

is the response function (static Green's function) for the perturbed lattice (Region II). It may be noted that $\underline{\underline{A}}^{-1}$ is just $\underline{\underline{G}}^{(o)}$, the response function for the perfect lattice (Region II filled with perfect Region I). $\underline{\underline{G}}$ and $\underline{\underline{G}}^{(o)}$ are related in the usual way via

$$\underline{\underline{G}} = \underline{\underline{G}}^{(o)} - \underline{\underline{G}}\,\underline{\underline{F}}^{(1)}\,\underline{\underline{G}}^{(o)} \quad . \tag{6.8}$$

Of course, these are only formal solutions. Practical methods of solution are of three kinds, which we may label as the Mott-Littleton (68), the Kanzaki (71) and the Tewary (72) methods. By Mott-Littleton methods we mean all those which employ the limiting dielectric and elastic continuum solutions to give the displacements in Region II; often the elastic part of the continuum solution is approximated by that for an isotropic solid and sometimes it is even omitted altogether. Provided Region I is large enough the errors so introduced do not often seem to be very serious. By contrast, the polarisation caused by charged defects is of long range ($\sim 1/r^2$) and should not be neglected even with the largest practicable Region I. In the Kanzaki method one avoids the difficulty presented by the inversion of the large matrix $\underline{\underline{A}}$ by making a Fourier transformation (the transform of $\underline{\underline{A}}$ is then the same 'dynamical matrix' as arises in the theory of lattice vibrations). Actually the Fourier transformation is applied to the equation

$$\underline{\underline{A}}\,\underline{\xi} = \underline{F}^{(o)} \tag{6.8}$$

to give Kanzaki's "first approximation" for $\underline{\xi}$ while the $\underline{\underline{F}}^{(1)}$ term is included in a subsequent real-space calculation to give his "second approximation" via

$$\underline{\xi} = (\underline{\underline{1}} - \underline{\underline{A}}^{-1}\,\underline{\underline{F}}^{(1)})^{-1}\,(\underline{\underline{A}}^{-1}\,\underline{F}^{(o)}) \quad , \tag{6.9}$$

the quantities $\underline{\underline{A}}^{-1}$ and $(\underline{\underline{A}}^{-1}\underline{F}^{(o)})$ having been obtained from the first calculation by Fourier transforming back again; this last equation is exactly the same as (6.7). In the Tewary method (72) one works entirely in real space, first finding $\underline{\underline{G}}^{(o)} \equiv \underline{\underline{A}}^{-1}$, i.e. the unperturbed lattice Green function, and then calculating the perturbed function

$$\underline{G} = (\underline{1} - \underline{G}^{(o)} \underline{F}^{(1)})^{-1} \underline{G}^{(o)} \quad (6.10)$$

which is just (6.7) and (6.9) again.

In comparing these methods, one notes that the Kanzaki and Tewary methods both use the full force constant matrix $\underline{A}$ i.e. they take full advantage of the harmonic assumption for Region II. Region I can therefore be quite small in these methods; in fact, it is often taken to be just the defect itself so that little or no further lattice relaxation remains to be done*. By contrast, the Mott-Littleton type of calculation assumes that Region II behaves as a continuum; thus a considerably larger Region I is required to ensure that the displacements are small enough in Region II for this assumption to be accurate. The main effort with these methods thus goes into the problem of minimising E w.r.t. the Region I variables, $\underline{x}$. Fortunately, progress in numerical minimisation methods in the last dozen years has now made it practical to handle quite large problems of this type economically (e.g. a few hundred variables). An example of an efficient program of the Mott-Littleton type is Norgett's HADES program (73). In addition to methods based on these advances in numerical analysis (search methods, conjugate-gradient methods, variable metric methods, etc. 74) we should also mention the dynamical method of minimising E (75); in this the actual atomic equations of motion are solved but with the kinetic energy periodically removed (every time it reaches a maximum). This remains a static lattice calculation though it has the advantage in principle of avoiding shallow secondary minima. Fully dynamical methods akin to 'molecular dynamics' studies of fluids are reviewed in Ref. 76.

There have now been many defect calculations using these methods (77). Broadly speaking we can say that the Kanzaki and Tewary methods are very suitable for defects of simple structure (e.g. vacancies, interstitials and impurity atoms) and high symmetry. They are convenient for localised forces (e.g. neutral defects) though the Kanzaki method has been extended and applied to charged defects, notably by Hardy et al. (78). However it is not clear that they have any general advantages over well-formulated Mott-Littleton methods (including dynamical relaxation methods)

---

*When the electronic states of the defect have also to be calculated (e.g. by minimisation of E w.r.t. the variational parameters $\underline{\lambda}$) then of course this lattice relaxation calculation has, in principle, to be repeated until the absolute minimum energy is determined. But simplifications are often possible (6).

which at present would seem to be the only practical approach to complex defects of low or unknown symmetry (though, of course, they are not limited to such cases).

6.2 Physical Models

The electronic theory of solids, despite its many successes, is still not well adapted to providing general potential energy functions of the sort required for applications of the lattice relaxation methods just described. As a result such calculations generally rely upon more or less empirical models based upon general ideas of the nature of interatomic forces in the different types of solid (79). We can only summarise these rather briefly here.

6.2.1 Rare-gas solids.
It has long been customary to discuss the forces between rare-gas atoms and the cohesion of the solid state in terms of pairwise interaction potentials of the Lennard-Jones type

$$V(r) = \frac{A}{r^n} - \frac{B}{r^m} \qquad (6.11)$$

generally, though not always, with $n = 12$ in the repulsive term and $m = 6$ in the attractive term (van der Waals attractions). The overlap repulsions between such closed-shell atoms are, however, better represented over a sizeable range of interatomic spacing r by the Born-Mayer form, $A \exp(-r/\rho)$, so that a somewhat better representation over this range is

$$V(r) = A \exp(-r/\rho) - \frac{B}{r^6} . \qquad (6.12)$$

Quantum mechanical calculations of the interaction energy of a pair of atoms can be used to obtain A and ρ, though these calculations will not generally include the particular correlation effects necessary for calculating the van der Waals term. For this one could use the classical London formula or later refined calculations (80). However, empirical determinations of $V(r)$ by analysis of data from atomic scattering and from gaseous and condensed phases show that these forms (6.11) and (6.12) are too simple to give an accurate description (81). Various, more complex functions for the pairwise potential $V(r)$ are now available which (supplemented by theoretical van der Waals' and three-body terms) accurately represent this data; qualitatively they resemble the corresponding Lennard-Jones functions though the depth of the minimum in all cases is somewhat greater (up to 20%) than that of the fitted Lennard-Jones potentials. From this work (81) it would appear that many-body terms in the total energy are not large and that they are adequately represented by the Axilrod-Teller 'triple-dipole' 3-body term*, though this conclusion has been disputed (82). Despite the undoubted success of these potentials in consistently fitting a wide range of physical properties of the rare gases one persistent difficulty remains; namely, that these, and most earlier rare-gas potentials as well, predict the stable solid structure to be hexagonal close-packed whereas it is actually face-centred cubic (Ne, Ar, Kr, Xe). This difficulty can seemingly be overcome by allowance for short-range 3-body terms coming from electron overlap and exchange effects (82) though the consequences of allowing for such terms in the determination of the empirical 2-body potential V(r) have not been consistently followed through. Lastly, we observe that these potential functions are all effectively rigid atom models. There is no allowance for the possible dependence of interaction energy upon atomic polarisation. The electrical polarisability of the atoms likewise must be separately determined.

6.2.2 <u>Ionic solids</u>. Empirical models of ionic solids have played an important part in solid state theory since the earliest work of Born. The cohesion of ionic solids is dominated by the Coulombic interactions between the ions (which are assumed to carry integral charges, $\pm e$, $\pm 2e$ etc.). Overlap repulsions are generally represented by a Born-Mayer form $A \exp(-r/\rho)$ often extended into a supposedly universal form by the idea of characteristic ionic radii. To represent their dielectric properties it is necessary to introduce ionic polarisabilities, often taken to be characteristic of the ions. However, it is now widely recognised that for the successful representation of both static and dielectric properties and of the lattice vibrations it is necessary to use shell-models, i.e. models in which each ion is given at least one internal degree of freedom corresponding to the displacement of the outermost electronic shell relative to the core of the ion; other degrees of freedom corresponding to the radial compression and the deformation of the electronic shell have also been introduced (93). Although these shell models were first introduced and analysed for the representation of harmonic properties, recent work, having defect calculations as its aim, has shown that one may embody the shell-model assumptions about the individual ions in a general Born model and that such models are very suitable for the alkali halides (84), the alkaline earth fluorides (85,86) and oxides such as UO_2 (85,87), FeO (22) MgO (88) and a number of others. In these, we retain the idea of Born-Mayer overlap repulsions (which may, in fact be independently calculated by Hartree-Fock methods)

---

*This is the 3-body analogue of the 2-body van der Waals polarisation forces, arising from the 3rd order terms in a perturbation treatment.

but assume that they act between the electron shells*. This has the effect of introducing a coupling between the overlap repulsions and the electric moments on the ions - which is important for the avoidance of those spurious divergences in the energy function which can occur with models which do not have this feature (90).

6.2.3 Covalent semiconductors.

It is generally recognised that the valence crystals (diamond, Si, Ge and related compounds with the zincblende structure) are distinguished by having strongly directional covalent bonding between nearest neighbours, with only much weaker interaction between more distant pairs of atoms. The traditional representation of the energy of a molecular covalent bond as a function of interatomic distance is the Morse potential

$$V(r) = D[\exp\{-2\alpha(r-r_o)\} - 2\exp\{-\alpha(r-r_o)\}] \qquad (6.13)$$

Such potentials and simple generalisations of them have been used for defect calculations in diamond, Si, Ge etc. The interaction is assumed to exist only between nearest neighbours so that r_o in (6.13) is then the nearest neighbour separation. The practical dificiency of this model is that it is purely a bond-stretching model and does not tell us how to predict the forces coming from a change in the angles between bonds. The lattice dynamics of these crystals confirms that these bond-angle forces are significant, and thus makes it unlikely that their neglect in some of these defect calculations can be satisfactory (91). A harmonic potential function for the solid, known as the valence force model (92), which is expressed in terms of the changes in bond lengths, Δr_{ij}, and the changes in bond angles, $\Delta \theta_{jik}$ (angle θ_{jik} between bonds i-j and i-k) has been shown to be convenient for the representation of lattice vibrations in these materials (12). This function is

$$V = \frac{1}{4} F_r \sum_{i=1}^{N} \sum_{j=1}^{4} (\Delta r_{ij})^2 + \sum_{i=1}^{N} \sum_{j=1}^{3} \sum_{k=j+1}^{4} \left\{ \frac{1}{2} r_o^2 F_\theta \Delta\theta_{jik}^2 \right.$$

$$\left. + f_{rr} \Delta r_{ij} \Delta r_{ik} + r_o f_{r\theta} (\Delta r_{ij} \Delta\theta_{jik} + \Delta r_{ik} \Delta\theta_{jik}) \right\}, \qquad (6.14)$$

where certain much smaller terms of second order in $\Delta\theta$ have been omitted. It implies that each bond of the perfect lattice is in equilibrium, something which is supported by the observation that

---

*Some very recent work (89) seems to show that, as in the rare-gas solids, the use of the form (6.12) for the non-Coulombic interaction of a pair of ions (overlap repulsion plus a van der Waals type of attraction) is too restrictive. As there, the use of a more elaborate form gives a better representation of the solid; it also leads to a smaller effective van der Waals term than is obtained with (6.12) and this has a small but significant effect on calculated defect energies.

the nearest-neighbour distances in diamond and Si are closely equal
to the mean single-bond length found in molecules of saturated hydrocarbons and similar compounds of Si (1.544Å in diamond compared to
1.537Å and 2.35Å in Si compared to 2.32Å). However (6.14) remains
a harmonic function and is thus inadequate for situations where we
break bonds, as in the formation of vacancies, or produce other
large strains. Accordingly, Sinclair and Lawn (93) proposed an
extension of (6.14) in the spirit of the Morse function (actually
they retained only the terms in F_r, F_θ and $f_{r\theta}$ which are the three
largest coefficients). They made the following replacements

$$\Delta r_{ij} \to \Delta_r(r_{ij}) \equiv \frac{1}{\alpha}[1 - \exp\{-\alpha(r_{ij}-r_o)\}]$$

$$\Delta\theta_{jik} \to \Delta_\theta(\theta_{jik}) \equiv \left(\frac{r_o}{-\sin\theta_o}\right)(\cos\theta_{jik} - \cos\theta_o) \times$$

$$\times \exp\{-\alpha(r_{ij}-r_o)\} \exp\{-\alpha(r_{ik}-r_o)\} \quad . \quad (6.15)$$

Here r_o is the nearest neighbour bond length and θ_o is the equilibrium tetrahedral bond angle ($\cos\theta_o = -1/3$). Their application of
this function to the modelling of cracks was quite successful and
indicated that (6.15) could yield a satisfactory model. In conclusion, however, two other things should be said. The first is
that these potential models are essentially rigid bond models;
they do not tell us anything about dielectric properties and thus
are incomplete if we want to describe charged defects. Shell
models, which do allow the representation of atomic polarisability,
have also been shown to permit good descriptions of the lattice
dynamics of these covalent crystals; indeed it was the discovery
of the need for coupling between fifth neighbours in Ge which first
led to their formulation (Cochran, (83)). However, the consistent
extension of these shell models beyond the harmonic regime does not
yet seem to have been tried for these materials. The second point
to be made is that the above models all refer to the interactions
of the bonded atoms. What happens when bonds are broken, as in
vacancy formation, must be separately considered. One may either
make intuitively founded assumptions (91) or attempt proper quantum
mechanical calculations (6).

6.2.4 <u>Metals</u>. Although this volume is devoted to non-metals,
very many defect calculations have been made for metals (9) and
many of these have more general implications; it is therefore appropriate to make brief reference to the potential functions which have
been employed. One important class of interatomic potentials are
those constructed on the basis of electronic 'pseudopotential' theory
which is a perturbation theory of those metals for which the interactions of the conduction electrons and the ion cores can be
treated as weak (e.g. the alkali metals). In a certain approximation the energy function for the solid can then be shown to be

the sum of two terms, one is independent of atomic arrangement but depends on the total volume of the solid while the other is a sum of two-body central-force interactions which, however, depend parametrically upon the total volume (94). There is therefore a close connection between electronic theory and the interatomic potentials in this case (15). Many metals of practical interest however may not be treated in the pseudopotential approximation (e.g. the noble and the transition metals). In these cases purely empirical potential models have been constructed, some of which neverless embody certain general features indicated by electronic theory, e.g. Born-Mayer repulsions between overlapping ion cores, the existence of a term in the energy dependent on total volume but independent of atomic arrangement, the fact that screening by conduction electrons can lead to spatially oscillatory screening densities and consequential effects in the interatomic potential, etc. (9,75).

6.3 Some Results

There is now a very large body of defect calculations made by using variations of the methods and models we have described (thus refs. 2, 6, 7, 9, 13, 17, 73, 75, 95, 96, although all books and reviews is far from being an exhaustive list). The aims of these calculations vary from case to case depending upon the extent of the corresponding experimental knowledge, but we can perhaps group them into three as follows.

(i) In cases where defect properties are well established experimentally one aims to verify that the relevant parts of solid state theory are adequate for calculating properties of the defect state as well as those of the perfect solid (e.g. cohesion, lattice vibrations, optical properties, etc.). Practical examples include the calculation of characteristic defect energies and their comparison with experimental values.

(ii) Another aim of calculations on experimentally well-understood situations may be to confirm theoretically that the observed defect populations and properties are those expected to be dominant (e.g. that, in the solid rare-gases, vacancies are present in much larger concentration than interstitials, that vacancy jumps to nearest neighbours are much more frequent than to next nearest neighbours, etc.).

(iii) In more complex situations or in less well studied systems the aim of the calculations will be to determine dominant defect types and processes, to limit the number of unknowns, to guide the choice of experiment, to help understand unusual or unexpected results, etc.

In this part we shall briefly survey some of the results which have been obtained to illustrate these various functions and the limitations of these calculations. In general however it may be remarked that, at present, progress would appear to be most rapid in calculations for ionic crystals and oxides. We limit ourselves to situations where the defect can be represented in terms of the models already described; situations where quantum mechanical calculations are needed for the defect are described in §7.

6.3.1 <u>Basic structures</u>. As we have already remarked in previous sections, the structure of vacancies is generally assumed to be close to that indicated by simply removing an atom from an otherwise rigid lattice. In defect calculations in particular the lattice relaxation around the vacancy is generally assumed to retain the point symmetry of the lattice site. This assumption has been surprisingly little tested theoretically, though it is now quite practicable to do so. Catlow (private communication) verified that the relaxation around the anion vacancy in a model of CaF_2 conformed to these expectations, namely that the four immediate cation neighbours relaxed outwards along <111> directions while the six anion first neighbours relaxed inwards along <100> directions. However it may not be safe to assume that this is always so in fluorite lattices (possibly not for example in cases such as PbF_2 where the cations are large and polarisable). 'Split-vacancy' defects (i.e. a pair of vacancies with an 'interstitial' atom midway between) have sometimes been suggested (see e.g. 23, 35) but have not been substantiated theoretically.

Interstitials, by contrast, have been looked at in greater detail especially in models of b.c.c. and f.c.c. metals (9,75) - perhaps because it is less obvious that the generally large distortions around an interstitial will not lead to a lowering of symmetry. The common conclusion seems to be that in the close-packed monatomic f.c.c. lattice the interstitial takes up the <100> - dumb-bell or split-interstitial form while in the b.c.c. lattice it is likely to take up the <110> - dumb-bell form (Fig. 3.1). However, although these appear commonly to be the configurations of lowest energy other configurations may have energies only a little higher with the result that activation energies for motion are low. Formation energies, on the other hand, are high, generally several times those of vacancies in these structures.

The stable configurations of interstitial ions in ionic crystals have been studied in a few cases. In the alkali halides (NaCl structure) the stable configurations appears to be the cube-centre position in most cases (97). The <111> oriented dumb-bell configuration appears as the lowest-energy saddle point configuration (corresponding to interstitialcy motion). In the fluorite compounds the stable interstitial anion position is at the centre of one of the alternate 'empty' cubes of eight lattice anions (Fig. 6.1) though in some

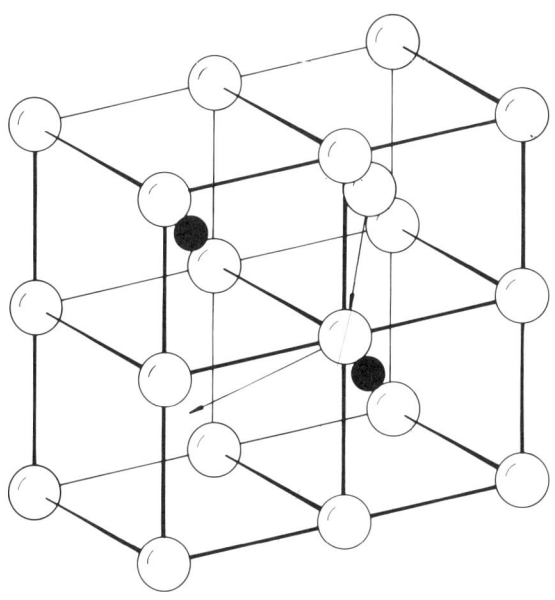

Fig. 6.1 A section of the fluorite lattice-open circles anions, filled circles cations. An anion interstitial is shown at the centre of the top right cube; the interstitialcy mode of displacement is as indicated by the arrows (7).

cases (e.g. UO_2) the exact location may be slightly off the exact cube-centre as a consequence of the Madelung potential there (85-87).   Interstitial anion movement occurs via an interstitialcy mechanism; the activation energy for direct movement from one cube-centre to another is much higher (85).

These statements about interstitial ions do not apply to interstitial halogen <u>atoms</u>. Thus interstitial halogen atoms in alkali halides effectively bond with a lattice anion to form an X_2^- molecule or dumb-bell occupying the lattice site, generally, though not always, in a <110> orientation, i.e. along the rows of anions ('H-centres' - see Schoemaker, this Volume).

These remarks have concentrated on the atomic structure of isolated vacancies and interstitials. The clustering of these defects can lead to complex structures and defect calculations can be particularly useful in these cases (22,75,85,87). We describe one or two successful such calculations separately below.

Table 6.1

Enthalpies of Vacancy Formation in Solid Rare Gases (cals/mole)

Substance	h_{fv} (expt.)	h_{fv} (calc.)
Ne	478	485
Ar	(1290)	1790
Kr	1780-1985	2520
Xe	(2480)	3510

The 'experimental' values in brackets have been obtained by applying a 'corresponding states' rule to the result for Kr but they are consistent with a number of other determinations from experiment in ways less direct than the Simmons-Balluffi experiment (eqn. 2.11) used to obtain the values for Ne and Kr. (After Chadwick and Glyde, (95)).

6.3.2 <u>Defects in solid rare-gases</u>. There have now been many calculations of the energies of formation and some of the energies of activation of vacancies in the rare gas solids, most, but by no means all, carried out in the static lattice approximation. There have been one or two calculations of interstitial formation energies (Cotterill and Doyama, (9)) which have shown these to be several times those for vacancies, in agreement with the experimental demonstrations that vacancy disorder is dominant (95). Most of these defect calculations have used pair potentials of the Lennard-Jones type (6.11). The more refined empirical potentials mentioned in §6.2 (8) have yet to be taken up in these defect calculations. The effects of zero-point vibrations and of 3-body interactions have, however, been studied. Attempts have also been made to improve upon the static lattice approximation and to avoid it altogether by molecular dynamics methods (76). All these vacancy calculations indicate small atomic relaxations around vacancies (e.g. nearest neighbours move inwards by only $\sim$ 1% of the mean nearest-neighbour distance in the perfect lattice) and correspondingly small energies of relaxation. The experimentally determined energies (enthalpies), with the exception of Ne, are substantially less ($\sim$ 30% less) than the sublimation energy, which points to much larger energies of relaxation (Table 6.1).

This is a disappointing result in view of the presumed simplicity of the electronic structure of the rare-gas solids and their position as the simplest members of the class of molecular solids.

The reason for the failure is far from clear. There can be little doubt about the mathematical accuracy of the calculations. Since many variations have been played on the theme of 'rigid' pairwise potential functions it might appear that the omission of atomic polarisability is to blame, but neither particular arguments nor the fact that the lattice vibrations appear to be well described by the rigid atom models give much support for the idea. To get a large atomic relaxation energy it is necessary for the forces on an atom in the perfect lattice to be made up of sizeable opposing terms from nearest and more distant neighbours. Then when a vacancy is created there is an unbalance of forces acting on the nearest neighbours. That the presently available potential models do not give a large enough force unbalance on these atoms, even though they agree with a wide range of other physical data, therefore seems to present a fundamental difficulty. The matter is made more puzzling by the fact the same calculations give close to the correct energy of activation for self-diffusion, i.e. for the sum of vacancy formation and migration energies.

6.3.3 Covalent semiconductors.

When we create a vacancy in the covalently bonded diamond lattice we break four bonds and are left with one unpaired electron in each 'dangling bond', as shown in Fig. 6.2. We should probably not expect to be able to describe the rebonding of these unpaired electrons in any simple intuitive way which is also useful; indeed it is very difficult to do so accurately, even by formal theory at present (6). Nevertheless there have been several calculations of the energies of formation of (neutral) vacancies in diamond and the Group IV semiconductors made by assuming (i) that the atoms rebond together in pairs (Fig. 6.2) and (ii) that the energy of these new bonds is given by the same bond-energy function as describes the perfect solid e.g. eqn. (6.13)). This particular rebonding assumption (to give an E-state) finds some justification in more fundamental calculations and in experiment (Watkins, this Volume). The results of these calculations (Table 6.2) are not unreasonable in themselves but seemingly neither have they been accurate enough to be of great help in the interpretation of diffusion in these materials (Dobson, this Volume). In the example of Si, the low activation energy of the vacancy V^o (0.3 eV) obtained from low-temperature e.p.r. studies (Watkins, this Volume) and the high activation energy for self-diffusion (5.1 eV) imply a formation energy higher than calculated. In Ge, where the energy of self-diffusion is 3.0 eV, the calculated energy seems reasonably good. Various ad hoc explanations of the Si result have been advanced (see e.g. the review by Casey and Pearson, (44)).

6.3.4 Schottky and Frenkel defects in ionic crystals.

Somewhat in contrast to these calculations of defect energies in solid rare-gases and in semiconductors, the calculations in ionic crystals now appear to be quite satisfactory. Indeed, the success of these

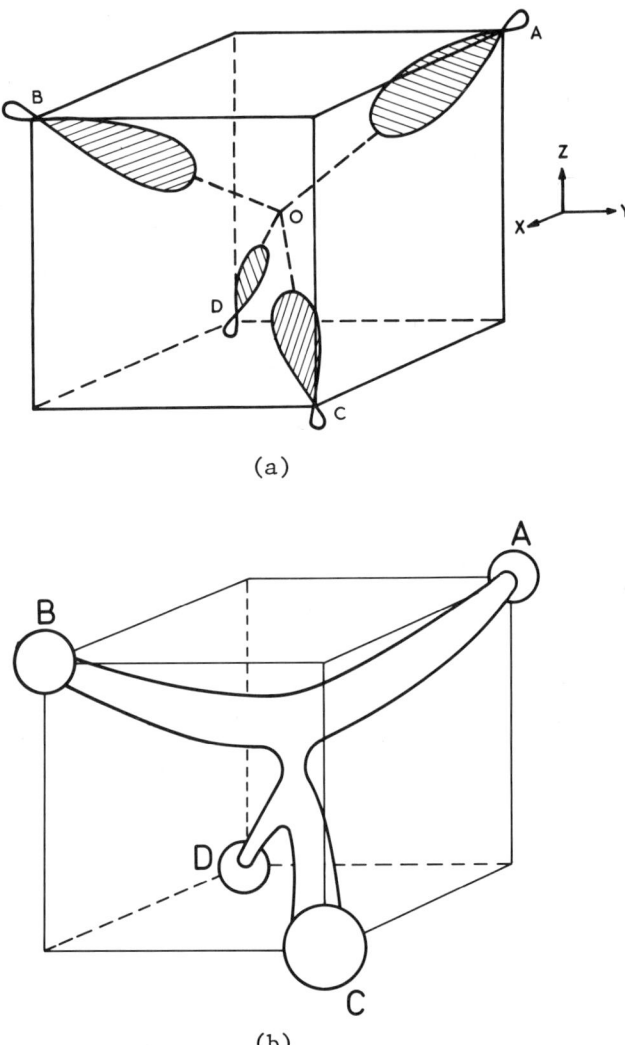

Fig. 6.2 Schematic diagrams of a vacancy in a diamond lattice (a) showing the four dangling orbitals which were paired with orbitals on the atom at O before it was removed and (b) showing one possible rebonded structure.

Table 6.2

Some Calculated Energies of Formation (in eV) of Vacancies in the Diamond Structure (91)

Substance	Calculated Formation Energy
Diamond	4.2
Si	2.3-2.8
Ge	2.1-2.5

computations of the energies of Schottky and Frenkel defects now enables us to tackle new and complex problems with considerable confidence (see below). To a large extent this success depends upon the dominance of Coulomb and electrostatic forces whose spatial variation is known and general. However, other factors are certainly important; more particularly, as a very large part of the relaxation energy associated with charged defects is polarisation energy, it is vital that the model used should correctly represent the static dielectric constant, ε_o. This is generally done with 'deformation-dipole' and shell models (§6.2) but has often not been done with polarisable point-ion models.

Some examples of the results of these calculations are given in Tables 6.3-6.6. These results were obtained either by the Kanzaki method (deformation dipole model, (78)) or by the Mott-Littleton method (shell models; (84,85,88)). They confirm that the dominant defects in the alkali halides are Schottky defects while those in the alkaline earth fluorides are anion Frenkel defects. Furthermore, they yield calculated formation energies of these majority defects which are in good agreement with experimental values. These results show that it is now possible to calculate accurate values of the energies of 'closed-shell' defects in these materials with relatively simple shell-model versions of the general Born ionic model. Extensions of this work to many other defects (e.g. simple clusters), other properties (e.g. activation energies) and other materials and structures (e.g. oxides, rutile) are now being rapidly made (see e.g. (22,84,85,88)). We describe two examples of these extensions which illustrate objectives of the third type described at the beginning of this §6.3.

6.3.5 <u>Structure of interstitial clusters in CaF_2/YF_3</u>. There is much experimental evidence to show that the F^- interstitials introduced into CaF_2 (and similar compounds) by doping with trivalent cations such as Y^{3+} and rare-earth ions may pair with these ions to

Table 6.3

Calculated Energies of Formation of Frenkel and Schottky Defects (eV) in Some Alkali Halides

Substance	Cation Frenkel	Anion Frenkel	Schottky
LiF	3.10	4.71	1.69-2.37
NaCl	2.88-3.21	3.85-4.60	1.79-2.34
NaBr	2.56-2.89	3.68-4.84	1.66-2.20
KCl	3.24-3.46	3.41-3.73	1.90-2.50
KBr	2.75-3.16	3.11-4.17	1.81-2.28
RbCl	2.96-3.71	2.94-3.52	1.90-2.36

This table gives the range of values calculated from deformation dipole models (78) and from shell models ((84); also (98) for Schottky energies only).

Table 6.4

Calculated and Experimentally-Determined Schottky Formation Energies (eV) in Some Alkali Halides

| Substance | Calculated | | Experimental |
	Deformation Dipole	Shell Models	
LiF	1.69-2.37	2.05-2.37	2.34-2.68
NaCl	1.79-2.34	2.22-2.32	2.18-2.50
NaBr	1.66-2.20	2.01-2.13	1.72-2.16
KCl	1.90-2.20	2.34-2.50	2.26-2.59
KBr	1.81-2.13	2.24-2.28	2.30-2.53
RbCl	1.90-2.36	2.36	2.04

The deformation dipole models were evaluated by Hardy et al. (78) while the shell model results were obtained by Faux (98) and by Diller (84). The ranges quoted result from variations in these two types of model. The experimental values are from the compilations in Refs. (12) and (17).

Table 6.5

Calculated Energies of Formation (eV) of Frenkel and Schottky Defects in 3 Different Shell Models of CaF_2

Defect	Formation Energy
Anion Frenkel	2.6–2.7
Schottky	7.0–8.6
Cation Frenkel	8.5–9.2

These results are due to Catlow and Norgett (85,86).

form nearest neighbour pairs having the expected tetragonal symmetry (see §5.3 and Ref. 7). In these pairs the interstitial F^- ion remains close to the 'empty' cube-centre position next to the substitutional cation (cf. Fig. 6.1). These defects are very eivdent at low impurity concentrations (say 0.1% or less). At high impurity concentrations (say several % upwards), however, there are equally definite experimental indications of interstitial F^- ions

Table 6.6

Calculated and Experimentally-Determined Frenkel Formation Energies (eV) in the Alkaline Earth Fluorides

Substance	Calculated	Experiment
CaF_2	2.6–2.7	2.2–2.8
SrF_2	2.2–2.4	1.7–2.3
BaF_2	1.6–1.9	1.9

The calculated values have been obtained for shell models (85, 86); the experimental values are taken from published compilations (7,12).

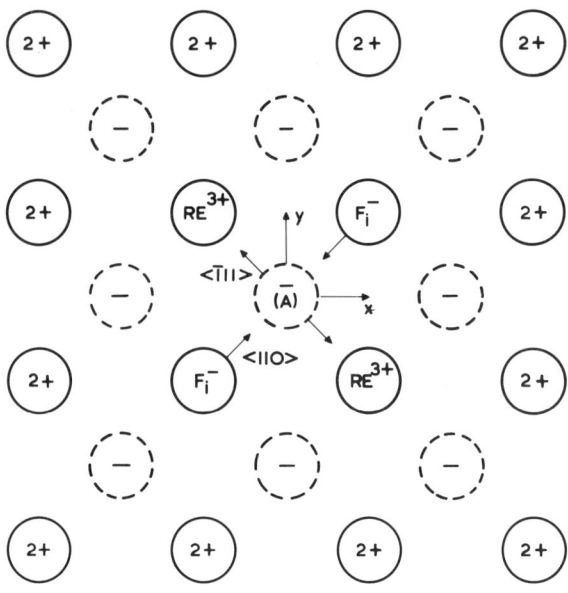

Fig. 6.3 A (100) section of a fluorite lattice showing the principal relaxations arising when two $Y^{3+} - F_i^-$ pairs are brought together 'head-to-tail' in the configuration shown. Ions initially lying in a (100) cation plane are shown as full circles while the anions in the planes above and below are shown as broken circles. The two anions above and below the cation plane relax towards the centres of the empty anion cubes above and below the two Y^{3+} ions.

in two different positions (neither of which is a cube-centre position) accompanied by unoccupied normal F^- sites (99), all three in roughly equal numbers. The two interstitial locations correspond to (i) ions displaced about $a_o/6$ in a <110> direction from a cube-centre position and about $a_o/4$ in a <111> direction from a normal F^- site (a_o being the lattice parameter, i.e. twice the edge length of the F^- cube). This seemingly complex situation was explained by Catlow (100) by modelling various defect clusters using the methods and potential models already described (Mott-Littleton method, shell model and HADES program). He showed firstly that isolated $Y^{3+} - F_i^-$ pairs did indeed have a positive binding energy and the expected tetragonal structure. Secondly, the more complex structures seen at high concentrations could be understood in terms of the stable relaxed structure taken up by an aggregate of two of these pairs. We may suppose them to be initially lying head to tail forming a square in a (100) plane (Fig. 6.3). However if the

Table 6.7

Energies of Formation (eV) of Intrinsic Defects in UO_2

Defects	Calculated	Experimental
Electrons and Holes	1.99	2.3-2.6
Anion Frenkel $- O_i''$ and O_V'	5.0-5.5	∿3
Anion Frenkel $- O_i'$ and O_V'	8.9	-
Schottky $- U_V''''$ and $2O_V''$	10.3	∿6
Cation Frenkel $- U_V''''$ and U_i''''	18.5	∿9.5

The calculated values were obtained by Catlow (85,87). The experimental values for electronic disorder are from Ref. 102, while the values for structural defects are taken from the assignments in Ref. 103. It should be noted that there is still substantial uncertainty over the latter. Also the theoretical values may be more sensitive to inherent limitations of the ionic model than was the case for the alkaline earth fluorides because the electrostatic forces are close to four times larger. Nevertheless the order of defect energies obtained in the two cases agrees.

lattice relaxation is constrained to retain this symmetry the binding energy relative to two separated pairs is relatively small. But when this constraint is removed, large symmetry-lowering relaxations occur (Fig. 6.3) which lower the energy considerably giving a sizeable binding energy relative to two separated pairs of some 1.2 eV in CaF_2. Furthermore the nature and large size of the relaxations is as observed; the two F_i^- are pulled towards one another along <110> by their attraction to the two Y^{3+} ions and the resultant repulsive forces on the two normal F^- ions above and below the mid-point push these ions off-site in <111> directions towards the centres of neighbouring empty cubes. These calculations thus provide an immediate explanation of the defect structures seen by the neutron diffraction experiments. They have been further elaborated beyond the example we have described to deal with still larger clusters. In conclusion, it may be remarked that these and some other calculations on model fluorite lattices point to a tendency for substitutional F^- ions to relax towards the centres of neighbouring empty F^- cubes, i.e. in <111> directions, rather than exactly in the direction of the force immediately acting on the ion. The fluorite lattice thus provides a situation where to assume the symmetry of relaxation may lead to serious error.

6.3.6 <u>Point defects in UO_{2+x}</u>. The oxide UO_2 also has the fluorite structure but rather readily takes up extra oxygen, giving oxygen interstitials and electron holes (p type electrical conductivity). The interstitial oxygen can give structures like those discussed above in CaF_2/YF_3; indeed the occurrence of such structures was first found by Willis in his neutron diffraction studies of UO_{2+x} in 1964 (101). The explanation for these structures and the extent to which the analogy with CaF_2 (pure and doped) could be relied on to provide statistical defect models for UO_2 remained in doubt for almost 10 years. Catlow extended the ionic shell model to give a satisfactory model of UO_2 (as made up of U^{4+} and O^{2-} ions) and employed the HADES program (73) to calculate many defect properties in this substance, thereby answering these long-standing questions (85,87). Here we shall touch upon just a few of the results.

Firstly in Table 6.7 we give the energy of formation of various defects in stoichiometric UO_2. The lowest of these is that for electron-hole excitation (interpreted as the formation of U^{3+} and U^{5+} ions), showing that stoichiometric UO_2 should be an intrinsic simiconductor. The lowest structural-defect formation energy is that for O^{2-} Frenkel defects (n.b. not O^-); apart from the electronic disorder UO_2 should thus be analogous to CaF_2. We would correspondingly expect that non-stoichiometric UO_{2+x} will contain O^{2-} interstitials and electron holes (U^{5+} states). That this is so is confirmed by Table 6.8 which shows that considerable energy would be required to transform these defects to various other possibilities. These and other results are important in showing which models to select for statistical analysis to provide theories for the thermodynamic and transport properties of this substance (87). Lastly the explanation of the interstitial structures inferred by Willis is similar though not identical to that obtained for CaF_2/YF_3. In UO_{2+x} it is only necessary to have one O_i^{2-} interstitial associated with two U^{5+} ions; one then obtains a defect structure made up of two O^{2-} ions displaced in <111> directions from normal sites and one O^{2-} ion displaced in a <110> direction from a cube-centre interstitial site. The relative proportions of <111> 'interstitials', vacant normal anion sites and <110> interstitials namely 2:2:1 appears to fit the experimental results on UO_2 better than would the equal proportions of the CaF_2/YF_3 cluster. Somewhat similar success has also been obtained recently in understanding the complex defect structures which occur in the system $Fe_{1-x}O$.

Table 6.8

Calculated Energies of Defect Reactions in UO_2 (eV)

Reaction	Energy Required
$2U_s^{\cdot} \rightarrow U_s^{\cdot\cdot} + U_s$	1.0
$U_s^{\cdot} + O_i'' \rightarrow O_i' + U_s$	3.8
$O_i'' + \frac{1}{2} O_{2,gas} \rightarrow 2O_i'$	2.7

These values were calculated by Catlow (85,87). All three reactions would therefore be endothermic.

7. ELECTRONIC THEORY OF DEFECT STATES

In the preceeding section we dealt with calculations of the atomic structure and energies of defects and their simple aggregates, but we confined ourselves to those cases where all the interatomic forces either derived from the potential energy function of the solid (e.g. "closed-shell" systems such as the rare-gas solids or the strongly ionic solids) or else could be regarded as given from some other calculation. Evidently, when dealing with colour centres for example, the forces which the defect exerts on the lattice depend upon the nature of the electronic state (e.g. whether it is localised or diffuse). We thus need to be able to calculate the electronic states and energy levels of colour centres and other localised electronic defects not only in order to make and to confirm the assignments of optical absorption bands and E.P.R. spectra, but also in order to describe the coupling of the centre to the lattice and its consequences for the structure and other observable properties of the defect. This is the subject of the present section.

Of course, the electronic states determine the forces on the surrounding atoms and thus the relaxation of these atoms and of the rest of the lattice; in turn this changes the potential in which the defect electrons move and thus their electronic states. What is needed therefore is a solution which is self-consistent in both electronic states and lattice relaxation. Unfortunately we do not have a theory which is both completely general and practical at the present time. We do, however, have theories which are practical and useful over particular areas where lattice relaxations either are small (e.g. shallow states in covalent semiconductors, ground states of F-like centres) or else are known well enough from experiment that certain broad features of the centre (such as its symmetry and the nature of its electronic states) can be assumed (e.g. trapped hole centres in alkali halides and in alkaline earth oxides and fluorides). In these cases the problems can be solved successfully because the theory has to encompass relatively small changes in structure. The nature of the difficulties arising when we do not have limitations of this sort are exemplified by the problems presented by the relaxed excited states of F-centres and by the vacancy and interstitial centres in diamond and Si.

In this section we shall describe some of the elements of the successful theories and also touch upon the nature of the difficulties presently encountered.

7.1 Shallow States in Semiconductors

By shallow levels we mean those lying close to a band edge in the 'forbidden gap'. The corresponding states are localised but diffuse. The perturbing potential which binds the electron (or hole) to the centre is weak. The formal theory (104) thus applies perturbation methods to the electronic Bloch states and, not surprisingly, shows that the localised electron wave-function is expressible simply in terms of the products of the Bloch functions at the band edge with an 'envelope' function. In the simplest case where the band minimum occurs at the centre of the Brillouin zone and is non-degenerate and the electronic band levels near $\underline{k} = 0$ are isotropic and define an effective mass m^*, the envelope function satisfies a Schrödinger equation in which m^* replaces the free electron mass and in which the perturbing potential enters instead of the full potential, i.e.

$$\left(-\frac{\hbar^2}{2m^*} + V_p(r) \right) \chi(\underline{r}) = E \chi(\underline{r}) \ . \qquad (7.1)$$

The total defect electron wave-function is $u_o(\underline{r}) \chi(\underline{r})$ where u_o is the band-edge Bloch function while the energy E given by eqn. (7.1) is the energy of the localised electron relative to the band edge. All this is very well known; indeed (7.1) was first presented and applied intuitively many years ago. However, the more formal analysis is necessary in order to derive the generalisations of these results to cases of degenerate band edges and non-spherical energy surfaces which are, in fact, often the cases of practical interest.

The systems for which such effective mass theories are most useful are the predominantly covalent semiconductors (Group IV

Table 7.1

Donor Impurity Levels (in meV) in Some Semiconductors with Conduction Band Minimum at $\underline{k} = 0$

Substance	$-E_{1s}$		$-E_{2s} = -E_{2p}$	
	Calc.	Expt.	Calc.	Expt.
GaAs	5.72	5.81–6.1	1.43	1.44
InP	6.86	7.28	1.76	–
CdTe	13.06	13.78	3.27	–
CdSe	18.48	–	4.62	–

The theoretical results have been obtained from eqns. (7.1)-(7.3). The experimental results and the values of m^* and ε_o are taken from the article by Bassani et al. (104). (N.B. the calculated values differ from those given in Table 1 of Ref. 104, apparently because Bassani et al. used an incorrect value of the Rydberg).

semiconductors, the III-V compounds, etc.) containing substitutional impurities having either one more or one less valence electron than the corresponding host atom, e.g. the Group III and V elements in Si and Ge. In the example of Group V elements in Si and Ge the fifth valence electron is only weakly bound and is describable by effective mass theory; except near the impurity the perturbing potential is just that due to the additional ion core charge, i.e.

$$V_p = -e^2/\varepsilon_o r \ . \tag{7.2}$$

Here ε_o is the static dielectric constant. We are thus allowing for the polarisation of the lattice by the attractive centre; only outside the electron orbit is the net charge (+e) screened by the trapped electron. If we substitute V_p in (7.1) by the Coulombic form (7.2) then we immediately see that the states and energy levels will be like those of a hydrogen atom in which m^* replaces the free electron mass, m, and e^2/ε_o replaces e^2. In particular, the energy levels relative to the band edge will be

$$E_n = -\frac{\text{Ryd}}{n^2}\left(\frac{m^*}{m\varepsilon_o^2}\right) \quad (n = 1,2,3 \ldots) \tag{7.3}$$

where Ryd denotes the Rydberg of energy (13.605 eV).

A number of semiconductors do have spherically symmetric non-degenerate conduction band minima at $\underline{k} = 0$ and these simple predictions agree rather well with experimental data on donor levels in these materials. Some examples appear in Table 7.1. However, the most completely studied semiconductors in the present connection are the elemental semiconductors Si and Ge which have conduction band minima away from $\underline{k} = 0$, the corresponding energy surfaces being ellipsoids of revolution about <100> and <111> directions respectively. This removes degeneracies associated with the magnetic quantum number, m_ℓ, as well as the accidental degeneracy of hydrogenic ns and np states, the only degeneracy remaining being that of $\pm m_\ell$ where the axis of quantisation is the axis of revolution of the ellipsoid. In addition, there is a small splitting arising from the coupling of states associated with different but equivalent minima, though this really lies outside the effective mass approximation. Calculated energy levels are compared with experimental values for Group V donors in Si and Ge in Table 7.2.

From these results, we can conclude that the effective mass theory of one-electron shallow donors is quite successful. It is indeed even more successful than we have indicated, since when the perturbing potential, V_p, is chosen so as to be accurate in the vicinity of the donor as well as away from it (so-called central cell corrections) the remaining small differences between different donors are accurately predicted in many cases.

One can set up a completely analogous theory for shallow acceptors, e.g. Group III elements in Si and Ge. In these cases the effective mass equation describes the motion of the localised hole. The effective mass which enters is that at the top of the valence band, while the energy levels are given relative to the valence band maximum. The wave-function is that for a hole in the otherwise filled set of valence states, the probability density $\psi^*\psi$ giving the spatial distribution of the missing electronic charge. However, in Si and Ge and in many III-V and II-VI compounds the top of the valence band, although at $\underline{k} = 0$ and having nearly spherical energy surfaces, is degenerate. (Though spin-orbit coupling partly removes the degeneracy this adds to the complexity of the problem except in the strong coupling limit). The required extension of the effective mass theory has been developed and worked out in detail. Unfortunately although many of the predictions are qualitatively correct these acceptor calculations cannot be said to be as quantitatively successful as those for donors. However, as the hole effective masses are generally substantially larger than those of the conduction electrons, the acceptor states will be less diffuse than the donor states and the specific central cell corrections will be relatively more important. The stronger localisation also means that the higher order corrections to the

Table 7.2

Energy Levels of Shallow Donor States in Si and Ge (in meV below the conduction band minima)

System	State							
	1S	$2P_0$	2S	$2P_1^{\pm}$	$3P_0$	3S	$3P_1^{\pm}$	$4P_0$
Si (theory)	31.27	11.51	8.83	6.40	5.48	4.75	3.12	3.33
Si (P)	45.5, 33.9, 32.6	11.45	–	6.39	5.46	–	3.12	3.38
Si (As)	53.7, 32.6, 31.2	11.49	–	6.37	5.51	–	3.12	–
Si (Sb)	42.7, 32.9, 30.6	11.52	–	6.46	5.51	–	3.12	–
Ge (theory)	9.81	4.74	3.52	1.73	2.56	2.01	1.03	1.67
Ge (P)	12.89, 9.88	4.75	–	1.73	2.56	–	1.05	–
Ge (As)	14.17, 9.96	4.75	–	1.73	2.56	–	1.04	–
Ge (Sb)	10.32, 10.01	4.74	–	1.73	2.57	–	1.04	–

The experimental ground levels are split as a result of the mixing of states from different but equivalent conduction band minima. This effect lies outside the effective-mass approximation and is unobservably small for the excited states (104).

effective mass approximation itself will be more significant. So far these effects have not been evaluated but there seems no reason to suppose that this approach to these shallow donor states is incorrect in any fundamental way.

This equivalence of localised donor and acceptor states can be demonstrated by a many-body formulation of the electron theory. It will also have been observed that lattice distortion does not play any part in the theory. There is an implied self-consistency in all this; thus when the electronic states are diffuse they exert only weak forces on the lattice and in turn there can be little effect of lattice relaxation upon the electronic state. On the other hand as we go towards more localised states we may rather suddenly enter a different regime in which the states are strongly localised. For example, from effective mass theory we would expect there to be a rather shallow donor state associated with substitutional N in diamond ($\sim$ 0.3 eV below the conduction band). In fact, the level is quite deep ($\sim$ 1.7 eV below the conduction band) and appears to derive from a T_2-type of molecular orbital function (centred on the N atom and its four C neighbours) which prompts a T_2-type of Jahn-Teller distortion so giving the centre a <111> axis (14). By itself effective mass theory cannot tell us when this

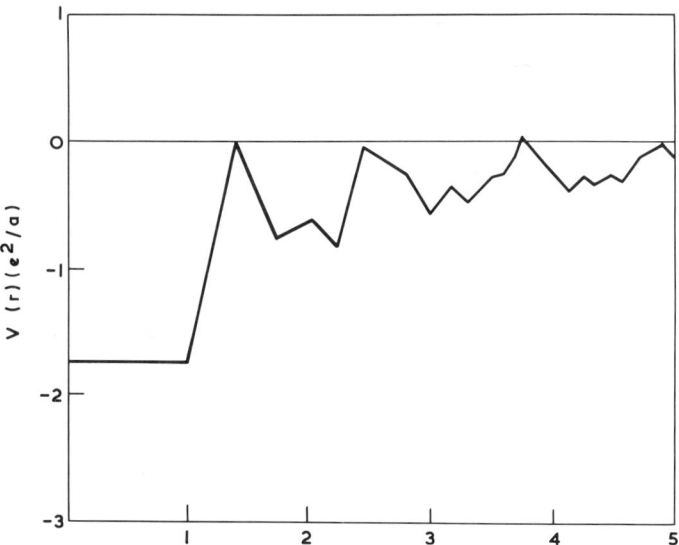

Fig. 7.1 The radial variation of the spherically symmetrical component of the point-ion potential of an F-centre in the rock-salt structure (abscissae in units of the anion-cation nearest-neighbour distance, a; ordinates in units of e^2/a). This spherically symmetrical part of V_{PI} dominates in the low-lying states.

sort of situation is likely to arise.

7.2 Electron Excess F-Like Centres: Point-Ion Models

Anion vacancies and their aggregates in the alkali halides, the alkaline earth fluorides and oxides and some other ionic crystals can trap electrons into strongly localised states which in many respects are in sharp contrast to the shallow donor states we have just described, (see Spaeth, von der Osten, d'Aubigne and Spinolo in this Volume). This trapping occurs as a result of the net positive charge carried by the anion vacancy; the levels are deep because the Madelung electrical potential inside the vacancy is strongly positive and rather flat (Fig. 7.1). Indeed, this Madelung potential plays a central role in the theory of these F-centres and their aggregates. This was suggested through the recognition many years ago of empirical relations between the energy of the F-band (and of the F-aggregate bands) and the lattice parameter of the form

$$E = \text{const.} \, a^{-n} \qquad (7.4)$$

valid for all materials of the same structure and valence (e.g. for the F-band in NaCl-structure, alkali halides n = 1.8, see Ref. 105). Formal expression was provided by Gourary and Adrian (106) who proposed the point-ion model in which all other terms were initially omitted. In other words the states and energy levels of the F-centre were calculated by solving the Schrödinger equation

$$\left(-\frac{\hbar^2}{2m}\nabla^2 + V_{PI}\right)\phi = E\phi \qquad (7.5)$$

in which V_{PI} is the potential which would be seen by an F-electron moving in a lattice of point-ions containing a vacancy. It was recognised soon afterwards that this approach can be regarded as a particular example of pseudopotential theory in which ϕ is the pseudo-wave function. The true wave-function for the F-electron ψ_F is then ϕ orthogonalised to the ion wave-functions ψ_c, i.e.

$$|\psi_F\rangle = |\phi\rangle - \sum_c \langle\psi_c|\phi\rangle|\psi_c\rangle \qquad (7.6)$$

while V_{PI} represents a first approximation to the corresponding pseudopotential.

Eqn. (7.5) has been solved for F-centres in all the alkali halides and a number of other materials as well. Although the accuracy of the individual solutions has been refined over the years the original, relatively simple variational solutions of Gourary and Adrian (106) have proved to be very accurate. In general, treating the A_{1g} ground state as spherically symmetric, i.e. s-like, and the first excited state (T_{1u}) as p-like proves to be a very good approximation; so that one commonly speaks of these two states as the 1s and 2p states-though, of course, they are not hydrogenic (cf. Fig. 7.1). The generalisation of (7.4) for aggregate centres containing more than one F-electron (e.g. $M \equiv F_2$, $R \equiv F_3$, $N \equiv F_4$ etc.) requires only the addition of a term for the Coulomb repulsions among the F-electrons and appropriate interpretation of V_{PI}. However, accurate solutions in such cases are much more difficult to obtain and variational methods which make use of the individual F-centre orbitals can be very useful and economical. We shall not go into these calculations and analyses in detail since extensive reviews are available elsewhere (e.g. 6.7). However it is appropriate to make the following summary remarks.

(i) Solving eqn. (7.5) in the rigid lattice approximation for the 1s and 2p states of the F-centre allows us to predict the F-band optical absorption energy fairly well in many ionic crystals (Table 7.3). When the point-ion wave-function is orthogonalised to the ion-core orbitals as in (7.6) then some other properties of the 1s and 2p states (e.g. hyperfine constants, spin-orbit coupling energy)

Table 7.3

Theoretical (Point-Ion) F-Band Energies (in eV) for Several Ionic Crystals

Substance	Point-ion 1s-2p	F-band (0K)
LiF	4.00	5.08
LiCl	2.75	3.26
NaF	3.24	3.70
NaCl	2.39	2.75
KCl	1.99	2.30
RbBr	1.71	1.85
CaF_2	3.19	3.30
SrF_2	2.91	2.85
BaF_2	2.61	2.03
MgO	4.79	4.95

The theoretical results, which were all obtained without allowance for ionic polarisation and relaxation, are taken from Refs. 106. Corresponding experimental results are from Refs. 105

can be obtained, again with fair success (Harker 107).

(ii) The discussions in §6 of lattice distortion and polarisation by defects immediately suggests that there will be corrections to the above results for these effects. In fact, because the Coulomb forces are so important in these structures and because the F-electron wave-functions are compact in both the ground state (1s) and in the state reached by absorption of a quantum of F-light (2p) these corrections often prove to be very small. Thus, in the alkali halides - with the exception of the fluorides - the corrections to the 1s → 2p transition energy for the F-centre appear to be <1/20th eV. The relaxation of the neighbours of the F-centre is correspondingly small - probably at most ∿ 10% of that for the empty anion vacancy. Accurate methods (Kanzaki; §6.1) have been used to calculate the relaxations theoretically but these prove to be sensitive to details of the physical model employed (lattice ions and F-centre), although the general concensus is that in the ground state the nearest neighbours (cations) relax outwards by ∿ 1% of the normal nearest-neighbour separation.

(iii) This success of the rigid lattice point-ion model of the simple F-centre has encouraged its use for F-aggregate centres, though perhaps small ionic relaxations should not be taken for granted when the defect comprises several vacant sites. Such

POINT DEFECTS IN NON-METALS

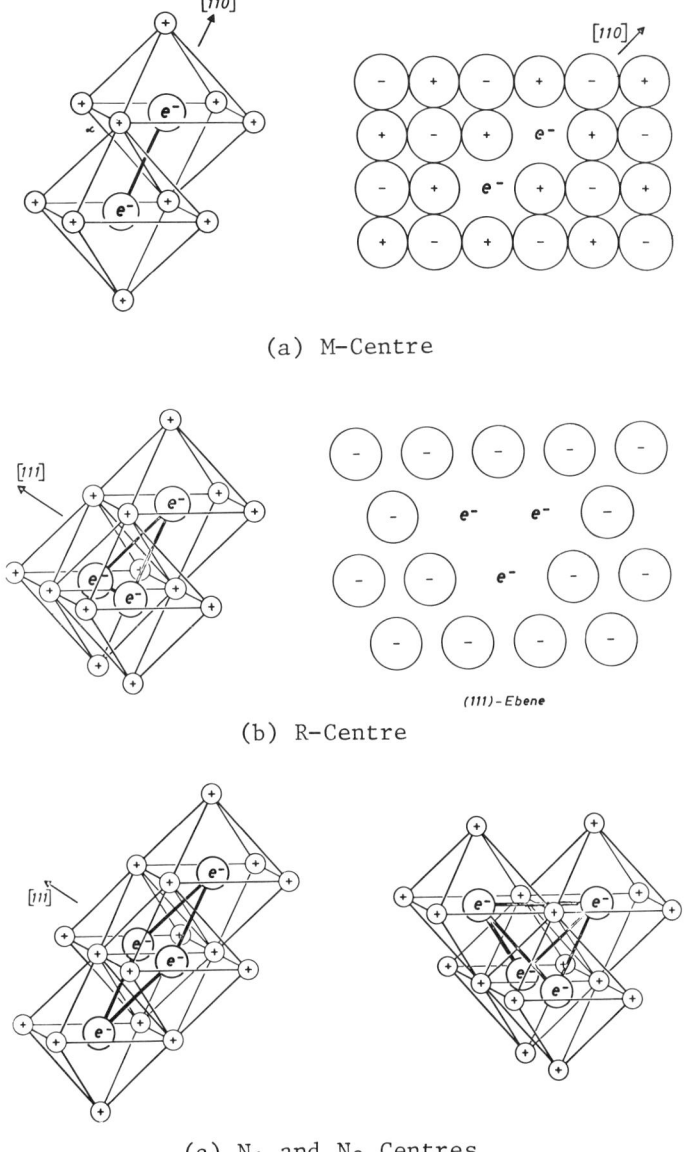

(a) M-Centre

(b) R-Centre

(c) N_1 and N_2 Centres

Fig. 7.2 Schematic diagram showing the structures proposed for M, R, N_1 and N_2 centres in alkali halides with the rock-salt structure (after Pick 111). There is now little doubt that the postulated M, R and N_1 structures occur and that these centres are responsible for the observed M-, R- and N_1- bands. The evidence in favour of the N_2- centre is mainly that from the calculations of Evarestov and Slonim (110).

Table 7.4

M-Centre Transition Energies as Calculated in the Point-Ion Approximation (108)

Substance		Transition energy		
		$^1A_{1g} \to {}^1A_{1u}$ <110>	$^1A_{1g} \to {}^1B_{1u}$ <100>	$^1A_{1g} \to {}^1B_{2u}$ <1$\bar{1}$0>
LiF	calculated	2.85	4.11	4.03
	experimental	2.79		
LiCl	calculated	2.01	2.96	2.91
	experimental	1.91		
NaF	calculated	2.26	3.43	3.35
	experimental	2.45		
NaCl	calculated	1.82	2.67	2.56
	experimental	1.71		
KCl	calculated	1.50	2.25	2.12
	experimental	1.55	2.31	2.27
KBr	calculated	1.44	2.18	1.99
	experimental	1.35		
RbBr	calculated	1.33		
	experimental	1.50		

The transition $^1A_{1g} \to {}^1A_{2u}$ corresponds to the M-band while the transitions $^1A_{1g} \to {}^1B_{1u}$ and $^1B_{2u}$ generally lie close to the F-band (so called M_F-bands).

extensions of the point-ion model have been made by several workers for M($\equiv F_2$)-centres (108), R($\equiv F_3$)-centres (109) N_1- and N_2-($\equiv F_4$) centres (110). Despite their simplifications these calculations have been useful in confirming and suggesting defect structures and spectral assignments. The structures for the M-, R- and N- centres shown in Fig. 7.2 have all been analysed. Evarestov, in particular, has used Heitler-London or 'valence-bond' wave-functions (71) constructed from simple, variational F-centre one-electron orbitals. Table 7.4 shows the nature of the agreement obtained for the M-centre. Unfortunately, the agreement with R-centre properties calculated this way is not so good; also the results obtained by Kern and Bartram (109, summarised in 6), who used more flexible wave-functions and

Table 7.5

N_1- and N_2- Centre Transition Energies (eV) as Calculated in the Rigid Lattice Point-Ion Approximation

Crystal	Experimental transition energies				Calculated transition energies				
					N_1 center			N_2 center	
	F band	R_2 band	N_1 band	N_2 band	$^1A_g \to {}^1B_u'$	$^1A_g \to {}^1B_u''$	$^1A_g \to {}^1A_u$	$^1E \to {}^1T_2$	$^1E \to {}^1T_1$
LiF	5.00	3.25	2.37	2.29	2.28	4.98	3.47	2.18	4.73
NaF	3.70	2.99	2.11	1.99	1.92	3.84	3.20	1.76	3.52
NaCl	2.77	2.07	1.50	1.44	1.47	2.94	2.04	1.34	2.61
KCl	2.31	1.70	1.27	1.20	1.22	2.48	1.90	1.13	2.06
KBr	2.07	1.56	1.14	1.09	1.16	2.18	1.07	1.07	2.00

These results were obtained by Evarestov and Slonim (110). The ground states are predicted to be spin-singlets in both centres (^1A_g in N_1 and ^1E in N_2), as is seemingly required experimentally, though in at least one case (N_1 in KBr) the lowest excited state, ^5B_g, lies sufficiently close in energy (~ 0.1 eV) that one might observe a proportion of centres thermally excited into this state. (N.B. the energies of the transitions $^1A_g \to {}^1A_u$ (N_1) and $^1E \to {}^1T_2$ (N_2) are given as 1.47 eV in the earlier reference; we presume this to be a misprint).

allowed ionic relaxation, do not lead to the same spectral assignments as were made by Evarestov. However, as the result of spectroscopic and resonance experiments, there is no doubt that the structure of the R-centre is that shown in Fig. 7.2. The valence-bond results for the F_4 centres, N_1 and N_2, appear to be much better (perhaps because the multi-centre integrals arising in this calculation were not approximated). Results for the N_1-centre (C_{2h} symmetry) and the N_2-centre (T_d symmetry) - see Fig. 7.2 - are given in Table 7.5. The lowest calculated transition energies for these structures agree well with the observed N_1- and N_2- bands respectively. As with the other aggregate centres there are also optical transitions in these centres with energies in the F-band region. In the case of N_1 there is an intermediate transition which is quite close to the R_2-band.

(iv) By suitable extension of these point-ion calculations it has also been possible to calculate other properties, e.g. the small singlet-triplet splittings of the ground states of M-centres (112) and related 2-electron centres (113), though in these cases it may be necessary to allow for lattice distortion and polarisation to get the best results. In one case, the so-called F_t-centre in the alkaline earth oxides, they showed that the measured singlet-triplet splittings were consistent with a linear trivacancy centre but not with a next-nearest neighbour divacancy centre (113). In summary, therefore we can say that these calculations with the point-ion model are not merely interesting because they reproduce transitions observable experimentally but that they have been useful in the identification of these transitions and the defect structures giving rise to them.

(v) We already noted that the point-ion potential, V_{PI}, is only a first approximation to the pseudopotential. Corrections to the predictions of the point-ion model coming from improvements to the potential are known as 'ion-size effects'. They have been calculated in different ways by different authors (see 6), but a convenient representation within the pseudopotential formalism was provided by Bartram et al. (114), even though it proves necessary to introduce an empirical correction factor ($\alpha = 0.53$) to all integrals of a certain type (A_v). The F-centre 1s → 2p transition energies differ significantly from those of the point-ion model when ion-size corrections are included (Table 7.6). In general, the agreement with observed F-band energies is much improved and systematic deviations (e.g. with ionic radii or with lattice parameter) largely removed. Thus ion-size effects in F-centres are generally more important than the corrections for ionic relaxation and polarisation. Whether this is so for F-aggregate centres, however, is largely undetermined so far.

Table 7.6

Ion-Size Effects in F-Band (1s→2p) Absorption Energies (eV)

Substance	Calculated		Expt. (0°K)
	Point Ion	Ion-Size Theory	
LiF	3.99	5.49	5.08
LiCl	2.76	3.34	3.26
LiBr	2.58	2.86	2.70
LiI	2.14	2.48	3.27
NaF	3.24	3.98	3.70
NaCl	2.38	2.75	2.75
NaBr	2.17	2.42	2.36
NaI	1.89	2.14	2.1
KF	2.60	2.93	2.85
KCl	1.99	2.18	2.30
KBr	1.83	1.98	2.06
KI	1.64	1.80	1.88
RbF	2.38	2.41	2.43
RbCl	1.87	1.93	2.05
RbBr	1.72	1.77	1.85
RbI	1.53	1.62	1.71
CsF	2.14	2.06	1.89
CsCl *	1.64	1.73	2.17
CsBr *	1.52	1.59	1.96
CsI *	1.37	1.50	1.68

* α-CsCl structure. These results were obtained by Harker (107). Minor differences between these point-ion results and those in Table 7.3 are due to small differences in the assumed form of wavefunction and choice of lattice parameter. The experimental values are from Dawson and Pooley (105).

(vi) The above remarks have largely centred on optical absorption from the ground state and other ground state properties. The large Stokes shifts between F-centre absorption and emission (e.g. in KCl from 2.3 eV to 1.0 eV), however, point to a large relaxation of the excited 2p state. Also there is direct experimental evidence from ENDOR studies on this excited state that the electronic wave-function in the relaxed excited state is <u>diffuse</u>, not compact as in the ground state and in the unrelaxed excited state (115). The correct theoretical description of this state poses a difficult

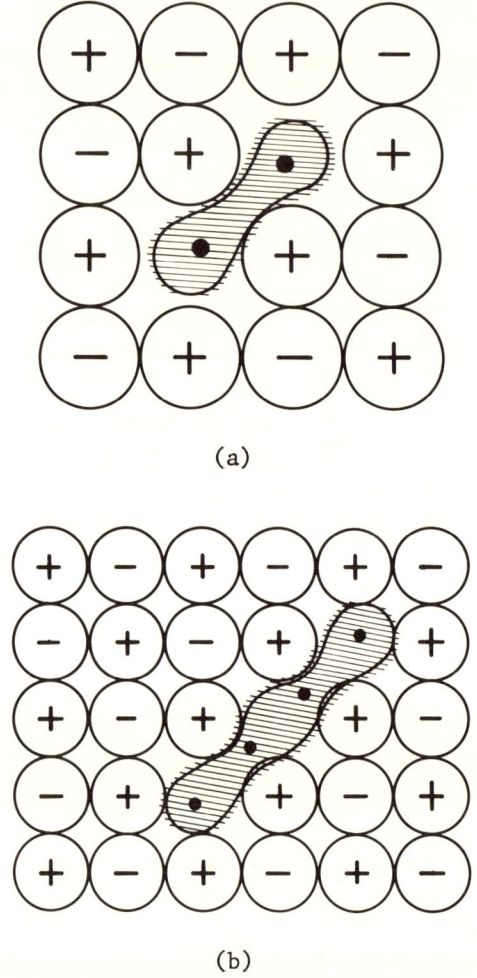

Fig. 7.3 Schematic diagrams showing the structure (a) of V_K-centres in alkali halides with the NaCl structure and (b) of H-centres as they occur in KCl, KBr and probably some other alkali halides with the NaCl structure (in LiF they are known to have <111> orientation rather than the <110> orientation shown here). Various V_K- and H-centres perturbed by the presence of other defects and impurities are also known. In all cases the essential part of the defect structure is the X_2^- molecule-ion.

problem, even for this one-electron centre, as a result of the requirement for self-consistency between the electronic state and the lattice relaxation and polarisation; and also because the 2s and 2p electronic states lie close to one another. For more detailed discussion see Spinolo this Volume.

7.3 Electron Deficient Centres

In §7.1 we noted that the theory for holes trapped at shallow acceptors was essentially the same as that for electrons trapped into shallow donor levels. One might therefore suppose that in the alkali halides there would be trapped hole centres analogous to the F-like centres discussed above (see e.g. Seitz 116). Quite a number of trapped-hole centres have now been identified in ionic crystals - both halides and oxides - but in none of them are the hole states analogous to the F-like states of the electron-excess centres. In general the hole is localised into quasi-molecular states which spread over only one or two atoms. Thus in the halides, several centres are all essentially a halogen molecule-ion X_2^- in different locations. The V_k-centre is X_2^- occupying two neighbouring anion lattice sites (Fig. 7.3). The interstitial H-centre is an X_2^- molecule occupying a single anion site (Fig. 7.3). In the oxides one finds holes trapped on single oxygen ions (giving O^-) next to vacancies. These identifications have been made by careful e.p.r. and optical work but the theory of these centres, regarded as 'quasi-molecular' centres, is quite successful.

We can illustrate this by the example of the V_k-centres. Thus Gilbert and Wahl (117) calculated the properties of the free X_2^- molecule while others applied these results to an X_2^- molecule subject to the interactions with an ionic lattice e.g. as implied by Fig. 7.3 for such a centre in a rocksalt structure. Table 7.7 compares the calculated and the observed V_k optical transition energies for LiF and KCl. The agreement is fairly good and would probably be improved if the lattice relaxation calculation were repeated with better models (shell model) and better methods (e.g. HADES); cf. §§6.1 and 6.2. Certainly the incorporation of these improvements in the extension of this work to the alkaline earth fluorides leads to no worse predictions (119). We can thus infer that the optical and some other related properties are consistent with the model of a V_k centre as an X_2^- molecule embedded in the crystal lattice.

The X_2^- model of the H-centre may not be quite so satisfactory as the E.P.R. work shows a small percentage ($\sim$ 5%) of the hole density to be on the two neighbouring <110> anions (KCl and KBr). While some previous calculations have predicted that the stable structure is one where the X_2^- molecule takes up a <111> orientation and not <110>, recent improved calculations by Diller (84) who used shell models in the HADES program predict that <110> is the stable orientation for KCl, RbCl and RbBr while <111> is stable for all the other alkali halides with the rocksalt structure. The predicted differences in energy in most cases are very small (only a few $\times$ 10^{-2} eV) and the omitted bonding with the <110> neighbours would possibly lead to a stronger preference for this structure (there

Table 7.7

Transition Energies (eV) of V_k-Centres and Free X_2^- Molecule-Ions

System	Calculated		Experimental	
	$^2\Sigma_u \to {}^2\Sigma_g$	$^2\Sigma_u \to {}^2\Pi_g$	$^2\Sigma_u \to {}^2\Sigma_g$	$^2\Sigma_u \to {}^2\Pi_g$
Free F_2^-	4.74	2.23	–	–
LiF : V_k	3.30	1.60	3.65	1.65
Free Cl_2^-	3.86	1.87	–	–
KCl : V_k	2.37	1.23	3.40	1.65

These results are selected from calculations for alkali fluorides and chlorides by Jette et al. (118). The molecular notation is used for the states; the crystal field splitting of the $^2\Pi_g$ states, which is not included in the calculations, in any case has not been resolved experimentally. The lattice calculation was relatively unsophisticated and used a Mott-Littleton method with a polarisable point-ion model. The molecule is stretched by being put in the lattice and this lowers the transition energies.

would be no such bonding in the <111> orientation). Since the structure of H-centres has only been identified by E.P.R. in KCl and KBr (<110> orientation) and in LiF (<111> orientation), it would clearly be desirable to have empirical information of some of the other alkali halides. However, important though the orientation is, the effects on the X_2^- molecule of going from <110> to <111> are small – as shown by the small energy differences in Diller's calculations. The model provides a satisfactory account of the optical properties of H-centres when evaluated in the same way as for the V_k-centre.

These calculations have also been extended for other H-centre properties, particularly their interaction energies, the structure of H-aggregates, etc. The reaction postulated by Hobbs et al. (120) by which two H-centres come together to 'dig their own hole' and displace an anion and a cation on to the edge of a dislocation loop as an explanation for dislocation loop growth under X-irradiation has been shown (121) to be exothermic ($\sim$ 3 eV), as required.

A rather similar success is obtained with the quasi-molecular calculations for trapped holes in oxides (6) though lattice relaxation and polarisation has not yet been included.

7.4 Conclusion

All this raises the question why we have to go to a strongly localised molecular model and, in particular, why the pseudo-potential theory of electrons, which leads to the point-ion model, does not also lead to an equivalent theory for holes. In other words why are there not electron-hole analogues of the F-centre as envisaged, e.g., by Seitz (116). Even when holes have been observed trapped in the vicinity of cation vacancies they are in perturbed X_2^- configurations rather than in any more extended F-like orbital. Part of the answer was indicated by Gilbert (see 6) in his consideration of the self-trapping of holes and electrons in KCl. He argued that self-trapping is favoured by (i) a narrow energy band (because this implies a small increase in band kinetic energy on forming a localised 'wave-packet' state) (ii) a polarisable lattice (i.e. large ε_o) and (iii) a large local rebonding. His estimates of the corresponding energy terms for electrons and holes in KCl are as follows:

		Hole	Electron
(i)	Band localisation	0.3 eV	1.9
(ii)	Polarisation	-0.5	-0.3
(iii)	Bonding	-1.5 (Cl_2^-)	-0.3 (K_2^+)
		-1.7 eV	+1.3 eV

Thus we expect self-trapping of the holes (to Cl_2^- or V_K-centres) but not of the electrons — as observed. However, 1.5 eV of the total energy lowering of 1.7 eV derives from the <u>assumed</u> rebonding; we have not predicted that this particular <u>rebonding</u> is what will occur merely that this is one possible way for self-trapping of holes to occur.

This example thus highlights one of the present limitations of the electronic theory of defects, namely that it cannot handle situations where the final electronic state and atomic configuration are both very different from the initial ones. When we know enough to choose the essential features of the initial state and configuration correctly then the electronic state and atomic configuration can often be calculated in detail quite satisfactorily. These remarks are illustrated not only by the examples from ionic crystals which we have already touched upon (excited F-centre relaxation and the X_2^- hole centres), but also by the example of vacancies and other defects in the covalent semi-conductors. The model of the vacancy indicated by Fig. 6.2 and similar models for vacancy pairs and other defects have provided a useful framework to represent the many detailed E.P.R. studies of these defects (Watkins, this volume). Watkins has been able to represent his results in terms of a 'Hartree' (independent electron) model of the centre states and known general results on the Jahn-Teller symmetry-lowering distortions of molecules and other localised systems. The same models have also been subject to a great deal of theoretical analysis by the sophisticated methods of molecular theory. Yet on the whole there is very poor agreement between the theoretically predicted levels, states, symmetries etc. and those obtained experimentally (see 6 and reviews presented at the series of conferences on Radiation Effects in Semiconductors, 122). Theory and experiment do, however, agree that lattice relaxations around the vacancy are large. This may be the cause of much of the difficulty, since the 'defect-molecule' calculations use the same basis orbitals for the relaxed as for the unrelaxed structure. Lattice relaxation and Jahn-Teller distortions are calculated within the same limitation while the variation of force with displacement is also neglected. These simplifications, while adequate for small relaxations may fail when the relaxations are as large as both the theory and experiment indicate. At the same time the defect-molecule calculations also indicate that the electron-electron interactions in the centre are very important in determining the final states and energy levels. This is in contrast to the representation of the experimental results by a Hartree model which would indicate that the interactions only enter in a non-specific average way. There have therefore also been some purely one-electron calculations which have tried to determine the one-electron orbitals and level structure and from them the atomic configuration (by evaluating and minimising the sum of the one-electron energies). These calculations have led to some interesting and suggestive results on structures, since as the electronic calculations are relatively simple their repetition for a variety of atomic configurations can be quite practical. (For a review which attempts to synthesise the results of these calculations with experimental indications and some speculations on defect structure in semiconductors see Ref. 35).

In conclusion therefore we can say that it is clear that we do not have one, single theory of defect structure, but instead a number of theories each spanning a certain area of the whole field of defect systems. We have seen that some of these are accurate and quantitatively reliable within their areas of validity (e.g. effective mass theory, point-ion models and their corrections). We also saw (in §6) that lattice relaxation calculations can now be performed very accurately even for complex defect structures when we are confident about the interatomic forces and the forces exerted by the defects upon the lattice. Difficulties arise at present when self-consistency in both electronic states and lattice relaxation involves large departures from the initial electronic state and atomic arrangement (e.g. excited F-centre relaxation, defects in the covalent semiconductors). It must be one of the aims of defect theory in the next few years to remove these limitations and to increase its ability to predict the structures and properties of defects in general with the same confidence as can now be done for 'potential-function' defects.

REFERENCES

1. See e.g. F. Seitz in "Imperfections in Nearly Perfect Crystals", Ed. by W. Shockley et al., (Wiley, New York, 1952) p.3.

2. C. P. Flynn, Point Defects and Diffusion (Oxford University Press, 1972).

3. F. A. Kröger, Chemistry of Imperfect Crystals (North Holland, Amsterdam, 1964).

4. Theory of Imperfect Crystalline Solids - Trieste Lectures 1970 (International Atomic Energy Agency, Vienna, 1971).

5. J. H. Crawford and L. M. Slifkin, Eds., Point Defects in Solids Vols. 1 and 2 (Plenum, New York and London 1972).

6. A. M. Stoneham, Theory of Defects in Solids (Clarendon Press, Oxford, 1975).

7. W. Hayes, Ed., Crystals with the Fluorite Structure (Clarendon Press, Oxford (1974).

8. A. V. Chadwick and J. N. Sherwood in Ref. 5, Vol. 2, Chap. 6.

9. See e.g. (i) R. M. J. Cotterill and M. Doyama, Phys.Letts. $\underline{25A}$, 35 (1967), (ii) R. A. Johnson and W. D. Wilson in Interatomic Potentials and the Simulation of Lattice Defects, Ed. by P. C. Gehlen, J. R. Beeler and R. I. Jaffee (Plenum, New York 1972) p.301 (iii) R. A. Johnson J.Phys.F. (Metal Phys.) $\underline{3}$, 295 (1973), (iv) R. A. Johnson in Diffusion (American Society of Metals, Metals Park, Ohio, 1973) p.25.

10. F. W. Schapink, Phil.Mag. $\underline{12}$ 1055 (1965).

11. A. R. Allnatt and E. Loftus, J.Chem.Phys. $\underline{59}$, 2541 and 2550 (1973); A. R. Allnatt and P. S. Yuen, J.Phys.C (Solid State Phys.) $\underline{8}$, 2199 (1975).

12. J. Corish and P. W. M. Jacobs in Surface and Defect Properties of Solids $\underline{2}$, 160 (1973). (The Chemical Society, Specialist Periodical Reports).

13. F. K. Fong, Prog.Solid State Chem. $\underline{3}$, 135 (1966).

14. G. Davies, in Chemistry and Physics of Carbon Ed. by P. L. Walker and P. A. Thrower Vol. 13 (1975).

15. N. H. March and J. S. Rousseau, Crystal Lattice Defects $\underline{2}$, 1 (1971).

16. R. W. Balluffi and P. S. Ho, in Diffusion (American Society for Metals, Metals Park, Ohio 1973) Chap. 4.

17. L. W. Barr and A. B. Lidiard, in Physical Chemistry - an advanced treatise Vol. X (Academic Press, New York and London 1970) Chap. 3.

18. G. D. Watkins and J. W. Corbett, Phys.Rev. 138, A543 (1965).

19. P. Kofstad, Non-Stoichiometry, Diffusion and Electrical Conductivity in Binary Metal Oxides (Wiley, New York, 1972).

20. H. Schmalzried, Solid State Reactions (Academic Press, New York, 1974).

21. See e.g. Defects and Transport in Oxides, Ed. by M. S. Seltzer and R. I. Jaffee (Plenum, New York, 1974).

22. C. R. A. Catlow and B. E. F. Fender, J.Phys.C. (Solid State Physics), 8, 3267 (1975).

23. G. G. Libowitz in Mass Transport in Oxides (U.S. National Bureau of Standards, Special Technical Publication 296, (1968) p.109.

24. R. G. Fuller, in Ref. 5, Vol. 1, Chap. 2.

25. J. K. Aboagye and R. J. Friauf, Phys.Rev. B 11 1654 (1975). But see also J. Corish and P. W. M. Jacobs, Phys.Stat.Solidi B 67, 263 (1975) and A. P. Batra and L. M. Slifkin, Phys.Rev. B 12, 3473 (1975).

26. J. A. A. Ketelaar, Trans.Far.Soc. 34, 874 (1938).

27. W. Jost, Diffusion in Solids, Liquids, Gases, (Academic Press, New York 1960).

28. J. Frenkel, Kinetic Theory of Liquids (Clarendon Press, Oxford, 1946).

29. T. Kurosawa, J.Phys.Soc. Japan 12, 338 (1957).

30. M. J. Rice, S. Strassler and G. A. Toombs, Phys.Rev.Letts. 32, 596 (1974).

31. B. A. Huberman, Phys.Rev.Letts., 32, 1000 (1974).

32. For an excellent review of the properties of AgI and related compounds see K. Funke, Prog. Solid State Chem., to appear.

33. See e.g. C. E. Derrington and M. O'Keeffe, Nature (Physical Science) 246, 44 (1973).

34. R. A. Huggins in Ref. 21 p.549.

35. J. W. Corbett and J. C. Bourgoin in Ref. 5, Vol. 2, Chap. 1.

36. It is difficult to recommend any single source for this field. For more extended introductions than are possible here see Refs. 2-4. The basic paper for classical behaviour is Ref. 37. A later review is that by H. Glyde, Rev.Mod.Phys. 39, 373 (1967). A more recent and comprehensive review (though one unfortunately marred by a rather pretentious style) is that by W. M. Franklin in Diffusion in Solids: Recent Developments, Eds. A. S. Nowick and J. J. Burton (Academic Press, New York, 1975) p.1. The recent work of C. W. McCombie and M. Sachdev, J.Phys.C. (Solid State Physics) 8, L413 (1975) on classical behaviour and of A. M. Stoneham Ber.Bunsenges.Phys.Chem. 76, 816 (1972) on the quantum behaviour of light interstitials in metals should also be studied. Many of the basic ideas in the subject derive from the theory of chemical reactions and the following reference is therefore also useful: N. B. Slater, Theory of Unimolecular Reaction Rates (Cornell Univ.Press, Ithaca, 1959). Lastly, the reader should perhaps be warned of the existence of papers in this field making somewhat extravagent claims.

37. G. H. Vineyard, J.Phys.Chem.Solids 3, 121 (1957).

38. Calculations in progress for the fluorite lattice, C. R. A. Catlow and M. J. Norgett (private communication).

39. J. R. Manning, Diffusion Kinetics for Atoms in Crystals (Van Nostrand, Princeton, 1968).

40. For a review see N. L. Peterson, in Diffusion in Solids: Recent Developments Ed. by A. S. Nowick and J. J. Burton (Academic Press, New York, 1975) p.715.

41. See for example; (a) R. E. Howard and A. B. Lidiard, Rep.Prog. Phys. 27, 161 (1964) (b) Y. Adda and J. Philibert, La Diffusion dans les Solides (Presses Universitaires de France, Paris, 1966), (c) C. P. Flynn Ref. 2, (d) A. B. Lidiard in Theory of Imperfect Crystalline Solids (I.A.E.A. Vienna 1971) p.339, (e) T. R. Anthony in Diffusion in Solids : Recent Developments Ed. by A. S. Nowick and J. J. Burton, (Academic Press, New York 1975) p.353.

42. S. Chandrasekhar, Rev.Mod.Phys. 15, 1 (1943).

43. A. D. LeClaire, Phil.Mag. 3, 921 (1958).

44. F. Benière in Physics of Electrolytes Ed. by J. Hladik (Academic Press, London 1972) Vol. 1, Chap. 6; R. G. Fuller in Ref. 5 Vol. 1, p.103; H. C. Casey and G. L. Pearson in Ref. 5, Vol. 2. p.163.

45. See e.g. F. W. Felix and M. Müller, Phys.Stat.Solidi, 46, 265 (1971).

46. A. D. LeClaire in Physical Chemistry - an Advanced Treatise Academic Press, New York 1970) Vol. 10 p.261.

47. A. Schoen, Phys.Rev.Letts. 1, 138 (1958).

48. K. Tharmalingam and A. B. Lidiard, Phil.Mag. 4, 899 (1959).

49. H. Bakker, Phys.Stat.Solidi, 44, 369 (1971).

50. See e.g. G. H. Wannier, Statistical Physics (Wiley, New York, 1966). Also Refs. 41.

51. W. D. Compton and R. J. Maurer, J.Phys.Chem. Solids 1, 191, (1956).

52. J. Corish and P. W. M. Jacobs, J.Phys.Chem.Solids 33, 1799 (1972).

53. W. J. Fredericks in Diffusion in Solids : Recent Developments Ed. by A. S. Nowick and J. J. Burton, (Academic Press, New York 1975) p.381.

54. T. R. Anthony, Ref. 41.

55. For comprehensive discussions see A. S. Nowick, Adv.Phys. 16, 1 (1967), and in Ref. 5, Vol. 1 p.151; also A. S. Nowick and W. R. Heller, Adv.Phys. 12, 251 (1963) and 14, 101 (1965).

56. See e.g. J. F. Nye, Physical Properties of Crystals (Clarendon Press, Oxford 1957).

57. R. W. Dreyfus, Phys.Rev. 121, 1675 (1961).

58. C. Bucci, R. Fieschi and G. Guidi, Phys.Rev. 148, 816 (1966).

59. H. Symmons, J.Phys.C (Solid State Phys.) 3, 1846 (1970).

60. G. D. Watkins, Phys.Rev. 113, 79 and 91 (1959).

61. A. D. Frankin, J. M. Crissman and K. F. Young, J.Phys.C. (Solid State Phys.) $\underline{8}$, 1244 (1975).

62. See e.g. I. M. Hoodless, J. H. Strange and L. E. Wylde, J.Phys.C. (Solid State Physics) $\underline{4}$, 2742 (1971).

63. See e.g. M. Tinkham, Group Theory and Quantum Mechanics (McGraw Hill, New York, 1964), or F. A. Cotton, Chemical Applications of Group Theory 2nd Edn. (Wiley, New York, 1971).

64. A. D. Franklin, A. Shorb and J. B. Wachtman, National Bureau of Standards J. of Res. $\underline{68A}$, 425 (1964).

65. G. Socino, R. de Batiste and R. Gevers, Proc.Brit.Ceram.Soc. No. 9, 73 (1967).

66. See e.g. G. S. Rushbrooke, Introduction to Statistical Mechanics (Clarendon Press, Oxford, 1949).

67. See e.g. T. Ninomiya, J.Phys.Soc. Japan $\underline{14}$, 30 (1959); I. M. Boswarva and A. D. Franklin, Phil.Mag. $\underline{11}$, 335 (1965).

68. N. F. Mott and M. J. Littleton, Trans.Far.Soc. $\underline{34}$, 485 (1938).

69. S. R. Tibbs, Trans.Far.Soc. $\underline{35}$, 1471 (1939).

70. H. B. Huntington and F. Seitz, Phys.Rev. $\underline{61}$, 315 (1942); H. B. Huntington, Ibid. $\underline{61}$, 325 (1942).

71. H. Kanzaki, J. Phys.Chem. Solids $\underline{2}$, 24 and 37 (1957).

72. V. K. Tewary, Adv.Phys. $\underline{22}$, 757 (1973).

73. M. J. Norgett, AERE Reports R7015 (1972) and R7650 (1974). Also A. B. Lidiard and M. J. Norgett in Computational Solid State Physics. Ed. by F. Herman et al., (Plenum, New York 1972) p.385.

74. R. Fletcher, Comp. Physics Commun. $\underline{3}$, 159, (1972).

75. See e.g. R. Bullough in Computing as a Language of Physics (I.A.E.A., Vienna, 1972) p.323.

76. C. H. Bennett in Diffusion in Solids : Recent Developments Ed. by A. S. Nowick and J. J. Burton (Academic Press, London, 1975) p.74.

77. See e.g. the surveys in Refs. 2, 6, 7, 9 and 12.

78. A. M. Karo and J. R. Hardy, Phys.Rev. $\underline{B3}$, 3418 (1971);
    P. D. Schulze and J. R. Hardy, Ibid. $\underline{B5}$, 3270 (1972)
    and $\underline{B6}$, 1580 (1972).

79. See e.g. (i) C. A. Coulson, Valence (2nd edn.) (Clarendon Press, Oxford, 1961) (ii) L. Pauling, The Nature of the Chemical Bond (3rd edn.) (Cornell Univ.Press, Ithaca 1960) and (iii) J. C. Slater, Introduction to Chemical Physics (McGraw Hill, New York, 1939); also Refs. 2 and 6.

80. J. N. Murrell in Orbital Theories of Molecules and Solids Ed. by N. H. March (Clarendon Press, Oxford, 1974) p.311.

81. J. A. Barker in Rare Gas Solids Ed. by M. L. Klein and J. A. Venables (Academic Press, New York, 1975) p.212.

82. See K. F. Niebel and J. A. Venables, Proc.Roy.Soc. $\underline{A336}$, 365 (1974) and other references cited there.

83. See e.g. W. Cochran, C. R. C. Reviews in Solid State Sciences $\underline{2}$, 1 (1971), or H. Bilz in Computational Solid State Physics, Ed. by F. Herman et al. (Plenum, New York, 1972).

84. See e.g. (i) A. B. Lidiard and M. J. Norgett Ref. 73, (ii) K. Diller, D.Phil thesis, University of Oxford (1975); also issued as AERE Report TP.642.

85. C. R. A. Catlow, D.Phil. thesis, University of Oxford 1973; also issued as AERE Report TP.566.

86. C. R. A. Catlow and M. J. Norgett, J. de Phys. (Paris) $\underline{34}$, C9-45, (1973) and J.Phys.C (Solid State Physics) $\underline{6}$, 1325 (1973).

87. C. R. A. Catlow and A. B. Lidiard in Thermodynamics of Nuclear Materials 1974 (I.A.E.A. Vienna 1975) Vol. II, p.27.

88. C. R. A. Catlow, I. D. Faux and M. J. Norgett, J.Phys.C. (Solid State Physics) in press; also issued as AERE Report TP.569.

89. C. R. A. Catlow and M. J. Norgett (personal communication).

90. I. D. Faux, J.Phys.C. (Solid State Physics) $\underline{4}$, L211 (1971).

91. See e.g. A. Seeger and M. L. Swanson in Lattice Defects in Semiconductors, Ed. by R. R. Hasiguti (Univ. of Tokyo and Pennsylvania State Univ. Presses, 1968) p.93.

92. See H. L. McMurry, A. W. Solbrig, J. K. Boyter and C. Noble, J.Phys.Chem. Solids 28, 2359 (1967) and also A. W. Solbrig, Ibid. 32, 1761 (1971).

93. J. E. Sinclair and B. R. Lawn, Proc.Roy.Soc. A329, 83 (1972).

94. See e.g. W. A. Harrison, Pseudo-potentials in the Theory of Metals (Benjamin, New York 1966).

95. A. V. Chadwick and H. R. Glyde in Rare-Gas Solids, Ed. by M. L. Klein and J. A. Venables (Academic Press, New York 1975) Chap. 20.

96. A. B. Lidiard in Orbital Theories of Molecules and Solids, Ed. by N. H. March (Clarendon Press, Oxford 1974) Chap. 4.

97. K. Tharmalingam, J.Phys.Chem. Solids 24, 1380 (1963) and 25, 255 (1964); K. Diller, Ref. 84.

98. I. D. Faux, Ph.D. thesis, University of London (1971). Also I. D. Faux and A. B. Lidiard, Z.Naturforsch 26a, 62 (1971). (N.B. the denominator of eqn. 7 of this paper should be 3 not 6; the following eqn. 8 and the numerical results have been obtained from the correct expression).

99. A. K. Chetham, B. E. F. Fender and M. J. Cooper, J.Phys.C. (Solid State Phys.) 4, 3107 (1971).

100. C. R. A. Catlow, J.Phys.C (Solid State Phys.) 6, L64 (1973): AERE Reports TP.617 and 618 (1975).

101. B. T. M. Willis, Proc.Brit.Ceram.Soc. 1, 9 (1964).

102. H. P. Myers, T. Jonsson and R. Westin, Solid State Commun. 2 321 (1964); J. L. Bates, C. A. Hinman and T. Kawada, J.Amer.Ceram.Soc. 50, 652 (1967).

103. H. Matzke, J.Phys. (Paris) 34, C9-317 (1973). See also J. R. Matthews, AERE Report M-2643 (1974).

104. See e.g. (i) F. Bassani, Ref. 4 p.265 (ii) F. Bassani, G. Iadonisi and B. Preziosi, Rep.Prog.Phys. 37, 1099 (1974).

105. See e.g. R. K. Dawson and D. Pooley, Phys.Stat.Solidi, 35, 95 (1969); A. E. Hughes and B. Henderson Ref. 5, Vol. 1, p381; Ref. 7 Chap. IV.

106. B. S. Gourary and F. J. Adrian, Phys.Rev. 105, 1180 (1957); J. C. Kemp and V. I. Neeley Phys.Rev. 132, 215 (1963); H. Bennett and A. B. Lidiard Phys.Letts. 18, 253 (1965).

107 A. H. Harker, AERE Report TP.621 (1974); in press.

108 R. A. Evarestov, Optika Spektroskp, 16, 361 (1964).
English version: Opt.Spectroscp. (USSR) 16, 198 (1964).

109 R. A. Evarestov, Spektroskp. tverd. Tela IV, 26 (1969);
R. C. Kern, Ph.D. thesis University of Connecticut (1972).

110 A. M. Stoneham, Proc.Phys.Soc. 88, 135 (1966); R. A. Evarestov and V. Z. Slonim, Phys.Stat.Sol. (b) 47, K59 (1971) and Optika Spektroskp 33, 910 (1972) (English version: Opt. Spectroscp. (USSR) 33 499 (1972)).

111 H. Pick, Z.Phys. 159, 69 (1960).

112 A. A. Berezin, Fiz,tverd.Tela 10, 2882 (1968). English version Sov.Phys.Sol.State 10, 2880 (1969).

113 M. J. Norgett, J.Phys.C (Solid State Phys.) 4, 1289 (1971).

114 R. H. Bartram, A. M. Stoneham and P. Gash, Phys.Rev. 176, 1014 (1968).

115 L. F. Mollenauer and G. Baldacchini, Phys.Rev.Letts. 29, 465 (1972); G. Baldacchini and L. F. Mollenauer, J.Phys. (Paris) 34, C9-141 (1974).

116 F. Seitz, Rev.Mod.Phys. 26, 7 (1954).

117 T. L. Gilbert and A. C. Wahl, J.Chem.Phys. 55, 5247 (1971).

118 A. N. Jette, T. L. Gilbert and T. P. Das, Phys.Rev. 184, 884 (1969).

119 M. J. Norgett and A. M. Stoneham, J.Phys.C (Solid State Phys.) 6, 223 and 229 (1973).

120 L. W. Hobbs, A. E. Hughes and D. Pooley, Proc.Roy.Soc. A332, 167 (1973).

121 C. R. A. Catlow, K. M. Diller and M. J. Norgett, J.Phys.C. (Solid State Phys.) 8, L34 (1975).

122 See (i) 7th Int.Conf. on the Physics of Semiconductors, Part 3, Radiation damage in semiconductors (Paris, Dunod 1965), (ii) Lattice Defects in Semiconductors, Ed. by R. R. Hasiguti (Univ. of Tokyo Press and Pennsylvania State University Press 1968), (iii) Radiation Effects in Semiconductors, Ed. by F. L. Vook (Plenum Press, New York 1968), (iv) Radiation Effects in Semiconductors, Ed. by J. W. Corbett and G. D. Watkins (Gordon and Breach, London, 1971), (v) Defects in Semiconductors, Ed. by J. H. Whitehouse (Institute of Physics, London 1973), (vi) Lattice Defects in Semiconductors, Ed. by F. A. Huntley (Institute of Physics, London, 1975).

POINT DEFECTS AND DIFFUSION

P. S. Dobson

Department of Physical Metallurgy
  and Science of Materials
University of Birmingham
Birmingham, B15 2TT, England

INTRODUCTION

Diffusion in solids occurs as a result of a large number of thermally activated atomic jumps from one lattice position to another. When two atoms simply exchange positions large distortions of the crystal are required to allow the atoms to squeeze past each other and the rate of this process is much too small to account for the experimentally observed diffusion rates. In practice the atomic movements involve the cooperation of point defects in the crystal and the diffusion coefficient can be related to the concentration and migration of these defects. Thus diffusion data represents a source of information on these basic point defect properties.

The diffusion of a particular species in a crystal is described by Ficks laws which treat the crystal as a continuum and state that the flux J of the diffusing species across a given plane in the crystal is proportional to the concentration gradient $\partial N/\partial x$ of the species across that plane.

$$J = - D\, \partial N/\partial x \tag{1}$$

The process is characterised by the diffusion coefficient D which is strongly dependent on temperature. The value of the diffusion coefficient can be determined by measuring the distribution of the species after it has been diffused in the crystal and then analysing the distribution or concentration profile using the phenomenological theory which is based on Ficks law. The details of the actual atomic mechanism of diffusion is contained within the parameter D and we can gain an insight into the physical significance of the diffusion coefficient by considering a simple derivation of Ficks first law.

Consider two parallel planes separated by a distance α in a one dimensional crystal of unit cross-sectional area. Suppose that a concentration gradient of solute atoms exists such that there are n solute atoms on plane 1 and $(n + \alpha\, \partial n/\partial x)$ on plane 2. If each atom makes Γ jumps per second, then the number of atoms jumping from plane 1 to plane 2 per second is $\tfrac{1}{2}\Gamma n$ where the factor $\tfrac{1}{2}$ allows for only one half of the total number of jumps being in the forward direction. There will also be a flux of $\tfrac{1}{2}\Gamma(n + \alpha\, \partial n/\partial x)$ in the backward direction and thus the net forward flux ϕ is given by:

$$\phi = -\tfrac{1}{2}\Gamma\alpha\, \partial n/\partial x = -\tfrac{1}{2}\Gamma\alpha^2\, \partial N/\partial x = -D\partial N/\partial x \qquad (2)$$

where $N = n/\alpha$ is the volume solute concentration. This is Fick's first law. A similar analysis is obtained for a three dimensional crystal apart from differences in the geometrical factor. For a cubic crystal.

$$D = \frac{1}{6} f\alpha^2 \Gamma \qquad (3)$$

where f is the correlation factor which is less than unity. This takes account of the fact that the direction of a jump may not be random but can depend on the direction of the preceeding jump.

The parameter of interest here is Γ, the number of jumps which an atom makes per second or the jump probability. It has already been pointed out that it is energetically unfavourable for the atoms to move by direct exchange and there are a number of possible diffusion mechanisms involving point defects which lead to higher values of Γ.

The vacancy mechanism involves an atom changing place with an adjacent vacant site (figure 1a). The crystal distortion involved in this transfer is very much less than that which occurs in direct exchange and thus the energy barrier is sufficiently low for a large number of thermally activated jumps to occur. We can now appreciate the significance of the correlation factor f since an atom which has exchanged sites with a vacancy has a higher probability of jumping back again to the vacant site than moving eventually to some other neighbouring site when this becomes vacant. Thus in the vacancy mechanism the atomic jumps are correlated rather than random.

The interstitial diffusion mechanism involves the transfer of an atom from one interstitial site to one of its nearest-neighbour interstitial sites. Clearly this mechanism will be more favoured by relatively small solute atoms which normally occupy interstitial positions and relatively open crystal structures. A related mechanism is the interstitialcy mechanism which can occur in cases where a relatively large atom such as a solvent atom occupies an interstitial position. (figure 1b). It will be difficult for this

POINT DEFECTS AND DIFFUSION

atom to move by the interstitial mechanism because of the large distortions involved but these are reduced if the interstitial atom pushes one of its nearest neighbour substitutional atoms into an interstitial site and then occupies the empty substitutional site.

To interpret eqt. 3 in terms of point defects, let us consider the case of an atom diffusing by the vacancy mechanism. It was stated above that an adjacent site must be vacant for a jump to occur and the probability of this is given simply by the atomic vacancy concentration c_v. When the atom and vacancy exchange positions the lattice is distorted and thus the atom (and vacancy) must surmount an activation barrier which is the vacancy migration energy E_m. If ν is the vibrational frequency of the atom, which can be considered as the rate at which the atom attempts to surmount the barrier, then we can write

$$\Gamma = c_v \nu \exp(-E_m/kT) = c_v \Gamma_v \quad (4)$$

where Γ_v is the jump probability of the vacancies. Since $c_v \sim \exp(-E_F/kT)$ where E_F is the vacancy formation energy we have

$$D = fc_v D_v = D_o \exp(-(E_F + E_m)/kT) \quad (5)$$

where D_v is the vacancy diffusivity and D_o is a constant. Thus by measuring the diffusion coefficient as a function of temperature, the sum of the vacancy formation and migration energies is obtained.

DIFFUSION IN IONIC SOLIDS

Point Defects in Ionic Solids

Because of the large electrostatic forces in ionic solids a diffusant is constrained to sites in either the cationic or anionic sublattices. The point defects exist in the form of anion vacancies,

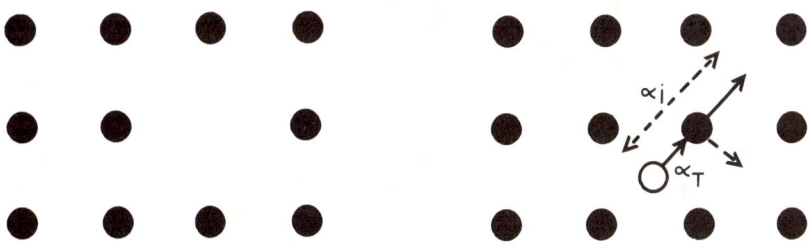

1a   Vacancy Mechanism   1b   Interstitialcy Mechanism

V_a and cation vacancies V_c (Schottky defects) or interstitials and vacancies in either sublattice (Frenkel defects). The concentrations of these defects as functions of temperature are given by

$$N_{Va} N_{Vc} = \exp(-g_s/kT) = K_s$$
$$N_{Ia} N_{Va} = \exp(-g_{Fa}/kT) = K_{Fa}$$
$$N_{Ic} N_{Vc} = \exp(-g_{Fc}/kT) = K_{Fc} \qquad (6)$$

For the case of pure material containing only Schottky defects for example then the constraint of electroneutrality requires there to be equal numbers of anion and cation vacancies. This is not true in general however where both Schottky and Frenkel defects exist and the concentrations of each species can be found from the above equations and the electroneutrality equation.

The relative proportion of anion and cation defects can also be perturbed if some of the matrix ions are replaced by ions of a different valence. For example if $MgCl_2$ is added to NaCl then in order to preserve electrical neutrality, one sodium vacancy is formed for every added magnesium atom. The concentration of each species can still be obtained from a combination of equation 6 and the electroneutrality equation which becomes:

$$N_{Va} + N_{Ic} + N_{Mg} = N_{Vc} + N_{Ia} \qquad (7)$$

This is a very useful effect since the defect concentrations can be varied in a known manner by deliberately doping the crystal.

Diffusion and Ionic Conductivity

In the absence of any electrical conduction by the tightly bound electrons, the conductivity of an ionic solid is due to the migration of the ions themselves and there is therefore a relationship between the electrical conductivity σ and the diffusion coefficient D. We can consider the mobile species to be either the matrix atoms of volume concentration N and self diffusion coefficient D_s or the appropriate point defects of volume concentration N_x, diffusivity D_x and correlation factor f. This relationship is given by the Einstein equation

$$\sigma/D_x = N_x q_x^2 / kT \qquad (8)$$

which for the vacancy mechanism is equivalent to

$$\sigma/D_s = N q_x^2 / fkT \qquad (9)$$

The relationship between electrical conductivity and diffusion

POINT DEFECTS AND DIFFUSION

coefficient is of great value and enables considerably more information about point defect properties to be obtained than either measurement would on its own. For example figure 2 shows an Arrhenius plot of the diffusion coefficient of sodium in sodium chloride obtained directly using a radiotracer diffusion technique and indirectly from measurements of the electrical conductivity. It can be seen that not only is each plot composed of two regions exhibiting different activation energies but that there are significant differences between the two sets of data.

In sodium chloride the bulk of the current is carried by the sodium ions and we will assume that these ions diffuse by the vacancy mechanism. This assumption will be justified below. For a crystal exhibiting Schottky disorder and containing a concentration N_D of divalent impurities we have from equations 6 and 7

$$N_{Vc} = (N_D/2) \{1 + (1 + 4K_S/N_D^2)^{\frac{1}{2}}\} \tag{10}$$

If $N_D \gg K_S$, the concentration of cation vacancies becomes independent of temperature and thus the activation energy of diffusion corresponds to the migration energy of the cation vacancy. This is the situation at low temperatures where K_S is small and thus the migration energy of the cation vacancy is 0.77eV. At high temperatures where K_S is large the behaviour is dominated by the intrinsic

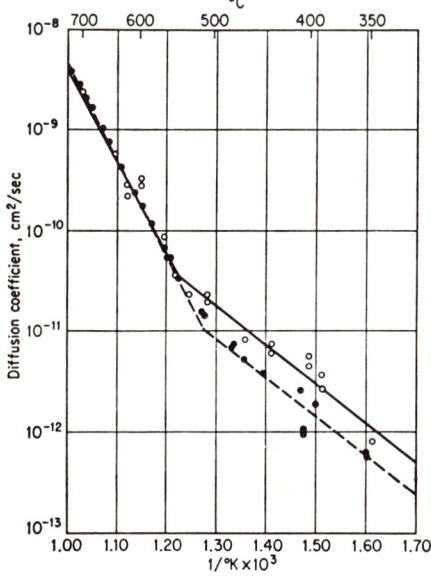

2. Log D versus 1/T for sodium in NaCl as determined with radioactive sodium (o) and as calculated from the conductivity (●). {Mapother, Crooks and Maurer, J. Chem. Phys., 18, 1231 (1950)}.

defects and thus the observed activation energy of 1.8eV corresponds to the sum of the migration energy and half the formation energy of a Schottky defect. The formation energy of the Schottky defect g_s is thus determined to be 2.06eV. More recent results using improved experimental techniques give the formation energy of a Schottky defect as 2.4 - 2.5eV and the cation vacancy migration energy as 0.65 - 0.72eV.

This interpretation can be checked by measuring the diffusion coefficient or conductivity as a function of the concentration of deliberately added divalent impurities. It was found that at low temperatures the conductivity increased with doping concentration in accord with the model. These experiments also justify the assumption that the cations migrate by the vacancy mechanism since the concentration of cation interstitials decreases with increasing doping concentration.

This description of diffusion in sodium chloride is somewhat historic but serves to illustrate how the identities and properties of point defects in ionic crystals can be determined. It is now known that in order to fully account for the electrical conductivity of sodium chloride, it is necessary to consider not only the cation vacancies but also a small contribution from anion vacancies together with effects due to vacancy-impurity pairs and divacancies. The presence of the pairs account for the observation that the directly measured diffusion coefficient is somewhat greater than that inferred from electrical conductivity measurements. The concentration of these pairs can be derived in terms of their binding energies using the same statistical method as is employed for the derivation of equation 6 and the experimental situation is such that the point defects can be fully characterised by a combination of electrical conductivity and diffusion measurements.

DIFFUSION IN SEMICONDUCTORS

Self Diffusion in Silicon and Germanium

In contrast to the case of ionic crystals, diffusion in the elemental semiconductors, silicon and germanium, is considerably less well understood. Studies of self diffusion in these materials using radiotracer techniques show that the diffusion coefficients are small and are about four orders of magnitude less than those of metals at the melting point. In silicon the measured activation energy is $\sim$ 4.9eV whereas the vacancy and interstitial migration energies as determined by EPR techniques are $\leqslant$ 0.3eV. These results imply a very large energy of point defect formation which means that the equilibrium point defect concentrations are very small. As a consequence it has not so far proved possible to detect

POINT DEFECTS AND DIFFUSION

directly the presence of equilibrium point defects, by for example, measuring the density and lattice parameter as a function of temperature. This combination of a high formation energy and a low migration energy also means that quenching experiments cannot be used to study the defects since they rapidly anneal out at internal sinks during the quench.

Apart from the high activation energy a second feature which characterises self-diffusion in silicon and germanium is the high value of the pre-exponential factor D_o. In silicon $D_o \sim 10^3$ cm^2 sec^{-1} (cf ~ 1 cm^2 sec^{-1} for metals). This factor includes the entropy terms for defect formation and migration and it has been suggested that the point defects are in fact extended defects. For example an extended vacancy is considered as a region in the crystal containing say nine atoms which would normally contain ten atoms.

Some information has been obtained by measuring the self diffusion coefficient as a function of the n or p-type doping concentration. This is analogous to the effect of impurities on diffusion in ionic crystals but, unlike the ionic crystals, it is not immediately obvious which type of point defect will be enhanced by a particular dopant impurity. This is because any changes in point defect concentration do not arise as a consequence of simple charge compensation but are due to the point defects having acceptor or donor states. For example if a vacancy has an acceptor level near the middle of the band gap, then in n-type material where the Fermi level is near the conduction band edge, the vacancy will accept an electron with a consequent energy release which effectively reduces the vacancy formation energy. Thus the total vacancy concentration is enhanced in n-type material and reduced in p-type material.

Self-diffusion in germanium is enhanced by n-type doping and retarded by p-type doping and the results are in reasonable quantitative agreement with the point defects having an acceptor level near the top of the valence band. It has been generally assumed that the point defects responsible for self diffusion in germanium are vacancies since on the basis of a hydrogenic model, the vacancy is expected to behave as an acceptor. In the absence of any detailed experimental data on the levels of the germanium vacancy however, this assumption must be regarded as unproven. For silicon the situation is less clear since the experimental evidence itself is partially conflicting. There are considerable experimental difficulties in carrying out self diffusion tracer studies because of the short half life of the ^{31}Si isotope which means that the experiments can only be conducted over a small temperature range and that the shallow diffusions may be influenced by surface effects. The diffusion coefficient was found to increase as the concentration of n-type dopant increased but the results for p-type doping are conflicting, both enhancements and retardations being observed by different workers.

Because of these experimental difficulties, McVay and Du Charme studied the diffusion of germanium in silicon and silicon-germanium alloys. They found that the diffusion profile in the near-surface region was influenced by some surface effect in the silicon rich alloys but considered that beyond this near-surface region the profile represented the true bulk diffusion. They obtained values for the activation energy and the pre-exponential factor as a function of alloy composition which were in good agreement with the self-diffusion data on silicon and germanium. When these parameters were plotted as a function of alloy composition, (figure 3) the plot showed a sharp change in slope at a germanium concentration of about 35%. This result was taken to imply a different diffusion mechanism in silicon and germanium.

These workers also found that the diffusion coefficient was increased by both n and p-type dopants. This result was taken as support for the interstialcy diffusion mechanism on the basis of a theoretical model of the interstitial which predicts that the interstitial is amphoteric, having an acceptor level in the upper half of the band gap and a donor level in the lower half.

Phosphorus Diffusion in Silicon

The diffusion of phosphorus in silicon is strongly influenced

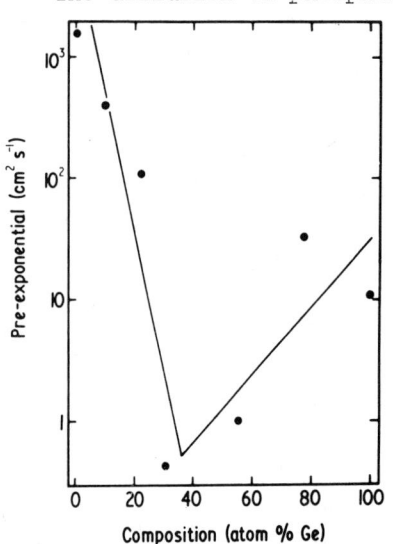 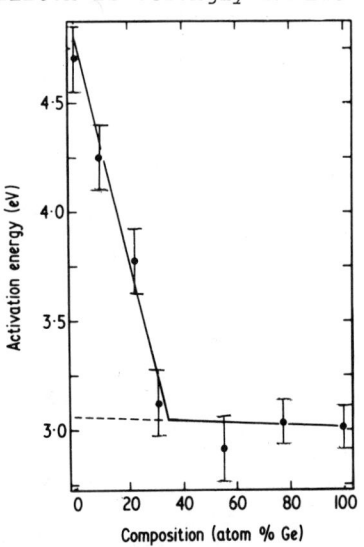

3. Variation of the activation energy and pre-exponential factor with composition for ^{71}Ge diffusion in Si-Ge alloys. {McVay and Du Charme, Lattice Defects in Semiconductors, p. 91, (1974, Institute of Physics, London)}.

POINT DEFECTS AND DIFFUSION

by surface effects, particularly oxidation, and exhibits anomalous behaviour if the phosphorus concentration exceeds the intrinsic carrier concentration at the diffusion temperature. Under this condition the diffusion profile is kinked (figure 4a) and is characterised by two diffusion coefficients rather than one unique value. These kinked profiles have been accounted for by the E-centre model which is based on the ionised phosphorus atoms forming E-centres with the negatively charged vacancies in the region of high phosphorus concentration near the surface. The first part of the profile is then governed by the diffusion of these centres. At greater depths in the specimen where the Fermi energy is less the E-centres dissociate releasing a large concentration of vacancies in excess of the equilibrium value. The second part of the profile is then governed by the simple vacancy exchange mechanism and is in fact characterised by a higher diffusion coefficient than that in the surface regions presumably due to a much enhanced non-equilibrium vacancy concentration.

Support for this type of model has been provided by diffusion experiments using marker layers. In these experiments a sandwich type silicon structure is fabricated which contains a thin marker layer of doped silicon lying about 10 microns from the surface. An oxide is then deposited on the surface and windows are etched in the oxide. Various diffusion treatments can then be carried out in the uppermost few microns of the exposed silicon and the expansion of the marker layer under the window can be compared with that under the protective oxide using a bevelling and staining technique. An example of this is shown in figure 5 which is actually a silicon

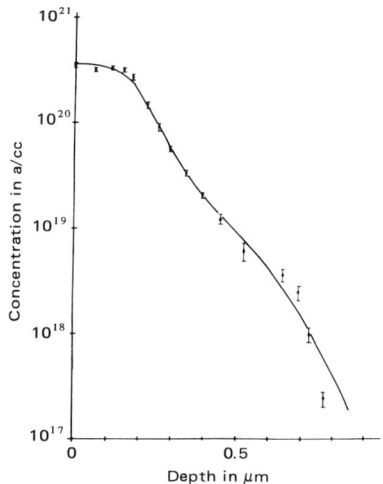
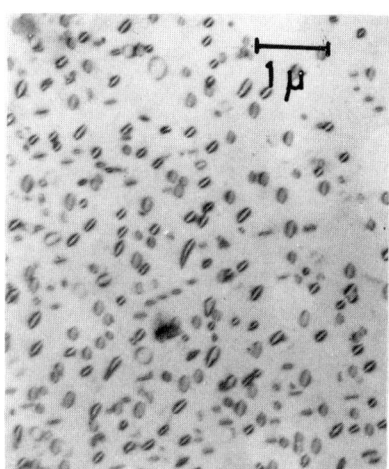

4a. Phosphorus profile after a diffusion from an oxide source at 938°C for 1 hour (courtesy of R. Francis). (b) Transmission electron micrograph of surface region of diffused slice after annealing for 1 hour at 700°C. (courtesy of P. Lewis).

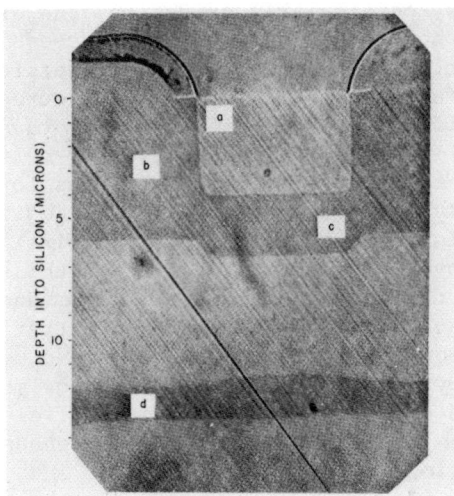

5. Section through transistor at 2° to the surface (a) phosphorus emitter, (b) boron base, (c) pushed-out base and (d) p-type marker layer. {Lee and Willoughby, J. Appl. Phys., 43, 245, (1972)}.

transistor but the enhanced diffusion of the p-type marker after the phosphorus emitter has been diffused through the window can clearly be seen. These results indicate that the shallow phosphorus diffusion produces an enhanced point defect concentration in the bulk of the silicon which is in general agreement with an E-centre type of model although it is not certain whether the complex involved is an actual E-centre or some other complex.

Interstitials have been detected in phosphorus diffused silicon by transmission electron microscopy. The surface regions of a diffused slice which had been subsequently annealed at a lower temperature ($\sim$ 700°C) contained a high density of dislocation loops. (figure 4b). These were found to be of interstitial character and thus consist of an extra plane in the lattice which is formed by the condensation of silicon interstitials. The depth distribution of these loops was determined by thinning the specimens perpendicular to the diffusion plane and it was found that the region containing the loops corresponded roughly to the first part of the diffusion profile. It is not clear at this stage whether the interstitial loops are produced by the condensation at the lower annealing temperature of a quenched-in interstitial concentration or whether the loops have effectively acted as a vacancy source to provide an increased equilibrium E-centre concentration at the lower annealing temperature.

It is clear from the above that in contrast to the ionic

crystals, the identity and properties of point defects in semiconductors are still the subject of much confusion. A more detailed discussion of some of the points mentioned above may be found in the following references.

References

"Diffusion in Solids", P.G. Shewman, (McGraw Hill, 1963).
"Point Defects and Diffusion", P. Flynn, (Oxford Univ. Press, 1970).
"Point Defects in Solids", Vols. I and II, edited by J.H. Crawford and L.M. Slifkin, (Plenum Press, 1972, 1975).
"Diffusion in Semiconductors", edited by D. Shaw, (Plenum Press 1973).
"Diffusion in Solids - Recent Developments" edited by A.S. Nowick and J.J. Burton, (Academic Press 1975).

FROM SPECTROSCOPY TO MICROSCOPY - THE PHOTON AS A PROBE

W. A. Sibley

Physics Department, Oklahoma State University

Stillwater, Oklahoma   74074

I.   INTRODUCTION

The title of this paper illustrates the considerable latitude of the subject matter to be covered. Photons range from gamma rays to very low energy infrared rays; thus, the scope is extremely broad. Using gamma rays from the Mossbauer effect Mullen[1] has studied impurity precipitates and vacancy-impurity pairs. Visible and infrared radiation can be used to study defect clusters through light scattering or microscopy and to study the effect of point defects on the vibration modes of OH molecules in crystals.[2] Because of the limited space available for this review only certain aspects of using the photon as a probe can be emphasized. Those areas which will be covered by other authors at this conference such as x-ray scattering, point defect optical absorption, magnetic circular dichroism, infrared absorption, etc. will be treated only briefly and the reader is referred to the more detailed papers in this volume. Research areas not covered by other participants in the conference will be emphasized but even so certain topics will have to be omitted. First, a review of optical transitions in pure and defect containing crystals is presented and then a treatment of the use of photons to study impurities, radiation damage and macroscopic defects is given.

II.   THE INTERACTION OF LIGHT WITH SOLIDS

A.   The Perfect Crystal

High energy photons are especially useful in studies of excitons and band to band transitions,[3] and Synchrotron radiation is an important tool in the investigation of these high energy transitions.[4] Figure 1 illustrates the optical absorption of MgF$_2$ over a wide range of photon energies.[5-8] At high energies exciton and band to band transitions dominate. Notice that the absorption coefficients for these transitions are of the order of 10^5 cm^{-1}. The absorption coefficient α is determined by the number of absorbing defects and their ability to absorb photons i.e. their oscillator strength, f. It is related to the incident photon flux $F_o(E)$ and the energy of the incident photons in the following way:[9,10]

$$F/F_o = [1-R]^2 \exp(-\alpha t) \tag{1}$$

where t is the sample thickness and R(E) is the reflection coefficient of the surface for normal incidence. The reflection coefficient for light at normal incidence is given by

$$R = (\eta-1)^2/(\eta+1)^2 \tag{2}$$

where η, the refractive index of the material, is changed very little by the introduction of impurities or radiation defects. The intensity of an optical absorption band is defined in terms of the probability of a transition occurring between the energy levels and as the area under the curve of the absorption coefficient, α, versus photon energy. That is[8,9,11]

$$I(E,T) = \int_{E_{min}}^{E_{max}} \alpha(T,E) dE = \sum_{i,f} |\langle \psi_{ai} | \vec{er} | \psi_{bf} \rangle|^2 B_{ai} \delta(E - E_{bf} - E_{ai}) \tag{3}$$

The subscripts a and b denote the ground and excited electronic states and ψ are the wave functions for these states. The thermal population factor for the ith vibrational level of the ground state is B_{ai} and the δ function locates the energy of the i→f transition. Assuming that an optical transition occurs more rapidly than the vibration of the lattice ions (adiabatic approximation) and that the total wave functions ψ can be expressed as a product of purely electronic wave functions, ϕ, and purely vibrational wave functions, χ, we find

$$|\langle \psi_{ai} | \vec{er} | \psi_{bf} \rangle|^2 = |\langle \phi_a | \vec{er} | \phi_b \rangle \langle \chi_{ai} | \chi_{bf} \rangle|^2 \tag{4}$$

As will be discussed later, the first term on the right hand side is proportional to the oscillator strength of the transition and the second term gives the band shape.

An electronic transition in alkali halides and alkaline earth halides for the energy range 5 to 20 eV may be viewed for most purposes as the removal of an electron from a halide-ion p-type orbi-

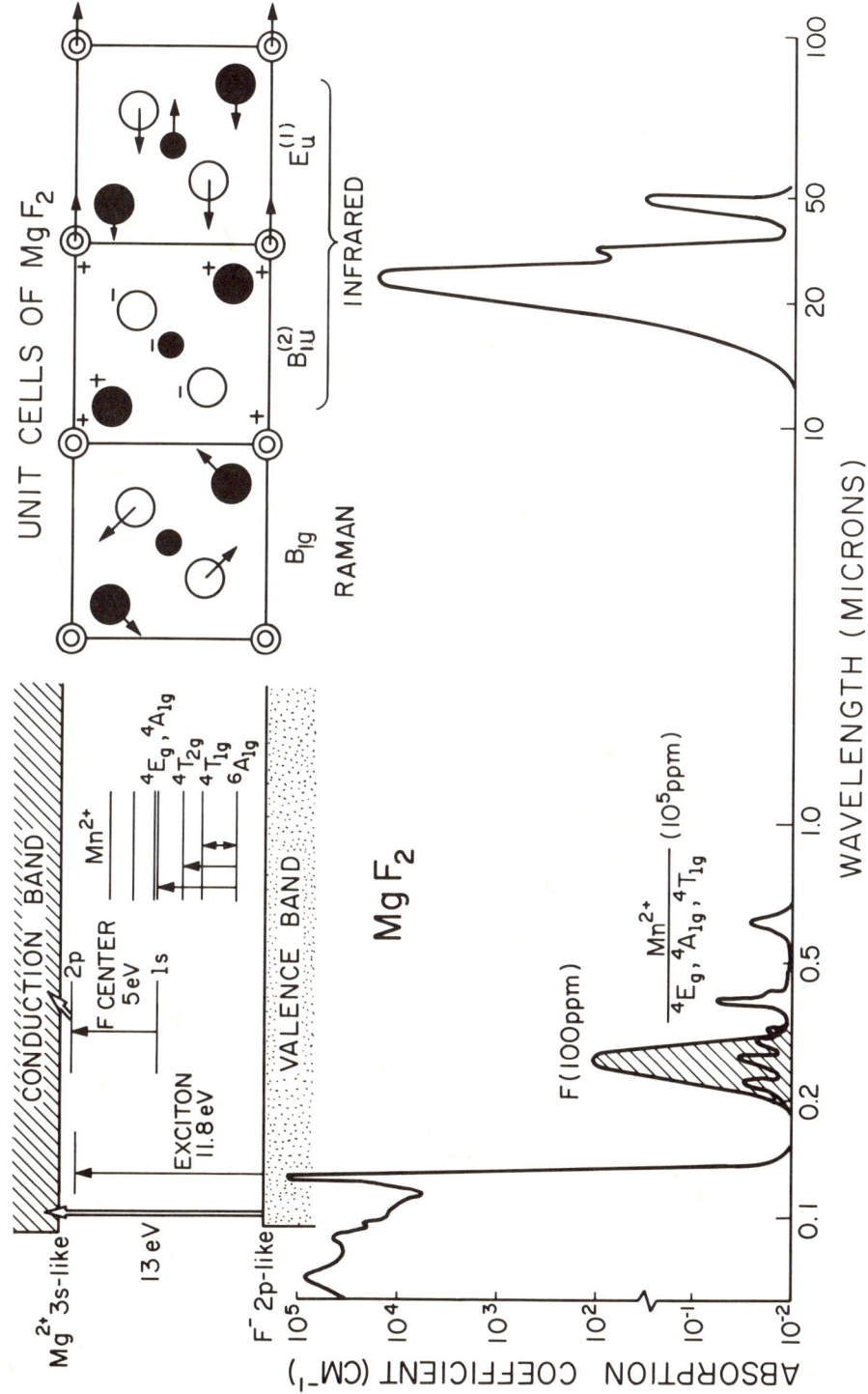

Fig. 1. The optical properties of MgF$_2$

tal. In the illustration shown in Fig. 1 the valence band for MgF_2 is made up of 1s, 2s and 2p-type orbitals of F^- and Mg^{2+}. The 2p-type orbitals for F^- dominate and the valence band is about 6 eV wide.[12] The conduction band consists of the higher order wave functions of Mg^{2+} and F^- with 3s-type Mg orbitals lowest. Given sufficient energy an electron may even be removed from the crystal, but the absorption shown in Fig. 1 at 11.8 eV is apparently due to a less energetic excitation in which the electron is raised to an orbital such that the lowest excited electronic states of the crystal result.[5,6] As can be seen in Fig. 1 the energy for these "exciton" transitions which consist of tightly or loosely bound electron-hole pairs is slightly less than that of band to band transitions where photoconductivity is observed. According to Williams et al.[6] the band at 11.8 eV in MgF_2 is due to excitons and band to band transitions start at about 13 eV. Theoretical calculations for MgF_2 have been made by Jouanin and Gout.[12] The large magnitude of the absorption coefficient for excitons and band to band transitions arises from the fact that 10^{22} ions/cm^3 are involved in the absorption, whereas for impurity or radiation defect absorption only about $10^{16} - 10^{19}$ defects/cm^3 are active. In absorbing regions of the spectrum the imaginary part of the high frequency dielectric constant $\varepsilon(\infty)$ is nonzero and both the real and imaginary parts of ε can be obtained from the experimental reflectivity and the use of the Kramers-Kronig relation.[13-16] Theorists solve the Schrodinger equation for excited states by various means to construct microscopic models which explain the behavior of the absorption coefficient in these high energy regions.[14-15]

Newman discusses infrared absorption of defects elsewhere in this volume and in an excellent review paper.[17] Intrinsic infrared absorption in insulators is due to lattice ion vibrations. In some transparent crystals such as diamonds the infrared absorption is not too intense, but in diatomic crystals such as MgF_2 the absorption is high.[13] This is because of the dipoles created by the vibration of oppositely charged ions in diatomic crystals (see Fig. 1). Moreover, the dispersion relation of frequency ω versus K, the wave vector, develops two branches known as the acoustical and optical branches. Longitudinal and transverse waves propagate in each branch giving rise to so-called longitudinal (LA) and transverse (TA) acoustical phonons and longitudinal (LO) and transverse (TO) optical phonons.[18] In a perfect crystal the coupling of these phonon modes to photons occurs at K≈0. The ratio of the longitudinal and transverse optical frequencies at K≈0 is related to the static ε_o and high frequency $\varepsilon(\infty)$ dielectric constants by[13,18]

$$\omega_L/\omega_T = [\varepsilon_o/\varepsilon(\infty)]^{\frac{1}{2}} \qquad (5)$$

In diatomic crystals the first order dipole interaction is restricted to the transverse optical mode; but even there only the vibration at K≈0 makes a large contribution since only then does the momentum vector of light match that of the lattice so that a nonvanishing macroscopic dipole can be sustained by the crystal. Thus, the infrared absorption spectra may be explained by the existence of a single oscillator i.e. the vibration of the anion lattice 180° out of phase with the cation lattice. As noted earlier, the probability of an optical transition occurring is proportional to integrals of the form (eq. 4)[19]

$$|\int <\psi_g|\vec{er}|\psi_e> d\tau|^2 \tag{6}$$

where the ψ's are wave functions of the system. In the case of infrared and Raman transitions these wave functions are those of the lattice ions, and a fundamental transition can occur only when the integrals in eq. (5) are nonzero. Since the ground state wave function for the ions is totally symmetric then for eq. 6 to be nonzero the excited wave function ψ_e must belong to the same symmetry representation as one of the dipole coordinates x, y or z. Thus, a simple rule for the activity of fundamental infrared absorption is: A fundamental lattice mode will be infrared active if the normal mode which is excited has the same symmetry representation as any one or several of the dipole coordinates. For Raman scattering, it is necessary that at least one integral of the type[19]

$$\int <\psi_g|\hat{P}|\psi_e> d\tau \tag{7}$$

be nonzero. In these integrals P is one of the quadratic functions of the cartesian coordinates i.e. x^2, y^2, z^2, xy etc. The above requirement means that there must be a change in polarizability when the transition occurs. Therefore, a fundamental transition will be Raman active if the normal mode involved belongs to the same symmetry representation as one or more of the components of the polarizability tensor of the system. In MgF_2 there are 15 optical vibrations which have the following group theory labels and vibrational frequencies in wave numbers (cm^{-1})[20,21,22] It may also be noted, at this point, that in a system with inversion symmetry, no Raman active vibration is also infrared-active and no infrared-active vibration can be Raman active.

A_{1g}	410	A_{2u}	438	B_{1g}	92	E_g	295
A_{2g}	325	B_{1u}	180, 450	B_{2g}	515	E_u	242, 298, 403, 404, 459, 615

In Fig. 1 the absorption coefficient for infrared absorption is shown to be about 10^5 cm^{-1} at 25 μm [7,8]. This absorption is due primarily to the E_u lattice modes. From the illustration in Fig. 1 of the movement of the lattice ions (shown immediately above the infrared absorption peak in Fig. 1) it is clear E_u modes are infrared active (a dipole is evident) whereas the B_{1g} mode is Raman active. It should also be noted that the B_{1u} and E_u modes have opposite polarization in MgF_2 and polarized absorption measurements could in principle be made. Since a number of modes are infrared active the absorption is normally broad and highly temperature dependent. As mentioned above, in perfect crystals only absorption at K≈0 can be observed in first order by optical techniques. Neutron scattering is a more powerful method of determining ionic movements across the whole Brillioun zone and for investigating phase transitions. Several good reviews of this method are available.[22,23] When defects are present in the crystal the possibility of local modes of vibration or resonance modes exists.[2] Obviously any change in the mass of one of the ions or in the force constant will give some change in the vibration frequency.

B. Impurities and Radiation Defects

In addition to intrinsic absorption Fig. 1 shows the absorption of Mn^{2+} impurities and radiation induced F centers (negative ion vacancies which have trapped electrons for charge compensation). For N centers each with oscillator strength f and a Gaussian band shape we find[9,10]

$$Nf = 0.87 \times 10^{17} \frac{\eta}{(\eta^2+2)^2} \alpha_m W \text{ cm}^{-3} \tag{8}$$

where W is the width of the band at half maximum and α_m is the maximum absorption coefficient. The oscillator strength for the F center is almost unity and the half-width of F bands is about 0.5 eV so the concentration of F centers in a crystal is about

$$N_F \approx 10^{16} \alpha_m \text{ cm}^{-3} \tag{9}$$

Although the oscillator strength is close to unity for F center transitions, for 3d impurity ions situated in octahedral, O_h, symmetry (as shown in Fig. 2) the oscillator strengths are 10^{-8} - 10^{-7}.[24,25] For O_h symmetry both ϕ_a and ϕ_b are even functions and the electric dipole r is an odd function; thus, in theory $f_{ED} = C_k <\phi_a|e\vec{r}|\phi_b>$ is zero. The transitions are then magnetic dipole allowed (the expression for a magnetic dipole oscillator strength[25] is $f_{MD} = C_\ell |<\phi_a|\hat{L}+2\hat{S}|\phi_b>|^2$ and all functions are even so that the integral is finite). In most cases the transitions are mixed dipole or are phonon assisted in which case eq. 4 should not be separated into two terms. In any case the oscillator strengths are extremely small. This difference in oscillator strengths between Mn^{2+} transitions and F center transitions is shown in Fig. 1. It is possible to change f by changing the site symmetry or by exchange inter-

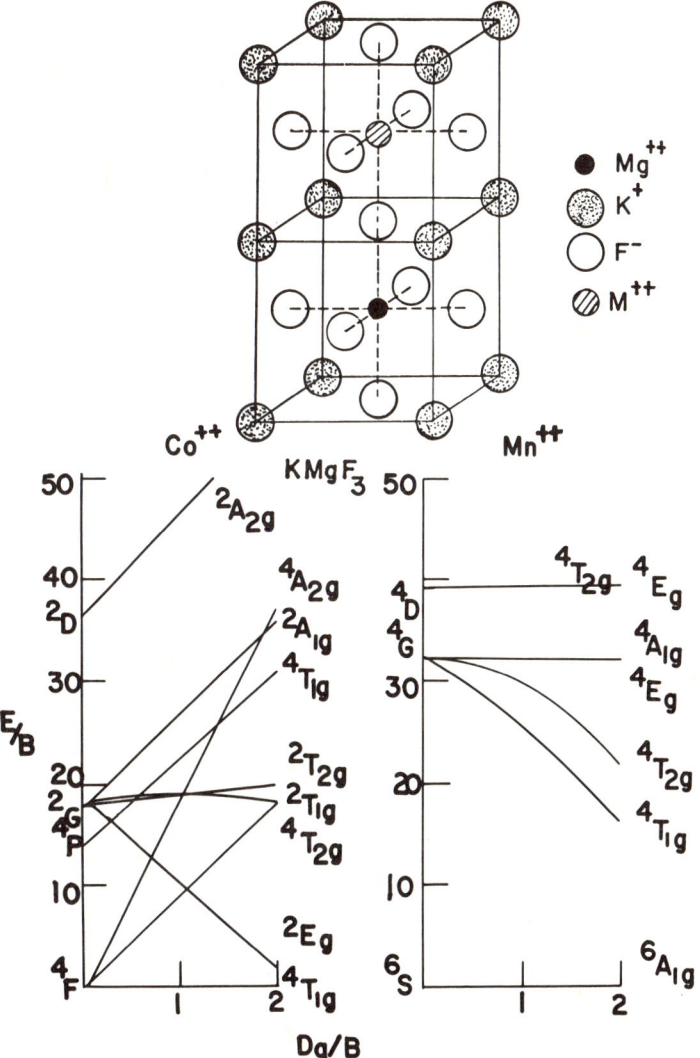

Fig. 2. Crystal structure of $KMgF_3$ showing a substitutional impurity ion in O_h symmetry. The lower part of the figure shows energy level diagrams after Tanabe and Sugano[34] for d^7 and d^5 configurations in O_h symmetry

actions, but usually if only the temperature of the sample is altered f and thus the area under the absorption or emission band remains the same.

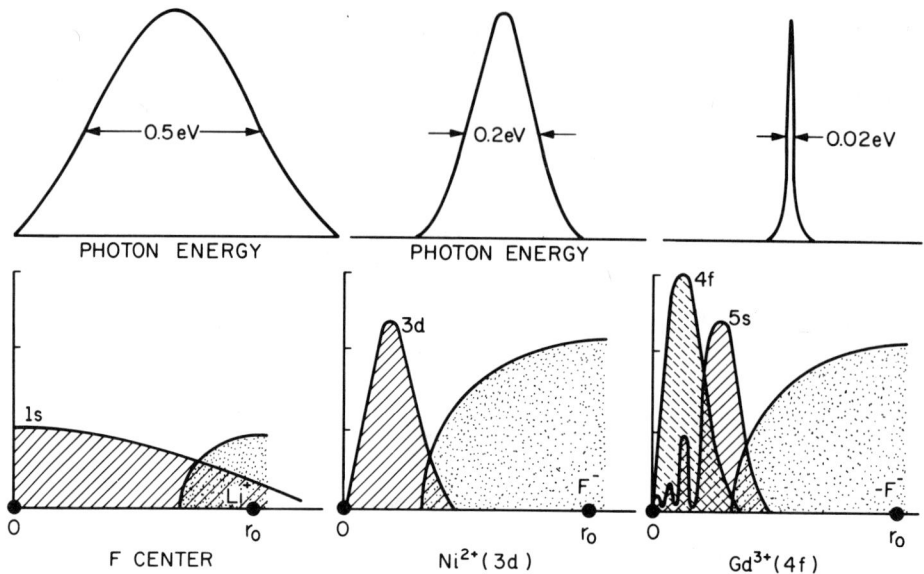

Fig. 3. Optical absorption and electronic overlap for defects in LiF

The change in band shape with temperature arises from the term $<\chi_{ai}|\chi_{bf}>$ in eq. 4. When the electronic wave function is spread out then the transition is extremely sensitive to the vibrations of the neighboring ions. This is portrayed in Fig. 3 which shows the overlap of the electronic charge distribution for an F center, Ni^{2+}, and Gd^{3+}. The 3d and 4f ions are relatively insensitive to neighboring defects and lattice vibrations whereas the F center is more strongly coupled to the host lattice and its transition energy is easily shifted by neighboring defects.[26,27] Rare earth ions are further shielded by outer electrons and have about the same absorption energy regardless of host lattice.[28] For broad bands the width varies with temperature according to the relation[9]

$$W^2(T) = W^2(0)[\coth(\hbar\omega/2kT)] \qquad (10)$$

where W(0) is the low temperature width of the band and $\hbar\omega$ is the energy of the single mode assumed to be interacting with the electron. In MgF_2 there are a number of vibrational modes interacting with the defects, but it can be assumed that one mode dominates the interaction. In fact, the observed vibration energy for ions around the F center is 300 cm^{-1}. Fig. 4 portrays a configuration coordinate model which gives a simple account of the vibronic structure observed on some of the narrower optical bands.[10] The basis of the model is the Born-Oppenheimer approximation which assumes that the electrons rapidly move to the lowest energy configuration of given symmetry appropriate to a given nuclear configuration. This is the same approximation that allowed a separation of eq. 4 into two terms and is not valid for phonon assisted transitions. The total energy of each electronic state will therefore depend on the nuclear configuration (as shown in the figure) having a minimum at the equilibrium configuration. Usually the configuration of minimum energy in an excited state is different from that for the ground state with the result that the most probable optical transition (the one directly upward in the figure from the configuration where the ground vibrational state wave function is a maximum) involves creating several, S, vibrational quanta. This is based on the assumption that only the lowest vibrational level of the ground electronic state is populated at very low temperatures. The probability of an optical transition in which vibrational quanta are created is[29]

$$\rho_n = S^n \exp(-S)/n! \qquad (11)$$

Although the configuration coordinate model implies strong coupling to a single vibrational mode, the expression for ρ_n is the same for weak coupling to a large number of equivalent modes. This latter case was given by Huang and Rhys[30] and thus, the coupling parameter S is referred to as the Huang-Rhys factor. For observable vibronic structure S must usually be less than 6, but Rolfe[31] has shown that in some cases for S<6 zero phonon lines are not observable. The width at low temperature of most zero phonon lines is usually determined by lattice strains. In fact in MgF_2 the zero phonon line associated with one defect is twice as broad in neutron irradiated samples as it is in electron irradiated specimens. Obviously, when zero phonon lines are present it is possible to use various perturbation methods to make defect structure analyses.[32] This type of experiment is discussed by Hughes in this volume.

Spectra from rare earth and transition metal ions arise from transitions between localized states.[28,13] In this case several effects are important in shifting energy levels or lifting degeneracies. There are spin-orbit interactions,[25,28] crystal field effects,[25,24] exchange and multipole interactions and Jahn-Teller effects.[33] The Hamiltonian for a multiple electron system can be written[13]

$$H(\vec{P}_i, \vec{r}_i) = \sum_i^n [\frac{P_i^2}{2m_i} - Z\frac{e^2}{\vec{r}_i} + \zeta(\vec{r}_i)\hat{s}_i\hat{\ell}_i] + \sum_{i>j=1}^n \frac{e^2}{\vec{r}_{ij}} + \sum_{ij} q_j \frac{e^2}{\vec{r}_{ij}} \qquad (12)$$

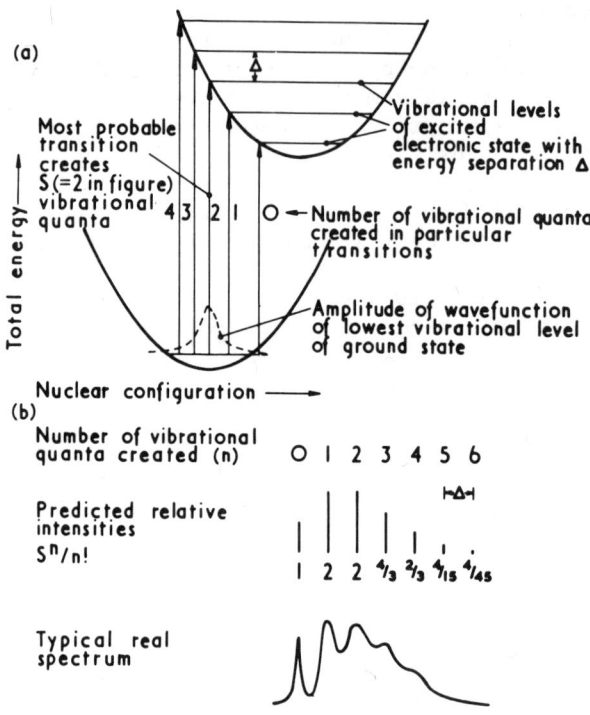

Fig. 4. Configuration coordinate diagram illustrating vibronic structure. After Sibley and Pooley[10]

In this equation the symbols have their usual significance and for the spin-orbit term

$$\zeta(\vec{r}_i) = \frac{h^2}{2m_i^2 c^2 \vec{r}_i^2} \left[\frac{\partial V(\vec{r}_i)}{\partial \vec{r}_i}\right] \quad (13)$$

The last term in eq. 12 refers to the crystal field effect. In the case of 3d ions the crystal field splitting of the energy levels is usually much larger than the spin-orbit splitting; however, for rare earth ions this is not the case. Fig. 5 illustrates the splitting of the d-electron levels in a cubic field for Fe^{2+} and shows the spin orientation for weak fields. In Fig. 2 the effect of the crystal field on these levels as calculated by Tanabe and Sugano[34] was shown. The energy levels for Mn^{2+} in Fig. 2 can also be compared with the transitions illustrated in Fig. 1. As the crystal field changes some of the levels are strongly effected and optical

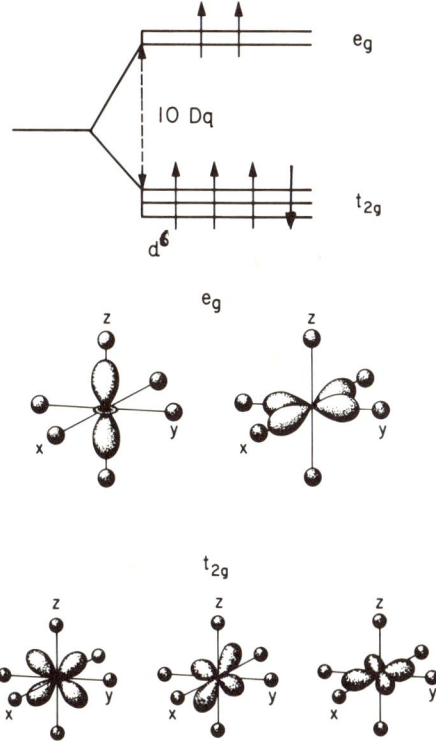

Fig. 5. Splitting of the d^6 configuration in cubic field

transitions to these levels are generally broader than those to the levels that are relatively independent of the crystal field. The Jahn-Teller effect also lifts the electronic orbital degeneracy of levels.[33] A nonlinear configuration of nuclei surrounding a defect will tend to distort to remove any electronic orbital degeneracy. If this interaction is strong, the defect will undergo a static Jahn-Teller distortion which lowers its symmetry and removes the degeneracy. If the interaction is weak and the distortion energy is comparable to the zero point energies of vibration then there is not stable, static distortion of the center, but rather a dynamic distortion involving coupled electronic and nuclear motion. A powerful tool for investigating these types of transitions is that of Magnetic Circular Dichroism (MCD) as will be discussed by Merle d'Aubigne in this volume.

C. Light Scattering and Microscopic Studies of Defects

Fröhlich and his collaborators[35] have shown with excellent radiography studies of impurity doped crystals that macroscopic defects are introduced into crystals during growth. These large size defects interact with electromagnetic radiation in a different manner than the atomic defects described above and other techniques are used to investigate their presence and behavior. When defects are greater than about 0.1 micron in size it is possible to use optical microscopy with visible light. The highest resolution of most optical microscopes is given by the expression:[36]

$$s = \frac{\lambda}{2\eta \sin i} \qquad (14)$$

where the product of $\eta \sin i$ is characteristic of a particular objective and η is the index of refraction of the oil which fills the space between the object and the objective lens when high magnification is sought. For 560 nm light $s \simeq 1.8 \times 10^{-5}$ cm. The resolution is less when working with infrared microscopes and can be increased with ultraviolet light. In either case, however, an imaging system or photographic techniques must be utilized. The optical microscope is also used in studies of dislocation etch pits and grain boundaries. In many instances these studies are on transparent materials and the intensity changes caused by the pits are not readily observable visually. The pits do give large differences in phase on the other hand and the phase-contrast microscope developed by Zernike[37] is a tremendous aid in studying transparent materials. In this system[38], interference occurs between the direct light which passes unaffected through the uniform parts of the material and the light which is diffracted by the irregular etch pits. The former consists of parallel beams, and is brought to a focus in the secondary focal plane of the objective while the latter is focused in the plane of the image conjugate to the object. By placing a quarter-wave plate in the secondary focal plane, the phase of the direct light which is spread uniformly over the image plane is altered in the proper way to produce an amplitide modulation in this plane. In this way details of the transparent material become visible.

Two other methods of investigating macroscopic defects in transparent solids are ultramicroscopy[39] and bulk light scattering.[40,41] Our concern in this paper will be with Rayleigh, Rayleigh-Gans and Mie scattering. The images seen in ultramicroscopy do not reveal the true size or shape of the scattering centers since the light is scattered and not reflected but calibration with particles of known size is possible.[39] The density and location of scattering centers can be determined by this method. Usually a laser or an intense arc lamp is used in conjunction with magnifications of 7X to 1200X. Since what is seen is light scattered by single spherical particles the Mie theory is applicable and several tables of Mie scattering are available for analysis of data.[40] The theory for thermal scattering, precipitate scattering, and cylindrical scattering units such as for impurities around dislocations has been available for sometime.[40] A convenient way to express the scattering power of a crystal is the Rayleigh ratio, R, which is defined in terms of the distance to the detecting system, r, and the polarization of the incident and scattered light. For example, the Rayleigh ratio $R_V = V_V r^2/I_o$ where I_o is the intensity of the incident light and V_V indicates that the incident and the scattered light are polarized in the vertical direction. Thus, the expressions of interest are:

$$R_{V_{Thermal}} = \frac{V^2 \pi^2}{2\lambda^4} (\eta^2-1)^2 kT\beta \qquad (15)$$

$$R_{V_{Rods}} = (\frac{2\pi^4 \eta^4}{\lambda^4}) NV^2 \sum_{j=1}^{3} (<\alpha_{zz}^2>_{AV})_j [(\frac{\sin(C_j Z^*/2)}{C_j Z^*/2})^2] \qquad (16)$$

$$R_{V_{Spheres}} = [(\frac{2\pi\eta}{\lambda})^4 NV^2 |m-1|^2 G^2 (2 x \sin \theta)]/8\pi^2 \qquad (17)$$

In the above equations V is the volume of the scattering region, β is the compressibility, k Boltzmann's constant, 2θ the scattering angle, η the index of refraction and m is the index of the precipitate relative to that of the medium. $x = \frac{2\pi\eta}{\lambda} a$, where a is the radius of the sphere and $G(u) = (9\pi/2u^3)^{\frac{1}{2}} J_{3/2}(u)$. In eq. (16) which specifies the scattering for a triad of orthogonal cylinders,[42] j is an index specifying the cylinder orientations, $<\alpha_{zz}^2>_{av}$ is the zz component of the polarizability of the centers, Z^* is the cylinder length and C_j defines the orientation of the cylinder with respect to the incident beam. From the Rayleigh ratio, the dissymmetry of the scattering (R_V at $2\theta = 45°$ to R_V at $2\theta = 135°$) and the depolarization ratios it is possible to determine the size and the orientation of the scattering centers. Unfortunately, only average quantities can be obtained since phase relations are not known. The beautiful recent work of Hobbs[43] with the electron microscope will enable workers to calibrate light scattering and ultramicroscope data and will lead to several advances in this field. Figure 6[42] which is a plot of scattering power versus scattering angle, 2θ, for thermal scattering, benzene, precipitate scattering, 0227, and rod-like scattering, 1203 gives an idea of the magnitude of the scattering. In the latter two cases the full symbols are for data taken with the incident light beam propagating along a <110> crystallographic direction, whereas the open circles are for <100> propagation.

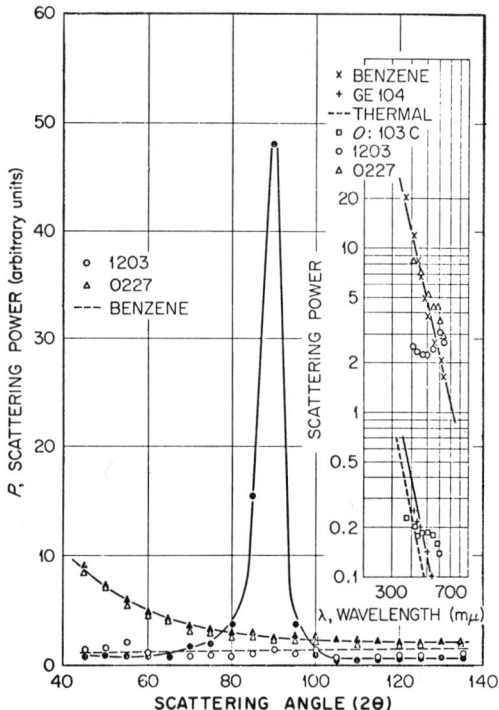

Fig. 6. Light scattering from several types of systems. After Sibley[42]

III. POINT DEFECTS AND SMALL CLUSTERS

Hughes and Von der Osten treat the optical properties of color center type defects in this volume. Therefore, the emphasis of this presentation will be on impurity defects and the effect of radiation induced defects on the optical transitions of impurities. Transition metal ions have been chosen as the means to illustrate the effects to be discussed here, but s^2 ions also have interesting optical properties and these have been reviewed in detail by Fowler.[9]

When Ni^{2+} is substituted for Mg^{2+} in MgO, $KMgF_3$, or MgF_2 it has O_h or for MgF_2, D_{2h} site symmetry which means that the optical transitions are parity forbidden.[24,25] Other transitions are also spin forbidden such as the ones from the $^3A_{2g}$ ground state to the ^1E_g or $^1T_{2g}$ excited states shown in Table I. The energy level

SPECTROSCOPY TO MICROSCOPY—PHOTON AS PROBE

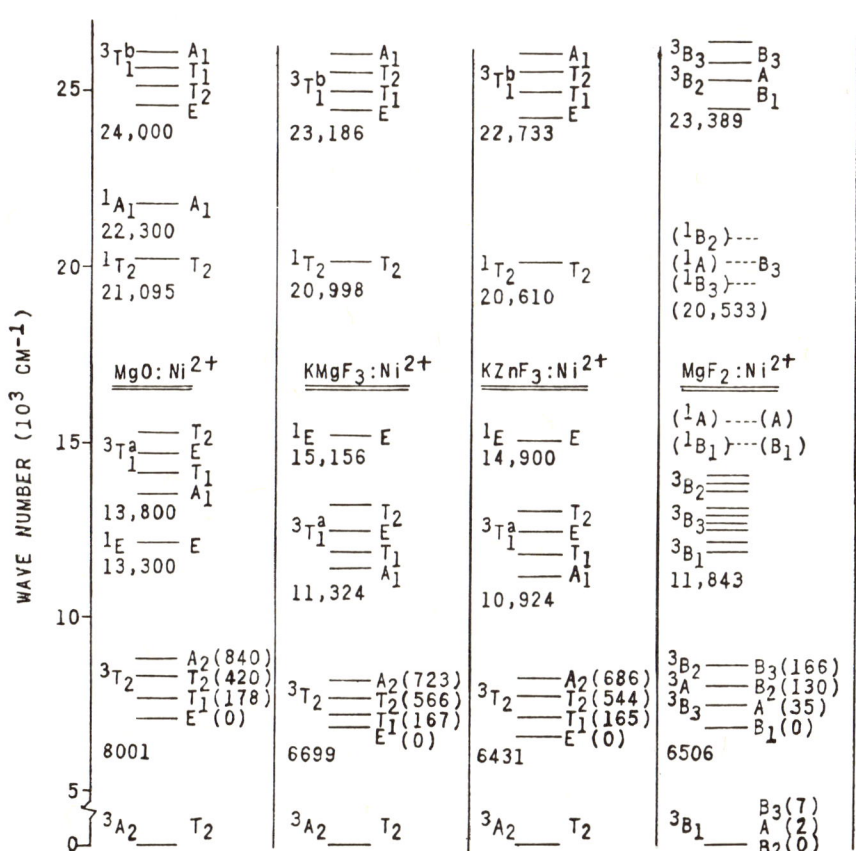

Table I. Energy-level schemes for Ni^{2+} ion in four host crystals. Group representations on the left are those for which spin-orbit coupling has been neglected. Spin-orbit components are shown on the right of each column.

assignments shown in the table have been taken from a number of sources.[25,44-46] The designations on the right side of each level are those for spin-orbit splitting with the splitting energy in wave numbers (8066 cm^{-1} = 1 eV). Studies have been made on the absorption and emission of Ni^{2+} in MgO and with the aid of Magnetic Circular Dichroism (MCD)[46] most of the observed bands have been assigned to the various energy levels. Until recently one assignment that was not clear was that of the 12,500 cm^{-1} (800 nm) emission band in MgO. The emission of this band and one at 20500 cm^{-1} (485 nm) is shown in Figure 7.[44] The peak marked "h" in the upper part of the figure increases with increasing temperature and is taken to be a "hot" band due to an electronic transition assisted by the annihilation of a phonon in the lowest possible mode. Since the magnetic dipole allowed zero phonon line for this transition is too weak to be observed the peak at 20872 cm^{-1} is taken to be a one phonon peak of 187 cm^{-1} energy (compare Fig. 4). Thus, the zero phonon transition should be at point 0 midway between the hot band and the first phonon peak. The figure has been drawn in such a way, by superimposing 0 over the sharp zero phonon line of the bottom emission band, as to illustrate that the phonon modes interacting with the Ni^{2+} are essentially the same for both transitions. The phonon energy is measured to be 187 cm^{-1} which agrees with the data in reference 46. In the lower band several sharp lines are evident (the first one marked with an arrow) which do not appear to be associated with phonons. Moreover, the intensity of the emission is such that it is most likely associated with a magnetic dipole transition. In a uniaxial crystal like MgF_2 the electric and magnetic dipoles can have their axes along either the basal plane directions or the c-axis. An optical spectrum can be determined in three different ways denoted by σ, α and π.[25] The α spectrum consists of unpolarized light propagating down the c-axis of the crystal and the σ and π spectra to plane polarized light propagating normal to the c axis with electric vector directions perpendicular and parallel, respectively to the c axis. If e_a and e_c are the two absorption coefficients or emission intensities for electric dipole transitions perpendicular and parallel to the c axis and the corresponding quantities for magnetic dipole transitions are m_a and m_c then the three possible spectra can be expressed in the following way:[25]

$$\alpha = e_a + m_a \qquad \pi = e_c + m_a \qquad \sigma = e_a + m_c$$

If $\alpha = \pi \neq \sigma$ the transition is purely a magnetic dipole one and if $\alpha = \sigma \neq \pi$ the transition is electric dipole. Figure 8 shows the emission bands for MgF_2. Note that in the lower portion of the figure $\alpha = \pi \neq \sigma$ for the zero phonon lines which indicates that the transition is magnetic dipole in nature. The broader bands are phonon assisted and thus are mixed electric and magnetic dipole in nature. Bird, et al.[46] using MCD have shown that the $^3A_{2g} \rightarrow {}^3T_{2g}$

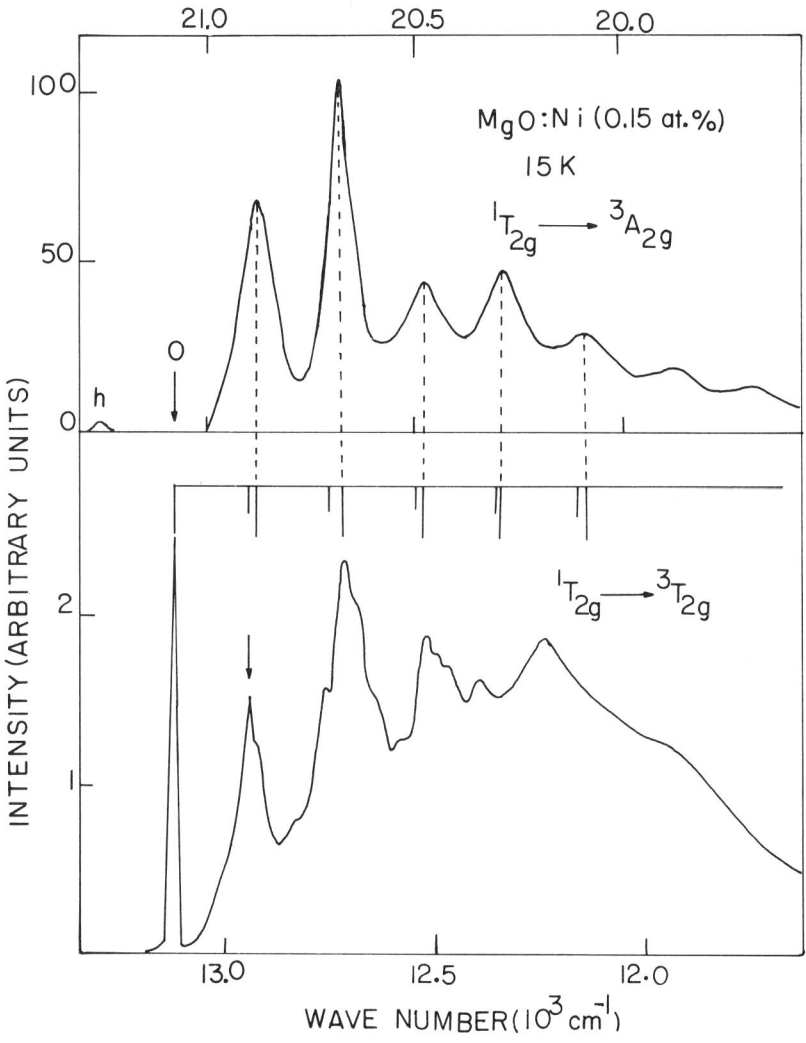

Fig. 7. Ni^{2+} emission from MgO. After Vehse et al.[44]

transition in MgO:Ni is purely magnetic dipole and that the other absorption bands are phonon assisted. The energy difference between the zero phonon peaks labeled m_1 and m_2 in Figure 8 is 36 cm^{-1}. A comparison of this number with the splittings shown in Table I indicates that the 13500 cm^{-1} emission band in MgF_2:Ni is due to a transition from the $^1B_{3g}$ level to the A_{1g} and B_{1g} spin-orbit split levels of $^3B_{3g}$. By analogy the emission in MgO:Ni arises from the $^1T_{2g}$ level and terminates at the spin-orbit split $^3T_{2g}$ level. The extra sharp line shown in Figure 7 is about 178 cm^{-1} from the zero phonon line and that is the spin-orbit splitting of the T_1 and E components of the $^3T_{2g}$ level. Some of the power of using optical tools to study defects becomes apparent from this illustration.

Since the optical transitions of the 3d ions in cubic sites are in many cases both symmetry and spin forbidden it would seem that

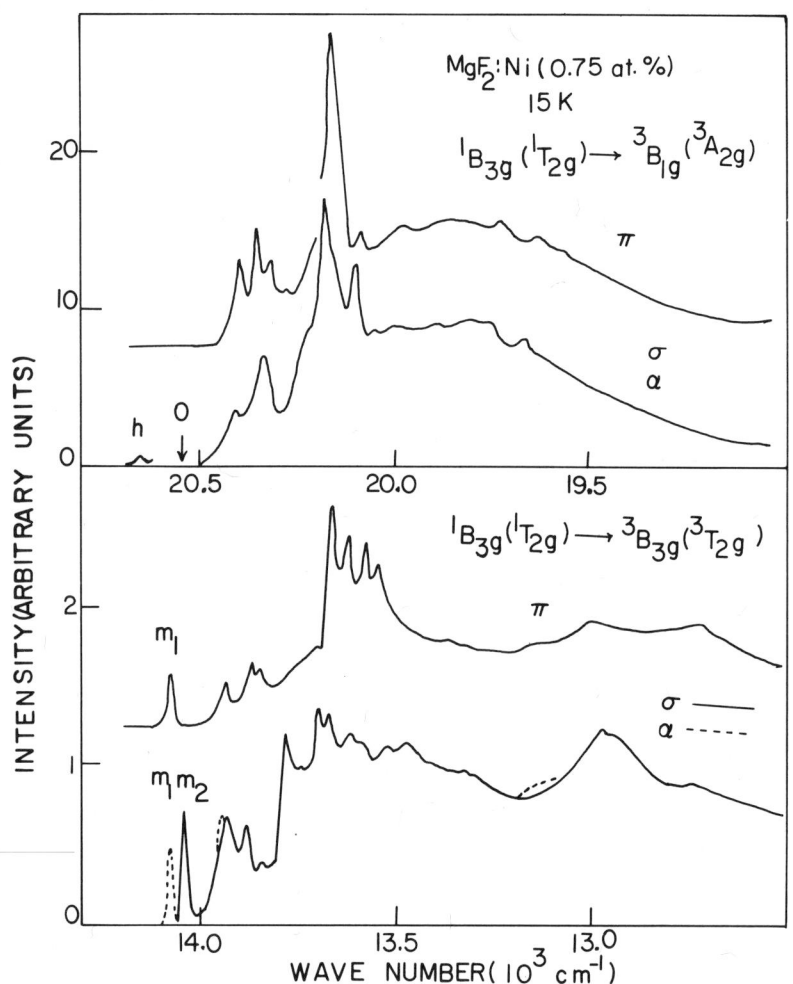

Fig. 8. Ni^{2+} emission from MgF$_2$. After Vehse et al.[44]

placing a defect beside them as a next neighbor might result in an oscillator strength change on two counts: (i) lifting the symmetry forbiddeness and, (ii) the possibility of exchange interactions.[25] Mn^{2+} is an ideal candidate for this type of investigation since there are no spin allowed transitions. When either MgF_2:Mn or $KMgF_3$:Mn crystals are irradiated with electrons two different types of Mn^{2+}-radiation defect centers are formed.[47] One of these centers gives rise to the emission band shown in Figure 9. The excitation spectrum, that is the spectrum taken by detecting the change in the intensity of the emission band as the crystal is excited with light ranging from 200 nm to 700 nm, is illustrated in Figure 10. Notice that although the peak positions are shifted in energy the spectrum has the typical Mn^{2+} absorption character[24,25] and the peaks can be fit to the energy level scheme shown in Figure 2 with $Dq = 955.2$ cm^{-1}, $B = 873.3$ cm^{-1} and $C = 3.45$ B. Prior to electron irradiation no emission from either this defect complex or

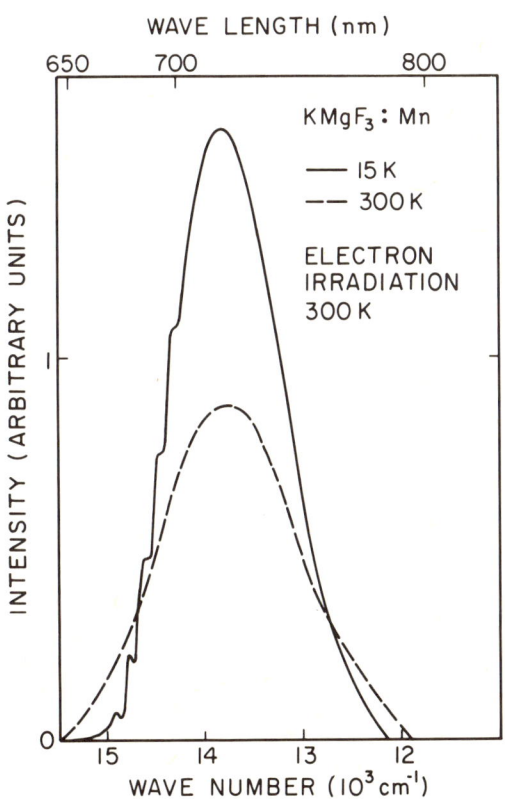

Fig. 9. Emission from irradiated $KMgF_3$:Mn. After Lee and Sibley[47]

the unperturbed Mn^{2+} ions can be seen; thus, the oscillator strength of the Mn^{2+}-defect complex must be at least 10^4 greater than that of the Mn^{2+}. In order to determine if the change in oscillator strength is due to the lifting of the symmetry forbiddenness or to exchange interactions between Mn^{2+} and F center electrons the samples were optically bleached with ultraviolet light at 15K. This removes the electrons from the F centers but leaves the vacancy next to the Mn^{2+} so that the symmetry is unchanged. The intensity is reduced by this treatment suggesting that exchange interactions play the dominant role in enhancing the oscillator strength. Halliburton et al.[48], using epr, have verified the existence of Mn^{2+}-vacancy complexes. From careful measurements of the absorption spectrum it is found that the ^4E_g and ^4A_g transitions which are almost accidently degenerate are separated by 117 cm^{-1} with ^4E_g lowest. Ferguson, et al.[49] using MCD on unirradiated heavily doped crystals find the two levels split by 100 cm^{-1} with ^4E_g lowest.

As mentioned above two types of Mn^{2+} perturbed centers are present after irradiation in MgF_2 and $KMgF_3$. The second type of complex <u>apparently</u> consists of an interstitial-F center-Mn^{2+} system. Other work has been done on Co^{2+} in these host materials and similar effects have been observed.

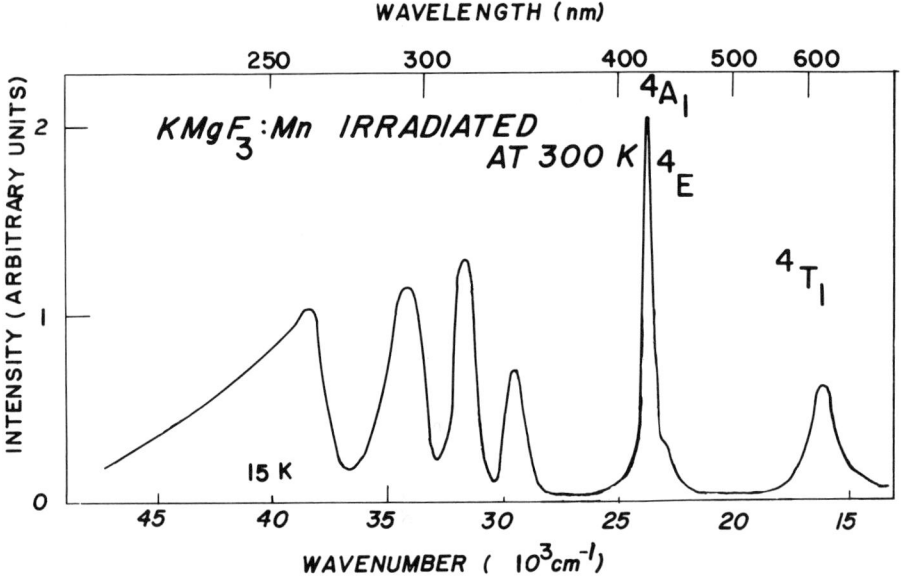

Fig. 10. Excitation spectra of the 720 nm emission band from electron irradiated $KMgF_3$:Mn. After Lee and Sibley[47]

IV. MACROSCOPIC DEFECTS

In this volume Hobbs has covered in some detail the aggregation of radiation defects and colloids in insulating materials from the point of view of electron microscope studies. Some years ago Amelinckx[50] and his coworkers were able to decorate dislocation networks with metal impurities and study the structure of these networks by means of microscopy. Of course, it would be preferable to study precipitates and defects in "as received" material in a nondestructive manner. Light scattering and ultramicroscope investigations give the promise of being able to accomplish this feat.[51,52] There are, however, a number of pitfalls along the way. For example, bulk light scattering is some 10^{-3} times less intense than surface scattering even when the sample is immersed in a liquid of almost the same index of refraction as the sample. Therefore, great care must be taken in obtaining data. Moreover, because of the resolution the ultramicroscope can sometimes "see" different defects than the bulk light scattering. Bansigir and Schneider[53] find no correlation between the number of scattering centers visible by ultramicroscopy and the intensity of scattering measured by a photomultiplier as a function of wavelength, whereas in heat treated NaCl Hauret and Girard-Nottin[54] do find correlations. Figure 11 is a photograph of scattered laser light from defects in a Ruby crystal.[51] From this work grain boundaries were observed to be oriented along $[11\bar{2}0]$ and $[10\bar{1}0]$ crystallographic directions in verneuil grown ruby. The same crystals showed large variations in light scattering intensity from point to point within the crystals. Although the scattering appears to come from several types of imperfections, it was possible to show that the scattering regions in the crystals tend to lie in $[10\bar{1}0]$ planes. Annealing studies showed that the scattering defects tend to line up at higher temperature and the scattering power goes through a maximum at about 440K and 595K. The French have done excellent work using both ultramicroscopic and light scattering techniques in research on precipitate formation and dissolution in NaCl. Taurel and Girard-Nottin[52] have shown, as have Hyder and Bansigir[55], that much of the observed light scattering from NaCl is due to impurity precipitates at dislocation lines while spherical impurity precipitates also contribute. It is possible to detect about 10^6 precipitates per cm^3 by ultramicroscopy and optical absorption.[56] On heat treatment of NaCl, precipitates form at around 275°C and dissolve at 395°C. Thus, kinetic studies of precipitate formation can be done. Recently Baltog[57], et al. have used light scattering in conjunction with ionic conductivity to study precipitates and their interaction with dislocations in NaCl. Light scattering peaks were observed at about 275°C and 480°C. The first peak agrees with other work and the second is attributed to the decoration of charged dislocations. These techniques can be extremely useful in observing the isoelectric temperature at which dislocations should be charged and have no imperfection cloud sur-

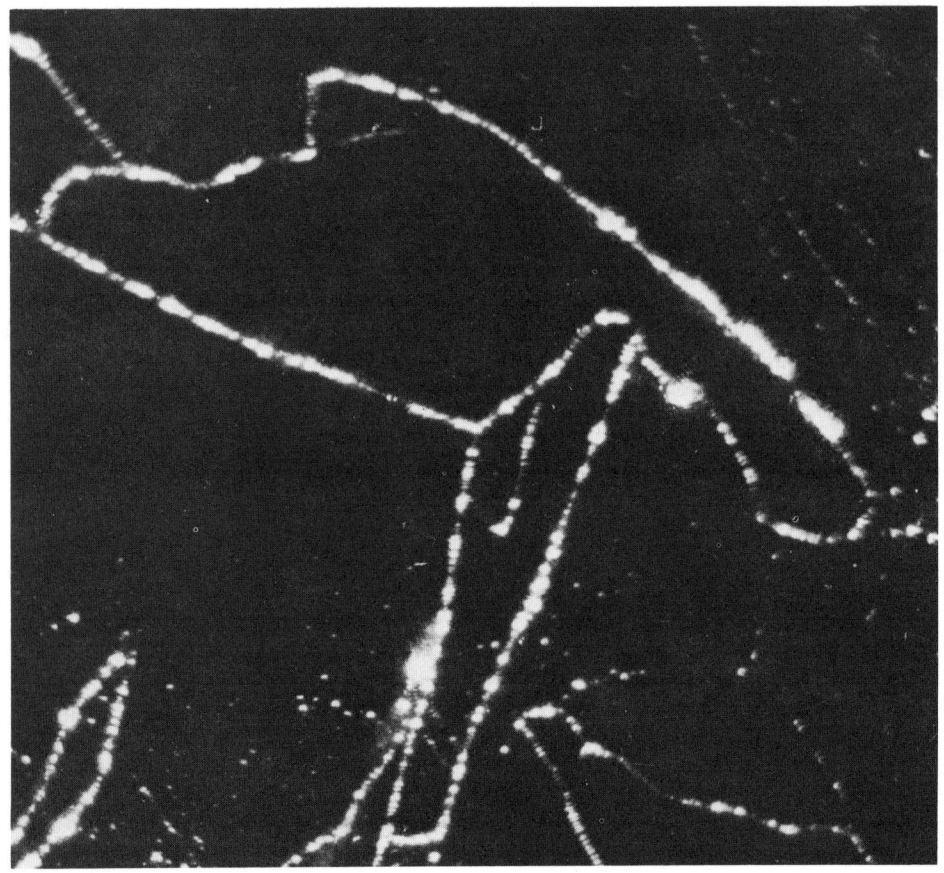

Fig. 11. Light scattering from Ruby. Courtesy of R. C. Powell (see ref. 51)

rounding them. It seems especially important to note that combination measurements of light scattering and ultramicroscopy with other experimental tools such as epr or ionic conductivity can be useful in unraveling the complex role of precipitation and solubility.

V. SUMMARY

In summary, it should be reiterated that the treatment given in this paper covers a wide range of topics and for this reason the reader is referred to the papers referenced for details. The areas emphasized here are those where it appears little work has been done to date and where much progress could be made. Some important problem areas are:

1. A comparison of photoconductivity, electron paramagnetic resonance and optical property work aimed at understanding radiation damage in oxides and fluoride compounds.
2. Theoretical and experimental investigation of exchange interactions between defects such as impurities and radiation defects. Hopefully these will compare the effects on different types of impurities such as s^2 ions, 3d ions, and rare earth ions.
3. Fast measurement techniques in the range 10^{-13} seconds must be developed to study two photon absorption and excited state relaxation processes.
4. More work on reflectivity must be done. This tool will be especially valuable as it becomes more sensitive.
5. A combination of ultramicroscopy, light scattering and optical absorption or electron microscope research on impurity kinetics and radiation defects should be undertaken.
6. Synchroton radiation studies of high energy band to band transitions.

REFERENCES

1. J. G. Mullen, Phys. Rev. 131, 1410 and 1415 (1963).
2. M. V. Klein in "Physics of Color Centers" (ed. W. B. Fowler), Academic Press, p. 429, New York (1968).
3. R. S. Knox and K. J. Teegarden, in "Physics of Color Centers" (ed. W. B. Fowler), p. 1. Academic Press, New York (1968).
4. F. C. Brown, Solid State Physics 29, 1 (1974).
5. W. F. Henson, E. T. Arakawa and M. W. Williams, J. Appl. Phys. 43, 1661 (1972).
6. M. W. Williams, R. A. MacRae and E. T. Arakawa, J. Appl. Phys. 38, 1701 (1967).
7. A. Duncanson and R. W. H. Stevenson, Proc. Phys. Soc. 72, 1001 (1958).
8. A. S. Barker, Phys. Rev., 136, 1290 (1964).
9. W. B. Fowler, in "Physics of Color Centers" (ed. W. B. Fowler) p. 53, Academic Press, New York (1968).
10. W. A. Sibley and D. Pooley, in "Treatise on Material Science and Technology", (ed. by H. Hermann), p. 46, Academic Press, New York (1974).
11. C. S. Kelley, Phys. Rev. B6, 4112 (1972) and B8, 1806 (1973).
12. C. Jouanin and C. Gout, J. Phys. C5, 1945 (1972).
13. M. Garbuny, "Optical Physics", p. 289, Academic Press, New York (1965).
14. R. C. Chaney, C. C. Lin, and E. E. Lafon, Phys. Rev. B3, 459 (1971).
15. A. B. Kunz, Phys. Rev. 162, 789 (1962).

16. D. J. Mickish, A. B. Kunz and T. C. Collins, Phys. Rev. B9, 4461 (1974).
17. R. C. Newman, Adv. in Physics 18, 545 (1969) and "Infrared Studies of Crystal Defects", Taylor and Francis, London (1973).
18. C. Kittel, "Introduction to Solid State Physics", Wiley New York (1971).
19. F. A. Cotton, "Chemical Applications of Group Theory", Wiley, New York (1963).
20. R. S. Katiyar and R. S. Krishnan, Can. J. Phys. 45, 3079 (1967).
21. R. Kahn, J. P. Trotin, D. Gribier and C. Benoit, Colloquium on Inelastic Neutron Scattering, SM-104/33, lAEA, Copenhagen 20-25 May, 1968.
22. B. N. Brockhouse, S. Hauteeler, and H. Stiller in "Interaction of Radiation with Solids" (ed. Struman et al.), North-Holland (1963).
23. G. Shirane, Rev. Mod. Phys. 46, 437 (1974).
24. D. S. McClure, in "Solid State Physics", 9, 399 (ed. F. Seitz and D. Turnbull) Academic Press, New York (1959).
25. J. Ferguson, in "Progress in Inorganic Chemistry", 12, 159 (ed. S. J. Lippard), Interscience, N.Y. (1970).
26. F. Luty, in "Physics of Color Centers" (ed. W. B. Fowler), p. 182, Academic Press, N. Y. (1968).
27. F. Luty, Surface Science 37, 120 (1973).
28. B. diBartolo, "Optical Interactions in Solids", Wiley (1968).
29. D. B. Fitchen, in "Physics of Color Centers, p. 294 (ed. W. B. Fowler), Academic Press, N. Y. (1968).
30. K. Huang and A. Rhys, Proc. Phys. Soc. A204, 406 (1950).
31. J. Rolfe and S. R. Morrison, Bull. Amer. Phys. Soc. 20, 431 (1975).
32. A. Hughes and B. Henderson, in "Defects in Solids" (ed. J. H. Crawford and L. Slifkin), p. 381, Plenum, N. Y. (1972).
33. M. D. Sturge, in Solid State Physics, 20, 91 (1967) (ed. F. Seitz and D. Turnbull) Academic Press, New York.
34. Y. Tanabe and S. Sugano, J. Phys. Soc. Japan, 9, 753 (1954).
35. H. Beleites and F. Frohlich, Kristall and Technik 7, 1329 (1972).
36. F. A. Jenkins and H. E. White, "Fundamentals of Optics", McGraw Hill, N. Y. (1957).
37. F. Zernicke, Physica 9, 686 and 974 (1942).
38. A. H. Bennett, H. Jupnik, H. Osterberg and O. W. Richards, "Phase Microscopy", Wiley, New York (1951).
39. V. Vand, K. Vedam and R. Stein, J. Appl. Phys. 37, 2551 (1966).
40. H. C. Van de Hulst, "Light Scattering by Small Particles", Wiley, New York (1957).
41. H. Z. Cummins, "Light Scattering in Solids", p. 3 (ed. Balkanski), Paris (1971).
42. W. A. Sibley, Phys. Rev. 132, 2065 (1963).
43. L. W. Hobbs, J. de Physique, 34, C9-227 (1973).

44. W. E. Vehse, K. H. Lee, S. I. Yun and W. A. Sibley, Journal of Luminescence 10 (1975).
45. J. E. Ralph and M. G. Townsend, J. Phys. Chem. 3, 8 (1970).
46. B. D. Bird, G. A. Osborne and P. J. Stephens, Phys. Rev. B5, 1800 (1972).
47. K. H. Lee and W. A. Sibley, Phys. Rev. (to be published).
48. L. E. Halliburton, M. A. Young and E. E. Kohnke, Private Communication and Bull. Am. Phys. Soc. 20, 328 (1975).
49. J. Ferguson, H. J. Gugganheim, and E. R. Krausz, Mol. Phys. 27, 577 (1974).
50. S. Amelinckx, Phil. Mag. 1, 269 (1956).
51. R. C. Powell, J. Appl. Physics 39, 3132 (1968).
52. L. Taurel and M. Girard-Nottin, J. de Physique 27, C3-25 (1966).
53. K. G. Bansigin and E. E. Schneider, J. Appl. Physics 533, 383 (1962).
54. G. Hauret and M. Girard-Nottin, Phys. Stat. Sol., 17, 881 (1966).
55. S. B. Hyder and K. G. Bansigir, Indian J. Pure and Appl. Phys., 2, 395 (1964).
56. M. Fayet-Bonnel, Phys. Stat. Sol.(b) 60, 713 (1973).
57. I. Baltog, C. Ghita and M. Giurgea, J. Phys. C7, 1 (1974) and Rev. Roum. Phys. 17, 1121 (1972).

OPTICAL TECHNIQUES AND AN INTRODUCTION TO THE

SYMMETRY PROPERTIES OF POINT DEFECTS

A. E. Hughes

Materials Development Division
Atomic Energy Research Establishment
Harwell, Didcot, Oxford

1. INTRODUCTION

The purpose of this contribution is to cover some of the groundwork in a subject which will be discussed in greater detail by other authors in this volume. In particular, the aim will be to give a broad outline of the use of optical techniques for studying point defects in non-metallic solids, emphasizing those aspects which relate to methods of determining information about defect structure and the properties of the electronic states of point defects. A brief introduction will also be given to some of the theory which is now often used freely by people active in this field, but which may be less familiar to many of those interested in the physical and chemical properties of the defect solid state. This includes the theory of vibronic structure of optical spectra (including the Jahn-Teller effect), the use of group theory in defect studies and an account of methods of extracting structural information from perturbation experiments.

It is first of all perhaps useful to recall why optical studies have played such a large part in defect studies in non-metals, particularly in ionic structures such as the alkali halides. The fundamental reason is, of course, that defects in insulators have electronic quantum states separated by energies of the order of a few electron volts, so that absorption and emission of electromagnetic radiation involves photons of visible or near visible light. The optical properties of defects have thus long been used to detect the presence of the defects and to label the models advanced for their structure. This latter point has resulted in a confusion of point defect nomenclatures which has not yet been totally rationalized (1). Since we shall be dealing with examples drawn from a number of different defect systems, it is as well to be reminded of the inter-relation of the various units used by optical spectroscopists as indicated in Table 1.

Table 1

Units and symbols used in optical spectroscopy

Wavelength λ : units : nm, μm, Å (10^{-8} cm)

Wavenumber $\bar{\nu} = 1/\lambda$: units : cm^{-1} (20,000 cm^{-1} ≡ 500nm)

Energy $E = h\nu = hc\bar{\nu}$: units : eV (1eV = 8066 cm^{-1})

There are some general points which can be made about the use of optical techniques in point defect studies, particularly absorption and luminescence spectroscopy which are of main interest for structural studies.

(i) Materials must be in a form which shows reasonable transparency in the spectral region of interest. This usually means that single crystals are required and this is essential for the perturbation experiments discussed later in section 6.

(ii) For point defects showing allowed electric dipole transitions (eg the F centre) the range of defect concentrations which may be studied in absorption is usually 10^{14}-10^{17} defects cm^{-3}. Larger concentrations of up to about 10^{20} defects cm^{-3} may be studied if thin (<0.1 mm) layers of materials are used. Luminescence is more sensitive and can detect as few as 10^9 defects cm^{-3}. These concentrations are usually comparable with the level of residual impurities in crystals, so one must always be aware of the possible role of impurities in defect structures. This point (and (i) above) has limited detailed defect studies to quite a small range of insulators, eg alkali halides, alkaline earth halides, alkaline earth oxides.

(iii) Optical studies can lead to information about the <u>symmetry</u> of defects and their electronic states. Detailed atomic <u>stucture</u> can only be reached by combination with other techniques, either 'spectroscopic' (eg EPR) or 'chemical' (eg formation kinetics or equilibrium).

2. INSTRUMENTATION

There are many textbooks on optical instrumentation (2) and only a brief summary will be given here of the types of instrument normally used by defect spectroscopists. These can be divided roughly into two classes according to the resolution $\Delta\lambda$ required. The two classes correspond to some extent to whether one is studying broad bands or the details of sharp zero-phonon lines (see section 3).

2.1 Low Resolution $\Delta\lambda > 0.1$ nm

Here there are many commercially available double beam absorption spectrophotometers whose principle is illustrated in figure 1. A continuous spectrum from a source (usually a tungsten filament for $\lambda > 400$ nm or a hydrogen/deuterium discharge for $\lambda < 400$ nm) is passed through a small monochromator (grating or prism) and the emerging monochromatic beam split so that it passes alternately through the sample and through a reference chamber. Both paths are recombined on a detector such as a photomultiplier tube and the signal, which is of the form shown in the insert, fed to integral electronics which allows the display of the required output on a recorder as the monochromator is scanned over the spectral region of interest. The most common outputs are sample transmission

$$T = I_{sample}/I_{ref}$$

or optical density (absorbance)

$$OD = \log_{10}(I_{ref}/I_{sample}).$$

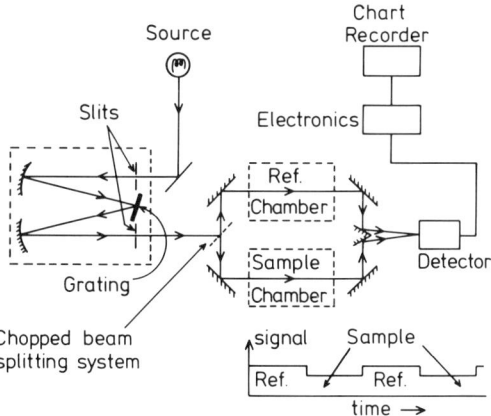

Fig. 1 Double beam absorption spectrophotometer.

The absorption coefficient $\alpha (\text{cm}^{-1})$ of the sample of thickness d may then be obtained from

$$\alpha = \frac{OD}{d} \times (\ln 10)$$

Commercial spectrophotometers are usually equipped with two or more sources and detectors so as to cover the wavelength range 200-2,000 nm, together with the possibility of modification so as to measure the spectral or diffuse reflectivity of samples instead of transmission.

2.2 High resolution $\Delta\lambda < 0.1\text{nm}$

Here one is usually searching for a resolution capabability of about $\Delta\lambda \simeq 0.01$ nm and the use of a fairly high quality monochromator is required. A typical choice might be a 1m focal length grating monochromator with a dispersion in first order of 0.8 nm mm^{-1} across the exit slit. Figure 2 shows a typical arrangement which could be used for single beam absorption measurements or for luminescence studies using a 90° geometry. Many variations are possible; eg use of a second monochromator in the excitation optics and the adaption to double beam operation as in figure 1. Where very low scattered light characteristics are required, as in Raman spectroscopy of defects(3), a double monochromator is necessary. The detection system may use a chopped beam and a phase sensitive detector or, for optimum signal to noise at low light levels, photon counting (4).

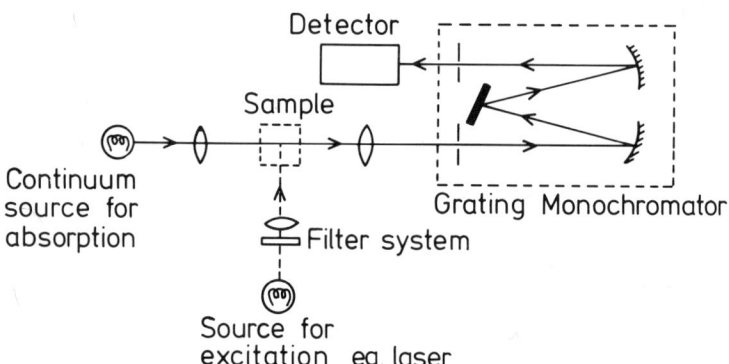

Fig 2 Single beam high resolution spectrometer.

3. THE FORM OF OPTICAL SPECTRA

The electronic states of defects or impurity ions in solids are sensitive to the positions of the nearby atoms or ions, so that the form of the optical absorption or emission spectrum of a defect depends on the equilibrium positions and vibrations of these ions. This is most conveniently discussed at a simple level with the aid of a configuration co-ordinate diagram, figure 3, which allows one to discuss the optical transitions within a system where the electronic states are coupled to vibrations of the surrounding ions. We suppose that the electronic energy of the system depends on only one mode of displacement Q of the surroundings, so that for each electronic state we can draw a curve which shows how the electronic energy depends on Q. In the spirit of the Born - Oppenheimer approximation (5) this then represents the potential energy for the vibration of the ions. In general the equilibrium position of the ions (the minimum in the curve) will not be the same in different electronic states, so that if we draw the curves for the ground and one excited state of a defect as in figure 3, their minima will be separated along the Q axis by some amount Δ.

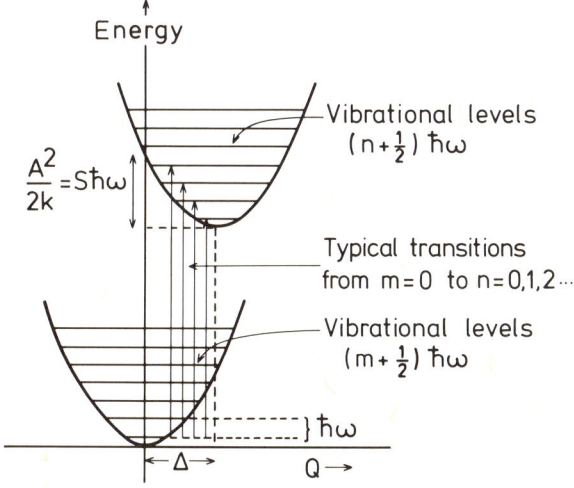

Fig 3 Configuration co-ordinate diagram.

In the harmonic approximation both curves are parabolic, and we usually make a <u>linear coupling</u> approximation so that the two curves are written

$$\text{ground state}: E_g = \tfrac{1}{2} K Q^2 \qquad (1)$$

excited state:
$$E_e = E_o + \tfrac{1}{2} K Q^2 - AQ$$
$$= E_o - \frac{A^2}{2K} + \tfrac{1}{2} K \left(Q - \frac{A}{K}\right)^2 \qquad (2)$$

In this case $\Delta = A/K$ and the curvature of both parobolae is the same; ie the vibration angular frequency ω will be the same in both the ground and excited states.

The full energy of the system will be quantized with the vibrational energy of the system in each electronic level given by $(n + \tfrac{1}{2}) \hbar\omega$ where n is an integer. Thus transitions between ground and excited states can occur as shown in figure 3 such that the photon energy is:

$$h\nu_{nm} = E_o - \frac{A^2}{2K} + n\hbar\omega - m\hbar\omega \qquad (3)$$

where n denotes the vibrational quantum number in the excited electronic state and m in the ground electronic state. The intensity of this transition, P_{nm}, will be determined by the wavefunctions in the two states. In the Born-Oppenheimer approximation these can be written as the product of an electronic and vibrational part:

$$\psi_g^m = \phi_g(\underline{r}, Q) \chi_g^m(Q) \qquad (4)$$

$$\psi_e^n = \phi_e(\underline{r}, Q) \chi_e^n(Q) \qquad (5)$$

Usually the ϕ's are slowly varying functions of Q, so that the transition matrix element for electric dipole transitions between the states can be broken up using the Condon approximation to

$$\langle \psi_e^n | e\underline{r} | \psi_g^m \rangle = \langle \phi_e | e\underline{r} | \phi_g \rangle \langle \chi_e^n(Q) | \chi_g^m(Q) \rangle \qquad (6)$$

The first, purely electronic, maxtrix element then determines the overall intensity (oscillator strength) of the transition, whereas the second vibrational matrix element contains information about the <u>shape of the band</u>, since it determines the intensity distribution over the various possible photon energies $h\nu_{nm}$. Since the wavefunctions $\chi(Q)$ are harmonic oscillator functions, the matrix elements can be readily evaluated (6). The most useful simple illustration of the implications of the theory is to take the case of absolute zero of temperature so that m = 0. We then find that

$$P_{n0} \propto |\langle \chi_e^n(Q) | \chi_g^o(Q) \rangle|^2 = \frac{e^{-S} S^n}{n!} \qquad (7)$$

where S, the <u>Huang-Rhys factor</u> is given by $S\hbar\omega = A^2/2K$.

Some examples of this Poisson distribution or <u>Pekarian</u> are shown in figure 4, where the P_{n0} are represented as delta functions at energies $n\hbar\omega$ from the origin E_o - $S\hbar\omega$. It can be seen that as S increases, the distribution gets broader and greater weight is attached to transitions with finite n i.e. transitions in which the number of vibrational quanta changes. These are known as <u>phonon assisted</u> transitions, and the transition n = 0 is known as the <u>zero-phonon</u> transition. Without going into details, we may highlight the following features of the theory:

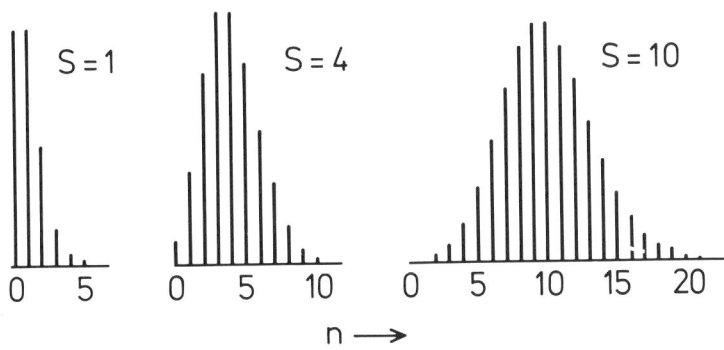

Fig 4 Pekarian bandshapes for various values of the Huang-Rhys factor S.

(i) The centroid of the distribution (which is close to the peak) occurs at n = S. Thus the centroid energy is E_Q i.e. occurs at an energy corresponding to Q = 0 and at a separation $S\hbar\omega$ from the zero-phonon line.

(ii) The second moment of the distribution is $S\hbar^2\omega^2$, so that the width of the distribution is proportional to $\sqrt{S}\,\hbar\omega$

(iii) The intensity of the zero-phonon line relative to the whole band is exp(-S). It thus becomes undetectable for high values of S; usually the limit is about S = 6.

(iv) At finite temperatures T > 0 levels with n > 0 are populated according to Boltzman statistics, $\bar{n}$ = $\{\exp(\hbar\omega/kT)-1\}^{-1}$. One then finds that the zero-phonon line intensity decreases like $\exp[-S(2\bar{n}+1)]$ and the second moment of the distribution increases like $S\hbar^2\omega^2(2\bar{n}+1)$. The centroid remains at E_o. Thus in practice the temperature dependence of the band features can be used to determine S and an effective value of ω.

(v) Finally, and most importantly, if we deal with coupling to more than one mode Q and a spectrum of vibrational frequencies $\hbar\omega$, then the zero-phonon line <u>remains sharp</u>, but all the phonon assisted transitions are <u>broadened</u>. The n-phonon band becomes the n-fold convolution of the spectrum of coupled modes. The sharpness of the zero-phonon line reflects the fact that it occurs between two well defined levels: the zero point vibrational states in each vibrational potential.

Full details of the theory approached in several different ways may be found in papers by Keil (6), Lax (7), Markham (8), Pryce (9), Maradudin (10) and many others e.g. refs (11-13). The crucial point in the theory is that it allows a classification of the different types of bands met in optical spectra of defects according to the <u>vibronic coupling</u> determined by the single parameter S.

It is convenient to define three regions of coupling strengths:

(i) Strong coupling $S > 10$. Here the zero-phonon line and details of the vibrational structure are too weak to be detected and only the broad envelope of the band is seen. Since $\hbar\omega \sim 200$ cm^{-1} the bandwidth is $> \sqrt{10} \times 200$ cm^{-1} i.e. > 0.1 eV. An example is the F band in the alkali halides where S is typically >20 (14, 15) and the halfwidths observed are 0.2 - 0.3 eV.

(ii) Intermediate coupling $1 < S < 6$. Here the zero-phonon line and structure can be resolved, but the multiphonon transitions dominate the integrated intensity. An example is the R_2 band in many alkali halides such as LiF shown in figure 5.

(iii) Weak coupling $S < 1$. In this case the zero-phonon line is the dominant feature and the vibrational sidebands are weak. The most extreme examples are transitions within the $4f^n$ configuration of rare earth ions in solids, where the shielding from the outer 5d orbitals makes the vibronic coupling very weak (16).

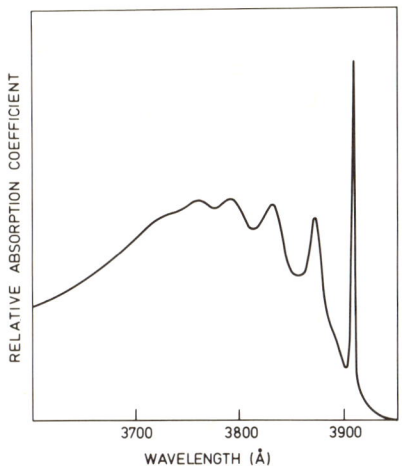

Fig 5 R_2 absorption band in lithium fluoride.

4. THE JAHN-TELLER EFFECT

The discussion in section 3 has assumed implicitly that the electronic states are non-degenerate, since only then can we in general write the wave-functions as Born-Oppenheimer products as in eqns. 4 and 5. Jahn and Teller showed (17, 18) that if a molecule is in a degenerate electronic state as a result of some high symmetry, then the molecule will be unstable with respect to a distortion which lowers the symmetry and removes the electronic degeneracy. The same applies to any localized system of electrons and nuclei such as we encounter at a defect or impurity in a solid (19-22).

A simple example which illustrates this point is shown in figure 6. Consider a square molecule or defect in an electronic state with two-fold degeneracy whose wave-functions are p-like functions as shown. The Jahn-Teller effect will lower the symmetry and remove the <u>electronic</u> degeneracy, as indicated in the second diagram of figure 6. If we denote the displacement in the distortion mode by Q, then the situation may be represented mathematically by the following secular matrix for the p-like states

$$\underset{\sim}{H} = \tfrac{1}{2}KQ^2 \begin{pmatrix} 1 & 0 \\ 0 & 1 \end{pmatrix} + AQ \begin{pmatrix} 1 & 0 \\ 0 & -1 \end{pmatrix} \tag{8}$$

whose eigenvalues are clearly

$$E = \tfrac{1}{2}KQ^2 \pm AQ \tag{9}$$

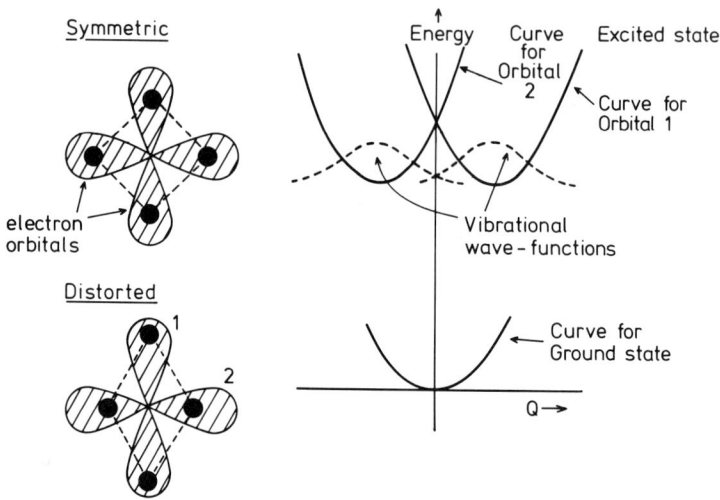

Fig 6 Simple illustration of the Jahn-Teller effect for a square molecule.

On a configuration co-ordinate diagram analogous to figure 3 we therefore have two branches as shown for the upper part of the energy versus Q plot in figure 6. There are two equivalent minima at $Q = \pm A/K$ each depressed by $E_{JT} = A^2/2K$ from the energy in the symmetric (Q =0) configuration. E_{JT} is called the Jahn-Teller energy, and is analogous to the quantity $S\hbar\omega$ discussed previously. It is important to note that although the electronic degeneracy has been removed, the system is still in a two-fold degenerate vibronic state, in the sense that distortions $Q = \pm A/K$ are degenerate in energy. It is a general property of Jahn-Teller systems of this kind that the inclusion of the Jahn-Teller coupling leads to a degenerate vibronic ground state of the same symmetry as the original degenerate electronic state.

Because of this, the presence of a Jahn-Teller effect in a particular system may not be immediately apparent. If the distortion is large enough ($E_{JT} \gg \hbar\omega$) then the physical properties of the system resemble those of a set of statically-distorted molecules or defects. On the other hand, for more modest distortion energies $E_{JT} \sim \hbar\omega$, Ham (23) showed that the main observable effect is the modification of certain measurable properties of the system. For example, consider the orbital angular momentum of the system shown in figure 6. In the absence of a distortion, the component of angular momentum

OPTICAL TECHNIQUES AND DEFECT SYMMETRY

perpendicular to the diagram, l_z, will have off-diagonal matrix elements between the p-like states of the form:

$$L = <\phi_1 | l_z | \phi_2> \tag{10}$$

In the presence of a Jahn-Teller effect, the wave-functions for the degenerate vibronic ground state will be Born-Oppenheimer products of the form:

$$\Psi_1 = \phi_1 \chi_1, \quad \Psi_2 = \phi_2 \chi_2 \tag{11}$$

where χ_1 is a vibrational wavefunction centred at $Q = +A/K$ and χ_2 is centred at $Q = -A/K$ as indicated in figure 6. Now we have

$$L = <\Psi_1 | l_z | \Psi_2> = <\phi_1 | l_z | \phi_2><\chi_1 | \chi_2>$$

$$= <\chi_1 | \chi_2 >L \tag{12}$$

The overlap integral $<\chi_1 | \chi_2>$ has the value $\exp(-2E_{JT}/\hbar\omega)$ so that we have

$$L = L \exp(-2E_{JT}/\hbar\omega) \tag{13}$$

The angular momentum is therefore reduced by the exponential factor, generally called the <u>Ham reduction factor</u>. It follows that the first order effect of, say, spin-orbit coupling will be reduced in the presence of a Jahn-Teller effect. The same applied to any operator off-diagonal in ϕ_1, ϕ_2, which includes the response of the Jahn-Teller system to uniaxial stresses along certain directions(23).

The optical properties of a Jahn-Teller system can depart significantly from those described by the simple configuration co-ordinate model of figure 3. If we imagine the p-like state in figure 6 to be an excited electronic state of the molecule, we may represent a non-degenerate ground electronic state on the same energy versus Q diagram as shown. In considering optical transitions between these two states we now clearly have to take account of both branches of the potential curves for the excited state. In the simple example considered in figure 6 this introduces nothing new since the two branches are mirror images of each other and the vibronic wave-functions for each branch are Born-Oppenheimer products as in eqn (11). However, some more complex situations exist in other Jahn-Teller systems e.g. molecules with threefold symmetry (20-22, 24) and it is then not possible to write the vibronic wave-functions as simple products of an electronic and vibrational part. The theory of section 3 then breaks down completely, and an analytical solution is generally not possible. Figure 7 shows some optical transition probabilities obtained by O'Brien (25) for the case of an s $\rightleftarrows$ p transition at an octahedral site where the p state is Jahn-Teller coupled equally to two different symmetry vibrational modes (E_g and T_{2g}, see section 5) of the octahedron. It can be seen that:

(i) the s ⟶ p transition has two 'humps' and the spacing of the vibronic levels is not equal

(ii) the p ⟶ s transition has a different, single-peaked shape and the spacing of the lines is equal since the vibrational levels in the terminal s state are involved.

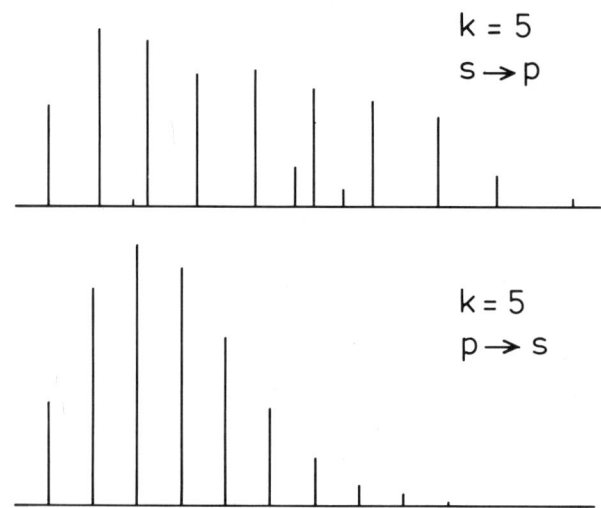

Fig 7 Optical bandshapes calculated for an s ⇄ p transition where the p state has equal coupling to E_g and T_{2g} modes (25). The parameter k is defined by $E_{JT}/\hbar\omega = 2k^2/15$.

An experimental example, the F^+ absorption band in CaO, is shown in figure 8 (26). This centre is an example of a system showing the type of Jahn-Teller effect considered by O'Brien (25, 26-30). The dashed curve shows a theoretical band shape obtained by smoothing the calculated s ⟶ p spectrum in figure 7 to allow for some coupling to symmetric vibrational modes ('breathing' modes) of the centre. It can be seen that the main features of the experimental curve are reproduced, particularly the double-hump shape.

In conclusion then, the main consequences of the Jahn-Teller effect for optical spectra are:

(i) Certain physical properties of the defect (eg spin-orbit splitting and response to external perturbations) may be modified by Ham reduction factors.

(ii) Optical transitions to Jahn-Teller states may give bands with multi-peaked structure.

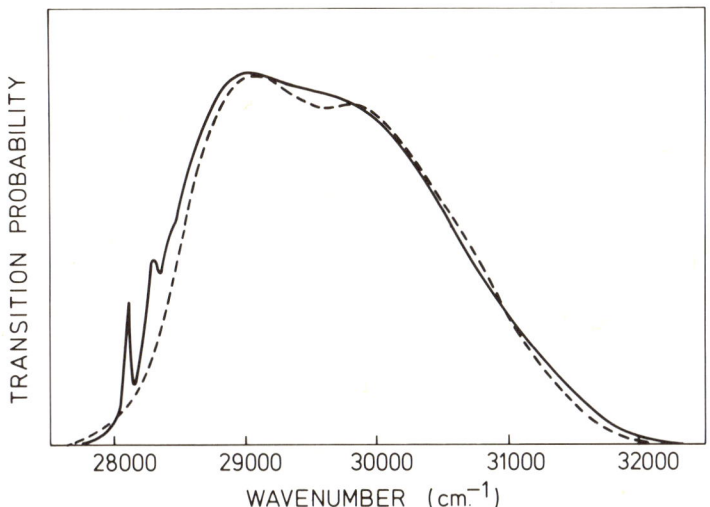

Fig 8 The F^+ absorption band in CaO compared with a smoothed version of the calculation for $E_{JT}/\hbar\omega = 3.3$ in figure 7. The smoothing corresponds to coupling to A_{1g} modes with $S \simeq 2$.

5. BASIC IDEAS OF GROUP THEORY

The purpose of this section is simply to provide some insight into the meaning of terms used in group theory to label electronic and vibrational states of defects (31, 32). The fundamental items of interest in the use of group theory are <u>irreducible representations</u> which are, for our purposes, simply labels which describe the symmetry properties of the wave-functions of an electron trapped at a defect or of vibrational normal modes of the system. Let us start by considering explicitly the labelling of wave-functions.

The presence of <u>symmetry</u> in a system implies that some electronic states will be <u>degenerate</u>. For example, it is obvious that the three-fold degeneracy of an atomic p-state is retained in cubic symmetry, since the three p orbitals have lobes pointing along the equivalent x, y and z axis of the cube. In fact, these p functions transform under symmetry operations of the cube in exactly the same way as the three simple functions x, y and z. Functions which have this transformation scheme belong to an irreducible representation of the cubic group O which is labelled T_1. There are, in fact, five irreducible representations of the group O which have the properties shown in Table 2.

Table 2

Irreducible representations of the cubic group O

Representation	A_1	A_2	E	T_1	T_2
Simple functions with same transformation rules	$(x^2 + y^2 + z^2)$	xyz	(x^2-y^2) $(3z^2-r^2)$	x y z	yz zx xy
'Degeneracy'	1	1	2	3	3

All electronic states of a defect of cubic symmetry will therefore be labelled according to one or other of these five irreducible representations. The advantage of knowing the symmetry of the wave-functions is that it allows one to use group theoretical rules to simplify the calculation of matrix elements between the wave-functions, as required, for example, in establishing optical selection rules. Note from Table 2 that there is no degeneracy higher than three. This means that an atom or ion in a state described by its angular momentum J, and therefore having degeneracy $2J + 1$, will have some of this degenracy removed if it is placed in a cubic environment. The way this occurs for some of the integral J states is shown in Table 3.

Table 3

Decomposition of free atom J states into irreducible representations of the group O

Free atom angular momentum J	Cubic group
0 (s)	A_1
1 (p)	T_1
2 (d)	$E + T_2$
3 (f)	$A_2 + T_1 + T_2$
4 (g)	$A_1 + E + T_1 + T_2$

The treatment of electron spin and half-integral values of J requires special methods which we cannot discuss properly here. Suffice it to say that three extra irreducible representations of the cubic group

O have to be introduced, which are labelled Γ_6, Γ_7 and Γ_8 (31). Γ_6 and Γ_7 are two-fold degenerate and Γ_8 four-fold. For many purposes it is convenient to treat electron spin in the spirit of the Russell-Saunders (L, S) coupling scheme in atoms, ie we label a state by its orbital symmetry with a superscript denoting the spin degeneracy 2S + 1. Thus a state with three-fold orbital degeneracy belonging to T_1 and a spin S = 1 would be labelled $^3T_1$.

Symmetry groups lower than the cube have irreducible representations of degeneracy less than three as shown in Table 4.

Table 4

Irreducible representations of low symmetry groups

Tetragonal (4-fold axis) and Trigonal (3-fold axis) groups:

    One dimensional representations A_1, A_2

    Two dimensional representations   E

Rhombic (2-fold axis) and lower symmetry groups

    One dimensional representations A_1, B_1, B_2, B_3.

As an example of the labelling of states of defects, figure 9 shows schematic energy level diagrams for the F centre and the F_2(M) centre in the alkali halides. The main optical transitions are also shown.

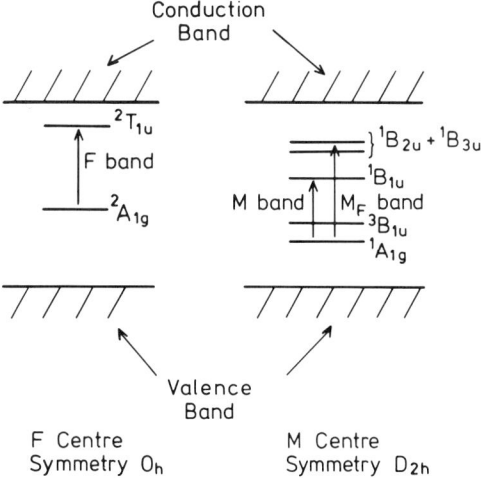

Fig 9   Schematic energy level diagrams for the F centre and the F_2(M) centre.

The vibrational normal modes of molecules and defects in solids may also be labelled by irreducible representations of the symmetry group of the system. Examples are given in refs 31 and 32.

6. METHODS OF ARRIVING AT STRUCTURAL INFORMATION

If the optical spectrum of a defect corresponds to relatively weak vibronic coupling (S < 6) then a sharp zero-phonon line will be resolved and the defect symmetry may be established by studying the splitting of the line under applied perturbations such as uniaxial stress, magnetic and electric fields. If the vibronic coupling is strong then only a broad band results, and it is generally not possible to deduce defect symmetry from perturbation experiments, although by sensitive techniques very useful information may be obtained about the properties of centres of known structure (33, 34). To establish defect symmetry from a broad band it is necessary to use the techniques of polarized bleaching or polarized luminescence (35-37). We shall discuss the latter briefly before giving a short description of the use of perturbation experiments on zero-phonon lines.

6.1 Polarized luminescence

Suppose we have a defect with a unique symmetry axis $\underline{p}$ in a cubic crystal. The defect may be of tetragonal or lower symmetry, and in general the symmetry axis $\underline{p}$ will be along one of the low index axes of the cube eg <100> for a tetragonal symmetry defect. The optical properties of such a centre will be anistropic: some optical transitions will be excited by $\underline{E}//\underline{p}$ and others by $\underline{E} \perp \underline{p}$. The probability of exciting the centre will be

$$I_{exc} \propto (\underline{E} \cdot \underline{p})^2 \propto \cos^2(\underline{E},\underline{p}) \tag{14}$$

in the former case and $I_{exc} \propto \sin^2(\underline{E},\underline{p})$ in the latter. If the centre, once excited, emits luminescence which is observed in a polarization $\underline{E}'$, then the observed intensity will be (for a transition allowed for $\underline{E}' // \underline{p}$).

$$I \propto I_{exc} \cdot (\underline{E}' \cdot \underline{p})^2 \propto \cos^2(\underline{E},\underline{p}) \cos^2(\underline{E}',\underline{p}) \tag{15}$$

A similar expression involving $\sin^2(\underline{E}',\underline{p})$ can be worked out for the case of transitions allowed for $\underline{E}' \perp \underline{p}$.

In a cubic crystal there will be several different equivalent orientations of the anisotropic defects so that the total luminescence intensity is found by summing expressions such as in eqns (14) and (15) over all equivalent orientations of $\underline{p}$ for the given polarization conditions of exciting and observing the luminescence. The result is that, for anisotropic centres, the intenstiy varies as the directions of polarization are varied. Full details of the

OPTICAL TECHNIQUES AND DEFECT SYMMETRY

calculations are given in refs 35-37, where it is shown how the polarization patterns may be used to deduce the symmetry axis of the defect, provided it is along a fairly simple direction like <100>, <110> or <111>. A typical geometry for excitation and emission polarization directions is shown in figure 10(a) and an experimental result for a defect in magnesium-doped CaO (38) is shown in figure 10(b). The highly polarized nature of the luminescence in figure 10(b) shows that the centre must have $\underline{p}$ //<100>, and it has been attributed to an F^+ centre perturbed by neighbouring magnesium impurities (38).

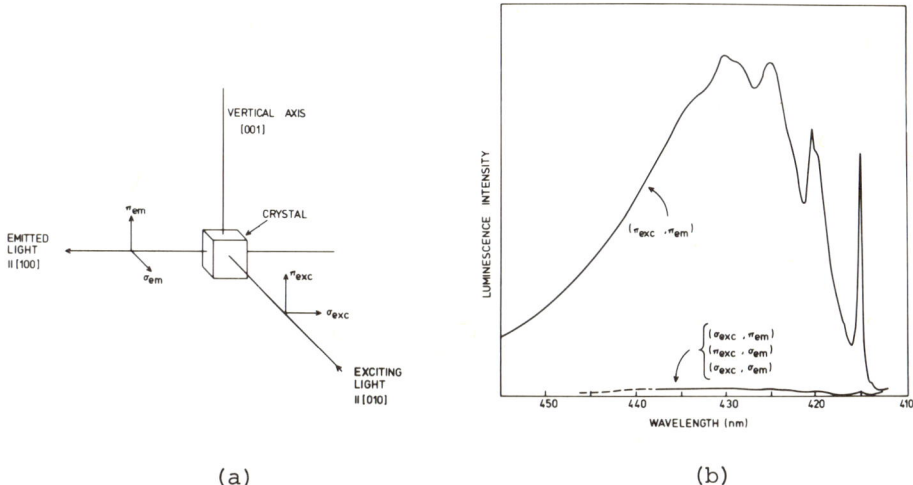

(a) (b)

Fig 10 (a) Right-angle geometry for polarized luminescence experiments.

(b) Experimental results on the 430 nm emission band in CaO:Mg taken with the geometry in (a).

6.2 Uniaxial stress

If a zero-phonon line can be detected, then its splittings under the application of uniaxial stresses directed along the <100>, <110> and <111> directions of a cubic crystal are probably the most powerful way of establishing to which symmetry class the defect belongs. The basic theory of uniaxial stress effects has been considered by Kaplyanskii (39, 40), Runciman (41), Hughes and Runciman (42) and reviewed by Fitchen (13). The defect symmetry can be deduced from the number of components in the splitting patterns and their polarization relative to the stress direction. Splittings are of the order of 10^4 cm^{-1} per unit strain, so that stresses high enough to produce strains of $>10^{-3}$ must be used to giving splittings of ~ 10 cm^{-1}. Since the width of a colour centre zero-phonon line is usually a few cm^{-1} (due to random internal strains of order 10^{-4} (43,44)), such a splitting can generally be resolved. Fitchen (13) and Von der Osten (45) give a number of examples.

The basic idea involved in a uniaxial stress experiment is shown in figures 11(a) and (b) (46). In figure 11(a), a uniaxial stress along [001] is applied to a centre of cubic symmetry having a state of T_1 symmetry (ie threefold degenerate). It is clear that the stress removes the degeneracy and will split the T_1 state into a singlet and doublet as shown. In figure 11(b) the situation is shown for the case of defects of tetragonal symmetry, where there is three-fold orientational degeneracy since the fourfold axis of the defect can be along [100], [010] or [001]. Again it is clear that the [001] stress will remove the degeneracy and result in a splitting of optical transitions within the centre. One important general feature of uniaxial stress experiments is that the perturbing Hamiltonian can be written to lowest order as a linear function of the stress or strain tensor components:

$$H = \Sigma_{ij} A_{ij} \sigma_{ij} = \Sigma_{kl} B_{kl} e_{kl} \qquad (16)$$

where the coefficients A_{ij} and B_{kl} depend on the properties of the defect. Since σ_{ij} and e_{kl} have even parity (σ_{ij} transforms like $x_i x_j$), a uniaxial stress experiment does not indicate whether the defect has a centre of symmetry or not.

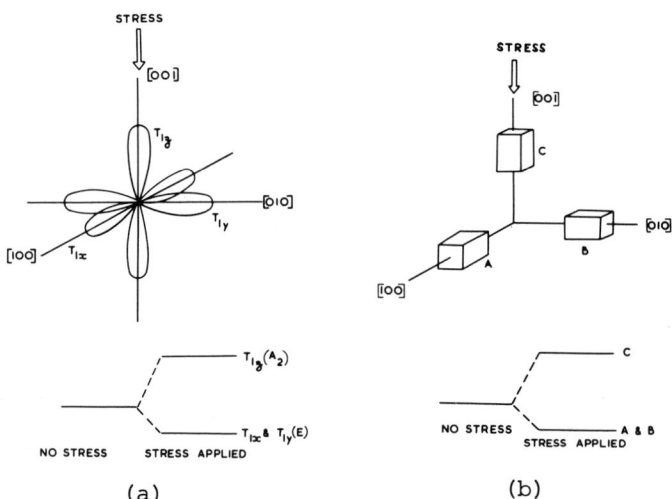

Fig 11 Illustration of uniaxial stress splitting: (a) removal of electronic degeneracy in the T_1 state of a cubic centre; (b) removal of orientational degeneracy of tetragonal centres.

6.3 Stark effect

The application of an electric field to the crystal can also, in principle, cause a splitting of a spectral line and determine the symmetry class to which a defect belongs. The theory is discussed by Kaplyanskii and Medvedev (47) and examples for colour centres and further theory may be found in refs 48-51. In this case the perturbing Hamiltonian is

$$H = \Sigma_i e \; \underline{r}_i \cdot \underline{E} \qquad (17)$$

Where $\underline{r}_i$ is the position of the ith electron in the defect. Since first order splittings (ie $\Delta\overline{\nu} \propto |\underline{E}|$) require matrix elements of the form $< \emptyset | H | \emptyset >$ to be non-zero, we note that such a splitting will only result if the defect has no centre of inversion symmetry. This is because $\underline{r}_i$ has odd parity. The presence of a linear Stark effect can therefore tell us this information, which, as explained above, cannot be obtained from a stress experiment. However, Stark splittings are often small ($e\underline{E} \times 0.1$ nm = 8cm^{-1} for $\underline{E} = 10^5$V cm^{-1}: close to breakdown) and the experimental application has been limited. A good example is the detection of a linear Stark effect for the F_3^+ (R^+) zero-phonon line (49, 51) which proves that this centre has no inversion symmetry. The quadratic Stark effect results from the coupling of different electronic states by the Hamiltonian eqn. (17), and can provide useful information about the level separations (34).

6.4 Zeeman effect

The application of a magnetic field $\underline{B}$ to a crystal may be described by a perturbing Hamiltonian

$$H = \beta \underline{B} \cdot (\underline{L} + 2\underline{S}) \qquad (18)$$

where β is the Bohr magneton. For a first-order splitting of a zero-phonon line we must have non-zero matrix elements of $\underline{L}$ and/or $\underline{S}$ in the electronic states of interest. For defect centres $\underline{L}$ is usually small and optical selection rules $\Delta M_S = 0$ can restrict the chance of seeing splittings due to $\underline{S}$ (52), so that direct Zeeman splittings are seldom observed. This is, however, not the case for impurity centres and the general theory is discussed in refs 53-56. Since $2\beta B \approx 10$ cm^{-1} for $B = 100$ kG (10T), very large magnetic fields would be required to give splittings much larger than the line-width. It is therefore usually much more useful for defect centres to detect any magnetic effects by differential techniques like magnetic circular dichroism (34) than to try to resolve direct Zeeman splittings.

REFERENCES

1. E. Sonder and W.A. Sibley in *Point Defects in Solids* (Vol. 1) eds. J.H. Crawford and L.M. Slifkin (Plenum Press, New York 1972) Chapter 4.

2. G.F. Lothian, *Absorption Spectrophotometry* (3rd edition, Adam Hilger Ltd., London, 1969).

3. *Light Scattering in Solids*, ed. G.B. Wright (Springer-Verlag, New York, 1969).

4. G.A. Morton, Applied Optics $\underline{7}$, 1 (1968).

5. L. Pauling and E.B. Wilson, *Introduction to Quantum Mechanics* (McGraw-Hill, New York, 1935).

6. T.H. Keil, Phys. Rev. $\underline{140}$, A601 (1965).

7. M. Lax, J. Chem. Phys. $\underline{20}$, 1752 (1952).

8. J.J. Markham, Rev. Mod. Phys., $\underline{31}$, 956 (1959).

9. M.H.L. Pryce in *Phonons* ed. R.W.H. Stevenson (Oliver and Boyd, Edinburgh, 1966) p. 403.

10. A.A. Maradudin in *Solid State Physics* ed. F. Seitz and D. Turnbull (Academic Press, New York, 1966) Vol. 18 p. 274.

11. K. Huang and A. Rhys, Proc. Roy. Soc. $\underline{A204}$, 406 (1950).

12. K.K. Rebane, *Impurity Spectra of Solids* (Translated by J.S. Shier, Plenum Press, New York, 1970).

13. D.B. Fitchen in *Physics of Color Centers* ed. W. Beall Fowler (Academic Press, New York 1968) Chapter 5.

14. W. Gebhardt and H. Kühnert, Phys. Letters $\underline{11}$, 15 (1964).

15. R. K. Dawson and D. Pooley, Phys. Stat. Sol. $\underline{35}$, 95 (1969).

16. G.H. Dieke, *Spectra and Energy Levels of Rare-Earth Ions in Crystals*, (Interscience, New York, 1968).

17. H. A. Jahn and E. Teller, Proc. Roy. Soc. $\underline{A161}$, 220 (1937).

18. H.A. Jahn, Proc. Roy. Soc. $\underline{A164}$, 117 (1938).

19. U. Öpik and M.H.L. Pryce, Proc. Roy. Soc. $\underline{A238}$, 425 (1957).

20. M.D. Sturge in *Solid State Physics*, ed. F. Seitz, D. Turnbull and H. Ehrenreich (Academic Press, New York, 1967). Vol. 20 p. 92.

21. F.S. Ham in *Electron Paramagnetic Resonance* ed. S. Geschwind (Plenum Press, New York, 1972), Chapter 1.

22. R. Englman, *The Jahn-Teller effect in Molecules and Crystals* (John Wiley, New York 1972).

23. F.S. Ham, Phys. Rev. 138, A1727 (1965).
24. H.C. Longuet-Higgins, U. Öpik, M.H.L. Pryce and R.A. Sack, Proc. Roy. Soc. A244, 1 (1958).
25. M.C.M. O'Brien, J. Phys. C. (Solid State Physics) 4, 2524 (1971).
26. C. Escribe and A.E. Hughes, J. Phys. C. (Solid State Physics) 4, 2537 (1971).
27. A.E. Hughes, J. Phys. C. (Solid State Physics) 3, 627 (1970).
28. A.E. Hughes, G.P. Pells, and E. Sonder, J. Phys. C. (Solid State Physics) 5, 709 (1972).
29. Y. Merle d'Aubigné and A. Roussel, Phys. Rev. B3, 1421 (1971).
30. J. Duran, Y. Merle d'Aubigné and R. Romestain, J. Phys. C (Solid State Physics) 5, 2225 (1972).
31. M. Tinkham, *Group Theory and Quantum Mechanics*, (McGraw-Hill, New York, 1964).
32. F.A. Cotton, *Chemical Applications of Group Theory*, (John Wiley, New York, 1971).
33. C.H. Henry, S.E. Schnatterly and C.P. Slichter, Phys. Rev. 137, A583 (1965).
34. C.H. Henry and C.P. Slichter in *Physics of Color Centers* ed. W. Beall Fowler (Academic Press, New York, 1968), Chapter 6.
35. P.P. Feofilov, *The Physical Basis of Polarized Emission*, (English translation: Consultants Bureau, New York, 1961).
36. C.Z. Van Doorn, Philips Research Reports Suppl. No. 4 (1962).
37. W.D. Compton and H. Rabin in *Solid State Physics* ed. F. Seitz and D. Turnbull (Academic Press, New York, 1964). Vol. 16 p. 121.
38. G.P. Pells and A.E. Hughes, J. Phys. C. (Solid State Physics) 8, 3703 (1975).
39. A.A. Kaplyanskii, Optics and Spectroscopy 16, 329 (1964).
40. A.A. Kaplyanskii, Optics and Spectroscopy 16, 557 (1964).
41. W.A. Runciman, Proc. Phys. Soc. 86, 629 (1965).
42. A.E. Hughes and W.A. Runciman, Proc. Phys. Soc. 90, 827 (1967).
43. A.M. Stoneham, Rev. Mod. Phys. 41, 82 (1969).
44. A.E. Hughes, J. Phys. Chem. Solids. 29, 1461 (1968).
45. W. Von der Osten, Z. Angew. Physik 24, 365 (1968).
46. A.E. Hughes, Proc. Brit. Ceram. Soc. No. 9 p. 51, (1967).

47. A.A. Kaplyanskii and V.N. Medvedev, Optics and Spectroscopy 23, 404 (1967).

48. A.A. Kaplyanskii, V.I. Kolyshkin, V.N. Medvedev and A.P. Skvortsov. Sov. Phys. Solid State 12, 2867 (1971).

49. A.A. Kaplyanskii, J. de Physique 28 (Suppl.) C4-39, (1967).

50. G. Johannson, F. Lanzl, H. Mödl, W. Von der Osten and W. Waidelich, Z. Physik 210, 1 (1968).

51. G. Johannson, W. Von der Osten, R. Piehl and W. Waidelich, Phys. Stat. Sol. 34, 699 (1969).

52. A.E. Hughes, D. Phil. Thesis, University of Oxford (1966).

53. G.F. Koster and H. Statz, Phys. Rev. 113, 445 (1959).

54. H. Statz and G.F. Koster, Phys. Rev. 115, 1568, (1959).

55. B.P. Zakharchenya and I.B. Rusanov, Optics and Spectroscopy 19, 207 (1965).

56. W.A. Runciman in Physics of Solids in Intense Magnetic Fields ed. E.D. Haidemenakis (Plenum Press, New York, 1969), p. 344.

MAGNETIC RESONANCE STUDIES OF VACANCY CENTERS IN IONIC CRYSTALS

J.M.SPAETH

Gesamthochschule Paderborn
Fachbereich 6 - Experimentalphysik
479 Paderborn
Pohlweg 55,W.Germany

1. INTRODUCTION

The application of magnetic resonance techniques to the study of paramagnetic centres has greatly improved our understanding of point defects. In fact ESR (Electron Spin Resonance) and especially ENDOR (Electron Nuclear Double Resonance) come close to what one may call an atomic scale microscopy of paramagnetic centres. In favourable cases one obtains a very detailed picture of the atomic structure of a defect (lattice site, neighbours, symmetry) including information about lattice distortions around the vacancy or the paramagnetic impurity, as well as detailed information about the electronic structure of the centre. This very precise picture of the paramagnetic defect is mainly the result of a careful determination of the strong hyperfine interaction (hf) between the unpaired centre electron and the centre nuclei as well as the weaker hf interaction with the nuclei of the surrounding lattice ions, (sometimes called "superhyperfine interaction", shf). In order to determine the latter, ENDOR measurements usually have to be performed, since they allow a much higher resolution of hf interactions compared to ESR measurements. Vacancy centres mostly contain no centre nuclei so that the shf interactions must be determined.

As early as 1949 ESR (1) and in 1961 ENDOR (2,3) measurements made decisive contributions to the determination of the structure of the F-centre in alkali halides, which consists of a single anion with one trapped electron, and finally confirmed the model originally proposed by De Boer (4). In the other ionic crystals like the alkaline earth oxides the identification of F type centres occurred first through ESR and ENDOR measurements, whereas their optical transitions were identified later.

It will be the purpose of the two lectures to discuss some results of ESR and ENDOR studies of a few vacancy centres with trapped electrons (sometimes also called electron excess centres or F-type centres) in some ionic crystals. It will not be possible within this limited scope to go into great detail and depth, nor will it be possible to present examples in all the ionic crystals where F-type centres have been studied. Somewhat arbitrarily most of the examples chosen are vacancy centres in the alkali halides. The discussion will be mainly concerned with the determination of the atomic structure of the vacancy centres. The question of the electronic structure can only be briefly touched. Relaxation phenomena and excited states will not be discussed.

For further information on the subject the reader is referred to review articles dealing with magnetic resonance of vacancy centres in the alkali halides (5), alkaline earth oxides (6,7) and alkaline earth fluorides (8).

2. ANALYSIS OF ESR AND ENDOR SPECTRA OF F-CENTRES

2.1. Spin Hamiltonian

F-like centres containing a simple electron in a vacancy have an orbitally nondegenerate ground state of mainly s-character with spin quantum number $s = \frac{1}{2}$. The electronic wave function is moderately delocalised and overlaps appreciably onto neighbouring ions. Since the electronic wave function is strongly exposed to the crystal field and no excited states are very close to the ground state, spin orbit effects are usually very small. Only through admixtures of neighbour ion orbitals because of the overlap of the F centre wave functions, does spin orbit coupling onto the neighbours show up in a small shift of the g-value relative to the free electron (5). This allows the use of a much simplified general spin Hamiltonian for the interpretation of the ESR and ENDOR spectra:

$$H = g\beta \vec{B}_o \vec{S} + \sum_1 (a_1 \vec{S} \vec{I}_1 + \vec{S} \tilde{B}_1 \vec{I}_1 - g_{I,1} \beta_n \vec{B}_o \vec{I}_1 + \vec{I}_1 \tilde{Q}_1 \vec{I}_1) \quad (1)$$

$$\underbrace{\phantom{g\beta \vec{B}_o \vec{S}}}_{H_{EZ}} \quad \underbrace{\phantom{a_1 \vec{S} \vec{I}_1 + \vec{S} \tilde{B}_1 \vec{I}_1}}_{H_{SHF}} \quad \underbrace{\phantom{g_{I,1} \beta_n \vec{B}_o \vec{I}_1}}_{H_{NZ}} \quad \underbrace{\phantom{\vec{I}_1 \tilde{Q}_1 \vec{I}_1}}_{H_Q}$$

H_{EZ} describes the electron-Zeeman-interaction (β is the Bohr magneton, $\vec{B}_o$ the applied magnetic field). H_{SHF} is the superhyperfine (shf) interaction between the F electron and the neighbouring lattice nuclei, H_{NZ} is the nuclear Zeeman interaction (β_n is the nuclear magneton, g_I the nuclear g-factor), and H_Q is the quadrupole interaction. The sum runs over all nuclei 1 of the lattice surroundings experiencing an interaction with the F-electron (in the alkali halides up to the 9th shell (9)) $\vec{S}$ and $\vec{I}$ represent the electron and nuclear spin operators. The hyperfine "contact" term a, or "isotropic shf constant" is proportional to the unpaired spin density at the nucleus of concern (index 1 omitted) (10):

$$a = \frac{2\mu_o}{3} g\beta g_I \beta_n |\psi(o)|^2 \qquad (2)$$

$\psi(r)$ is the wave function of the F electron, $\psi(o)$ its value at a particular nucleus of the lattice surrounding. The anisotropic shf interactions can be visualised as the classical dipole-dipole interactions of electron and nuclear moments averaged over the density distribution $|\psi|^2$ of the unpaired electron. $\tilde{B}_1$ is a traceless tensor with the elements

$$B_{ik} = \frac{\mu_o}{4\pi} g\beta g_I \beta_n \int_V (\frac{3}{r^5} x_i x_k - \frac{1}{r^3} \delta_{ik}) |\psi(\vec{r})|^2 dV \qquad (3)$$

$\vec{r}$ means the radius vector from the nuclear site of concern (origin) where the origin is spared in the integral Equ.3.
$\tilde{Q}$ is the traceless quadrupole interaction tensor with the elements

$$Q_{ik} = \frac{eQ}{2I(2I-1)} \frac{\partial^2 V}{\partial x_i \partial x_k} \bigg|_{r=o} \qquad (4)$$

$\tilde{Q}$ contains the electrical field gradient at the nuclear site due to the total charge distribution in its neighbourhood.

For the analysis of spectra it is important to note that each neighbour nucleus has its own principal axes for the tensors of the shf (x', y', z') and quadrupole interaction (x", y", z"). The point symmetry of the nucleus may determine some or all of the principle axes (2). The nearest neighbours of the F-centre in alkali halides, for example, have their connection line to the

F centre as one principle axis, whereas the other two are equivalent (axial symmetry) and can be chosen in any direction perpendicular to it. Instead of the principle values of the tensors B and Q, the following quantities are often used:

$$b = \frac{1}{2} B_{z'z'} \qquad b' = \frac{1}{2}(B_{x'x'} - B_{y'y'}) \qquad (5)$$

$$q = \frac{1}{2} Q_{z''z''} \qquad q' = \frac{1}{2}(Q_{x''x''} - Q_{y''y''}) \qquad (6)$$

b and q respresent the axially symmetric part of the shf and quadrupole-tensors respectively, z' and z'' are taken as the principle axes with the largest interaction. b' and q' describe the deviations from axial symmetry.

2.2. ESR Transitions

Using first order perturbation theory with the assumptions of high field ($g \beta B_0 \gg a$) and small anisotropic interaction ($B_{ik} \ll a \pm g_I \beta_n B_0$) the eigenvalues of Equ. (1) are given by

$$E = g\beta B_o m_s - g_I \beta_n B_o m_I + m_s m_I (a + b(3\cos^2\gamma - 1)) + \frac{1}{2}q(3\cos^2\gamma' - 1)(3m_I^2 - I(I+1)) \qquad (7)$$

γ and γ', respectively, are the angles between the principal axes (z') and (z'') and the magnetic field B_o (z). For simplicity only the shf interaction with one neighbour nucleus is considered in Equ.7. ESR transitions occur at magnetic field values B_{res} because of the selection rules $\Delta m_S = \pm 1$, $\Delta m_I = 0$ according to the following equation:

$$h\nu_{ESR} = g\beta B_{res} + m_I(a + b(3\cos^2\gamma - 1)) \qquad (8)$$

Since m_I has $2I + 1$ values, one obtains a shf splitting of the ESR spectrum. The lines are equally spaced with a field separation of $(g\beta)^{-1}(a + b(3\cos^2\gamma - 1))$. In the case of F centres the electron interacts with many nuclei, e.g. 6 nearest neighbour cations and 12 nearest neighbour anions in the alkali halides. For certain field directions or negligible anisotropic interactions the nuclei of one neighbour shell may be equivalent. Then Equ.8 becomes

$$h\nu_{ESR} = g\beta B_{res} + \sum_1 M_I(1)\left(a(1) + b(1)(3\cos^2\gamma - 1)\right) \qquad (9)$$

where the total spin quantum number $M_I(1)$ of the 1 th shell of equivalent nuclei is given by

$$M_I(1) = \sum_{\beta=1}^{N} m_{I,\beta} = NI, NI - 1, \ldots, - NI \qquad (10)$$

N is the number of equivalent nuclei in the 1 th shell. The intensity of the M_I - shf - line depends, however, on the different statistical weights, corresponding to the number of possible combinations of the m_I to form M_I (5).

2.3. ENDOR - Transitions

In the ENDOR experiment one measures the nuclear spin resonance of the lattice nuclei which are coupled to the unpaired electron. The nuclear resonance is detected through the desaturation of the partially saturated ESR transition (2,11). ENDOR transitions usually occur in the range between 0,5 - 100 MHz in vacancy centre problems. The selection rule for ENDOR transitions is: $\Delta m_s = 0$, $\Delta m_I = \pm 1$.

In the first order perturbation theory, that is with the conditions

$$|B_{ik}|, |Q_{ik}| \ll |a \pm \frac{1}{m_s} g_I \beta_n B_o|$$

the ENDOR frequencies obtained from Equ.(1) for $S = \frac{1}{2}$ (as in F centres) are:

$$\nu_{ENDOR}^{\pm} = \left| \frac{1}{2h} W_{shf} \mp \nu_h \pm \frac{1}{h} m_q W_Q \right| \qquad (11)$$

with the following abbreviations:

$$W_{shf} = a + b(3\cos^2\gamma - 1) + b'\sin^2\gamma \cos 2\delta \qquad (12)$$

$$W_Q = 3\{q(3\cos^2\gamma' - 1) + q'\sin^2\gamma'\cos 2\delta'\} \qquad (13)$$

γ,δ and γ',δ' are the polar angles of B_o in the shf and quadrupole principal axis, respectively.

$$\nu_n = \frac{1}{h} g_I \beta_n B_o \qquad (14)$$

ν_n is the Larmor frequency of a free nucleus in the magnetic field $\vec{B}_o$

$$m_q = \frac{m_I + m'_I}{2} \qquad (15)$$

m_q is the average between the two nuclear spin quantum numbers which are connected by the transition. If no quadrupole interaction is present each interacting neighbour nucleus of the F centre gives according to Equ. (11) a pair of ENDOR lines. The two lines are separated by $2\nu_n$ if $\frac{1}{2h}W_{shf} > \nu_n$ and by $\frac{1}{h}W_{shf}$ if $\nu_n > \frac{1}{2h}W_{shf}$.

In case of non-vanishing quadrupole interactions each of the lines is split into quadrupole multiplets. Very often nuclei have nuclear spin I = 3/2, so that a symmetrical quadrupole triplet is observed. The "quadrupole lines" have a frequency separation of $\frac{1}{h}W_Q$ from the "shf-line". The simple first order solutions are very often not adequate to explain the details of the ENDOR spectra. Especially for larger anisotropic shf interactions and quadrupole interactions of the same order as shf interactions as well as stronger deviations from axial symmetry of the tensors, higher order perturbation theory or even a full diagonalisation of the Spin Hamiltonian Equ.(1) must be taken into account. For the F-centres in alkali halides second order perturbation theory is sufficient (9). One of the clearly visible effects of higher order corrections is e.g. that quadrupole triplets are no longer symmetric and the "shf-lines" are shifted by the quadrupole interactions. For further details of the analysis of ENDOR spectra the reader is referred to the literature (2, 12, 13, 14, 15, 16).

Basis for the assignment of ENDOR lines to certain neighbour nuclei is the careful analysis of the dependence of the ENDOR spectrum on the orientation of the magnetic field. An ENDOR spectrum is understood if the orientation of the principal axes and the interaction constants determined from some ENDOR lines describe fully the measured angular dependence. For the identification of a particular sort of neighbour nuclei the following criteria can also be used:

(i) the pairwise appearance of "shf" ENDOR lines according to Equ. (11) with separation $2\nu_n$ or $\frac{1}{h}W_{shf}$

(ii) repeated appearance of the ENDOR lines in the presence of several isotopes with a frequency ratio equal to the ratio of nuclear g-factors

(iii) further splittings because of quadrupole interactions.

The analysis of an ENDOR spectrum is usually an iterative process. The ENDOR method has certainly become a very powerful tool in determining the atomic structure of a point defect such as a vacancy centre. In practice, however, one first assumes a model for the defect and then tries to explain the ENDOR spectra on the basis of that model.

3. F CENTRES

3.1. ESR Spectra

Fig. 1 shows the ESR spectrum for F centres in CaF_2 for $B_o||(110)$. The 7 equidistant shf lines with a field separation of 61 Gauss have within 5% an intensity ratio of 1:6:15:20:15:6:1. Such a shf structure is explained by 6 equivalent nuclei of $I = \frac{1}{2}$. With this spectrum a model for the F centre is consistent, where the unpaired electron occupies an F^- vacancy and the observed shf structure is due to 6 second nearest F neighbours at the corners of a regular octahedron, provided the anisotropic shf interaction is small compared to the isotropic interaction. The latter is noted in only a small angular dependence of the shf splitting. The nearest 4 Ca^{++} cations are not seen, since the concentration of Ca^{++} ions with nuclear moments (^{43}Ca) is too small.

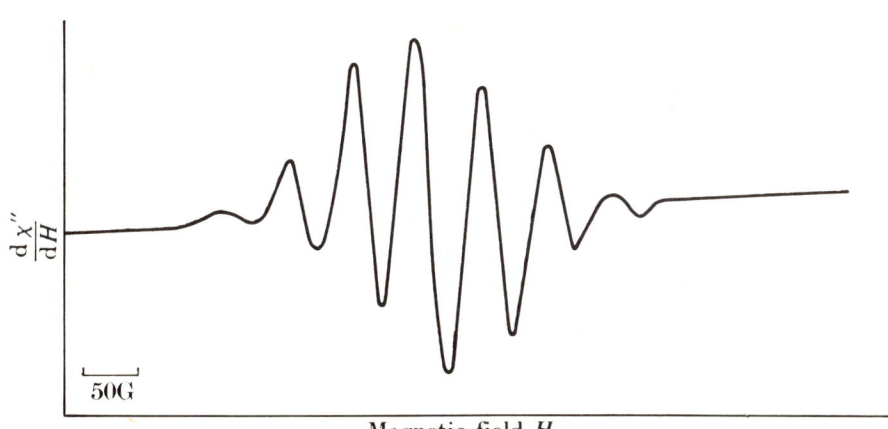

Fig. 1. ESR spectrum of F centres in CaF_2 at 300 K. $B_o||(110)$. (after Arends (17)).

The ESR spectra of F centres in most alkali halides consist only of a single, broad line of Gaussian shape without further structure. The half width of the ESR line varies between about 50 to 750 G. In some cases (LiF, NaF, RbCl, CsCl) the line shows some shf structure, the analysis of which is not straightforward, however, so that from the ESR spectrum alone no safe conclusion on the centre model can be drawn. (5). The broad line arises because of "inhomogeneous broadening". The F electron interacts with so many neighbour nuclei that the shf structure is not resolved any more (5). The g factors are in all cases slightly smaller than the g-value of the free electron. The g-shift varies from 1 to about 500×10^{-4}, the shift increasing on going to crystals with heavier ions (19).

Analogous vancancy centres in the alkaline earth oxides show a nicely resolved shf structure. In these materials vacancy centres with one trapped electron are called F^+ centres. Fig.2 shows the ESR-spectrum in SrO for $B_0||(111)$. The alkaline earth oxides (except BeO) have the face centred cubic rocksalt structure and dominant ionic binding. The unpaired electron occupies an oxygen vacancy. In SrO it has 6 Sr^{++} ions as nearest neighbours. Only the ^{87}Sr has a magnetic moment and $I = \frac{9}{2}$. The strong line at $g = 1.9845$ is due to these 65% F^+ centres with only nonmagnetic ^{86}Sr and ^{88}Sr neighbours. For $B_0||(111)$ all nearest Sr nuclei are equivalent (there is only an isotropic shf interaction, since $\cos\gamma = \frac{1}{\sqrt{3}}$ in Equ.8). If one Sr neighbour is ^{87}Sr, one expects 10 equidistant shf lines due to $I = 9/2$ which are observed, (line separation about 15 Gauss). About 5% of the centres will have two equivalent ^{87}Sr neighbours, which leads to 19 shf lines with the same splitting and a certain intensity ratio which can be clearly seen in the outer region of the spectrum. Between the lines just described further strong lines appear which were identified as forbidden transitions ($\Delta m_s = \pm 1$ and $\Delta m_I = \pm 1$) due to strong quadrupole interactions (7,20). The many details of the shf structure of the ESR spectrum allow a rather safe conclusion on the centre model and rather precise determination of the shf and quadrupole interactions.

3.2 ENDOR Spectra

Results of ENDOR investigations will briefly be discussed for F centres in alkali halides. Fig.3 shows the model of the F centre, which, of course, is the result of such an investigation, and Fig.4 shows the complete ENDOR spectrum of F centres in KBr for $B_0||(100)$: (2).

In the spectrum the shf and quadrupole interactions with 8 shells of neighbour nuclei are resolved in the frequency range between 0,5 and 30 MHz. All lines can be unambiguously assigned to nuclei

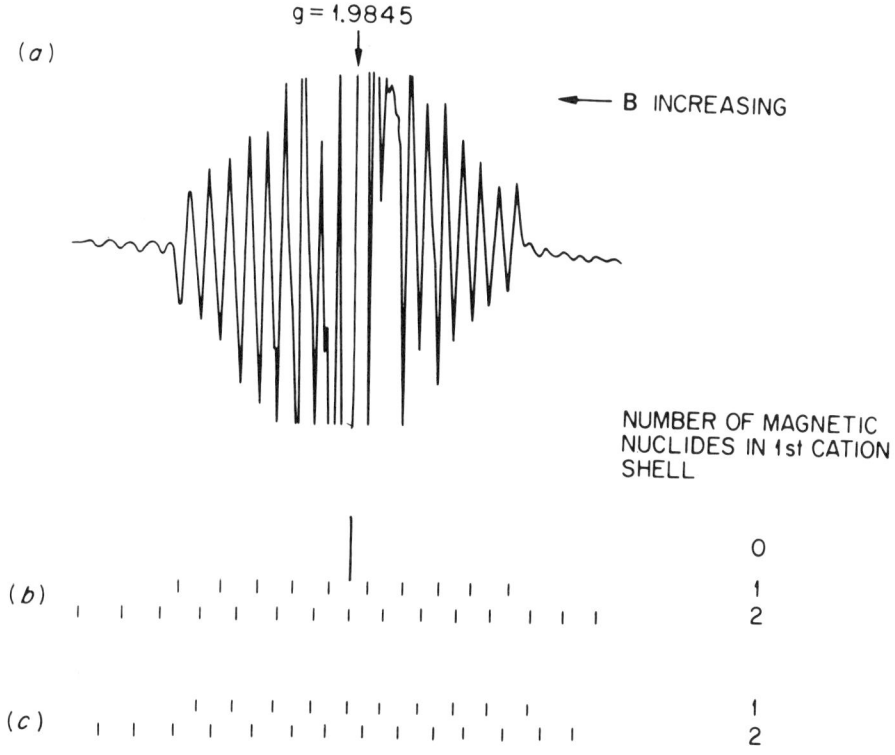

Fig. 2. a) ESR spectrum of F^+ centres in SrO for $B_0 || (111)$
b) reconstruction of spectrum with 0, 1 and 2 87Sr neighbours
c) forbidden transitions due to strong quadrupole interaction. The separation between vertical lines in b) and c) is about 15 G (after Culvahouse et al. (20)).

in the 8 shells. Following the principles briefly outlined in chapter 2 the spectrum can be analysed very precisely. The angular dependence of the first shell K neighbours is particularly simple and shown schematically in Fig. 5.

The angular dependence of ENDOR lines of higher shell nuclei is far more complicated. For a precise and unambiguous analysis of the spectra the magnetic field must be rotated in several planes. Fig. 6 shows as an example the angular dependence of the ν^- ENDOR lines (i.e. for $m_s = -\frac{1}{2}$) of shells V, VI, VIII and IX a (nuclei 003) in KCl (9). The solid curves are calculated using

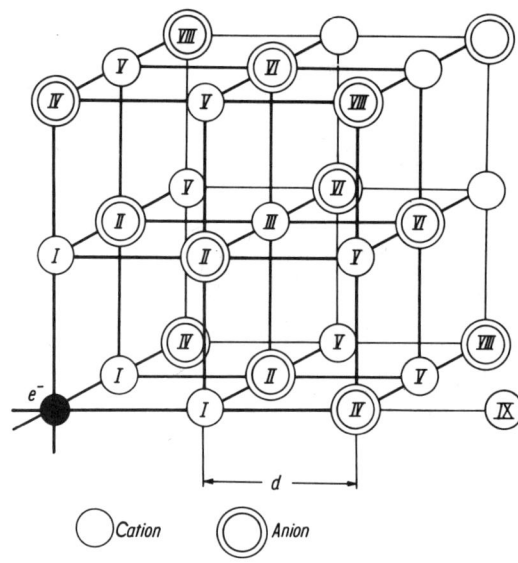

Fig. 3   Environment of the F centre with shells I - IX.

second order perturbation theory with the best set of interaction constants.

The results of the ENDOR analysis do not only confirm beyond any doubt the model of the F centre in the alkali halides as an electron trapped in an anion vacancy, but also yield very detailed information on the electronic structure by means of the shf and quadrupole interaction constants. Table 1 reproduces the results for F centres in KBr (2). Noteworthy is the high precision with which the interaction constants can be determined.

The quantitative results show that in the lower shells the shf interaction is predominantly isotropic. On going from shell I to about shell V there is a sharp drop in the shf interaction by about two orders of magnitude which tends to become more constant towards the higher shells (2,9). The interaction tensors have only very small deviations from axial symmetry where on pure symmetry arguments the axial symmetry is not required (shells II,V,VI,VII). Very detailed ENDOR investigations were also made on F centres in the alkaline earth fluorides (8) and on F^+ centres in the alkaline earth oxides (7).

Table 1 : shf interaction constants of the F centre in KBr in MHz (T = 90°K) (after Seidel(2)).

Shell	nucleus	a/h	b/h	3q/h
I	^{39}K	18.33	0.77	0.20
II	^{81}Br	42.85	2.73+	0.23
III	^{35}Cl	0.274	0.022	
IV	^{81}Br	5.70	0.410	0.105
V	^{39}K	0.16	0.02	
VI	^{81}Br	0.838	0.082+	
VIII	^{81}Br	0.583	0.068	

+ There is a small deviation from axial symmetry.

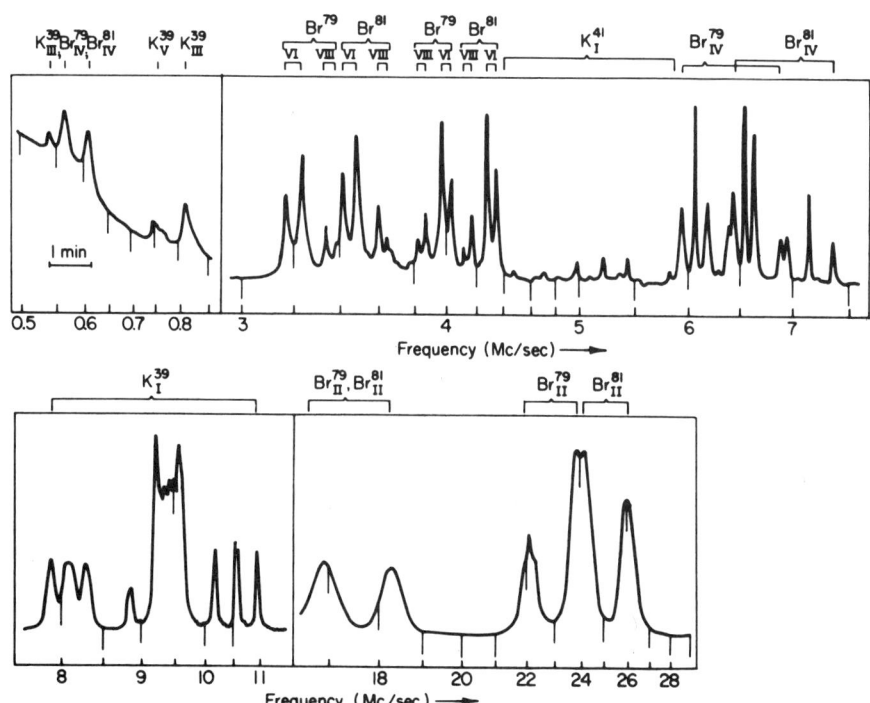

Fig. 4 ENDOR spectrum of F centres in KBr for $B_0 \| \langle 100 \rangle$, T = 90°K, with lines from shell I to VIII. ν_{ESR} = 9.5 GHz (after Seidel (2)).

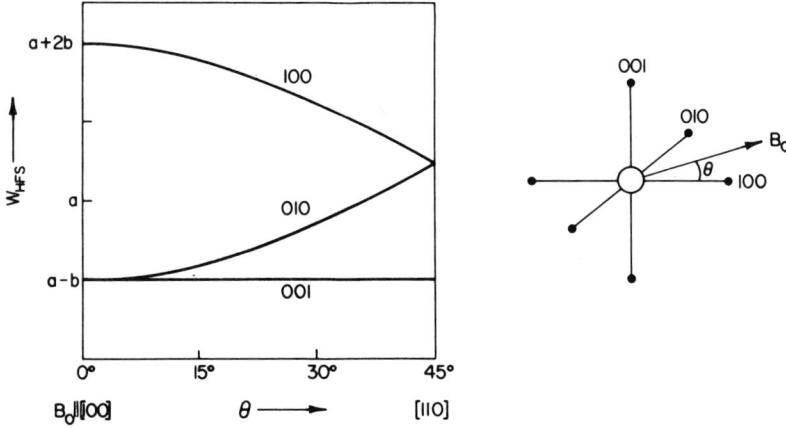

Fig. 5 Schematic representation of the angular dependence of the shf interaction of first shell nuclei in F centres. B_0 is rotated in a (001) plane. (after Seidel(2))

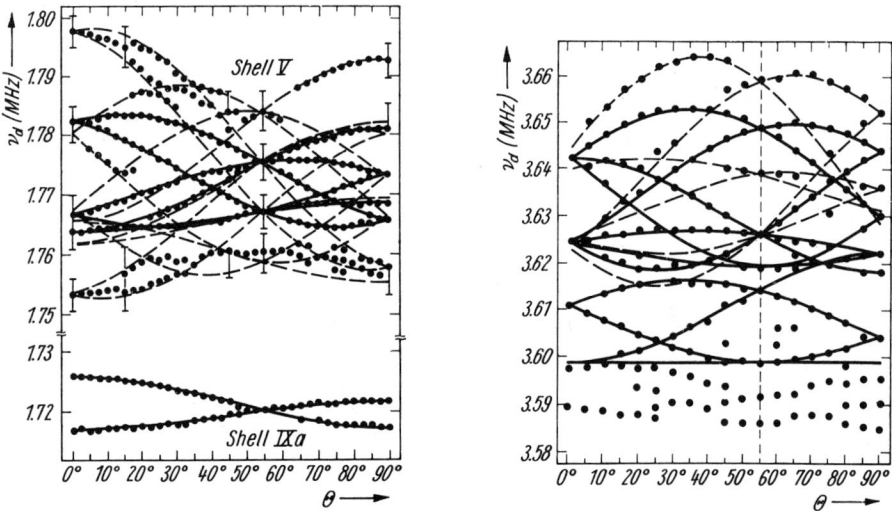

Fig. 6 Angular dependence of higher shell ν_{ENDOR}^- lines for F centres in KCl for B_0 rotated in a $(1\bar{1}0)$ plane. $\theta = 0° \hat{=} (001)$. $\nu_{ESR} = 24$ GHz, T = 77°K. Right side: upper curves belong to shell VI and lower curves belong to shell VIII.

3.3. Electronic Structure

According to Equ.(2) and (3) the shf interaction constants furnish an extraordinarily exact determination of the wave function $\psi(r)$ of the unpaired electron in its ground state. The isotropic shf constant is a direct measure of the density at the various sites of neighbour nuclei. For a theoretical interpretation of the results one has thus to discuss the electronic wave function of the unpaired electron. If one wants to do this from first principles this is quite a formidable task, which involves very difficult questions of molecular and solid state quantum theory, which cannot be discussed here. However, this theoretical part of the investigations is of great importance since the ENDOR measurements provide so much detailed information that a very good test of any theoretical model can be made.

In many ionic crystals the experimental results of vacancy centres are discussed in some form of a pseudopotential theory. In its simplest form a smoothly varying, vacancy-centred envelope function ψ_F is determined as a one-electron solution of an appropriate Schrödinger equation employing the potential of the centre and a simplified potential of the lattice where the detailed structure of the ions is neglected. This envelope function is then orthogonalised to the core wave functions ψ_i of the neighbour ions. (7,18,21,22,23,25). The centre wave function thus obtained is:

$$\psi = N \left(\psi_F - \sum_i \langle \psi_F | \psi_i \rangle \psi_i \right) \qquad (16)$$

N is a normalisation constant which generally is practically equal to 1.

In this "method of orthogonalised wave function" the centre electron is described by the one particle wave function ψ which is then used to calculate the interaction parameters according to Equ.(2) and (3).

The agreement between experiment and theory in the alkali halides obtained with this type of calculations is reasonably good for the first two or three shells of neighbour nuclei. The agreement for the isotropic shf constants for higher shell nuclei is still rather poor, the discrepancies being about two orders of magnitude for shells VIII and XI (5,9).

The theoretical interpretation of the quadrupole interaction is still more difficult. The main problems arise in connection with the proper treatment of the Sternheimer antishielding. For F centres in KCl the quadrupole interactions were discussed in detail in connection with the lattice distortion around the vacancy (26).

4. VACANCY CENTRES IN CRYSTALS WITH IMPURITIES

In many ionic crystals perturbed F-type centres can be produced if the crystals contain impurity cations. For example, if one bleaches optically at room temperature F-centres in potassium chloride containing Li^+ or Na^+ ions, perturbed F-centres are produced which are called F_A centres. As revealed by an ENDOR investigation, F_A centres are F centres where one of the 6nearest K-cations is replaced by Li^+ or Na^+. Fig. 7 shows the ENDOR spectrum for randomly oriented $F_A(Li)$ centres in KCl for $B_o || (100)$ (27).

The spectrum shows clearly the ENDOR lines of the Li-impurity. The line denoted by $Li_{||}$ is due to those Li nuclei parallel to the field direction, that denoted by $Li_\perp$ due to those perpendicular to B_o. The potassium lines appear in two groups that are slightly shifted with respect to each other. This is because in the F_A centre the four K nuclei in the symmetry plane perpendicular to the centre axis, which is the connection line between the vacancy and the Li^+ neighbour, are no longer equivalent. The ESR spectrum of the F_A centres differ only in the line width from that of the F centre (28).

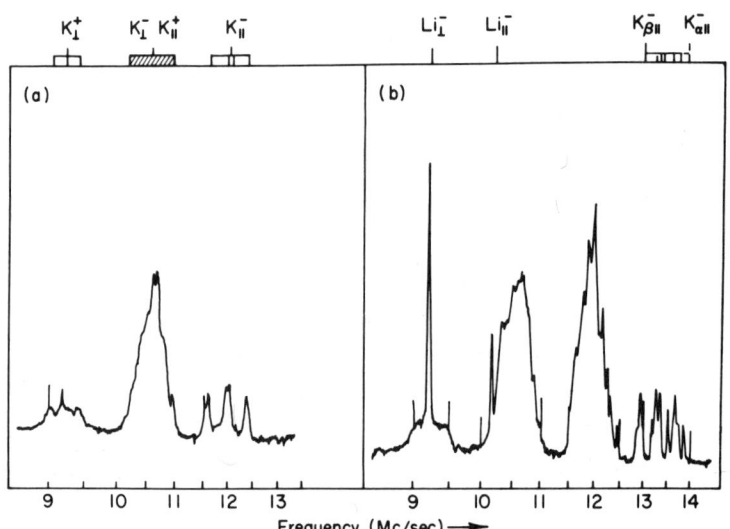

Fig. 7 ENDOR lines of first shell nuclei of F and F_A centres in KCl: Li, $B_o || (100)$, $T = 90°K$, $\nu_{ESR} = 9.38$ GHz.
a) F centres in the quenched crystal
b) F_A centres after five minutes irradiation by 5460 Å light at room temperature. (after Mieher (27)).

Magnetic resonance studies of Z centres in alkali halides, which are formed in crystals with divalent cation impurities, were less successful. For instance as a result of the ENDOR investigation of Z_1 centres the atomic structure of the Z_1 centres could not be established unambiguously. Only two possible configurations could be given. Very many ENDOR lines could be measured, but a complete analysis could not be achieved (29).

Another interesting example of a perturbed F centre in an alkali halide is the H_2O^--centre in KCl. It is produced photochemically from OH^--centres. The ESR-spectrum is very similar to that of the F centre. In the ENDOR spectrum below 15°K ENDOR lines from protons can clearly be seen. From the analysis the model shown in Fig.8 could be derived: a H_2O molecule occupies an F centre with its C_{2v} molecular axis along a $\langle 111 \rangle$ direction. The shf interactions with the nearest lattice neighbours are very similar to those of the F centres. Therefore the centre was sometimes called the "wet F centre". (30).

5. F AGGREGATE CENTRES

When F centres in alkali halides are exposed to light in the F-band at room temperature aggregation of F centres occurs. Part of the F centres form aggregate centres such as M and R centres, others form loose aggregates. The F centres in the loose aggregates are somewhat perturbed. Their ESR linewidth is decreased compared to normal F centres, the line shape altered in the central part of the ESR line, and the relaxation times are shorter (5,31). The ENDOR spectrum, however, shows no difference from the unperturbed F centre. (31). The changes in the ESR spectra are explained as due

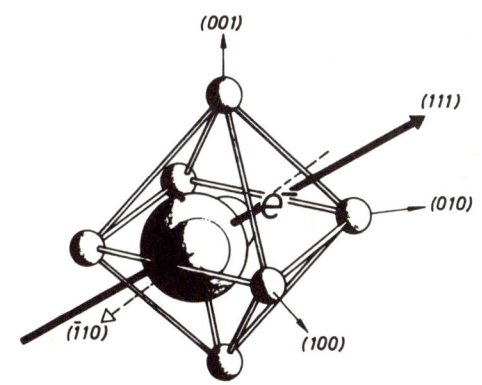

Fig. 8 Model of the H_2O^--centre in KCl. (after Rusch (30)).

to exchange effects between loosely aggregated F centres (31).

M centres have a diamagnetic ground state (32) and the ESR of R centres can only be observed under special conditions (33). The models for the M and R centre in KCl are shown in Fig.9.

They were definitively confirmed through the ESR and ENDOR measurements in their metastable multiplet states with lifetimes of many seconds into which they can be brought optically (34,35).

Their electronic states can be described in a very good approximation as Heitler-London combinations of two respectively three F centre wave functions. For the M centre triplet state in KCl this is immediately seen in the ENDOR spectrum of the nearest neighbouring nuclei shown in Fig.10.

New lines appear with the M excitation. The lines in the range of 18 to 21 MHz are due to the K_α^I nuclei (see Fig.9) and have approximately twice the frequency of the corresponding F centre K^I lines. The electron density of K_α^I is approximately twice the value of the F centre due to the two M centre electrons. (34).

For magnetic resonance investigation of multiplet states of M centres in the alkaline earth fluorides the reader is referred to (8).

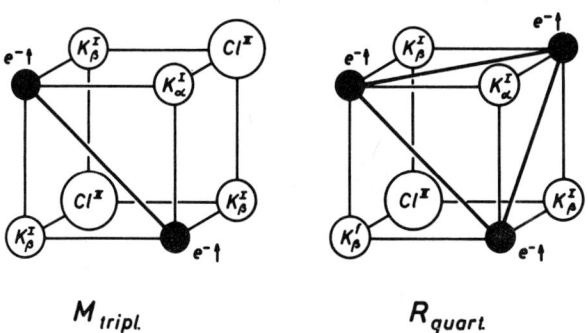

Fig. 9   M and R centre in KCl. Spin configurations are shown for the metastable multiplet states. (after Seidel (36)).

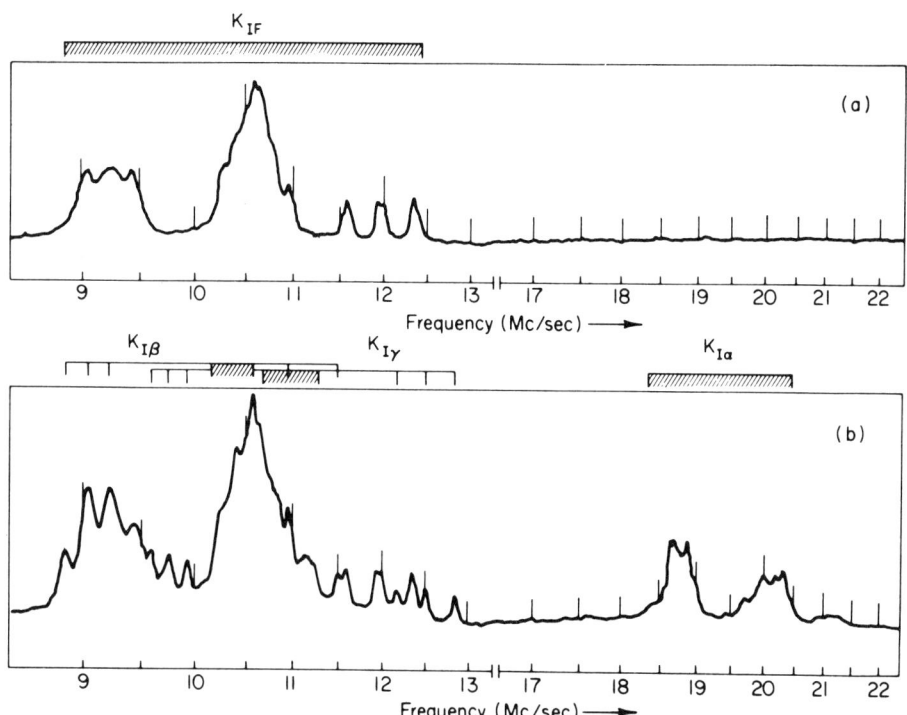

Fig. 10 ENDOR spectra of F and M centres in KCl. $B_0 || (100)$, T = 90°K, ν_{ESR} = 9.38 GHz.
a) without illumination (F centres, first shell)
b) during excitation with 3650 Å. M centres in the triplet state (after Seidel (34)).

REFERENCES

1. C. A. Hutchison, Phys. Rev. 75, 1769 (1949)
2. H. Seidel, Z. Physik, 165, 218, 239 (1961)
3. W. T. Doyle, Phys. Rev. 126, 1421 (1962)
4. J. H. De Boer, Rec. Trav. Chim. 56, 301 (1937)
5. H. Seidel and H. C. Wolf, in "Physics of Color Centers," ed. by W. Beall Fowler (Academic Press, New York, 1968) p.538
6. B. Henderson and I. E. Wertz, Advan. Phys. 17, 747 (1968)
7. A. E. Hughes and B. Henderson, in "Point Defects in Solids", Vol. 1, ed. by J. H. Crawford, Jr. and L.M. Slifkin (Plenum Press, New York-London, 1972), p. 381
8. W. Hayes, in "Crystals with the fluorite structure", ed. by W. Hayes, p. 185 (Clarendon Press, Oxford 1974)
9. R. Kersten, phys. stat. sol. 29, 575 (1968)
10. E. Fermi, Z. Physik 60, 320 (1930)
11. G. Feher, Phys. Rev. 114, 1219, 1249 (1959)
12. W. C. Holton and H. Blum, Phys. Rev. 125, 89 (1962)
13. U. Ranon and J. S. Hyde, Phys. Rev. 141, 259 (1966)
14. J. M. Spaeth, Z. Physik 192, 107 (1966)
15. H. Seidel, "Habilitationsschrift" Stuttgart (1966)
16. J. M. Spaeth, "Habilitationsschrift" Stuttgart (1966)
17. J. Arends, phys. stat. sol. 7, 805 (1964)
18. A. M. Stoneham, W. Hayes, P. H. S. Smith and J. P. Stott Proc. R. Soc. A 306, 369 (1968)
19. D. Schmid, phys.stat. sol. 18, 653 (1966)
20. J. W. Culvahouse, L.V. Holroyd and J. L. Kolopus Phys.Rev. 140, 1181 (1965)
21. B. S. Gourary and F. J. Adrian, Phys. Rev. 105, 1180 (1957)
22. B. S. Gourary and F. J. Adrian, Solid State Physics 10, 127, (1960)
23. J. K. Kübler and R. J. Friauf, Phys. Rev. 140, A 1742 (1965)
24. W. B. Fowler in "Physics of Color Centres" ed. by W.B. Fowler, p. 54 (Academic Press, New York 1968)
25. R. H. Bartram, A. L. Harmer and W. Hayes, J. Phys. C. Sol.St. Phys. 4, 1665 (1971)
26. R. Kersten, Solid State Comm. 8, 167 (1970)
27. R. L. Mieher, Phys. Rev. Letters 8, 362 (1962)
28. H. Ohkura, K. Murase and H. Sugimoto, J.Phys.Soc.Japan 17, 708 (1962)
29. J. C. Buschnell, Thesis, Univ. of Illinois, unpublished (1964)
30. W. Rusch and H. Seidel, Solid State Comm. 9, 231 (1971)
31. M. Schwoerer and H. C. Wolf, Z. Physik 175, 457 (1963)
32. H. Groß, Z. Physik 164, 341 (1961)
33. D. C. Krupka and R. H. Silsbee, Phys. Rev. 152, 816 (1966)
34. H. Seidel, Phys. Letters 7, 27 (1963)
35. H. Seidel, M. Schwoerer and D. Schmid, Z. Physik 182, 398 (1965)
36. H. Seidel, Colloque Ampère XV, North Holland, Amsterdam, p. 141 (1969)

INTERSTITIAL CENTRES :

OPTICAL ABSORPTION AND MAGNETIC RESONANCE

DIRK SCHOEMAKER

Physics Department

University of Antwerp (U.I.A.) 2610 Wilrijk, Belgium

Ionising radiations (x-,γ - rays, electrons, neutrons) produce defects in solids. The defects can be varied and complex and depend not only on the type of crystal but also on the type of radiation, the temperature, pretreatment (thermal, mechanical), impurity content etc. No wonder then, that detailed investigations on the structure and properties of defects have centered primarily around simple solids : ionic crystals, elemental semiconductors, simple metals.

In the simplest of ionic crystals, namely the alkali halides, the knowledge of the geometric and electronic structure of the defects (often called "colour centres" or just plain "centres") is probably the most developed. This is not only because of their simple crystal structure and easy availability, but also because they are transparant for electromagnetic radiation ranging from radio- and microwaves up to the far U.V. The defects in these ionic crystals have usually well localised ground-and excited states which can be studied by a broad spectrum of spectroscopic techniques whose results often lead to a thourough understanding of the structure and the physics of these defects.

Even in such simple materials as the alkali halides the number and kind of defects that can be produced is quite staggering. However most of these are derived from a few basic or primary defects. If one excludes neutron irradiated crystals (few of which have been studied anyway) one can state as an experimental observation that the ionising radiations only affect the halogen sublattice and the primary defects that are created are the following :
        electrons
        positive holes
        interstitial halogen ions

interstitial halogen atoms
halogen ion vacancies

It is convenient for our discussion to view these as the primary defects and to ascribe to the first four of them as having, or not having, depending on the temperature, a certain mobility and range in the crystal. It is doubtful that the primary defects move or exist as bare entities in the crystal. However, the undoubtedly very complicated combination of electronic and ionic processes that take place during and after the absorption of an x or γ photon or after optical excitation, are not yet fully understood and fall outside the scope of these lectures.

The primary defects can be stabilised in a great variety of ways. An electron can be trapped by a negative ion vacancy and the well known F centre is then created. At sufficiently low temperatures (e.g. below 20 K in KCl) the interstitial halogen ion is stable and immobile. It is called the I center. It possesses absorption bands in the U.V. close to the fundamental absorption of alkali halide. It is not paramagnetic and cannot be studied directly with ESR techniques. We will not treat it in these lectures, but refer to a recent discussion by Itoh[1].

At sufficiently low temperatures (below 120 K in KCl) the hole is stabilised in the lattice without help from any defect or impurity and forms what is known as the V_K centre. The interstitial halogen atom is also stabilised at sufficiently low temperatures (below 40 K in KCl) and forms what is known as the H centre.

In these lectures we will give an introduction to the H centres and centres derived from it. Although fundamentally different from the V_K centre, both centres possess an unpaired electron spin and exhibit great similarities in their electronic structure. In fact they are often confused with one another when first encountered. For this reason, and also because the knowledge about the V_K centre is more advanced in many respects, we will start by giving a short survey of the properties of this centre. It will become clear that the two experimental tools of optical absorption and ESR have played an important role in the unraveling and understanding of the properties of these centres. The ESR spectroscopy especially has been indispenseble in identifying and studying the V_K-, and H-type centres. The reason is essentially that the ground states of these centres are very well localised states and the interaction of the unpaired electron spin with the nucleair spin of the halogens results in very characteristic anisotropic ESR spectra from which oftentimes the essential aspects of the geometric and electronic structure can be derived. These notes are not meant to constitute a review paper, and no extensive data are presented. Rather they aim to be an introduction to the field of hole and interstitial centres in ionic materials, and KCl will be used in most cases as a representave alkali halide. For recent critical review papers giving extensive data and literature references we refer to papers of Itoh[1] and Kabler[2].

THE V_K- OR "SELF TRAPPED HOLE" CENTRE

ESR Spectrum and Model

In most alkali halides the V_K centre is stable at 77 K and can be produced by x- or γ irradiation at this temperature. Its ESR spectrum in KCl observed at this temperature is shown in Fig. 1. Though the spectrum may look complex at first glance it is really an example of a clean well resolved ESR spectrum. Just about every line can be accounted for.

The dominant feature of the spectrum is a set of seven strong equally spaced hyperfine lines with intensity ratios 1:2:3:4:3:2:1. This shows that the unpaired spin interacts equally with the nuclear spins of two chlorine nuclei which are equivalent with respect to the magnetic field $\vec{H}$ used in the ESR measurements. Indeed, there are two Cl isotopes, ^{35}Cl (75% abundant) and ^{37}Cl (25% abundant) and both have nuclear spin 3/2. The dominant seven lines arise from a ^{35}Cl ^{35}Cl combination which has a probability of occurence of (3/4) x (3/4) = 9/16. The weaker lines in the spectrum can all be accounted for by the presence of the ^{35}Cl ^{37}Cl and ^{37}Cl ^{37}Cl combinations which have, respectively, a probability of 6/16 and 1/16.

A second important aspect of the ESR spectrum is its anisotropy, i.e., the hyperfine separation between the seven lines, and the average position of the spectrum, change when the crystal is rotated in the magnetic field. The anisotropy is determined by the symmetry of the centre. The analysis is not difficult and we refer to the original papers [3][4]. The conclusion is that the centre has to a very good approximation axial symmetry, with the symmetry axis oriented exactly along a <110> direction of the crystal.

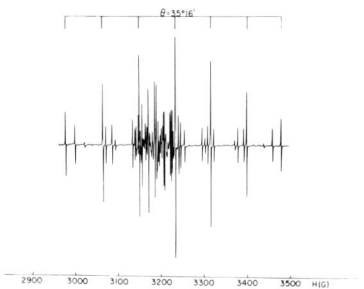

Fig. 1. ESR spectrum of the Cl_2^- V_K centre in KCl for $\vec{H}$ ∥<111>.

The hyperfine separation is maximum when $\vec{H}$ is parallel to the symmetry axis and minimum when perpendicular to it. There are six <110> directions in KCl and each has an equal number of V_K centres in it. These six orientations are indeed observed separately in the ESR spectrum and this is useful in experiments where the equality of populations is disturbed.

The quantitative analysis of the ESR spectrum is done in terms of the following spin Hamiltonian (usual notation) :

$$\frac{\mathcal{H}}{g_0 \beta} = \frac{1}{g_0} [g_\parallel H_z S_z + g_\perp (H_x S_x + H_y S_y)] \qquad (1)$$

$$+ A_\parallel S_z I_z + A_\perp (S_x I_x + S_y I_y)$$

with $\vec{I} = \vec{I}_1 + \vec{I}_2$ and in which the symmetry axis z is parallel to <110>. One finds (4) :

$g_\parallel = 2.00145$ $A_\parallel (^{35}Cl) = 101.31$ gauss
$g_\perp = 2.0435$ $A_\perp (^{35}Cl) = 12.13$ gauss†

The axial symmetry around <110>, the interaction with two Cl nuclei and the fact that all experimental indications point to a trapped hole centre lead to the model for the V_K centre as depicted schematically in Fig. 2 : It is a Cl_2^- molecule ion, occupying two negative ion sites and whose internuclear axis is oriented along <110> . Note that this centre has an effective positive charge, a fact that will be exploited later on.

Electronic Structure

At this stage one can ask : How good is the description of the V_K centre, both in the ground and the excited states, in terms of a Cl_2 molecule?

It is instructive to turn to the neutral Cl_2 molecule which is treated in just about every book on molecular structure. The molecule can be looked upon as being composed out if two Cl atoms and each of these contribute five p valence electrons to the molecular bond. Spectroscopic studies have shown that these 10 electrons are accomodated in five molecular orbitals in the following fashion :

$$\sigma_g^2 \pi_u^4 \pi_g^4$$

giving a $^1\Sigma_g^+$ ground state. The σ_g orbital has rotation symmetry around the molecular axis and its molecular orbital, expressed in terms of atomic p orbitals centered on each of the two nuclei, is

INTERSTITIAL CENTRES: OPTICAL AND MAGNETIC RESONANCE

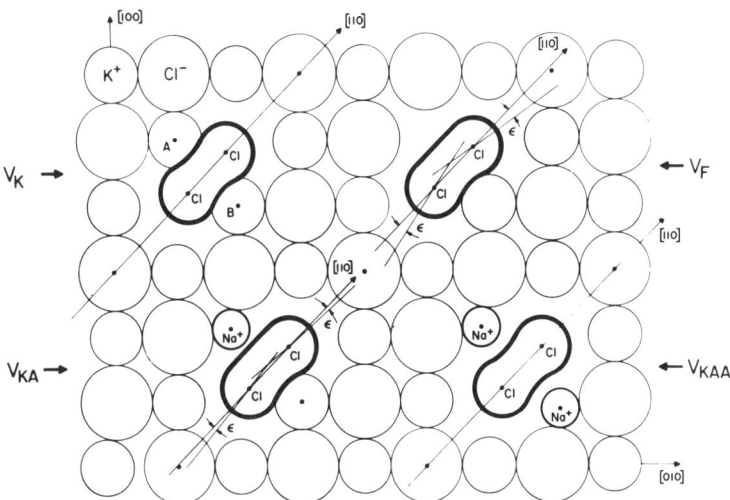

Fig. 2. Model of the Cl_2^- V_K centre in KCl. Also shown are models for the V_F, V_{KA} and V_{KAA} centres [4] which are not discussed in the text.

(without normalisation constant) : z_1-z_2. The π orbitals are doubly degenerate and can be expressed as $\pi_{u,x} = x_1 + x_2$ and $\pi_{u,y} = y_1 + y_2$ while $\pi_{g,x} = x_1 - x_2$ and $\pi_{g,y} = y_1 - y_2$. The indices u and g indicate the uneveness and eveness under inversion.

The first unoccupied orbital of Cl_2 is a σ_u type orbital of the form $z_1 + z_2$ and one expects therefore that the Cl_2^- valence configuration will be :

$$\sigma_g^2 \pi_u^4 \pi_g^4 \sigma_u^1$$

giving a $^2\Sigma_u^+$ ground state.

The configuration and ground state has been confirmed by the SCF-MO calculations of Wahl and Gilbert [5] who also showed that the binding energy of this molecule is about 1 eV. The spin Hamiltonian parameters are in very good qualitative and quantitative accord with this ground state and configuration. The demonstration of this falls outside the scope of these notes and we refer to the analysis given by Schoemaker [4]. Recently Lagendijk and Schoemaker [6], employing gauge invariance arguments of a special kind, were able to determine from the spin Hamiltonian parameters the internuclear V_K centre distances. They found, e.g., for Cl_2^- in KCl that $R = 2.61$ Å which is in very good agreement with the 2.6 Å value obtained by Gilbert and Wahl[5].

Optical Transitions

Because the electric dipole operator is uneven under inversion, simple symmetry considerations suggest that the Cl_2^- centre should possess the following two optical transitions :

$$\sigma_g^2 \pi_u^4 \pi_g^4 \sigma_u^1 \; (^2\Sigma_u^+) \rightarrow \sigma_g^1 \pi_u^4 \pi_g^4 \sigma_u^2 \; (^2\Sigma_g^+)$$

and

$$\sigma_g^2 \pi_u^4 \pi_g^4 \sigma_u^1 \; (^2\Sigma_u^+) \rightarrow \sigma_g^2 \pi_u^4 \pi_g^3 \sigma_u^2 \; (^2\Pi_g).$$

These total energy states plus the $^2\Pi_u$ excited state which is not involved in optical transitions, are shown in Fig. 3 as a function of the internuclear distance R. This Fig. is based on the SCF - MO calculation of Wahl and Gilbert.[5] Note that both excited states $^2\Sigma_g^+$ and $^2\Pi_g$ involved in the optical transitions are dissociative states, i.e., when excited to these states the Cl_2^- molecule dissociates into $Cl°$ and Cl^- fragments with kinetic energy determined by their excess energy over the binding energy. This is an important remark.

The $^2\Sigma_u^+ \rightarrow \, ^2\Sigma_g^+$ transition is, in the parlance of molecular spectroscopy, a charge-transfer transition and is strongly allowed.

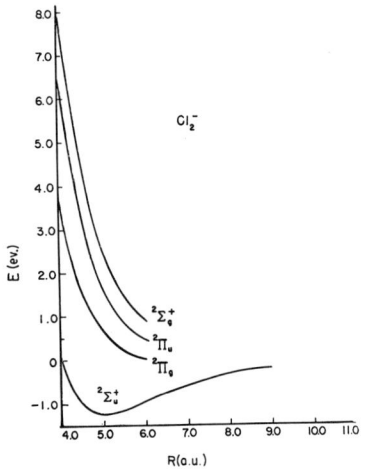

Fig. 3. Energy levels of the V_K centre as function of the internuclear distance R.

The second transition $^2\Sigma_u^+ \rightarrow {}^2\Pi_g$ is expected to be much weaker. These two transitions were indeed observed at 77 K very early in the V_K centre history by Delbecq, Smaller, Hayes and Yuster[7]. For Cl_2^- in KCl, the strong $^2\Sigma_u^+ \rightarrow {}^2\Sigma_g^+$ transition is situated at 365 nm and the much weaker (by about two orders of magnitude) $^2\Sigma_u^+ \rightarrow {}^2\Pi_g$ at 750 nm. Fig. 4 gives the 365 nm absorption spectrum. These results substantiate the qualitative validity of the description of the V_K centre in terms of the Cl_2^- configuration given above.

The V_K centre absorption band is not obtained by a straightforward optical absorption measurement. Usually there are interfering absorptions in the neighbourhood. Use is made of the fact that the two transitions are anisotropic. The $^2\Sigma_u^+ \rightarrow {}^2\Sigma_g^+$ transition is a σ - polarised transition which means that it is allowed only for polarised light whose electric vector is parallel (or has a component parallel) to the molecular axis. The $^2\Sigma_g^+ \rightarrow {}^2\Pi_u$ transition should be π polarised (i.e. allowed for light polarised perpendicular to the molecular axis).

Experimentally one finds indeed that the strong U.V. transition is σ polarised[7]. This and the fact that the excited $^2\Sigma_g^+$ is dissociative allows one to produce an anisotropy in the absorption by inducing a non-equal distribution among the six equivalent <110> V_K centre oroentations. Fig. 5 shows a cube on which are drawn the six <110> orientations. Suppose one irradiates a sample with an intense beam of [0$\bar{1}$1] polarised light in the 365 nm absorption band. The V_K centres in orientation 2 (which makes an angle of 0° with the electric vector) and the V_K centres in orientations 3, 4, 5 6 (which make 60° angle with the electric vector) will be excited to the dissociative $^2\Sigma_g^+$ state. Since the Cl_2^- molecule is trapped in the solid the excess energy will be given off to the lattice as phonons and the hole will move several lattice spacings in the process. It will come to rest as a Cl_2^- molecule whose orientation is uncorrellated with the original Cl_2^- orientation. If it comes

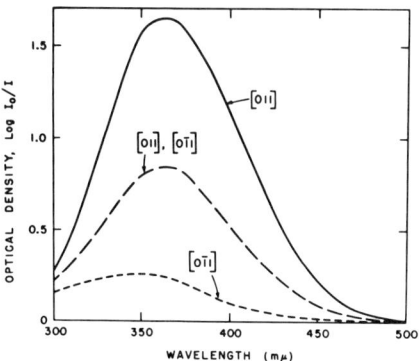

Fig. 4. Anisotropic absorption spectrum of Cl_2^- V_K in KCl. Substraction of curve [0$\bar{1}$1] from [011] gives the absorption spectrum.

to rest in the [011] orientation it will be hidden from the [0̄1̄1] polarised light and will not be excited again. In any of the other five <110> orientations it will in due time be excited again. Eventually most Cl_2^- will be in the [011] direction. Measuring the absorption now with [0̄1̄1] and [011] polarised light will give respectively a low and a high optical density as shown in Fig. 4. The difference of these curves gives the 365 nm anisotropic absorption of Cl_2^- assuming that any other underlying band is isotropic. In this fashion the optical absorption of the V_K centres (and similar centres) have been determined.

Thermal Reorientation and Diffusion

In the foregoing section we have implicitly assumed that a Cl_2^-, once it comes to rest after optical excitation, is immobile. This is not always so and in fact it depends on the temperature. Experimentally this is easily seen by investigating the effect of temperature on a crystal having preferentially oriented Cl_2^- centres such as shown by curves [0̄1̄1] and [011] in Fig. 4. If one follows these absorption curves as a function of temperature above 77 K one observes that if one approaches - 100°C curve [0̄1̄1] rises and curve [011] drops in intensity. The rate of change is fastest at T_D = -100°C in KCl and this temperature is called the disorientation temperature. Around -90°C all optical anisotropy has disappeared. This experiment implies that the Cl_2^-'s are thermally excited out of their preferential orientation into new orientations when the temperature is sufficiently high. In order to discuss the mechanism of the V_K centre

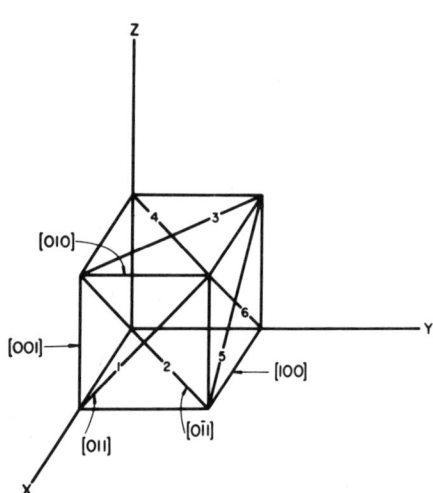

Fig. 5. The six <110> orientations of the V_K or H centre

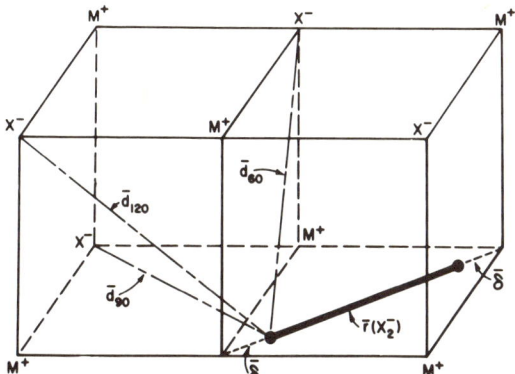

Fig. 6. Three dimensional model of the V_K centre used in the discussion of the thermal reorientation mechanism.

reorientation we refer to the three dimensional representation in Fig. 6. The Cl_2^- is indicated by a heavy line ending in two heavy points indicating the Cl nuclei. Because the Cl_2^- internuclear distance is smaller than the normal $Cl^- - Cl^-$ separation of the lattice, the two nuclei are a distance δ away from the normal lattice site. The lines $d_{60°}$, $d_{90°}$ and $d_{120°}$ indicate distances from one Cl_2^- chlorine nucleus to a few of the nearest neighbour Cl^- ions. If the Cl_2^- nucleus would be on the normal lattice site ($\delta = 0$) then these three distances would be equal to each other, but now $d_{60°} < d_{90°} < d_{120°}$.

At increasing lattice temperatures the vibrational amplitude of the Cl_2^- and the neighbouring Cl^- ions increase and at one point three specific Cl nuclei (from the Cl_2^- plus one neighbouring Cl^-) are in such a configuration that the unpaired hole can be on any of the three Cl^- ions. An instant later this favourable configuration is broken and a Cl_2^- is formed again. This can be the same Cl_2^- but it could be with equal probability another Cl_2^- which would have only one Cl nucleus in common with the original Cl_2^-.

The Cl^- ions most likely to be involved in a favourable configuration are the ones closest to the Cl_2^-. Looking at Fig.6 one sees that the four Cl^- ions connected to the Cl_2^- by $d_{60°}$ are the closest ones. A careful study of the reorientation kinetics by Keller, Murray, Abrahams and Weeks[8] has indeed shown that the V_K centre reorients exclusively by "60° jumps", proving indeed that these four Cl^- ions are dominant in the thermal reorientation process. A "60° jump" means of course that the new Cl_2^- makes a 60° angle with the Cl_2^- in the previous orientation.

The reorientation process is a normal thermally activated one. For Cl_2^- in KCl an activation energy of 0.54 eV, and a frequency factor of about 1×10^{13} Hz were found [8].

It is important to point out that, because of the nature of the V_K centre, the thermal reorientation process is identical to the diffusion process. Indeed, looking at a succession of reorientation processes one sees that with each reorientation jump the Cl_2^- (or better : the hole) has moved a distance in the lattice; the reorientation process is not confined to one lattice site (as is the case e.g. for the H centre). Thus when a Cl_2^- containing KCl crystal is heated 20 to 30 degrees above the disorientation temperature of -100°C, the reorientation rate, and thus the diffusion rate, is very high, and the probability of annihilation by recombining with a trapped electron (or an impurity) will increase. Indeed, the Cl_2^- in KCl is observed to decay rapidly at the decay temperature of -65°C.

This concludes the summary of the more important V_K centre properties. It will be useful to compare and contrast these with those of the H centre which will now be discussed.

THE H - OR INTERSTITIAL HALOGEN ATOM CENTRE

ESR Spectrum and Model

The H centre is in essence an interstitial halogen atom centre. Compared to the V_K centre it has a distinctly lower thermal stability. In KCl it decays around 40 K, and thus the H centre must be produced by ionising radiation below this temperature. Fig.7 shows the H centre ESR spectra obtained at 23 K (because the signal happens to be strongest at this temperature). The second derivative of the absorption is presented. The spectra, first obtained by Känzig and Woordruff [9], look complicated, much more so than the V_K centre resonance of Fig. 1, but is not nearly as bad as it may seem.

Ignoring the finely spaced lines (called superhyperfine lines) for a moment, one notices that the basic ESR spectrum consists again out of seven hyperfine lines with intensity ratios 1:2:3:4:3:2:1. Furthermore, the separation between these lines is very comparable to those of the V_K centre. One concludes immediately that the basic constituent of the H centre is also a Cl_2^- molecule ion. Looking now carefully at the finely spaced superhyperfine structure lines one finds again a 1:2:3:4:3:2:1 seven line pattern, indicating that the unpaired electron spin interacts further, but weakly, with two other equivalent Cl nuclei (all Cl isotope effects can again be resolved within this superhyperfine structure).

For the quantitive analysis spin Hamiltonian (1) is used, but the following term is added to account for the superhyperfine interaction with Cl nuclei 3 and 4 :

INTERSTITIAL CENTRES: OPTICAL AND MAGNETIC RESONANCE 183

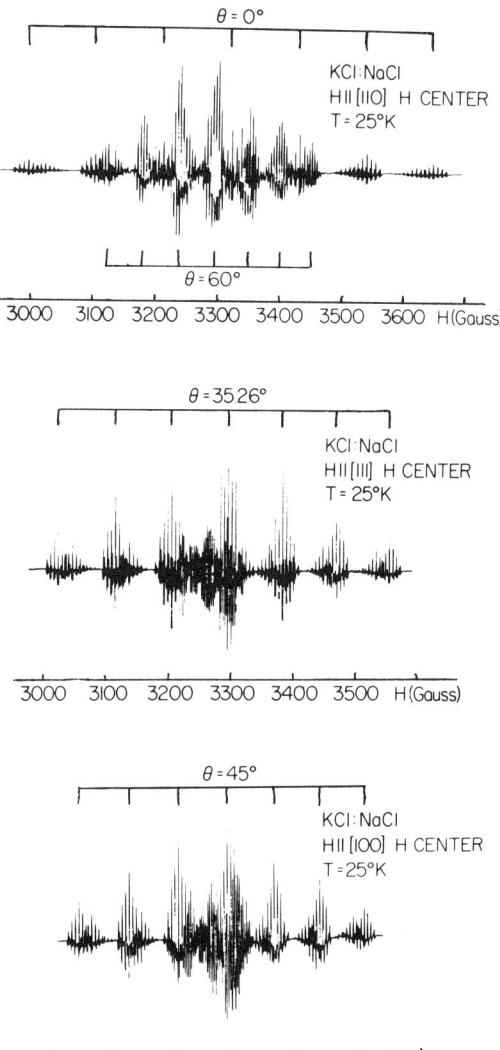

Fig. 7. ESR spectrum of the H centre in KCl for three orientations of the crystal.

$$a_\parallel S_z I'_z + a_\perp (S_x I'_x + S_y I'_y)$$

with $\vec{I}' = \vec{I}_3 + \vec{I}_4$

One finds [9][10]:

$$g_\parallel = 2.0018$$
$$g_\perp = 2.0224$$
$$A_\parallel = 108.6 \text{ gauss} \qquad a_\parallel = 7.4 \text{ gauss}$$
$$A_\perp = 16 \text{ gauss} \qquad a_\perp = 2.7 \text{ gauss}.$$

These parameters are very similar to those of the V_K centre indicating that one may consider the basic H centre entity to be a Cl_2^-. The $a_\parallel$ and $a_\perp$ parameters are quite small but their existence is important as we will see shortly.

There are various experimental indications pointing to the fact the H centre must be an interstitial centre and the model for the H centre that is derived from the ESR results is shown in Fig.8: it is a Cl_2^- molecule ion, exactly <110> oriented and occupying a single negative ion site. Furthermore, as a result of its crowded nature it possesses two weak molecular bonds with the two adjoining substitutional Cl^- ions 3 and 4.

<110> and <111> oriented H-centres

The Cl_2^- of the H centre is obviously crowded into the negative ion vacancy. Looking more closely into the various possibilities of squeezing a Cl_2^- into a negative ion vacancy one could be surprised that the Cl_2^- chooses the <110> orientation. Indeed, there is more room along a <111> direction. A theoretical calculation by Dienes, Hatcher and Smoluchowski[11] which did not consider the possibility of molecular bond formation, yielded indeed the <111> orientation as the most favourable one for an H centre.

It must be then that the two weak molecular bonds of the Cl_2^- with Cl^- ions 3 and 4 lock the H centre into the <110> orientation, at least in KCl and KBr where this symmetry has been observed. That this need not always be so is shown by the ESR and ENDOR experiments of Chu and Mieher[12] who determined that the F_2^- of the H centre in LiF lies along a <111> direction. In this case release of strain by orienting in the <111> direction obviously wins over the reduction in energy resulting from the two weak molecular bonds in the <110> direction. Thus in the alkali halides the H centre can have two

INTERSTITIAL CENTRES: OPTICAL AND MAGNETIC RESONANCE

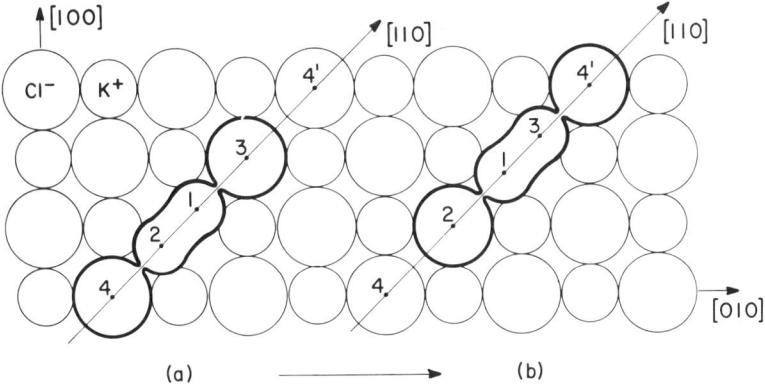

Fig. 8. Model for the H centre in KCl, and basic step in the diffusion or crowdion motion along <110>.

orientations depending on the crystal. Whether these two centre configurations can exist in the same crystal at different temperatures is a question which has not yet been investigated.

Electronic Structure and Optical Absorption

Ignoring the small molecular bonds with ions 3 and 4, the ESR results indicate that the H centre is basically a Cl_2^- molecule ion. The same question that was asked for the V_K centre can again be posed : How good is the Cl_2^- description, especially for the excited states. The answer is again : Quite good, at least qualitatively. Indeed, the optical absorption spectrum determined by Delbecq et al.[13] gives two absorption bands, a strong one at 336 nm and a much weaker one at 522 nm. These, as we did for analogous transitions of the V_K centre at 365 nm and 750 nm, are ascribed to, respectively the

$$^2\Sigma_u^+ \rightarrow {}^2\Sigma_g^+$$

and

$$^2\Sigma_u^+ \rightarrow {}^2\Pi_g$$

transitions derived from the $\sigma_g^2 \pi_u^4 \pi_g^4 \sigma_u^1$ ground configuration of Cl_2^-.

The fact that the H centre transitions are at higher energies compared to those of the V_K centre, is attributed to a spreading out of the Cl_2^- energy level scheme as a result of a slightly reduced internuclear Cl_2^- distance (see Fig. 3) caused by the squeezing of the Cl_2^- in a single vacancy. The g and A components of the H centre are in qualitative accord with this mechanism. However, no convincing quantitative analysis exists as yet for the H centre spin Hamiltonian parameters.

Thermal Reorientation and Diffusion

The strong $^2\Sigma_u^+ \rightarrow {}^2\Sigma_g^+$ transition should be σ polarised. This was confirmed by the measurements of Delbecq et al.[13] who used this property to obtain the H centre optical absorption spectrum. When the Cl_2^- of the H centre is excited to the anti-bonding $^2\Sigma_g^+$ state, it wants to dissociate into a Cl^- ion and a Cl atom. The Cl^- ion is likely to remain on the lattice site because of the coulomb attraction, but the Cl atom will take off as an interstitial, moving through the lattice and in the process, getting rid of its kinetic energy. The atom will eventually come to rest on a different lattice site, and form an H centre with a substitutional Cl^-. There is very likely no correlation between the orientations of the original H centre and the new H centre at rest. Thus if an H centre containing KCl crystal is excited at 4.2 K with [0$\bar{1}$1] polarised light in the 336 nm band, most of them will, and this is indeed experimentally verified[13], be pumped into the perpendicular [011] orientation. This results in a spatial anisotropy of the H centre orientations and thus in an optical anisotropy which is shown in Fig.9. This anisotropy persists indefinitely if the crystal is kept at 4.2 K, indicating that the H centres exhibit no motions and are frozen into their orientations.

To see whether the H centre, like the V_K centre, can reorient thermally, the optical anisotropy is followed while the sample is warming up. The optical anisotropy does indeed disappear and the disorientation temperature in KCl is found to be T_D = 10.9 K corresponding to an activation energy of 0.031 eV[13][14].
The kinetics of the H centre reorientation were studied by Bachmann and Känzig[10] using uniaxial stress. They found that similar to the V_K centre, the H centre in KCl reorients exclusively by "60° jumps" of the Cl_2^- among the six <110> orientations.

This looks all very similar to the V_K centre, but it is important to point out one fundamental difference : The nature of the H centre is such that after thermal reorientation the Cl_2^- is still at exactly the same lattice site, and the nuclei of the Cl_2^- in the new direction are identically the same as in the original Cl_2^-.

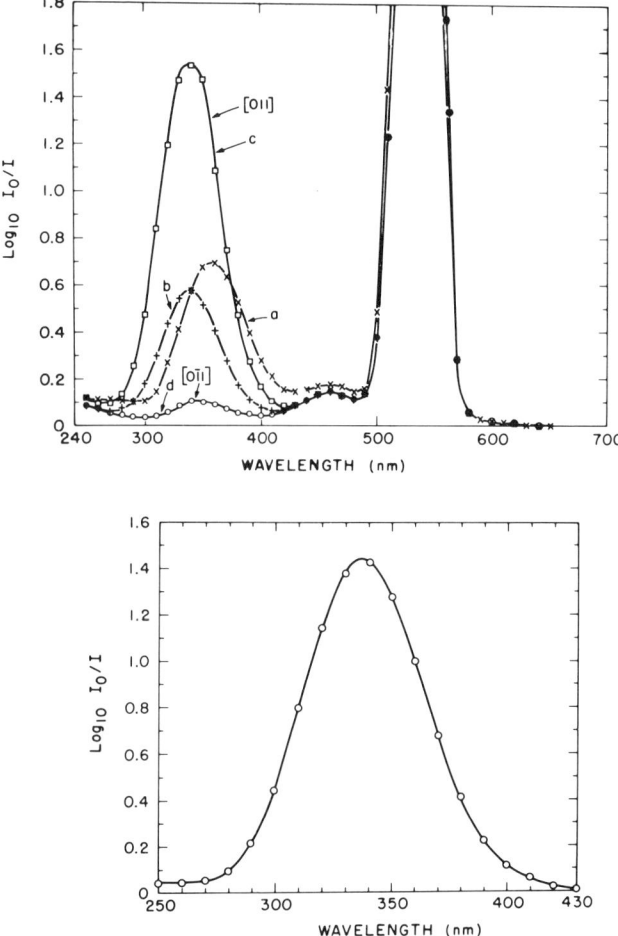

Fig. 9. Anisotropic optical absorption of the H centre in KCl. The anisotropic absorption (lower curve) is obtained by subtracting curve [0̄11] from [011] in the upper figure.

To put it still differently : When the H centre reorients thermally it does not diffuse through the lattice as is the case for the V_K centre.

Indeed one can warm KCl, containing H centres, to 35 K and no decay is observed although the reorientation rate of the Cl_2^- at this temperature is very high.

The H centre in KCl is observed to decay between 40 and 45 K, indicating that now the interstitial Cl atom must have a certain thermally induced mobility in the lattice. The basic diffusion step is indicated in Fig. 8 and the mechanism can be described as follows : With increasing temperature the vibrational amplitude of the

Cl_2^- increases and at a certain point three Cl nuclei,1-2-3 (or 1-2-4),are in such a configuration that the Cl_2^- breaks up and a Cl_2^- with nuclei 1 and 3 (or 2 and 4) is formed on the neighbouring site. One could also say that the interstitial Cl atom has jumped to a neighbouring site along a <110> direction. There is no doubt that the jump along a <110> direction is determined by the fact that the H centre has small molecular bonds with the two neighbouring substitutional Cl^- ions along <110>.

Two remarks are in order. First there is a similarity with the V_K centre reorientation process in the sense that the new Cl_2^- (in the same orientation though!) has only one Cl nucleus in common with the Cl_2^- in the previous configuration. Second, although a given jump is along a given <110> direction, the successive jumps are not confined to this particular <110> direction. Indeed, this interstitial jump takes place at and above 40 K where the reorientation rate of the basic Cl_2^- in the negative ion vacancy is high (~10^9 Hz). Thus when the interstitial Cl jumps to a neighbouring site, the new Cl_2^- starts immediately reorientating among the six <110> orientations and this is a much faster process than the jump to the next site, the diffusion process.

When the temperature of the KCl sample is raised above 40 K the diffusion rate of the interstitials increase exponentially and the interstitials move rapidly through the crystal. A substantial fraction will recombine with F centres and the perfect lattice is restored. One says that the F centre and the H centre are antimorph to one another.

However, the history of a diffusing H centre need not necessarily be that drastic and short. The interstitial may be trapped and converted into other centres in a very large number of ways depending on the purity of the crystal. In particular a large number of H-type centres may be formed by using specifically doped crystals.

The remainder of these notes will deal with several examples showing the varied ways in which interstitial halogen atoms, are stabilised by impurities. This gives rise to new centres, H-type centres , whose existence and properties vary from the merely amusing to the quite fascinating. A large number of impurities and defects stabilise interstitials but two types of impurities are of particular importance, namely foreign ions and foreign alkali ions. These will be discussed in the following sections.

THE <111> ORIENTED FX^- CENTRES IN KCl

KCl can be doped with small amounts of substitutional F^- ions

INTERSTITIAL CENTRES: OPTICAL AND MAGNETIC RESONANCE

by adding KF to the KCl melt. When such KCl:F⁻ crystals are irradiated at 77 K for a few hours with x or γ rays, one detects a strong ESR spectrum at 77 K[15][16]. The analysis show that it originates from an axially symmetric FCl⁻ molecule ion whose axis is oriented exactly along <111>. That it is an interstitial centre is supported by the observation that it is a neutral centre : Electrons released into the crystal (e.g. by optical excitation of F centres) do not influence its intensity to any extent. The positively charged V_K centres which are also present are wiped out by this treatment. The FCl⁻ is very likely produced by the following mechanism. At 77 K, far above the H centre decay temperature of 40 K, the interstitial Cl atom produced by the ionising radiation is quite mobile. When it encounters a substitutional F⁻ impurity, it sees a chance to relieve its strain by forming a FCl⁻ molecule, which is smaller than the Cl_2^- molecule of the H centre because the F⁻ ion is considerable smaller than the Cl⁻ ion.

The FCl⁻ centre has the expected <111> orientation for interstitials. Apparently no molecular bonds of sufficient strength are formed with the Cl⁻ ions along <110> to make this the low energy direction. The FCl⁻ interstitial centre has a much higher thermal stability than the H centre : It decays at 170 K . There is evidence that the centre decays by releasing an interstitial F atom leaving a substitutional Cl⁻ ion behind. Indeed, if the KCl:F⁻ crystal also contains Br⁻ (or I⁻) as an impurity it is observed that the <111> oriented FBr⁻ (or FI⁻) is formed when FCl⁻ decays thermally. This can be understood by the trapping of the mobile interstitial F atom by a substitutional Br⁻ ion impurity. However, one could also argue that there is a preferential association of F⁻ and Br⁻ in adjoining substitutional sites and that the FBr⁻ is formed by the trapping of a mobile interstitial Cl atom by this pair. This mechanism would imply that FCl⁻ dissociates by releasing an interstitial Cl atom. Thus the thermal decay mechanism of FCl⁻ is not settled with certainty. The configuration of the FX⁻ centres can be written as $\sigma_1^2 \pi_1^4 \pi_2^4 \sigma_2^1$ in analogy with the X_2^- centres. The strongly allowed σ polarised transition of FCl⁻ has been found at 305 nm and the FCl⁻ disorientation temperature is 45 K. At and above this temperature the FCl⁻ reorients among the eight <111> directions.

The stabilisation of an interstitial Cl atom by a substitutional F⁻ ion was ascribed to the small size of the F⁻ ion compared to the Cl⁻ ion. This suggest another way to trap a mobile interstitial Cl atom : Replace the alkali ion, K⁺ in KCl, by a smaller alkali ion, such as Na⁺ or Li⁺, and the interstitial may be stabilised in the vicinity of this ion. Such is indeed the case and the resulting trapped H centres are called H_A centres, where the A stands for Alkali.

The symmetry of an H_A centre depends on the alkali halide and the foreign alkali ion, and in fact several geometries have been observed. Some of these H_A centres have quite interesting physical properties and they will be discussed at some length.

THE $H_A(Na^+)$ CENTRE (V_1 CENTRE) IN KCl:Na^+

ESR and Model

The Na^+ ion in KCl can stabilise an H centre in its immediate vicinity and forms what is called the $H_A(Na^+)$ centre. Its production is straightforward : The KCl:Na^+ sample is irradiated with x or γ rays at 77 K where the interstitial Cl atoms are mobile, and these are quickly trapped by the Na^+ ions.

Before discussing its structure and properties it is of some interest to point out that the $H_A(Na^+)$ centre has a long history. It was one of the first known V-type centre and was called the V_1 centre when observed by optical absorption in KCl. Its electronic and geometric structure was a matter of speculation for almost two decades, until it became clear that it was an H centre next to a Na^+ impurity[17][18].

The detailed $H_A(Na^+)$ structure was elucidated by the discovery of its ESR spectrum which is only strongly observable in a narrow range of temperatures around 35 K. The spectrum is complex and we refer for its analysis to the original papers[17][18].

The $H_A(Na^+)$ model derived from it is depicted in Fig. 10. The Cl_2^- lies in a {100} plane next to a substitutional Na^+ impurity. The Cl_2^- internuclear axis makes a 5.7° angle with the <110> direction and the two nuclei are inequivalent. Comparison with Fig. 8 shows the great similarity with the H centre. As in the case for the H centre there is also a small superhyperfine interaction with nuclei 3 and 4 and these nuclei are also inequivalent. For comparison we give the ESR results :

$g_\parallel$ = 2.0018 $A_\parallel$ (1) = 109.1 gauss $A_\parallel$ (2) = 101.1 gauss
$g_\perp$ = 2.025 $A_\perp$ (1) = 14 gauss $A_\perp$ (2) = 14 gauss

$a_\parallel$ (3) = 13.7 gauss $a_\parallel$ (4) = 3.1 gauss
$a_\perp$ (3) = 4.6 gauss $a_\perp$ (4) = 1.4 gauss.

These are similar to the parameters of the H centre. The optical absorption measurements[17] yield a strong $^2\Sigma_u^+ \to {}^2\Sigma_g^+$ absorption at

357 nm and a much weaker $^2\Sigma_u^+ \to {}^2\Pi_g$ at 560 nm.

Thermal Reorientation Motions

The reorientation motions of the $H_A(Na^+)$ centre have been studied[17], first through of the optical anisotropy that can be produced at 4.2 K with polarised 357 nm light but also by a study of the effect of temperature on the ESR spectrum. The decay of the optical anisotropy as a function of temperature gives a single disorientation temperature T_D = 16.8 K. Although the kinetics of this motion have not yet been studied, the experimental data and comparison with other H-type centres (see further) strongly suggest that mechanism for thermal reorientation is the following.

In its unperturbed position, the Na^+ impurity and the negative ion vacancy, which the Cl_2^- occupies, define a particular <100> direction with which the Cl_2^- axis makes a 45° - 5.7° = 39.3° angle in a {100} plane. There are four {100} half-planes passing through this particular <100> direction and the Cl_2^- could be positioned in any one of them. The reorientation motion corresponding to T_D = 16.8 K is ascribed to a thermal activation of the Cl_2^- among those four equivalent orientations. These four orientations around a given <100> define a pyramid of C_{4v} symmetry and therefore this reorientation motion has been christened the Pyramidal Motion, P.M., for short.

Note that this P.M. is essentially the same as the reorientation motion of the H centre. In the latter there is no impurity to restrict the motion around a given axis, and the Cl_2^- jumps among all six <110> directions performing "60° jumps". For $H_A(Na^+)$ this motion is restricted to the four orientations around a given <100> and the "60° jumps" becomes "53.2° jumps" because of the 5.7° tilting away from <110>.

Motions, if sufficiently rapid (>10^6 Hz), are also visible in the ESR spectrum as a broadening, and sometimes subsequent at higher temperatures, motional narrowing of the lines. The $H_A(Na^+)$ centre exhibits such effects (broadening at 37 K, motional narrowing at 50 K) and a detailed analysis[17] has shown that these effects are caused by another motion than the P.M. This second motion is depicted schematically in Fig. 10 : Cl number 1, closest to the Na^+ ion and forming a Cl_2^- with nucleus number 2, can jump over to the neighbouring site and form a Cl_2^- with Cl^- ion number 3 while Cl number 2 now plays the role of number 3 before the jump. The Cl_2^- orientation changes in this jump by 2 x 5.7° = 11.4°. Thus in the second motion the movement of the interstitial is confined to those two equivalent positions, and this motion is therefore called the Restricted Interstitial Motion, R.I.M., for short. Comparison with Fig. 8 of the H centre shows the great similarity of this R.I.M. with the

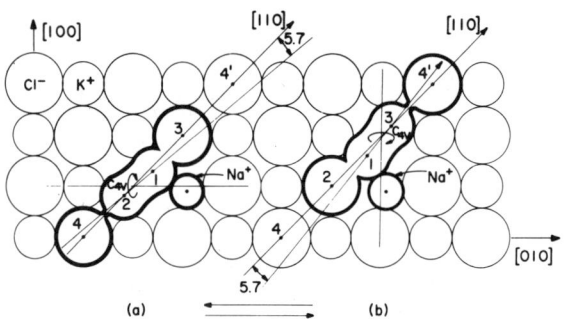

Fig. 10. The $H_A(Na^+)$ centre model in $KCl:Na^+$, and the two orientations involved in the Restricted Interstitial Motion. The C_{4v} axis of the Pyramidal Motion is also indicated.

diffusion motion of the H centre along <110>. However for the H centre there is no impurity to restrict the motion of the interstitial, and as a result it diffuses through the crystal.

The $H_A(Na^+)$ centre decays at 113 K. The interstitial Cl atom breaks away from the Na^+ impurity and moves as a highly mobile H centre through the lattice.

THE $H_A(Li^+)$ CENTRE IN $KCl:Li^+$

ESR and Model

If Na^+ stabilises an interstitial Cl atom in KCl, one expects the still smaller Li^+ to do the same. It turns out that it does. The mobile interstitial Cl atoms produced by x or γ irradiation at 77 K are stabilised by the Li^+ ion and $H_A(Li^+)$ centres are formed[19].

The $H_A(Li^+)$ properties are in several respects quite different from those of the $H_A(Na^+)$. First the ESR signal hardly shows any sign of saturation at 4.2 K indicating an anomalously short spin-lattice relaxation time. This has the result that the $H_A(Li^+)$ ESR signal is easily and strongly observed at 4.2 K, a feature which facilitates many experiments. The analysis of the ESR-spectrum at 4.2 K shows that the $H_A(Li^+)$ centre consists of a Cl_2^- molecule (with two slightly inequivalent Cl nuclei) whose internuclear axis lies in a {110} plane and makes a 26° angle with <100>. A three dimensional model is presented very schematically in Fig. 11(a) and Fig 12 gives a two dimensional model in a {110} plane. The ESR analysis also indicates that the Cl_2^- wave function extends slightly over the substitutional Cl^- ions 3 and 4 and this is indicated schematically

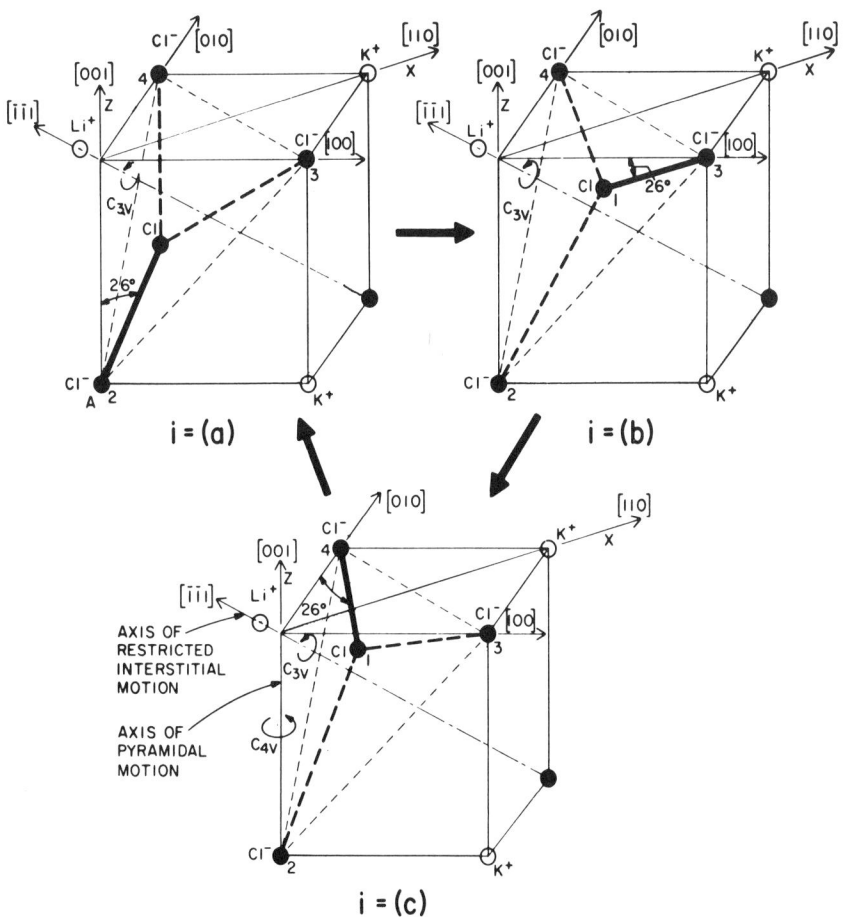

Fig. 11. (a) Very schematic three dimensional model of the $H_A(Li^+)$ centre in $KCl:Li^+$; (a), (b) and (c) present the three orientations involved in the Restricted Interstitial Motion (R.I.M.).

by the two heavy broken lines in Fig. 11. These two Cl⁻ ions play an important role in the reorientation motion as we shall shortly see.

Anisotropic optical absorption measurements [19] presented in Fig. 13 show that $H_A(Li^+)$ has a strong σ polarised $^2\Sigma_u^+ \rightarrow {}^2\Sigma_g^+$ transition at 354 nm and a much weaker transition at 618 nm ascribed to $^2\Sigma_u^+ \rightarrow {}^2\Pi_g$. Optical anisotropy produced at 4.2 K persists at this temperature and a warm up yields a single disorientation temperature at T_D = 23.5 K. At this temperarure all optical anisotropy decays.

This T_D is associated with the onset of a Pyramidal Motion of the Cl_2^- among the four {110} half-planes passing through a given <100> direction[20].

A study of the temperature dependence of the ESR lines[20] shows motional broadening above 29 K indicating the presence of a second motion which must be distinct from the P.M. Indeed, motional broading becomes observable only when the reorientation rate exceeds 10^6 Hz. The reorientation rate of the P.M. at 23.5 K is only 0.01 Hz and cannot possibly increase to 10^6 Hz at a temperature only about 6 K higher. At 75 K this second motion is so fast that a motionally averaged ESR spectrum is observed and from its analysis it was possible to deduce the symmetry of this second motion. It is shown schematically in Figs. 11(a),(b) and (c). The interstitial Cl number 1 breaks and reestablishes molecular bonds with the three substitutional Cl⁻ ions 2,3 and 4 that surround it and the motion takes place around <111>. This is again a R.I.M. very similar to the $H_A(Na^+)$ R.I.M. depicted in Fig. 10. Note that these R.I.M.'s are similar to the V_K centre reorientation process in the sense that the Cl_2^- in a new orientation has only one Cl in common with the Cl_2^- of the previous configuration.

This R.I.M. of $H_A(Li^+)$ has an interesting property. It is an Arrhenius type motion around 29 K and if this is extrapolated to a reorientation rate of ~ 0.01 Hz in order to obtain the disorientation temperature T_D of this R.I.M. one calculates that $T_D(R.I.M.) \approx 11$ K. However, experimentally [see Fig. 13(b)] there is no decay of optical anisotropy at 11 K or at any temperature between 4.2 K (where the optical anisotropy was produced) and 23 K. One concludes that the R.I.M. does not freeze in but goes over into a tunnelling motion at low temperature[20].

The tunnelling nature of the $H_A(Li^+)$ R.I.M. has been further investigated with uniaxial stress measurements at 4.2 K. In experiments in which uniaxial stress parallel to <110> was applied[21] it was verified that the stress lifts the degeneracy of the three R.I.M. tunnelling orientations. The three tunnelling orientations can be

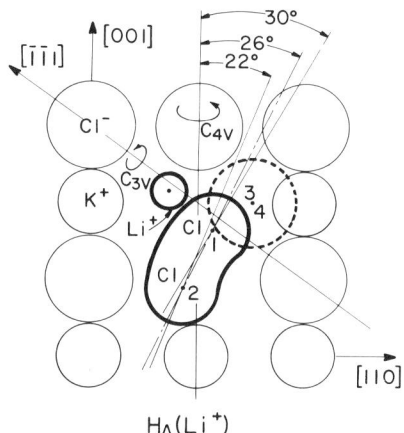

Fig. 12. Schematic two dimensional model of the $H_A(Li^+)$ centre in KCl, in a {110} plane.

observed seperately in the ESR spectrum because the $H_A(Li^+)$ spectrum is anisotropic. The lifting of the degeneracy is seen as a change in intensities (certain lines increase and others decrease) of lines corresponding to the three tunnelling orientations. The populations of these orientations are governed by the Boltzmann distribution.

The foregoing described a $\vec{\sigma} \parallel <110>$ experiment. When a $\vec{\sigma} \parallel <100>$ experiment was performed at 4.2 K something quite unexpected was observed[22]: The uniaxial stress changed the geometry of the centre in a continuous fashion and this effect was so strong that it completely overwhelmed any effect that the uniaxial stress could have had on the lifting of the three R.I.M. tunnelling orientations. The continuous change in geometry is indicated schematically in Fig. 14 : The stress pushes the Cl_2^- out of the {110} plane and, as the stress increases, it moves all the way to the {100} plane perpendicular to the stress. The ESR spectra show that the Cl_2^- describes an octant of a cone around <100> and that the 26° angel with <100> hardly changes.

At the highest stresses that the crystal tolerates at 4.2 K, the Cl_2^- is within a few degrees of the {100} plane. If the crystal is warmed up from 4.2 K while high uniaxial stress is applied it is observed that the Cl_2^- returns to the {110} plane, where it arrives at about 50 K.

This change of geometry is not yet completely understood, but this and the tunnelling nature of the $H_A(Li^+)$ R.I.M. illustrate that

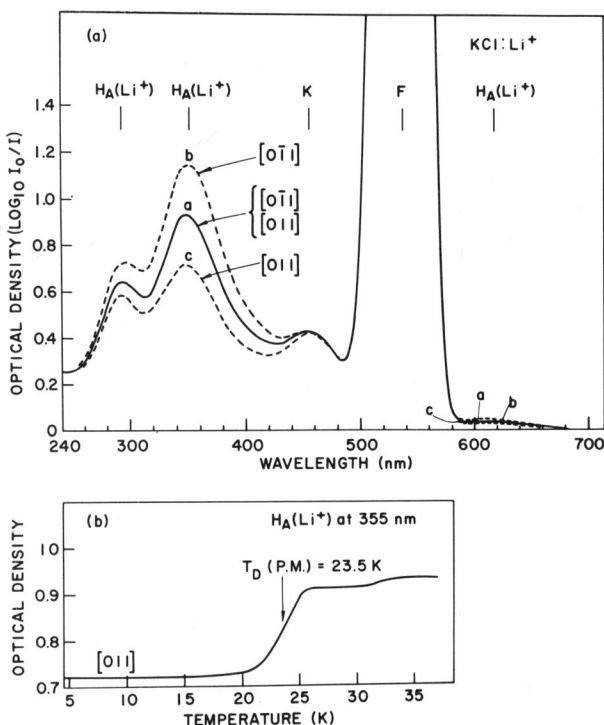

Fig. 13. (a) Anisotropic optical absorption of $H_A(Li^+)$ in $KCl:Li^+$; (b) Decay of optical anisotropy.

sometimes the non-primary centres may possess remarkable physical properties uniquely their own which are worthwhile investigating.

$H_A(Na^+)$ CENTRE IN $LiF:Na^+$

In the foregoing we presented examples showing that small impurity ions stabilise interstitials because the strain is reduced compared to the H centre. This may push one into thinking that e.g., alkali impurity ions larger than those of the host lattice cannot stabilise interstitials. The existence of the $H_A(Na^+)$ centre in $LiF:Na^+$ proves that such a conclusion is wrong.

When the ESR spectrum of this centre was first found by Känzig and Woodruff[9] it was interpreted as originating from the H centre. This interpretation was almost inevitable at the time because the spectrum had (as far as one could tell) exactly the same symmetry

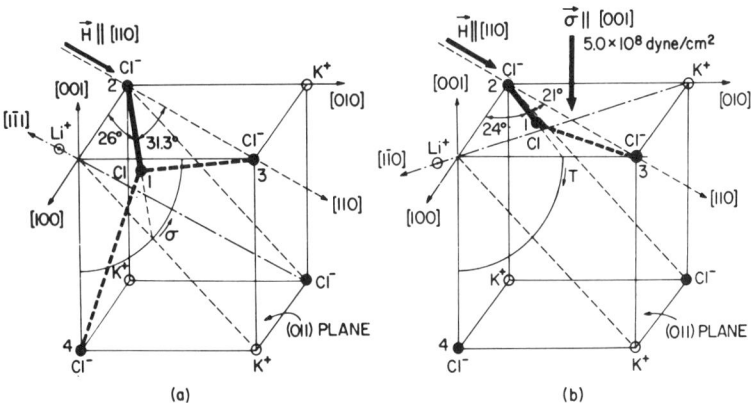

Fig.14. (a) Effect of $\vec{\sigma} \parallel [001]$ uniaxial stress at 4.2 K on the structure of $H_A(Li^+)$ and (b) the effect of increasing temperature on the centre under high uniaxial stress.

as the H centre spectra in KCl and KBr : An F_2^- molecule with equivalent F nuclei, exactly <110> oriented and exhibiting supplementary weak superhyperfine interactions with two other equivalent F nuclei along <110>. Furthermore, it was not realised that Na^+ is a common impurity in LiF crystals. It is only after (a) Dakks and Mieher[23] established through a study of the ENDOR spectrum that this centre showed the presence of a Na^+ impurity along a <100> direction perpendicular to the F_2^- <110> axis and, (b) Chu and Mieher[12] found the real H centre as being a <111> oriented F_2^-, that the situation was cleared up.

No optical absorption or reorientation studies have been performed as yet, but if the $H_A(Na^+)$ centre possesses a reorientation motion it must be a 90° jump of the F_2^- molecule confined to a {100} plane.

THE $H_A(Li^+)$ CENTRE IN NaF:Li^+

The foregoing examples of H_A centre geometries do not exhaust all possibilities. Indeed, Plant and Mieher[24] have found in NaF:Li^+ a $H_A(Li^+)$ centre which possesses no symmetry elements : The F_2^- on the negative ion site next to the Li^+ is neither oriented along a crystal axis, nor does it lie in any symmetry plane of the crystal. Uniaxial stress measurements by Schoemaker and Lagendijk have shown that this centre possesses complicated tunnelling motions at 4.2 K.

THE H_{AA} AND $H_{A'A}$ CENTRES

If one small impurity ion can stabilise an interstitial to higher temperatures, it would seem that two neighbouring small alkali ions should do even better. Experimentally this is confirmed. Interstitial Cl atoms are stabilised by pairs of Li^+ or pairs of Na^+ impurity ions which are nearest neighbours or next nearest neighbours of one another. These centres have been called, respectively, H_{AA} and $H_{A'A}$ centres[20]. Their structure as derived from the ESR spectra are shown in Figs. 15, 16 and 17. The Cl_2^- of the $H_{AA}(Na^+)$ or $H_{AA}(Li^+)$ centre is exactly <110> oriented and is parallel to the Na^+-Na^+ or Li^+-Li^+ axis of two nearest neighbour Na^+ or Li^+ ions. In the $H_{A'A}(Na^+)$ centre the Cl_2^- is tipped 11.5° away from the <100> direction defined by the two Na^+ ions which are next nearest neighbours of one another. The tipping for the $H_{A'A}(Na^+)$ centre is in the {100} plane, and this makes it similar to the $H_A(Na^+)$ centre. In the $H_{A'A}(Li^+)$ centre the Cl_2^- is tipped by 18.5° way from <100> in the {110} plane and this makes it similar to the $H_A(Li^+)$ centre.

The H_{AA} centres are rigid centres, i.e., there are no other equivalent orientations of the Cl_2^- with respect to the two nearest neighbour alkali ions, than the orientation shown in Fig. 15.

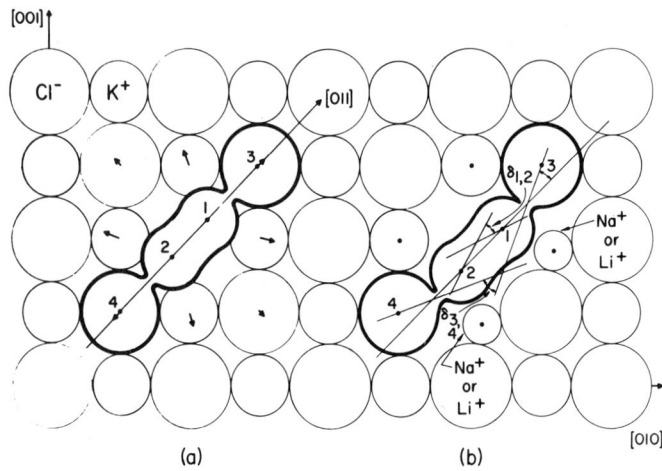

Fig. 15. (a) H centre; (b) $H_{AA}(Na^+)$ or $H_{AA}(Li^+)$ centre in KCl.

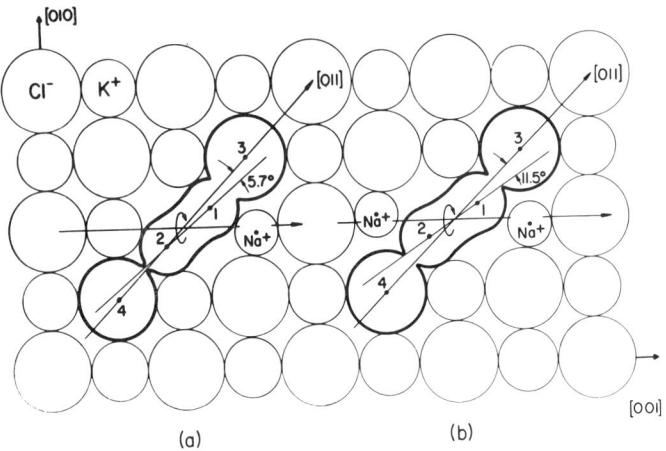

Fig. 16. (a) $H_A(Na^+)$ centre; (b) $H_{A'A}(Na^+)$ centre in KCl.

The $H_{A'A}$ centres on the other hand do exhibit a Pyramidal Motion around <100> at sufficiently high temperature. This P.M. is among four {100} half-planes for $H_{A'A}(Na^+)$ and among four {110} half-planes for $H_{A'A}(Li^+)$. The direct experimental observation of these Pyramidal Motions via the ESR spectra was important in establishing the P.M.'s for the $H_A(Li^+)$ and $H_A(Na^+)$ centres.

It should be clear by now from the foregoing examples that interstitials are readily stabilised by suitable impurities or agregates of impurities. In the latter case the impurities need not be of the same kind. We cite only the $H_A(Li^+)$ type $BrCl^-$ centre formed by the trapping of a mobile interstitial Cl atom by a (Br^-, Li^+) agregate and the $H_A(Na^+)$ $BrCl_2^{--}$ centre formed by the stabilisation of an interstitial Cl by a (Br^-, Na^+) pair[25]. The impurities need not be foreign alkali or halogen ions. An other important class of impurities are the divalent cations (Ca^{++}, Ba^{++}, Mg^{++}, Pb^{++} etc...). It is well known, mainly from optical absorption measurements[1], that divalent ions with their associated positive ion vacancies can trap interstitials. One could call these H_D centres. An early ESR investigation by Hayes and Nichols[26] has essentially substantiated this. However, a large amount of unpublished ESR work by Schoemaker on H_D centres has shown that the situation is much more complex than the work of Hayes and Nichols suggests. The complexity of the spectra and the fact that the interstitial can be trapped by the cation or the vacancy or both in different configurations make an interpretation of the data very difficult.

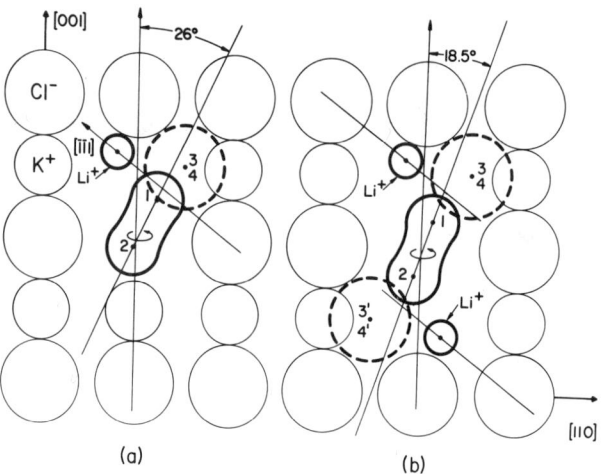

Fig. 17. (a) $H_A(Li^+)$ centre; (b) $H_{A'A}(Li^+)$ centre in KCl.

DIINTERSTITIAL CENTRES

In the foregoing sections the trapping of interstitials by substitutional impurities was amply demonstrated. There is also strong evidence now that interstitials may be trapped by other interstitials. An H centre in particular can be stabilised by another H centre. If the association is intimate a diinterstitial neutral Cl_2 molecule is formed. Because this is not paramagnetic one cannot use ESR techniques to verify its structure. The existence of this diinterstitial centre, called H' centre, is based on careful optical absorption measurements. The H' band is situated at about 245 nm in KCl and irradiation studies show that the heigth of this H' band increases with the square of the H band. Actually, the situation is more complex because the H' band seems to have more than one component at low temperature. It is possible that several di-H centre configurations are stable at low temperatures. For a discussion we refer to Itoh[1].

Quite recently, paramagnetic diinterstitial centres have been observed. Plant[27] observed one in NaF:Li^+ crystals. The ESR spectrum has the characteristics of a F_3^{--} molecule ion and presumably the centre is formed by the stabilisation of an interstitial F atom and a interstitial F^- ion next to a Li^+ impurity. The two interstitials then combine to form a F_3^- with a substitutional F^- ion.

A diinterstitial $BrCl_2^{--}$ centre of a similar nature was recently observed with ESR in KCl by Schoemaker[28]. The KCl crystal contained both Br^- and Na^+ impurities and the $BrCl_2^{--}$ is thought to be formed as a result of the trapping of both an interstitial Cl atom and Cl^- ion by a substitutional (Br^-, Na^+) pair. Other diinterstitial centres will undoubtedly be found in the future.

V_K and H centres are not unique to the alkali halides. They have by now been observed and analysed in a large number of other ionic materials such as the ammonium halides, the alkaline earth halides etc... The latter in particular have been studied thouroughly by Hayes and his coworkers at Oxford[29]. The centres also manifest themselves primarily as diatomic halogen molecule ions and many of the concepts presented in these notes may be applied. On the other hand because the crystal structure etc... is different sometimes unique differences with the alkali halides are observed.

Finally, neutron irradiated crystals are being studied more frequently recently[30]. Since neutrons presumably also effect the alkali sublattice this may give rise to centres whose existence and structure may not fit into the viewpoints presented in these notes.

References

1. N. ITOH, Crystal Lattice Defects $\underline{3}$, 115 (1972)
2. M.N. KABLER, in Point Defects in Solids edited by J.H. CRAWFORD and L.M. SLIFKIN (Plenum Press, N.Y.,1972)
3. T.G. CASTNER and W. KÄNZIG, J.Phys.Chem.Solids $\underline{3}$, 178 (1957); T.O. WOODRUFF and W. KÄNZIG, ibid, $\underline{5}$, 268 (1958)
4. D. SCHOEMAKER, Phys.Rev. $\underline{7B}$, $\overline{786}$ (1973)
5. T.L. GILBERT and A.C. WAHL, J.Chem.Phys. $\underline{55}$, 5247 (1971)
6. A. LAGENDIJK and D. SCHOEMAKER, Phys.Rev. $\underline{11B}$, 4030 (1975)
7. C.J. DELBECQ, B. SMALLER and P.H. YUSTER, Phys.Rev. $\underline{111}$, 1235 (1958); C.J. DELBECQ, W. HAYES and P.H. YUSTER, Phys.Rev. $\underline{121}$, 1043 (1961)
8. F.J. KELLER and R.B. MURRAY, Phys.Rev.Letters, $\underline{15}$, 198 (1965); F.J. KELLER, R.B. MURRAY, M.M. ABRAHAM and R.A. WEEKS, Phys.Rev. $\underline{154}$, 812 (1967)
9. W. KÄNZIG and T.O. WOODRUFF, J.Phys.Chem.Solids $\underline{9}$, 70 (1958)
10. D. SCHOEMAKER, Phys.Rev. $\underline{3B}$, 3516 (1971)
11. G.J. DIENES, R.D. HATCHER and R. SMOLUCHOWSKI, Phys.Rev. $\underline{157}$, 692 (1967)
12. Y.H. CHU and R.L. MIEHER, Phys.Rev. $\underline{188}$, 1311 (1969)
13. C.J. DELBECQ, J.L. KOLOPUS, E.L. YASAITIS and P.H. YUSTER, Phys. Rev. $\underline{154}$, 866 (1967)
14. K. BACHMANN and W. KÄNZIG, Phys.Kondensierten Materie $\underline{7}$, 284 (1968)
15. J.W. WILKINS and J.R. GABRIEL, Phys.Rev. $\underline{132}$, 1950 (1963)
16. D. SCHOEMAKER, Phys.Rev. $\underline{149}$, 693 (1966)
17. C.J. DELBECQ, E. HUTCHINSON, D. SCHOEMAKER, E.L. YASAITIS and P.H. YUSTER, Phys.Rev. $\underline{187}$, 1103 (1969)
18. F.W. PATTEN and F.J. KELLER, Phys.Rev. $\underline{187}$, 1120 (1969)
19. D. SCHOEMAKER and J.L. KOLOPUS, Phys.Rev. $\underline{2B}$, 1148 (1970)
20. D. SCHOEMAKER and E.L. YASAITIS, Phys.Rev. $\underline{5B}$, 4970 (1972)
21. D. SCHOEMAKER, Bull.Amer.Phys.Soc. $\underline{18}$, 305 (1973)
22. D. SCHOEMAKER, Phys.Rev. $\underline{9B}$, 1804 (1974)
23. M.L. DAKKS and R.L. MIEHER, Phys.Rev. $\underline{187}$, 1067 (1969)
24. W. PLANT and R.L. MIEHER, Phys.Rev. $\underline{B7}$, 4793 (1973)
25. D. SCHOEMAKER and C.T. SHIRKEY, Phys.Rev. $\underline{B6}$, 1562 (1972)
26. W. HAYES and G.M. NICHOLS, Phys.Rev. $\underline{117}$, 993 (1960)
27. W. PLANT, Solid State Comm. $\underline{11}$, 1219 (1972)
28. D. SCHOEMAKER, Proc.Int.Conf. on Color Centers in Ionic Crystals, Sendai, Japan, abstract 197 (1974)
29. "Crystals with the Fluorite Structure", W. HAYES, editor, (Oxford University Press, 1975)
30. D. RADOUX, Chem. Solids $\underline{36}$, 359 (1975); Y. KAZUMATA, J. Phys. Soc. Japan $\underline{35}$, 1442 (1973)

EPR STUDIES OF LATTICE DEFECTS IN SEMICONDUCTORS

G. D. Watkins

Department of Physics
Lehigh University
Bethlehem, Pa.   18015, USA

I.   INTRODUCTION

Lattice defects have been studied by EPR more thoroughly in silicon than in any other semiconductor material. As a result, these lectures will tend to concentrate in large measure on the results found in this material. No discussion will be given specifically to the other elemental materials, although some EPR work has been reported in germanium,[1] and quite a bit has been reported in diamond.[2] The state of understanding in these materials is still relatively poor and relies heavily upon the insight gained from the silicon studies.

No EPR studies of lattice defects in III-V materials exist at present. However, recent successful inroads have begun to be made into the II-VI materials. We will also review selected results from these studies.

In these lectures, "lattice defects" refers to the <u>intrinsic</u> defects, i.e. lattice vacancies or interstitial host atoms, and their interaction with other defects. In all the studies described, lattice defects are introduced by high energy particle irradiation (electrons, neutrons, etc.).

II.   SILICON

A.   Low Temperature Irradiation

Consider the following experiment:[3] We take p-type high

purity vacuum floating zone silicon (aluminum-doped $\sim 10^{16}/cm^3$) and irradiate it with 1.5 MeV electrons to a dose of $\sim 10^{17}$ el/cm^2. The irradiation is performed at liquid hydrogen temperature, 20.4K.

1. <u>Interstitial aluminum</u>.[4,5] Figure 1 shows the spectrum observed at 20.4K immediately after the irradiation. Most prominent are six equally intense lines. These have been identified as arising from a single <u>interstitial</u> aluminum atom, the 2(I+1)=6 hyperfine lines reflecting the 100% abundant ^{27}Al isotope, I=5/2. Its interstitial configuration has been confirmed by study of the satellite structure on each line (see blowup of one of the lines in Fig. 1): These satellites arise from hyperfine interaction with 4.7% abundant ^{29}Si(I=1/2) at sites surrounding the defect. Angular dependence study[4,5] shows that the structure is consistent with the neighboring silicon shells around the interstitial site but not the substitutional one. ENDOR studies[5] have revealed additional hyperfine interactions further confirming the interstitial assignment.

The aluminum hyperfine interaction is large (440.6 x 10^{-4}cm^{-1}) and isotropic. The g-tensor is also isotropic (2.0019 $\pm$ .003) and very close to the free electron value (2.0023). This tells us that we are dealing with a single electron in an atomic S-state. We deduce, therefore, that the EPR signal arises from Al^{++}, with configuration (1s^2 2s^2 2p^6 3s^1).

In this straightforward way, we have determined that interstitial aluminum atoms have been created by the irradiation. They are double donors and are observed in their doubly ionized state (Al$_i^{++}$) in this p-type material.

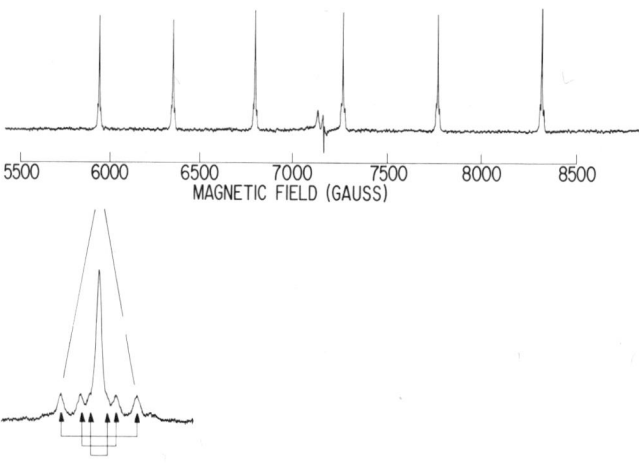

Fig. 1. Spectrum in silicon (Al$\sim 10^{16}$cm^{-3}) at 20.4K, ν_o=20 GHz, after 20.4K irradiation.

LATTICE DEFECTS IN SEMICONDUCTORS

The production of interstitial aluminum at this low temperature is a remarkable and unanticipated result. We will return to consider possible mechanisms for this later. In the meantime, let us look more closely for other spectra.

2. <u>Lattice vacancies</u>.[4,6,7] In Fig. 1, weak structure near 7200 gauss can also be seen. This can be greatly enhanced by shining light on the sample. Two separate spectra are revealed and are shown with expanded magnetic field scale in Figs. 2a and 2b.

The spectrum in Fig. 2a is identified as arising from the isolated lattice vacancy in its singly positive charge state (V^+).[4,6] The model is shown in the inset, the unpaired electron spread equally over the four silicon neighbors surrounding the vacancy. This identification comes primarily from the ^{29}Si hyperfine satellites (a' and a" for component a, b' and b" for b) which, in their angular variation, reflect the symmetry of these four neighboring sites and, in their magnitude, account for most (~65%)

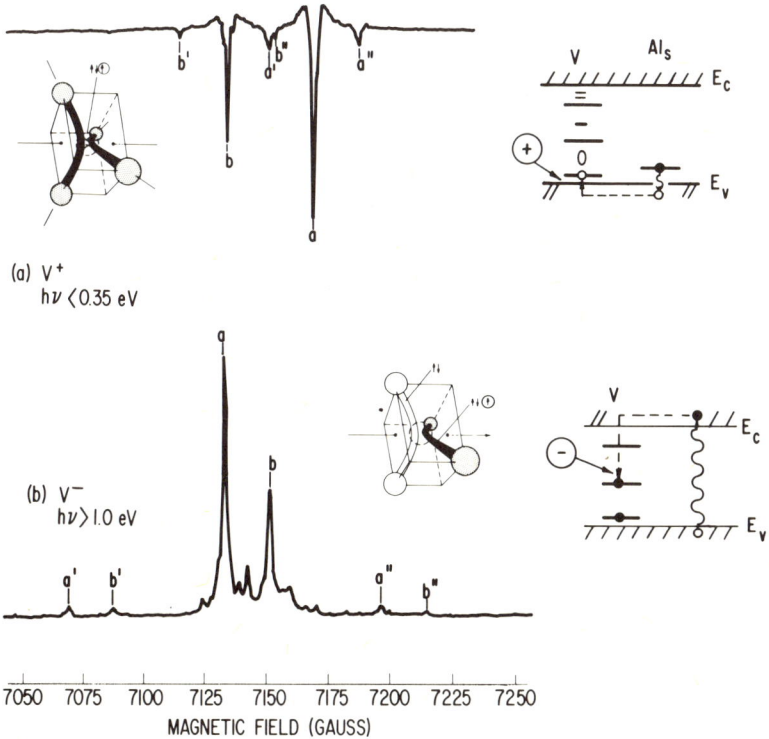

Fig. 2. Vacancy spectra, $H \parallel \langle 100 \rangle$, ν_o=20 GHz. (a) V^+ at 4.2K; (b) V^- at 20.4K.

of the wavefunction. A tetragonal distortion, illustrated as a pairing by twos in the figure, is detected as a small asymmetry of the g-tensor ($\sim 1/2\%$) as well as in slight tilting of the ^{29}Si hyperfine axes. This spectrum is optimized by light $h\nu < 0.35$ eV. Such light, in this p-type material, produces free holes which can be trapped by the neutral vacancy to produce V^+. This is illustrated in the inset of the figure.

Band gap light, on the other hand, can produce free electrons which, when trapped by the neutral vacancy, can give V^-. This is illustrated in Fig. 2b along with the spectrum identified with V^-.[6,7] In this identification the ^{29}Si satellites again serve as the main diagnostic element. In this case, the vacancy takes on the interesting configuration as shown, in which the unpaired electron sloshes to one side and is spread between only two atoms.

In the insets of Fig. 2, an additional double minus state is also indicated. This is deduced from studies in n-type material where light also is required to generate the EPR charge states.[7] The vacancy, therefore, can assume four charge states in the gap. In semiconductor language, it can act as a single (V^-) or double ($V^=$) acceptor and as a single donor (V^+).

The spectra of Figs. 1 and 2 are the dominant EPR spectra observed after such a low temperature irradiation.

B. The Electronic Structure of the Vacancy

The EPR results have suggested a very simple linear combination of atomic orbital-molecular orbital (LCAO-MO) model[4,7] for the various charge states of the vacancy. This is illustrated in Fig. 3. Here the atomic orbitals are the "broken bonds" (a,b,c,d) of the four atoms surrounding the vacancy. In the symmetry T_d of the undistorted vacancy a singlet (a_1) and triplet (t_2) set of molecular orbitals are formed. For V^+ (Fig. 3a), two electrons go into the a_1 orbital, paired off, and the third goes into the t_2. Because of the degeneracy, a tetragonal Jahn-Teller distortion results, as shown, lowering the symmetry to D_{2d}. The resulting orbital (b_2) is spread equally over the four atoms as seen in the ^{29}Si hyperfine interactions for the V^+ spectrum. The axial $\langle 100 \rangle$ anisotropy in the g-tensor reveals the tetragonal distortion.

In forming V^o, Fig. 3b, the next electron goes into b_2, paired off, further enhancing the tetragonal Jahn-Teller distortion. In this state the defect is diamagnetic and no EPR is observed.

In forming V^-, Fig. 3c, the added electron goes into the degenerate e orbital and a further Jahn-Teller distortion occurs.

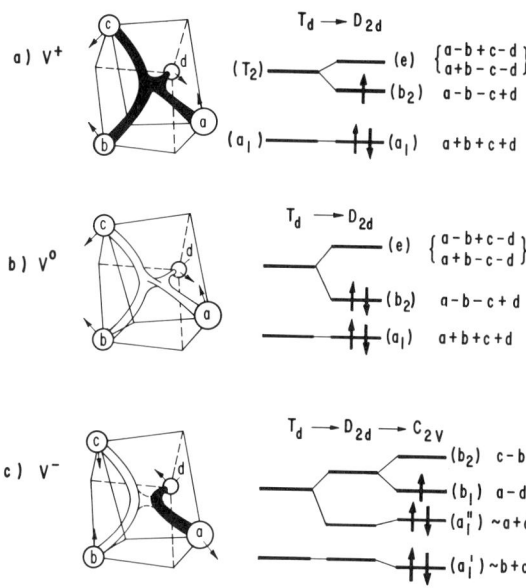

Fig. 3. Simple LCAO-MO treatment for the lattice vacancy.

The distortion is of b_2 symmetry (in D_{2d}) with atoms b and c pulling together and a and d separating slightly. The unpaired electron is now localized on only two of the four atoms, consistent with the observation in the V^- spectrum.

We conclude the following: The concept of a "defect molecule"[8] is valid. Approximately 65% of the wavefunction is accounted for as distributed between the "dangling bonds" on the atoms adjacent to the vacancy.[9] The dangling bond character is evidenced from the ^{29}Si hyperfine interactions which reflect both the 3s (from the isotropic part) and the $3p_{(111)}$ (from the anisotropic part) character expected for a hybrid 3s-3p orbital pointing into the vacancy from each atom. The g-values can also be shown to be consistent with those expected for broken silicon bonds.[9]

The success of these one-electron models implies that the electron-electron interactions that tend to favor parallel spin coupling (Hund's rules for atoms, etc.) are small. We are, in effect, in the strong crystal field regime. We fill each level before going to the next. When degeneracy occurs, a Jahn-Teller distortion results which imposes a new crystal field, decoupling the electrons again.

C. Vacancy Mobility

In p-type material, the vacancy disappears upon annealing at 150-180K.[6] As the vacancy disappears, new EPR centers emerge that are identified as vacancies trapped by impurities and other defects. The annealing is, therefore, the result of long-range migration of the vacancy. In p-type material the stable charge state of the vacancy is neutral (V^o), see Fig. 2. The measured activation energy for the annealing, 0.33 ± 0.03 eV,[6] is, therefore, identified as the activation energy of migration for V^o.

In n-type material, the vacancy disappears at 70-80K, again with the emergence of trapped vacancy spectra.[7] The activation energy 0.18 ± 0.02 eV for this process is identified with the migration energy for $V^=$, the stable charge state in n-type material (see Fig. 2).

The activation barrier for migration, therefore, depends strongly upon the charge state of the vacancy. This has been confirmed by the observation of enhanced low temperature annealing in p-type material in the presence of light[10] or electrical minority carrier injection[11] which allows the formation of metastable negative charge states for the vacancy.

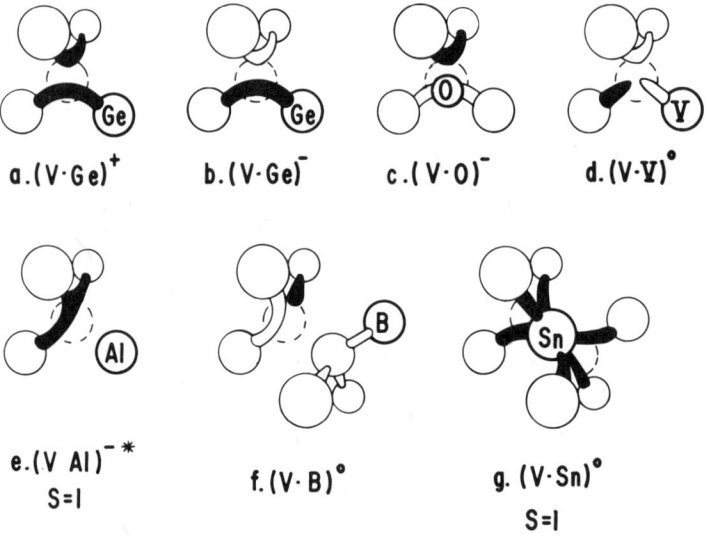

Fig. 4. Vacancy-impurity pairs identified in silicon.

D. Vacancy-Impurity Interactions

A large assortment of vacancy-impurity pairs emerge when the vacancy anneals. The principal ones that have been identified by EPR are illustrated in Fig. 4. In Fig. 4a and 4b the vacancy is shown trapped next to substitutional germanium. Both the Ge·V^+ and Ge·V^- charge states have been studied.[3] The distribution of the wavefunction is altered slightly between the four neighbors from that for V^+ and V^- (Fig. 2), but otherwise, the electronic structure, Jahn-Teller distortions, etc. are very similar. The presence of germanium is evidenced by the (2I+1)=10 hyperfine lines for ^{73}Ge (I=9/2, 7.6% abundant).

The vacancy-oxygen pair results from the trapping of a vacancy by interstitial oxygen, a common impurity in silicon. Shown (Fig.4c) is the stable configuration for $(V \cdot O)^-$.[12,13] The excited S=1 state of $(V \cdot O)^0$ has also been studied[14,15] (generated by light) as have two metastable S=1/2 configurations[4,7] which are low-temperature precursors to the formation of the final configuration shown in the figure. Vacancies trapped by the group V atoms P,[9] As,[16] Sb[16] take on the configuration shown in Fig. 4d. A photo-excited S=1 state of the aluminum-vacancy pair has been observed,[17] Fig. 4e, as has the S=1/2 state of a boron-vacancy pair,[18] Fig. 4f. The boron atom is deduced to reside at the next nearest neighbor site, as shown. The tin-vacancy pair,[19] Fig. 4g, takes on the interesting configuration in which the tin atom resides halfway between two atom sites and bonds partially with all six resulting dangling bonds. For it the ground state is S=1.

For all of these defects, simple one-electron molecular orbital models similar to those of Fig. 3 provide an adequate description of the structure. For most, Jahn-Teller distortions occur as evidenced in the figure by bond reformation by atom pairs which, in turn, produce minimum multiplicity for the ground state (i.e., S=1/2 or 0), as for the isolated vacancy. The only exception is the tin-vacancy pair. (The S=1 state for the aluminum-vacancy pair is an excited state and evidence has been cited that the ground state is S=0 with a Jahn-Teller distortion similar to that for the group V atom-vacancy pair.[17]) In the figure the localization of the unpaired spin is indicated in black. Hyperfine interactions and g-shifts again are consistent with the atomic dangling 3s-3p hybrid bond character for the wavefunction, with ~65% localized in each case on the near atoms to the vacancy.

In Fig. 5 the stability of the various vacancy and vacancy-impurity pairs is schematically indicated as would be observed in ~15 min isochronal annealing studies.

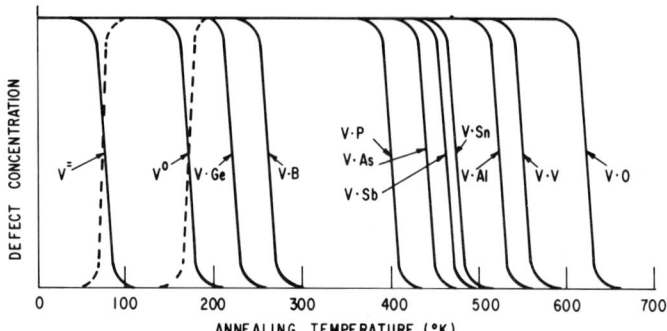

Fig. 5. Schematic of vacancy and vacancy-impurity pair annealing stages.

E. Vacancy Aggregates

The simplest vacancy aggregate is the divacancy shown in Fig. 6. This has been studied extensively by EPR[20-22] and ENDOR[23] in both its single positive and single negative S=1/2 charge states. A Jahn-Teller distortion occurs in which the bonds reform by pairs, as shown, and the unpaired electron is spread equally between the dangling bonds on the two opposite ends of the defect. The neutral state has not been seen by EPR and is deduced to be S=0, consistent again with molecular orbital arguments similar to that for the other vacancy-associated centers.

There are several S=1 EPR centers, particularly those observed in neutron irradiation experiments, that are believed to be associated with higher aggregates of vacancies. Their symmetries suggest chain-like arrays in a {110} plane as if simply building upon and continuing the divacancy as shown in Fig. 6. Assuming that the two electrons are located one at each end, the expected fine structure splitting, D_1, can be calculated by averaging the dipole-dipole interaction between the spins over the two-electron wavefunction as deduced from typical [29]Si hyperfine interactions.

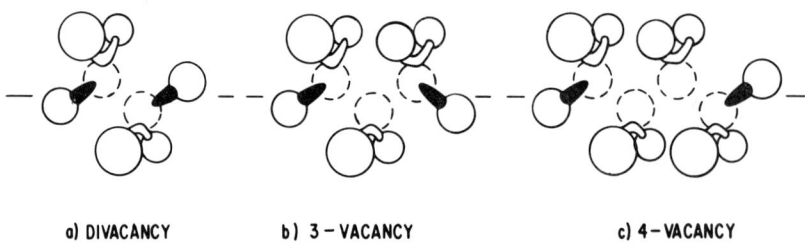

a) DIVACANCY   b) 3-VACANCY   c) 4-VACANCY

Fig. 6. Models for simple vacancy aggregates.

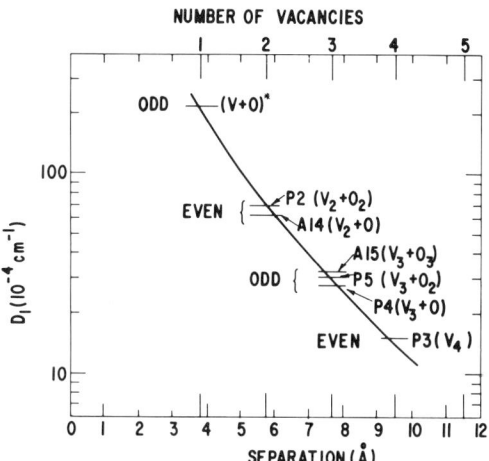

Fig. 7. Calculated and experimental values of D_1 vs separation of the end silicon atoms for $\{110\}$ vacancy chains.

The result[24] for 1-,2-,3-, and 4-vacancy chains is given in Fig. 7. The experimental values for several of the S=1 EPR centers are also included. The centers can be further classified as to whether they have even or odd numbers of vacancies from their EPR symmetries, as shown. In this way it has been deduced that the spectra labeled P2,[25] and A14[26] involve two vacancies, A15,[26] P5,[25] and P4[25] involve three, and P3[25] involves four.

The presence of oxygen has been found necessary to form all except P3 and an assignment has been suggested[24,26] as shown in the figure with one, two, three, etc. of the pair-bonds bridged by oxygen, as in the simple oxygen-vacancy pair, Fig. 3c. The P3 center has been identified[24] as the 4-vacancy center with no oxygen present.

Assigning S=1/2 centers to aggregate centers is more difficult but some assignments are beginning to be made. A center labeled P1[27] has recently been assigned to a non-planar pentavacancy cluster[28] and two other centers,[29] A4 and A3, tentatively to a trivacancy and tetravacancy cluster, respectively.[30] These centers are not believed to contain oxygen. As experience is gained in these identifications and cross checks can be made, the prospects in the future are good that the hierarchy of simple vacancy aggregates will be sorted out.

F. Interstitial Defects

We noted earlier that interstitial aluminum ions are produced by low temperature irradiation. Interstitial boron[31] has also been identified under similar irradiation conditions. Annealing studies reveal further clues about interstitials, as shown schematically in Fig. 8. The loss of interstitial Al_i^{++} is accompanied by the growth of an EPR spectrum identified as $(Al_i^{++}) \cdot (Al_s^-)$ pairs.[4] These are presumably formed when the Al_i^{++} interstitials diffuse through the lattice and pair with substitutional Al_s^- atoms. In irradiated gallium-doped material, $(Ga_i^{++}) \cdot (Ga_s^-)$ pairs have also been observed to emerge in the same temperature region, confirming that gallium interstitials must also have been formed.[4,32] $(C_i) \cdot (C_s)$ pairs[33] have also been detected. They emerge at $\sim 300K$[34] presumably when interstitial carbon, produced by the original damage event, migrates and pairs with substitutional carbon.

We deduce, therefore, that interstitial carbon and group III atoms are formed by irradiation at low temperature. No spectrum has been observed that can be unambiguously identified with the interstitial silicon atom.

G. The Damage Mechanism

The dominant defects identified after a low temperature irradiation appear to be isolated lattice vacancies and <u>interstitial impurities</u>. In p-type material, the interstitial group III atom concentration appears to be roughly equal to the vacancy concentration. The model suggested for this[4] is shown in Fig. 9. Here the interstitial silicon atom produced in the primary event is mobile and migrates until it is trapped by a substitutional group III (or carbon) impurity. In the trapped state, the silicon atom replaces the impurity, ejecting it into the interstitial site. The mechanism of its migration at these low temperatures (as low as 4.2K from EPR studies,[4] 1.6K from electrical measurement studies[35])

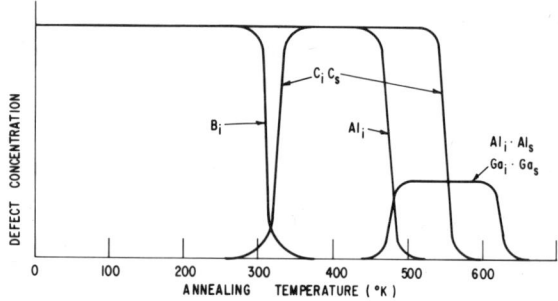

Fig. 8. Schematic of interstitial impurity annealing stages.

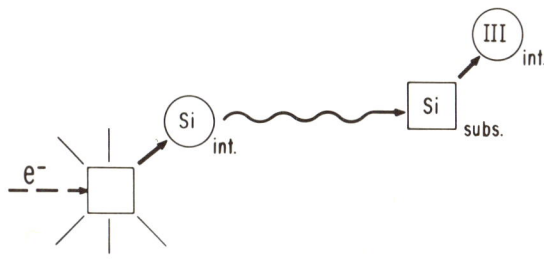

Fig. 9. Model for low temperature damage.

is not understood, but it is believed that the ionization accompanying irradiation may be playing a role.[36] Mechanisms under current consideration involve electron-hole pair recombination at the defect or simply lattice position alternations accompanying charge state change.

E. Summary

The structure of lattice vacancies and their interactions with simple impurities are fairly well understood. Divacancies have been identified and small vacancy aggregates are beginning to be sorted out. For most of these, the concept of a defect molecule made up of dangling silicon bonds is valid. Jahn-Teller effects are important.

Interstitials are less well understood. Interstitial silicon seems to be mobile in p-type material at very low temperatures under the conditions of irradiation. The mechanisms are not understood but it is believed that ionization is playing a role. There is indirect evidence that interstitial silicon may be less mobile in n-type material (annealing ~140K)[10] but this has not been definitely established.

Many problems remain. There are numerous EPR centers that have been observed but not identified.[4,29,37] Irradiations at temperatures >100K, where both vacancies and interstitials are mobile, have been less studied and appear to give complex and often different defect structures.[38] Ionization is believed to play an important role in a defect's stability, mobility and its reactions with other defects. However, little is understood, experimentally or theoretically, in this area. And, there remains the enticing and formidable problem of the theoretical treatment of the electronic structure of the known defects.

III. II-VI COMPOUNDS

A. The Chalcogen Vacancy

In neutron-irradiated ZnS, an EPR center has been observed that has been identified[39] as a single electron trapped at an isolated sulfur vacancy. The spectrum is isotropic and resolved hyperfine interactions with 4.1% ^{67}Zn reveal that the wavefunction is spread equally over the four zinc nearest neighbors. No Jahn-Teller distortions are evident. The center is thus highly similar to the F-center in alkali halides or the F^+-center in the alkaline earth oxides. As in these ionic materials, the center can also be produced by additive coloration (cooking in liquid zinc at ~1100°C).

Similar centers have also been reported in electron-irradiated ZnO[40] and neutron-irradiated BeO.[41] In the latter study, EPR and ENDOR of the near neighbor Be hyperfine interactions confirmed the identification. In all three of these systems the chalcogen (group VI atom) vacancy is stable to well above room temperature.

B. The Metal Vacancy

The metal vacancy has been observed by EPR in its single negative charge state (S=1/2) in ZnSe,[42,43] ZnS,[43] ZnO,[44,45] BeO,[46,47] and CdS.[48] For all, a static trigonal Jahn-Teller distortion is observed. This is illustrated in Fig. 10, which shows a simple molecular orbital model for the defect, constructed from the "dangling orbitals" on the four chalcogen neighbors.

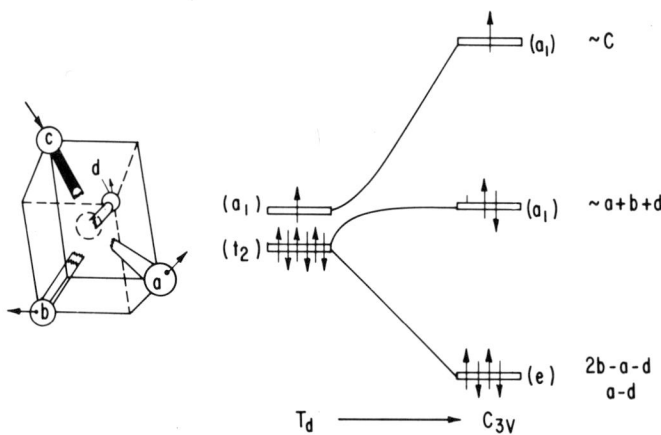

Fig. 10. Molecular orbital model for the zinc vacancy in ZnSe (single negative charge state).

From studies of ^{73}Se hyperfine interactions in ZnSe, it is determined that the unpaired electron is localized primarily in a p-orbital on the single Se neighbor which is on the axis of the trigonal distortion, as shown in the figure. In all materials, the axially symmetric g-value is consistent with that expected for a p-orbital on the corresponding chalcogen atom.

In ZnSe,[42,43] the zinc vacancy anneals at ~150°C with an activation energy 1.26 ± .06 eV. As the vacancy disappears new centers are observed which are identified as zinc vacancy-impurity pairs. The 1.26 ± .06 eV has, therefore, been identified as the activation energy for migration of the zinc vacancy. Vacancy-impurity pairs that have been identified are vacancies trapped by sulfur,[43] tellurium,[43] and chlorine.[49] For these, the impurity replaces one of the near neighbor selenium atoms. In each case, a simple molecular orbital model serves to describe the structure, with Jahn-Teller distortions again playing an important role.

In ZnS,[43] the zinc vacancy anneals at ~100°C. The activation energy, 1.04 ± 0.07 eV, has been correspondingly assigned to the vacancy migration energy in this material. Zinc vacancy pairs with chlorine, bromine, aluminum, and gallium have also been identified[50] in ZnS.

In BeO[51] and ZnO,[45] an S=1 center has been observed that has been identified with the neutral metal vacancy. A model consistent with EPR and ENDOR[52] results is that of two exchange coupled holes, one on each of two of the four oxygen atoms neighboring the vacancy. The triplet S=1 state appears to be the ground state. This would not have been predicted from the molecular orbital model of Fig. 10.

C. Interstitials

No EPR spectrum has been observed that can be identified with either the metal or chalcogen <u>isolated</u> interstitial atom. However, recently, zinc vacancy-zinc interstitial close-pairs have been detected in ZnSe.[43,53] They are observed directly after 1.5 MeV electron irradiation at 20.4K, and appear as zinc vacancy S=1/2 spectra that are perturbed by the presence of the nearby interstitial zinc ion. Four discrete pairs have been detected. From the symmetry of the spectra, their behavior under uniaxial stress, and the anisotropy of their production vs orientation of the irradiating electron beam, specific models for three of the pairs have been deduced. These are shown in Fig. 11. The pair (V') in Fig. 11a is the dominant one formed in the primary event, and results when the zinc atom is displaced into the nearest accessible interstitial site in the [$\bar{1}\bar{1}\bar{1}$] easy-displacement direction. The V" pair in Fig. 11b results upon annealing of V' (~60K), as the interstitial zinc ion moves into the equidistant site <u>behind</u> a

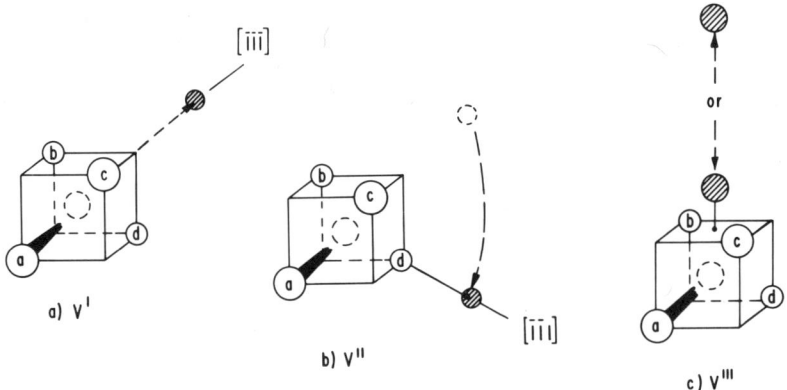

Fig. 11. Models for zinc vacancy-zinc interstitial close-pairs in ZnSe.

selenium near neighbor, apparently a more stable configuration. V''' in Fig. 11c has the interstitial in a $\langle 100 \rangle$ direction, and the fourth pair, the interstitial in a low symmetry direction from the vacancy, not specifically determined. The rearrangement and ultimate destruction of these Frenkel pairs occur in discrete annealing stages from 60-170K. This suggests a relatively low activation energy for migration of the zinc interstitial.

D. Low Temperature Damage Mechanism

Careful study of the production rates of isolated vacancies and Frenkel pairs at 20.4K and with 1.5 MeV electrons revealed the following interesting result:[43,53] The initial production of V' close-pairs is linear with fluence (φ). Isolated vacancies are also produced but the initial production rate is zero. They appear to grow in as φ^n where n is between 2 and 3. This reveals that the isolated vacancy production is at least a two-step process and that the first step is probably the close-pair production.

IV. CONCLUSIONS

Certain patterns are beginning to emerge for defects in semiconductors:

(1) Vacancies in group IV semiconductors and metal vacancies in the compound semiconductors tend to exhibit large Jahn-Teller distortions. Although we have not treated the magnitude of the distortion in these lectures, estimates have been made both from optical absorption studies[54] and the effects of external stress on

the EPR spectra.[7] The results indicate that the Jahn-Teller energies are $\gtrsim 1$ eV. The negative ion vacancy, on the other hand, tends to be like an F-center, with no distortions. [We note that this pattern also exists in the alkali halides, with the neutral alkali atom vacancy (V_F center[55]) undergoing a comparable Jahn-Teller distortion (~ 1 eV).[54]]

(2) A simple one-electron molecular orbital treatment based upon the dangling atomic orbitals on the atoms surrounding a vacancy appears to provide a satisfactory framework for understanding the electronic structure of the defects, their Jahn-Teller distortions, etc. The success of the one-electron treatments is in substantial measure a result of the large Jahn-Teller distortions which serve to "decouple" the electrons.

(3) In the compound materials, the vacancies are less mobile than in the elemental semiconductors. This is reasonable in the sense that diffusion in the elemental materials involves the nearest neighbor moving into the vacancy. In the compound materials, it is the next nearest neighbor which must make the move, the nearest neighbor serving to impede its motion.

(4) Much less is known about interstitial atoms. In silicon, the interstitial is believed to be mobile at < 1.6K. Current speculation tends to favor some type of athermal mechanism involving the ionization accompanying irradiation. Interstitialcy, bonded-like, configurations are believed to be the stable configurations. In ZnSe, the interstitial metal ion has been observed indirectly by its effect on the zinc vacancy. The symmetry of the resulting spectra indicates that the zinc ion sits in the center of a normal interstitial space in the lattice. It is mobile at ~ 170K.

(5) In contrast to the alkali halides, ionization does not appear to produce damage by itself. The lattice atom must be elastically displaced by collision with a bombarding particle. However, the evidence is strong that ionization plays an important secondary role in a defect's stability, mobility, and its reactions with other defects. This is an area of considerable current activity and interest.

REFERENCES

1. For EPR studies in germanium, see J. A. Baldwin, Jr., J. Appl. Phys. **36**, 793, 2079 (1965), D. L. Trueblood, Phys. Rev. **161**, 828 (1967).

2. For EPR studies in diamond, see the review by J. Owen, in Physical Properties of Diamond, edited by R. Berman

(Clarendon Press, Oxford 1965) p. 274. See also J. F. Lomer and A. M. A. Wild, Phil. Mag. 24, 273 (1971) and Rad. Effects 17, 37 (1973); Y. M. Kim, Y. H. Lee, P. Brosius and J. W. Corbett, in *Radiation Damage and Defects in Semiconductors* (The Inst. of Physics, London 1973) p. 202.

3. G. D. Watkins, IEEE Trans. on Nucl. Science NS-16, 13 (1969).

4. G. D. Watkins, in *Radiation Damage in Semiconductors* (Dunod, Paris 1965) p. 97.

5. K. L. Brower, Phys. Rev. B1, 1908 (1970).

6. G. D. Watkins, J. Phys. Soc. Japan 18, Suppl. II, 22 (1963).

7. G. D. Watkins, in *Lattice Defects in Semiconductors, 1974* (The Inst. of Physics, London 1975) p. 1.

8. C. A. Coulson and M. J. Kearsley, Proc. Roy. Soc. A 241, 433 (1957).

9. G. D. Watkins and J. W. Corbett, Phys. Rev. 134, A 1359 (1964).

10. G. D. Watkins, in *Radiation Effects on Semiconductor Components* (Journees d'Electronique, Toulouse 1967), Vol. 1, paper A1.

11. B. L. Gregory, J. Appl. Phys. 36, 3765 (1965).

12. G. D. Watkins and J. W. Corbett, Phys. Rev. 121, 1001 (1961).

13. G. Bemski, J. Appl. Phys. 30, 1195 (1959).

14. K. L. Brower, Phys. Rev. B4, 1968 (1971).

15. K. L. Brower, Phys. Rev. B5, 4274 (1972).

16. E. L. Elkin and G. D. Watkins, Phys. Rev. 174, 881 (1968).

17. G. D. Watkins, Phys. Rev. 155, 802 (1967).

18. G. D. Watkins, Phys. Rev., to be published.

19. G. D. Watkins, Phys. Rev., in press.

20. G. D. Watkins and J. W. Corbett, Phys. Rev. 138, A543 (1965).

21. J. W. Corbett and G. D. Watkins, Phys. Rev. 138, A555 (1965).

22. C. A. J. Ammerlaan and G. D. Watkins, Phys. Rev. B5, 3988 (1972).

23. J. G. deWit, C. A. J. Ammerlaan, and E. G. Sieverts, in Lattice Defects in Semiconductors, 1974 (The Inst. of Phys., London 1975) p. 178.

24. K. L. Brower, in Radiation Effects in Semiconductors, edited by J. W. Corbett and G. D. Watkins (Gordon and Breach, New York 1971) p. 189.

25. W. Jung and G. S. Newell, Phys. Rev. 132, 648 (1963).

26. Y. H. Lee and J. W. Corbett, to be published.

27. M. Nisenoff and H.Y. Fan, Phys. Rev. 128, 1605 (1962).

28. Y. H. Lee and J. W. Corbett, Phys. Rev. B8, 2810 (1973).

29. Y. H. Lee, Y. M. Kim, and J. W. Corbett, Rad. Effects 15, 77 (1972).

30. Y. H. Lee and J. W. Corbett, Phys. Rev. B9, 4351 (1974).

31. G. D. Watkins, Phys. Rev., to be published.

32. G. D. Watkins and J. W. Corbett, Disc. Faraday Soc. 31, 86 (1961).

33. K. L. Brower, Phys. Rev. B9, 2607 (1964).

34. G. D. Watkins, unpublished.

35. R. E. McKeighen and J. S. Koehler, Phys. Rev. B4, 462 (1971).

36. J. C. Bourgoin and J. W. Corbett, in Lattice Defects in Semiconductors, 1974 (The Inst. of Physics, London 1975) p. 149.

37. N. Almeleh and B. Goldstein, Phys. Rev. 149, 687 (1966).

38. G. D. Watkins, in Radiation Effects in Semiconductors, edited by F. L. Vook (Plenum, New York 1968) p. 67.

39. J. Schneider and A. Räuber, Solid State Comm. 5, 779 (1967).

40. J. M. Smith and W. H. Vehse, Phys. Letters 31A, 147 (1970).

41. R. C. DuVarney, A. K. Garrison, and R. H. Thorland, Phys. Rev. 188, 657 (1969).

42. G. D. Watkins, in Radiation Effects in Semiconductors, edited by J. W. Corbett and G. D. Watkins (Gordon and Breach, New York 1971) p. 301.

43. G. D. Watkins, "Intrinsic Defects in II-VI Compounds," Report No. ARL TR75-0011, available from N.T.I.S., Clearinghouse, Springfield, Va. 22151, USA.

44. A. L. Taylor, G. Filipovich, and G. K. Lindberg, Sol. St. Comm. 8, 1359 (1970).

45. D. Galland and A. Herve, Phys. Lett. 33A, 1 (1970); Sol. St. Comm. 14, 953 (1974).

46. A. Herve and B. Maffeo, Phys. Lett. 32A, 247 (1970).

47. B. Maffeo, A. Herve, G. Rius, C. Santier, and R. Picard, Sol. St. Comm. 10, 1205 (1972).

48. A. L. Taylor, G. Filipovich, and G. K. Lindberg, Sol. St. Comm. 9, 945 (1971).

49. W. C. Holton, M. deWit, and T. L. Estle, in International Symposium on Luminescence-The Physics and Chemistry of Scintillators (Munich 1965), edited by Riehl and Kallmann (Verlag Karl Thiemig K. G., Munich 1966) p. 454.

50. J. Schneider, A. Räuber, B. Dischler, T. L. Estle, and W. C. Holton, J. Chem. Phys. 42, 1839 (1965).

51. B. Maffeo, A. Herve, and R. Cox, Sol. St. Comm. 8, 2169 (1970).

52. B. Maffeo and A. Herve, to be published.

53. G. D. Watkins, Phys. Rev. Letters 33, 223 (1974); in Lattice Defects in Semiconductors, 1974 (The Inst. of Physics, London 1975) p. 338.

54. G. D. Watkins, in Radiation Damage and Defects in Semiconductors (The Inst. of Physics, London 1973) p. 228.

55. W. Kanzig, J. Phys. Chem. Solids 17, 80 (1960).

INFRARED STUDIES OF DEFECTS

R.C. Newman

J.J. Thomson Physical Laboratory

University of Reading, Berks., U.K.

GENERAL DISCUSSION

Semiconductor crystals such as silicon are opaque in the visible region of the spectrum and to investigate defects by optical means it is usually necessary to use long wavelength radiation. A difference between silicon and the compound semi-conductors is that there is no intrinsic one phonon absorption (Reststrahl-band) because of the symmetry of the diamond structure. Consequently, apart from relatively weak absorption features due to two and three phonon processes[1], silicon and germanium are transparent from zero frequency up to their fundamental electronic edges, at about $10^4 cm^{-1}$ ($\lambda \sim 1\mu m$) for silicon, with an energy gap of $1 \cdot 1 eV$, and a somewhat lower frequency for germanium with a gap of $0 \cdot 65 eV$. It follows that the presence of extrinsic absorption in the form of discrete bands arising from defects is relatively easy to detect providing any continuum of electronic absorption due to free holes or electrons is negligible. This means that the material that can be examined may contain only a very small concentration of shallow donors or acceptors, or alternatively, there must be a high degree of electrical compensation. In the latter case either one or even both types of defects could be intrinsic in nature rather than impurity atoms. Fortunately, as discussed below, it is possible to render samples transparent subsequent to their growth either by diffusion or by irradiation treatments, even if they are initially very highly conducting. It is then possible to examine defects optically but it must be remembered that the compensation treatment may itself lead to large changes in the original arrangement of the defects. These comments are also applicable to compound semiconductors such as gallium arsenide or gallium phosphide,[2] which is actually transparent in the visible region of the spectrum because of its relatively large energy gap of 2·25 eV. The compounds have a strong Reststrahl - band, typically in the wavelength range 25-35μm. and it is not possible to detect extra weak absorption from defects in a region on either side of this band. In addition, the strengths of the multiphonon combination bands in such materials are greater and sometimes show much sharper features [3] than those in silicon or germanium which imposes a further limitation on the sensitivity for detecting extrinsic absorption. For certain compounds, of which gallium phosphide is an example, where one type of atom has a much greater mass than that of the other there is a gap in the allowed vibrational frequencies between the acoustic and optical modes and the materials are relatively transparent in the corresponding spectral region [4]. An interesting example of extrinsic absorption in such a gap arises from a vibrational mode of the F-centre (negatively charged iodine vacancy) in the ionic compound potassium iodide. This will be discussed briefly below and compared with the corresponding situation in covalent compounds such as gallium phosphide.

At this stage it would seem sensible to classify the types of defect that may be present and then to comment briefly in a general way on the types of extrinsic absorption (electronic or vibrational) that may ensue. Five varieties of defect may be envisaged:

(1) Single vacancies or self interstitials that are far removed spatially from each other or from any other defects.
(2) Aggregates of intrinsic defects such as divacancies or a cluster of self-interstitials.
(3) Impurity atoms far removed from each other and other defects.
(4) Aggregates of impurity atoms which may or may not all be of the same type.
(5) Aggregates of impurity atoms with intrinsic defects; some further clarification would seem desirable here which is best given by way of examples. An impurity atom displaced from a substitutional site into an interstitial site as a result of combining with a mobile self-interstitial should be regarded as a complex, since it may not be clear which atom, if either, occupies a true interstitial site; this is another way of stating that locally there may be large spatial displacements of atoms from normal lattice sites. Likewise, the oxygen-vacancy complex (A-centre) in silicon[5] should not be considered simply as a substitutional oxygen atom because the symmetry of the centre is C_{2V}, whereas that of a substitutional site is T_d.

Although there is no intention of discussing impurity absorption[4] in its own right in this article, it is necessary to have a knowledge of such absorption. This is because it has been more than amply demonstrated that intrinsic defects interact with an exceedingly wide range of impurities in silicon.[6] It is natural to ask whether this is also the case in other materials; some comments on this point will be given below.

All the categories of defect listed above, together with others such as dislocations, will disturb the translational symmetry of the perfect crystal lattice. It follows that $\underline{k}$ and $\underline{q}$, the wavevectors of electrons and phonons respectively, may no longer be used as 'perfect quantum numbers' to determine selection rules for forbidden optical transitions. Thus one phonon absorption is found to be present in both silicon[7] [8] and germanium[9] containing high concentrations of point defects and it has also been reported that similar absorption is present in silicon containing a high concentration of dislocations[10]. The frequencies of the normal modes of vibration will also be changed, perhaps just up or down, as for example when germanium impurities are incorporated in silicon, but in certain cases localized modes may occur either above the maximum lattice frequency or in a gap if one exists. The two latter types of mode which will in general lead to absorption are of most interest in studies of point defects. This is because isotope effects can give valuable information about the atoms of which the defect is constituted; it should be remembered that the frequency of an oscillator is simply proportional to the inverse square root of its reduced mass. In addition, it is possible to measure optical dichroism for defects of low symmetry giving sharp absorption lines and in some cases these results can be compared with corresponding EPR data. An example which embraces all these concepts is afforded by the oxygen-vacancy complex in silicon.[5] A major advantage of making measurements of the vibrational states of a defect is that its charge state is not especially important in determining the magnitude of the absorption and it is possible to measure effects from isoelectronic impurities such as carbon in silicon and boron in gallium arsenide or gallium phosphide.[4] The absolute magnitude of the integrated absorption in a localized vibrational mode is given approximately by :

$$\int \alpha(\omega) d\omega = \frac{2\pi^2 D}{ncM'} \eta^2$$

where α is the absorption coefficient, D is the concentration of defects, n is the refractive index of the crystal, c is the velocity of light, M' is the mass of the impurity and η is what Leigh and Szigeti [11] have termed an apparent charge. η is in fact just the dipole moment per unit displacement in the normal mode (the localized mode) being considered and so has the dimensions of charge. For many impurities in a large number of host crystals the magnitude of this charge is close to e, the electronic charge, and so it is wrong to apply a Lorentz local field correction which would be very large in most homopolar crystals. A disadvantage of examining vibrational absorption is also now apparent because its magnitude is inversely proportional to M' and so is relatively small compared with that arising from allowed electronic transitions where M' is effectively reduced to m, the mass of an electron (the latter absorption is of course itself reduced somewhat by an f-number which is less than unity). In practice, localized vibrational absorption can be detected fairly readily for most impurities and defects when their concentrations exceed about 10^{17} atoms cm^{-3}.

Similar arguments apply to the electronic states and absorption by the defective lattice. The disturbance to the crystal will lead to changes in the distribution of electronic energies including a smearing effect close to the edges of the valence and conduction bands. Optical absorption is therefore expected at frequencies below that of the fundamental edge. Various mechanisms including actual spatial displacement of atoms from their normal sites and internal electric fields produced between compensating donor and acceptor centres have been considered and are reviewed by Fischer.[12] In practice a continuum of absorption is indeed found in irradiated crystals of silicon[13] (this is apparent in fig.2 which is discussed in section 2), germanium[14], gallium arsenide[15] and gallium phosphide; the latter material actually becomes quite opaque in the visible region due to this process and this has some important consequences in the manufacture of arrays of light emitting diodes. Like the one phonon continuum absorption produced in germanium and silicon however, measurements of this edge absorption are not particularly valuable because there is no obvious way of correlating the absorption with specific defects. That is, several different defects may make contributions with similar spectral forms.

In addition to changes in the bands of allowed electronic levels as described above, or following the substitution of certain impurities, discrete levels may be produced in a normally forbidden gap; these are in some sense analogous to the localized vibrational mode frequencies. A simple example of the production of a localized electronic level is when a shallow donor such as phosphorus is substituted into a silicon lattice. An electron bound to such a level at low temperatures produces a paramagnetic centre which also gives rise to strong sharp infrared absorption lines due to electronic transitions to excited states (in the region of 30 μm). However, if a more complicated defect is introduced, such as a trivacancy or a self-interstitial, the situation is theoretically much more complicated and it is usually not obvious whether localized electronic states should be produced and if so whether they would be expected to give rise to appreciable electronic absorption. An interesting and most important defect which does satisfy these criteria is the divacancy in silicon, which will be discussed in the following section. However, a vast number of localized levels have been detected by electrical measurements in a whole variety of host crystals but without corresponding reports of optical absorption although in some cases photoconductivity has been reported.

Following these general remarks, the remainder of the article will be devoted to some specific examples of the uses of infrared absorption to study the behaviour of intrinsic defects. Electronic absorption will be discussed first including a few references to absorption which appears to arise from impurities complexed with intrinsic defects. Vibrational absorption will then be considered but limitations on space will mean that selected examples only can be considered.

ELECTRONIC ABSORPTION

Silicon

The author is unaware of any optical absorption which has been ascribed to isolated vacancies or self-interstitials. Indirect evidence obtained from both EPR[16] and other optical measurements[4] suggest that the latter defects are present in irradiated n-type material provided the temperature is never raised above $150°K$, while direct EPR spectra show that vacancies are similarly stable up to $160°C$ in p-type material. It is not certain that a self-interstitial gives rise to localized electronic levels, but the EPR evidence for the vacancy appears quite unambiguously positive.[6] It must be concluded therefore that any optical absorption associated with the latter defects is either very weak, or alternatively is very diffuse otherwise characteristic bands would have been observed in samples irradiated at $77°K$.

Irradiation of silicon with high energy particles (electrons, neutrons, etc.) would be expected to produce divacancies by the direct displacement of two neighbouring lattice atoms, a process which has been confirmed from EPR measurements made on samples after irradiations at temperatures where isolated vacancies are not mobile.[17] The structure of the divacancy is shown in fig.1 and the simplest model may be considered as six broken bonds, three from each of the nearest neighbour atoms to each vacancy. Because pairs of electrons go into bonding orbitals it is clear that three bonds could form as shown and the V_2- centre would then be neutral and non-paramagnetic. However, one bond is extremely extended and it would not be surprising if the charge state of the defect could be modified if the Fermi level ϵ_F were moved away from the centre of the band gap, so that the supply of available electrons was either increased or decreased. According to the EPR data this model is sensible and paramagnetic V_2^- and V_2^+ - centres are found in n and p-type material respectively[17]; there is also evidence for

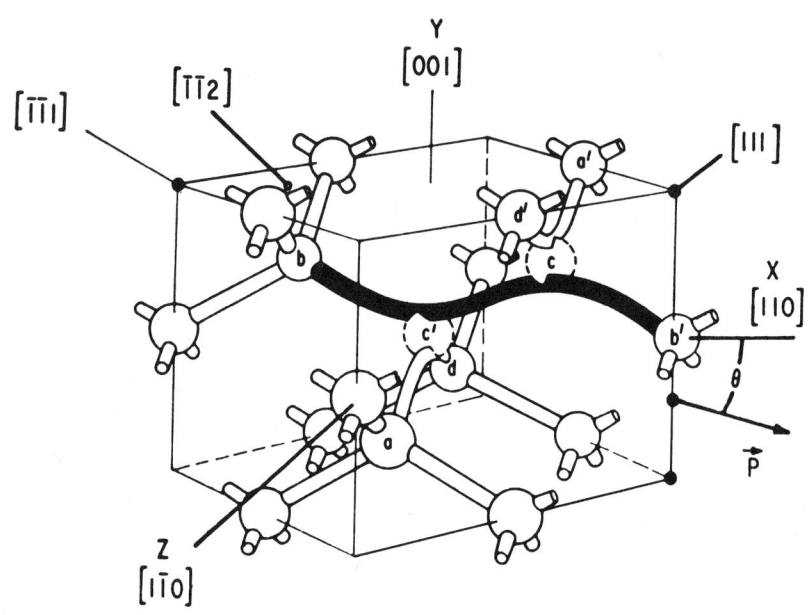

FIGURE 1. The structure of the divacancy in silicon (from Cheng et al.[18])

V_2^{2-} - centres (non paramagnetic) in highly doped n-type material (see for example ref.(18)).

Before proceeding two important points have to be made. The first is the obvious one that there are four equivalent $\langle 111 \rangle$ axes for the divacancy. The second concerns the symmetry of the defect which would be D_{3d} in the absence of rebonding. However, a bond forms between two of the neighbours to a particular vacancy and this can occur in three different ways. This is also true for the three neighbours to the other vacancy. This process does not occur randomly and the inversion symmetry is preserved so that there are only three possible configurations for the extended bond shown in fig.1. Thus the divacancy has a total of twelve different but equivalent configurations which are all distinct in terms of anisotropic EPR spectra.[17] The different configurations can be made nonequivalent by the application to the sample of a uniaxial stress, since some arrangements then have lower energies than others. Two types of process can be envisaged. The first simply changes the local bonding arrangement so that the long bond in fig.1 is between different pairs of atoms. The activation energy for this process is found to be 0.06eV which is very low and so non-equilibrium distributions of centres can be frozen into the crystal (i.e. when the stress is removed after cooling the sample) only if the temperature is kept below 20°K. At much higher temperatures (150°C), the axis of the centre can change from one [111] direction to another; this is effectively a diffusion process and has an activation energy of 1.25eV so that non-equilibrium distributions can easily be frozen in at room temperature.[17]

Absorption bands at 1.7, 3.3 and 3.9 μm have been attributed to divacancies. Normally this absorption is isotropic because a crystal will contain a random distribution of V_2- centres and the lattice has cubic symmetry. If the distribution is modified by subjecting the crystal to a uniaxial stress, the absorption coefficient will be different according to whether the electric vector of incident polarized light is parallel or perpendicular to the direction of the stress which itself can be oriented along different crystallographic axes. The ratio of the two absorption coefficients $\alpha(E_\perp)/\alpha(E_\parallel)$ is called the dichroic ratio. These ratios have been measured in detail by Cheng et al.[18] under two sets of conditions. Firstly, samples with random distributions of V_2- centres were stressed at room temperature and measured without removing the stress. This perturbation produced electronic reorientation (i.e. changes in the bonding configurations) but NO orientational changes because the temperature was too low. Secondly, the stress was applied at a high temperature (160°C) and the samples allowed to cool, whereupon the stress was removed. These samples would then be expected to have preferential geometrical alignment of V_2- centres, but no preferred bonding configuration of these centres; an example of the measurements showing the dichroism is shown in fig.2. Clearly by heating such samples subsequently, it was possible to determine the rate at which a random distribution of centres was re-established and moreover the activation energy for the process was determined. This energy and all the measured dichroic ratios were found to be in excellent agreement with those deduced from the earlier EPR data. It was therefore established beyond reasonable doubt that the optical bands arose from electronic transitions of divacancies.

A point which is of paramount importance is that it was the EPR measurements of g-shifts and hyperfine interactions with ^{29}Si nuclei which established the structure of the divacancy. Thus, although measurements of dichroic ratios will give information about the symmetry of a defect, they will not in themselves allow the atomic configuration to be determined. For example, although Chen et.al[19] have found absorption bands in irradiated and annealed silicon that show dichroism consistent with a defect having a ⟨111⟩ symmetry axis, the actual structure of the centre is still open to speculation. It is therefore important to assess the value of measurements of optical dichroism rather carefully before embarking on the actual experiments.

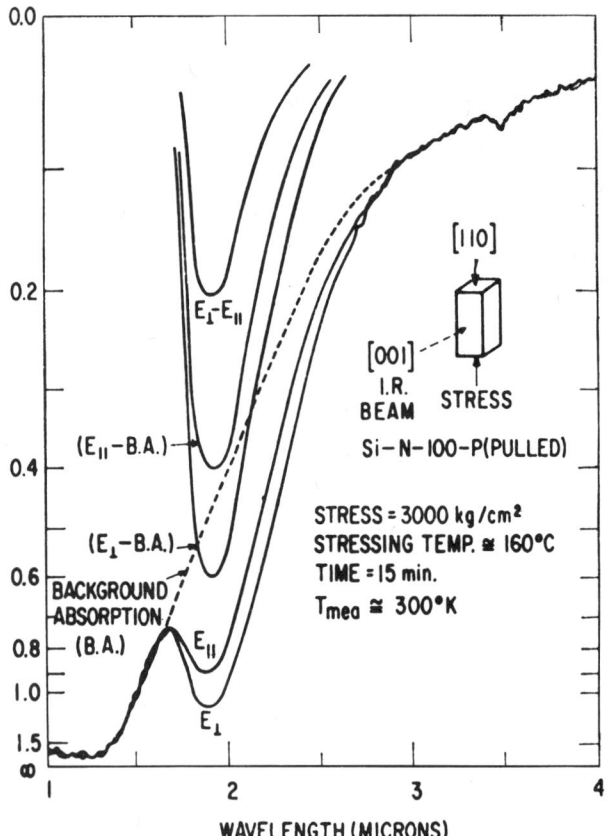

FIGURE 2. The V_2^- centre absorption at 1.7 μm and edge absorption in irradiated silicon showing dichroism quenched in from 160°C (from Cheng et.al.[18]).

It is pertinent to return briefly to the V_2^- centre. The band at 3.9 μm is found only in highly doped p-type silicon, the 1.7 μm band is found for a wide range of ϵ_F lying anywhere from the valence band to well within the top half of the energy gap, while the 3.3 μm band can be produced by photoexcitation and is particularly strong in n-type material. Other measurements of associated photoconductivity have been made but no entirely satisfactory energy level diagram has yet been produced (see refs. (20)(21)).

A large amount of space has been devoted to the divacancy because of its outstanding importance: it is very much easier to measure absorption at 1.7 μm in the near infra-red and estimate the number of V_2^- centres present from the calibration of Cheng et.

al.[18] than to carry out EPR measurements; indeed, the latter measurements may fail if the V_2- centres are in their neutral charge state as they are then non-paramagnetic. A knowledge of the concentration of divacancies can be most invaluable. If impurities are present and trap vacancies at temperatures where these defects are mobile, the production of V_2- centres will be reduced; if at a higher temperature the impurity vacancy complexes dissociate, extra V_2- centres will form. It was shown by these means that Ge or Sn impurities in silicon captured vacancies although there was no optical absorption detected from the impurity complexes.[22] Alternatively, an impurity may trap self-interstitials which will reduce the degree of vacancy-interstitial recombination, leading to an enhanced production of divacancies[4]; large effects of this type are found in carbon doped silicon.

There is little more to add except to list a few other references. A series of absorption lines found in neutron irradiated and annealed pure silicon are described by Mordkovich et.al.[23]; lines attributed to centres containing oxygen or carbon have been discussed by Bean et.al[24], while several papers have dealt with special effects when lithium is diffused into crystals.[25]

Germanium

The introduction of defects into germanium by irradiation at 300 or 77°K leads to carrier removal in lightly doped n-type material so that it eventually becomes intrinsic and then progressively more p-type. Thus n-type samples given a critical dose of radiation can always be rendered transparent. Recently, it has been found that irradiation of more highly doped n-type material ($5 \times 10^{18} cm^{-3}$) removes the free carriers but does not cause it to become p-type.[9] Previous publications dealing with the induced electronic absorption have been reviewed elsewhere[26]. A continuum of near edge absorption, similar to that shown in figure 2, is always found. For lightly doped n-type samples irradiated at 77°K there is a superposed band at 2.4 µm of electronic origin, similar in appearance to the 1.7 µm band due to V_2- centres in silicon (fig.2). Stein suggested that this band which anneals around 200°K is due to V_2- centres in germanium. There is now some supporting evidence for this view from variations in the annealing of the electrical properties of samples irradiated at different electron energies;[27] such differences would be expected to modify the production rate of divacancies relative to single vacancies. It should not be inferred that the 2.4 µm band is now identified with V_2- centres. There are no EPR data with which comparisons can be made and it has not even been established that the centres giving the band have ⟨111⟩ symmetry axes.

Consistent with these results is the observation that e^--irradiation of the heavily doped samples at 300°K does not produce a band. However, n^o- irradiation produces a different band at 2.78 µm, irrespective of whether the dopant is P, As or Sb; this band is stable up to about 150°C. It is now thought unlikely that this latter band is due to V_2- centres but this is purely a subjective view. To re-iterate a point, very few EPR spectra have been reported for defects in germanium and none identified unambiguously with any specific defect involving a vacancy or an interstitial.

III-V Compounds

Apart from absorption arising from band tailing, there appear to be no reports of discrete bands due to intrinsic defects. The situation regarding EPR spectra is worse than for germanium because no spectra are observed except from Mn^{2+} and Fe^{3+} impurities and electrons in the conduction band.

VIBRATIONAL ABSORPTION

To obtain spectroscopic details about defects it is very desirable that they should give rise to sharp absorption lines. For vibrational modes this condition is seldom satisfied unless the defect involves an impurity atom. We shall consider a substitutional impurity of mass M' in a compound crystal where the masses of the host atoms are M and m respectively, with M > m. In homopolar crystals where the local force constants are usually similar to those of the perfect lattice, high frequency localized modes are expected for M' << M or M' << m, while gap modes can occur for M' < M or M' > m; for M' > M no localized or gap modes have been reported. These expectations are well illustrated for B_{Ga}, C_P, As_P and Cd_{Ga} in GaP (fig.3). For a complex where the point symmetry is lower than T_d, there will be more than one associated vibrational mode, but it may happen that only one of these falls in the local mode region $\omega > \omega_{MAX}$, where ω_{MAX} is the maximum perfect lattice frequency. A problem limiting the detection of absorption from local modes is that they usually fall in the region $\omega_{MAX} < \omega_L < 2\omega_{MAX}$ where there is strong two-phonon absorption. For interstitial impurities there will be changes of force constants depending upon whether the impurity is bonded or not: for unbonded $^7Li^+$ in $^{28}Si, \omega < \omega_{MAX}$ (520 cm^{-1}), whereas for bonded ^{16}O in ^{28}Si, the frequency (1136 cm^{-1}) is actually much greater than for lighter substitutional ^{12}C(608 cm^{-1}). For Si, Ge and GaA there are of course no gap modes because of the densities of states of the intrinsic phonons. All these topics are discussed elsewhere[4].

Impurities which give localized modes in silicon are substitutional carbon and boron, and interstitial oxygen. Oxygen atoms can capture vacancies to form [O(i) - V] defects, or A-centres which act as electron traps with a level at E_c - 0.17eV. This centre is paramagnetic in the negative charge state (i.e. in n-type material), shows two well resolved local mode absorption lines from A^o and A^-, the A^o line shows optical dichroism and is probably the best under-

INFRARED STUDIES OF DEFECTS

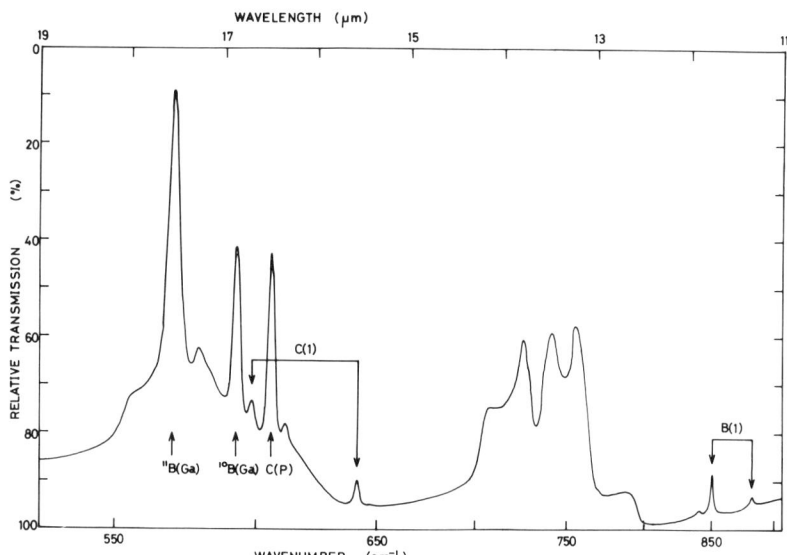

FIGURE 3. Differential transmission spectrum of irradiated GaP containing boron and carbon versus undoped and unirradiated sample. Unlabelled peaks due to residual intrinsic two phonon absorption; note lines due to irradiation damage centres ^{12}C(1) involving carbon and ^{11}B(1) and ^{10}B(1) involving boron.

stood impurity centre in a semiconductor. A full review of these observations has been presented previously [4] and will not be repeated here. The only comment to be made is that the vibrational frequency of the A-centre is lower than that of isolated oxygen. This is explicable because the Si-O bonds are lengthened which causes them to become weakened.

Irradiation of carbon or boron doped silicon again leads to a reduction in the strengths of the lines due to the unperturbed impurities, but new lines which grow are, by contrast to the discussion above, at higher frequencies. These new lines have been attributed to carbon and boron respectively in interstitial sites, but the symmetries for the two impurities must be different. The interstitial carbon centre shows two lines per isotope and is probably axial, whereas only one line per isotope is found for boron, indicating either a tetrahedral site or a low symmetry site like that of oxygen. These observations are important because localized vibrational modes may be observed from impurities in germanium and III-V compounds where EPR measurements have not so far been successful. It must be made clear that the almost trivial observation of a progressive decrease in the strength of a line with an increasing dose of irradiation must mean that the impurity is trapping either vacancies or interstitials (we are assuming here that the impurity is either isoelectronic, neutral or an ionized shallow donor or acceptor so that variations in e_F do NOT change the magnitude of the absorption). If a high frequency line appears simultaneously, this is almost certainly due to the formation of an interstitial defect because the presence of a vacancy will always "soften" the lattice locally and so reduce any vibrational frequencies. The rest of the discussion will therefore be devoted to Ge, GaAs, GaP and a few comments about KI.

N-type germanium highly doped with P, P+Ga, P+In, B+As, O or Si has been irradiated at room temperature with either 2MeV electrons or fast neutrons[9][26] to render it transparent. Apart from complexes produced with oxygen, which are now well known, we have found NO positive evidence for the formation of any other impurity complexes, in spite of the fact that very high concentrations of free carriers ($\sim 10^{19}$ cm^{-3}) were removed in some samples. These observations are quite different from those found in silicon and probably account somehow for the differences found in the electrical properties of irradiated germanium compared with silicon.

The behaviour of GaAs and GaP are very similar to each other, and intermediate to the behaviour of Ge and Si. It appears that light atoms like boron and carbon, that also have small covalent radii, must interact with mobile interstitials. The self-interstitital concerned is probably gallium. Thus $B_{Ga} + Ga_i \rightarrow B_i + Ga_s$, and the local mode absorption from the B_i is to be associated with the B(1) lines from ^{11}B and ^{10}B respectively shown in fig.3. The correlation between the two types of "boron" centre is shown in fig.4. If a gallium interstitial were captured by a carbon impurity, displacement of the impurity would not be expected because the gallium atom would then be located on a group V site. Hence a complex with axial symmetry is expected showing two lines per isotope (see fig.3) like B-Li pairs in a silicon host lattice.[4] It is believed that this is the interpretation of the C(1) lines due to ^{12}C shown in fig.3.[28] On the basis of these results it might have been expected that $Ga_{(i)}$ would pair with say Si_{As} acceptors in GaAs but there is no evidence for this; likewise there is no evidence for defect associations with other acceptors, donors or isoelectronic impurities, except for nitrogen in GaP.[29]

In general, localized modes from defects not involving an impurity are not expected. Exceptions are found in GaP where a pair of lines[4][28] is thought to be associated with an interstitial phosphorus defect, and in KI where a gap mode has been ascribed to the F-centre (negatively charged iodine vacancy).[30] The latter result can be understood by comparison with the normal modes of the U-centre (I$^-$ replaced by H$^-$); i.e. it is assumed that the mass of

INFRARED STUDIES OF DEFECTS

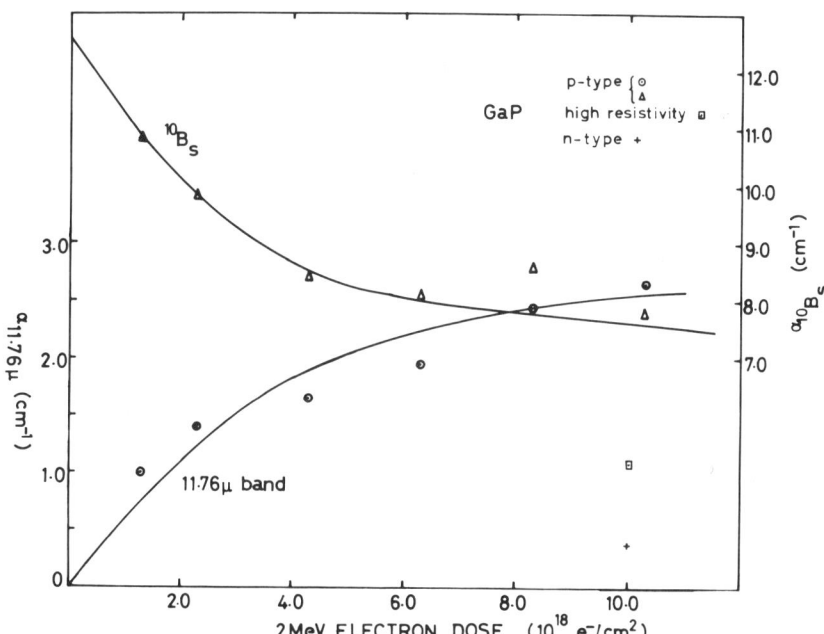

FIGURE 4. Removal of substitutional boron from gallium sites in GaP with increasing doses of electron irradiation and the growth of the B(1)-centre (11.76 μm band).

the H^- ion tends to zero. This is not unreasonable because the negative charge is still present in the ionic crystal, and the only real change is the removal of the local repulsive potential of the H^- ion. For a covalent crystal such as GaP the validity of this procedure is much less certain because of local rebonding. However two uncorrelated gap modes, apparently unassociated with impurities, have been found at 293 and 298 cm^{-1} in e^--irradiated GaP.[31] As a final comment it is therefore interesting to speculate that these modes may be due to vacancy centres.

ACKNOWLEDGMENT

The author wishes to thank The Science Research Council for financial support for some of the work reported here.

REFERENCES

1. F.A.Johnson, Proc.Phys.Soc., $\underline{73}$, 265, (1959).
2. S.R.Morrison and R.C.Newman, J.Phys.C.:Solid St.Phys.,$\underline{7}$, 619, (1974).
3. D.A.Kleinman and W.G.Spitzer, Phys.Rev.,$\underline{118}$, 110, (1960).
4. R.C.Newman, Infra-red studies of Crystal Defects (Taylor & Francis, London, 1973).
5. G.D.Watkins and J.W.Corbett, Phys.Rev. $\underline{121}$, 1001, (1961); ibid $\underline{121}$, 1015, (1961).
6. G.D.Watkins, Radiation Effects in Semiconductors (Dunod, Paris) p.97, (1965).
7. M Balkanski and W.Nazarewicz, J.Phys.Soc.Japan, $\underline{18}$, suppl.II, 37, (1963).
8. J.F.Angress, A.R.Goodwin and S.D.Smith, Proc.Roy.Soc.A, $\underline{308}$, 111, (1968).
9. M.R.Brozel, K.Laithwaite, R.C.Newman and D.H.J.Totterdell, Sol.St.Comm. in the press.
10. V.I.Vettegren, E.G.Kuzminov, V.V.Baptizmanskii and I.I.Navak Sov.Phys.Solid State, $\underline{15}$, 770, (1973).
11. R.S.Leigh and B.Szigeti, Proc.Roy.Soc.A, $\underline{301}$, 211, (1967).
12. J.E.Fischer, Phys.Rev., $\underline{181}$, 1368, (1969).
13. L.J.Cheng and J.Lori, Phys.Rev., $\underline{171}$, 856, (1968).
14. H.J.Stein, Rad.Damage and Defects in Semiconductors (IOP, London) p315, (1973).
15. R.Coates and E.W.J.Mitchell, J.Phys.C.:Sol.St.Phys., $\underline{5}$, L113, (1972).
16. G.D.Watkins, Sym.on Rad.Effects on Semicond.Components (Toulouse) A1, 1, (1967).
17. G.D.Watkins and J.W.Corbett, Phys.Rev., $\underline{138A}$, 543, (1965).
18. L.J.Cheng, J.C.Corelli, J.W.Corbett and G.D.Watkins, Phys.Rev., 152, 761, (1966).
19. C.S.Chen, R.Vogt-Lowell and J.C.Corelli, Rad.Damage & Defects in Semiconductors (IOP, London) p210, (1973).
20. M.T.Lappo and V.D.Tkachev, Sov.Phys. - Semiconductors $\underline{4}$, 1882, (1971).
21. J.C.Corelli, R.C.Young and C.S.Chen, I.E.E.E. Trans. Nucl.Sci. NS17, 128, (1970).
22. A.Brelot, Rad.Damage and Defects in Semiconductors (IOP,London) p191, (1973).
23. V.N.Mordkovich, S.P.Solovev, E.M.Temper and V.A.Kharchenko, Sov.Phys-Semiconductors, $\underline{8}$, 666, (1974).
24. A.R.Bean, R.C.Newman and R.S.Smith, J.Phys.Chem.Solids, $\underline{31}$, 739, (1970).
25. L.C.Kimerling, P.J.Drevinsky and C.S.Chen, Rad.Damage and Defects in Semiconductors (IOP London) p.182, (1973).
26. R.C.Newman and D.H.J.Totterdell, Lattice Defects in Semiconductors (IOP, London) p.172, (1975).
27. A.Jaworowski and H.Rzewuski, ibid. p.221, (1975).

28. F.Thompson, S.R.Morrison and R.C.Newman, Radiation Damage and Defects in Semiconductors (IOP, London) p371, (1973).
29. S.R.Morrison and R.C.Newman, J.Phys.C: Solid St.Phys.,$\underline{6}$, L223, (1973).
30. D.Bäuerle and R.Hübner, Phys.Rev.B.$\underline{2}$, 4252, (1970).
31. S.R.Morrison, Ph.D.Thesis, University of Reading (1973).

VACANCY AGGREGATE CENTERS IN IONIC CRYSTALS

W. von der Osten

Experimentalphysik I

Technische Hochschule Darmstadt

1. INTRODUCTION

The purpose of this chapter is to review briefly the optical properties of vacancy aggregate centers and how these are related to the electronic structure of the defects. While the F center is still the best studied and understood defect, considerable progress has been made through the years in understanding the more complex centers. The atomistic structure of the F_2(M) and F_3(R) center and their positively or negatively charged species (F_2^+, F_3^+ and F_2^-(M'), F_3^-(R')) is well established today. They are known to be complexes of two or three F centers at neighboring anion sites (for models see Pick, 1965). Aggregate centers are produced under various conditions not discussed here and give rise to optical absorption and emission spectra on the long wavelength side of or even beyond the F band. In other cases of more complex centers like the N centers we are still far from knowing even only the atomistic structure, although many efforts were undertaken to find it out.

Theoretical investigations of F-aggregate centers have been fairly limited to date (see e.g. Fowler, 1968; Inoue et al., 1970; Stoneham, 1972 and ref. therein). A simple model with which to treat the F_2 and F_3 type centers is that of a hydrogen molecule imbedded in a dielectric medium. Scaling according to the dielectric constant works surprisingly well in understanding the electronic states of the centers in analogy to the corresponding hydrogen configurations (Silsbee, 1965; Aegerter and Lüty, 1970, 1971). Qualitative statements about the degeneracies of the states of a center can be made using the molecular orbital treatment of molecules, giving at least a rough estimate of the ordering of levels (Hughes, 1966).

In the following experiments are discussed which indeed reveal the optical structure and the symmetry of F-aggregate centers by studying their optical spectra. Of key importance for the application of the described methods is the anisotropy of the centers with regard to their physical and especially their optical properties. F-aggregate centers are orientationally degenerate in the lattice and therefore macroscopically their anisotropy is cancelled. To reveal it the crystal is subjected to an anisotropic external action, that selectively affects individual groups of centers responding differently. This idea is behind the experiments using preferential bleaching and polarized luminescence as discussed in section 2 They allow us to determine the directions of the transition dipole moments without, however, revealing the center symmetry. The occurrence of narrow zero-phonon lines due to purely electronic transitions brought about a new class of experiments: the investigation of these lines in external fields. These are discussed in section 3 and it will be shown there that very detailed information is extracted about the symmetry and the symmetry properties of the defects involved.

As will be explained in section 4 the optical spectra do not only contain information on the electronic structure. In most cases where zero-phonon lines are observed more or less resolved sidebands occur. These can be analyzed with regard to the vibrational properties and the interaction of the centers with the lattice. Based on more or less empirical methods though, these possibilities will be briefly mentioned. Finally in section 5 the implications of a Jahn-Teller effect for the optical spectra are discussed, the importance of which in interpreting optical properties of orbitally degenerate localized states in solids is beyond doubt. The close analogy to non-linear molecules in which this effect occurs make certain aggregate centers good candidates to observe the expected features.

It will, of course, be impossible to cover the field completely. Instead of giving many details I have tried to discuss the physical idea and to present a few selected and representative examples for it. Most of these refer to aggregate centers in alkali halides but similar and even more detailed work was done in the alkaline earth oxides and fluorites. The references quoted are far from being complete, but they should provide a good starting point for the reader interested in a more detailed study. Several reviews treating optical properties of F-aggregate centers in alkali halides have been published. These include books and articles by Schulman and Compton (1962) , Compton and Rabin (1964), Pick (1965), Lanzl et al. (1966), Hughes (1967), Kaplyanskii (1967), Fitchen (1968) and von der Osten (1968). Optical spectra of defects in other crystals and topics related to the present article are reviewed by Henderson and Wertz (1968), Hughes and Henderson (1972) and Hayes and Stoneham (1974).

2. POLARIZATION EFFECTS IN ABSORPTION AND EMISSION

The methods treated first have since long been used to establish the "optical structure" of F-aggregate centers, i.e. to determine the orientation of the dipoles associated with various optical transitions. These traditional methods include

i. preferential bleaching and
ii. polarized luminescence.

Ueta (1952) was the first to demonstrate the anisotropic nature of the F_2 center in KCl by partial bleaching of the main (M_1) absorption band. The basic idea of his experiment was to introduce some anisotropy by using M_1 light polarized along [011], say. He then measured the M_1 absorption with light polarized in the [0$\bar{1}$1] and [011] directions. The different optical absorption for differently polarized light (due to the preferential destruction of the [011] dipoles) proved the M_1 transition to be most probably associated with an optical dipole moment along <110>. Later work (for ref. see Compton and Rabin, 1964; Fowler, 1968) demonstrated that the primary F_2 transition is excited by light polarized along the <110> center axis (compare inset of Fig. 1) and corresponds in the language of molecular chemistry to a $^1\Sigma_g^+ \rightarrow {}^1\Sigma_u^+$ transition. Also higher electronic states of the F_2 center underlying the F band were uncovered with transition moments along <100> and <110> perpendicular to the center axis. Using difference absorption and excitation spectra the absorption in the triplet state of the F_2 center could be detected in KCl (Haarer and Pick, 1967). Similar work was reported for the F_3^+ and F_3 center (Elsässer and Seidel, 1971). Since these measurements more sensitive experimental techniques were developed. In particular, extremely small dichroic absorptions can now be studied with high precision by using rotating polarizers and phase-sensitive detection (Schnatterly, 1965). Another method was recently applied by Engstrom (1975). He used a piezo-optical birefringence modulator (Kemp, 1969; Jasperson and Schnatterly, 1969) to modulate the light polarization and study F_2 center singlet and triplet states with enhanced sensitivity.

Even without producing anisotropic absorption, polarized luminescence allows one to identify the orientation of optical dipoles. Assuming that there occurs no bleaching or reorientation of dipoles, the degree of polarization obtained for polarized light depends on the dipole direction. Both preferential bleaching and polarization of luminescence are reviewed in articles by Schulman and Compton (1962) and Compton and Rabin (1964). They have tabulated quantitatively the changes in optical absorption and the degree of polarization for various dipole orientations and differently polarized light.

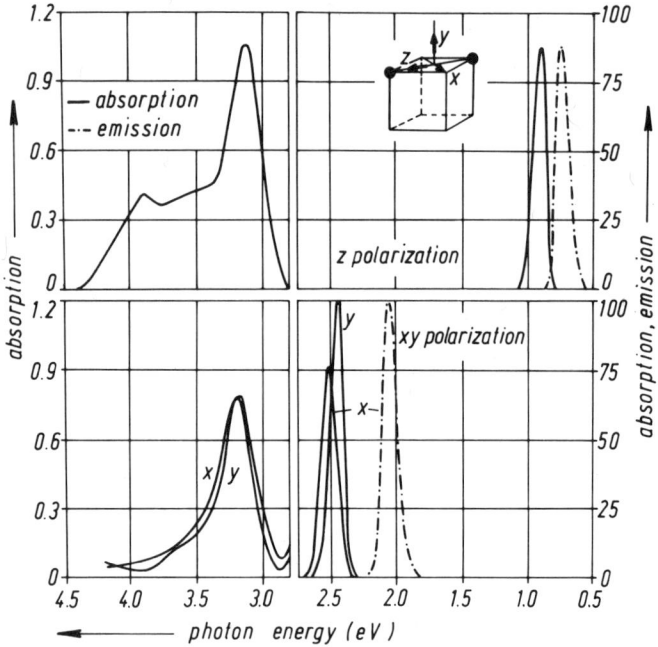

Fig. 1. Optical transitions of F_2^+ centers (Aegerter and Lüty, 1971). Dipole moments indicated in the model.

A common feature of earlier bleaching experiments is that only partial alignment of the anisotropic centers was achieved, frequently by preferential destruction of centers. In the case of F_2 type centers, our understanding of the reorientation mechanism* has improved considerably in recent years. Progress in this field allows measurements with ideally aligned systems resulting in completely dichroic spectra.

This possibility is of key importance in revealing new and weak optical transitions and completing the energy level systems of various F-aggregate centers. For example, Aegerter and Lüty (1970, 1971) performed an analysis of the optical properties of F_2^+ centers in KCl, using an almost perfectly aligned center system. Fig. 1 illustrates the observed absorption and emission bands in the near UV

---

* As first demonstrated for KCl by Schneider (1970) F_2 centers reorient via an F_2^+ mechanism. The F_2 centers are ionized through absorption in the M_F transition underneath the F band to form F_2^+ centers. They in turn absorb light in the same spectral region and reorient during nonradiative deexcitation (for ref. see Collins, 1973) resulting in aligned F_2^+ or - after capture of an electron - F_2 centers.

to IR region and their corresponding polarizations along the principal center axes (compare inset). To obtain complete dichroism in the ionized center system they started out from aligned F_2 centers and introduced extra electron traps into the crystal by X-raying at LHeT to produce F_2^+ centers. Using three different geometries for incidence and polarization of the light it was possible to identify strongly overlapping absorption bands by their different polarization. The aligned system also allowed to study polarized emission, even as a function of the polarization angle of the exciting light.

This system closely resembles a H_2^+ molecular ion embedded in a dielectric medium with a dielectric constant k_0. It beautifully demonstrates to what extent this analogy can be used to derive and interpret the electronic states of a F-aggregate center. Using a theoretical treatment of Herman et al. (1956) the energy positions, level degeneracies, polarizations and oscillator strengths of 8 absorption transitions were found in close agreement to the calculated states of the H_2^+ molecular ion corrected for a single dielectric constant and the proper separation between the two effective positive charges. In Fig. 2 the correspondence of states of the H_2^+ ion and the F_2^+ center is represented and the details of the fit are indicated. The best fit is obtained for the internuclear distance R = 1.70 Å which determines the value of k_0 and the distance r_{AB} between the two anion vacancies of the F_2^+ center.

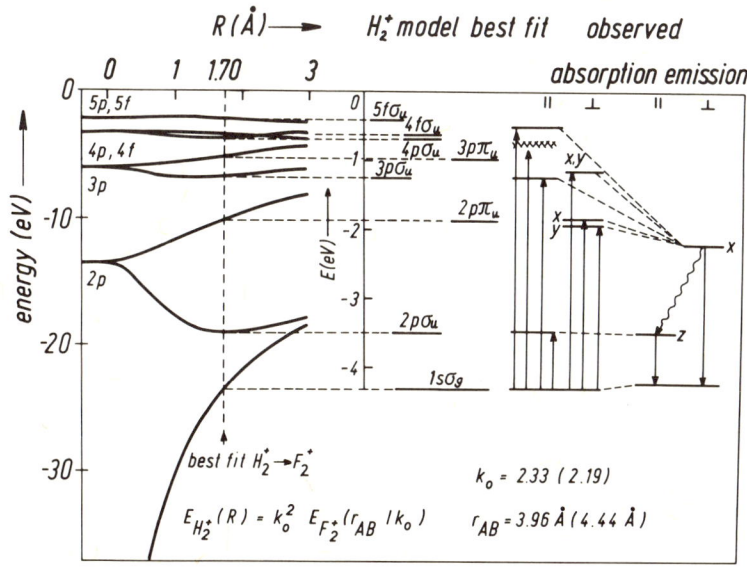

Fig. 2. Energy levels of the H_2^+ molecule versus distance R of protons, compared to F_2^+ energy levels (Aegerter and Lüty, 1971).

The techniques described in this section clearly can demonstrate the optical anisotropy of any F-aggregate center. Although they provide detailed information about the principal axes of the dipole moments, it is not possible to infer the symmetry of the center from them. Centers of different point symmetry might have their dipole moments along the same direction. The Seitz model of the "M" center and the model proposed by van Doorn (Compton and Rabin, 1964) were both expected to have a <110> transition moment. In spite of their different symmetries (C_{2v} and D_{2h}, respectively) they were not distinguishable in the early days of color centers.

3. UNIAXIAL STRESS AND ELECTRIC FIELD EFFECTS

With the discovery and interpretation of narrow zero-phonon lines in the optical spectra of defects it became possible to measure directly small shifts that are due to applied external fields. These effects are naturally difficult to detect in the case of broad bands. The most profitable results were obtained from studying zero-phonon lines in uniaxial stress and high electric fields. These investigations allow determination of the symmetry of the defects involved and their electronic states.

The basic concept is to remove various types of degeneracy of the center by applying the external perturbation. As a consequence, in general, a zero-phonon line associated with the center splits. There are two types of degeneracy that are present either separately or together:

i. orientational degeneracy associated with the equivalent orientations of an anisotropic center in the lattice and

ii. the electronic degeneracy that occurs in centers with three- and fourfold symmetry axes, for which degenerate energy levels are expected.

3.1 Orientational Degeneracy

The detailed theory of <u>uniaxial stress</u> is a straightforward application of group theory and first order perturbation theory and has been widely discussed in the literature (for comprehensive reviews see Kaplyanskii, 1964; Hughes 1966; Lanzl et al. 1966; Fitchen, 1968; von der Osten, 1968). It is not intended to go into the details here. Three pieces of information are provided by the experiments:

i. the number of components into which a line splits for stress along various directions,

ii. the energetic shifts with respect to the original line position, and

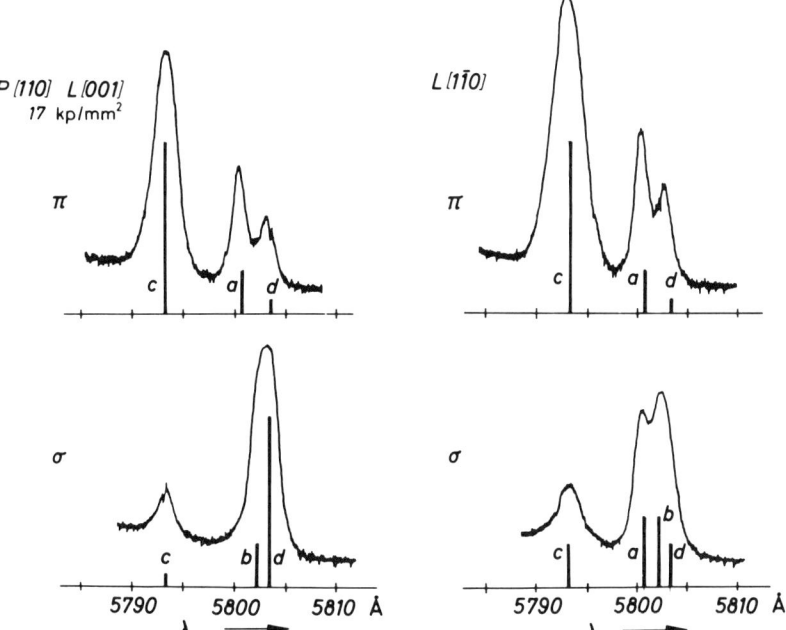

Fig. 3. Splitting of the 5803 Å (N_1) line in NaF under [110] stress, for different viewing directions. Predicted intensities and positions are indicated (Johannson et al., 1967b).

iii. the polarization of the splitting components depending on the orientation of the transition dipole moment at the center.

From these the symmetry of the defect and of the electronic states involved in the optical transition can be inferred. Commonly stress along different directions is applied and the spectra are examined with light polarized parallel (π) and perpendicular (σ) to the stress axis.

From the wealth of results obtained with this technique one example is illustrated in Fig. 3. It is left to the reader to examine the agreement between experimental results and theoretical predictions with regard to the number and relative intensities of the components for the derived model. The line shifts are linear with applied stress. The stress perturbation can therefore be written as

$$H' = \sum_{ij} A_{ij}\, \sigma_{ij}$$

where σ_{ij} and A_{ij} are the components of the stress tensor and of the

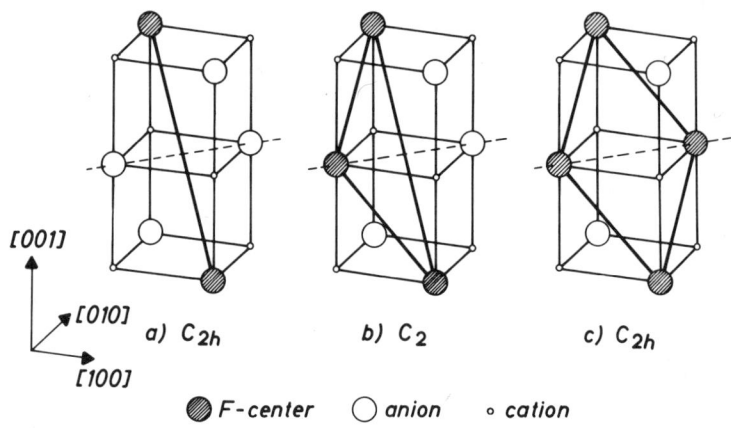

Fig. 4. N_1 center models with monoclinic symmetry. Possible transition moments along [112] and [110] (Johannson et al. 1965).

"piezospectroscopic" tensor. The number of independent parameters A_{ij} depends on the center symmetry. In Fig. 3 the positions of the expected lines are calculated using the four independent parameters which are obtained from the energy shifts as function of applied stress.

The example shown corresponds to the N_1 zero-phonon line in NaF and corresponding studies of this line in LiF and NaCl have been reported. The center associated with this line has monoclinic symmetry with a twofold axis along <110> and the transition moment along <112> perpendicular to the rotation axis. In Fig. 4 three possible models are shown. However, the true atomistic structure of the defect involved is not yet kown.

Precisely, uniaxial stress measurements provide the symmetry system to which a color center belongs which still includes several point groups. Stark measurements (for a survey see Kaplyanskii, 1967) in <u>electric fields</u> further can restrict the number of possible point groups within a certain symmetry system. They distinguish between centers with and without inversion. The perturbation operator for a center in an electric field $\vec{E}$ is written as

$$H' = \vec{E} \cdot \vec{d}$$

where $\vec{d}$ is the electronic dipole moment operator of the defect. The first order change in energy caused by $\vec{E}$ is calculated for the state n from the matrix element $\langle n|H'|n \rangle$. Due to the odd parity of H', for a center containing inversion these matrix elements vanish and no linear energy shift of the levels is produced. If the center does <u>not</u> contain inversion, external electric fields can produce a linear

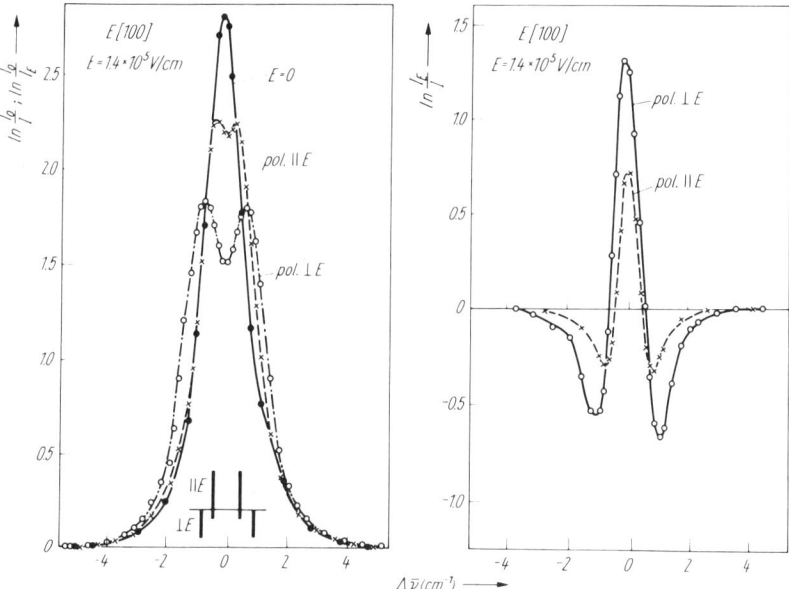

Fig. 5. Stark and pseudo-Stark effect of the F_3^+ (A→E) zero-phonon line (5456 Å) in NaF at 4.2 K (left); field-induced changes of optical density (right) (Johannson et al. 1969).

energy shift. Taking into account the orientational degeneracy in these cases a splitting of zero-phonon lines due to this linear "pseudo-Stark" effect occurs. As an example, Fig. 5. (left) illustrates the Stark splitting of the F_3^+ zero-phonon line in NaF compared to the line without electric field. The experimental information taken from the spectra is entirely analogous to the uniaxial stress case. It should be realized, however, that lack of inversion does not guarantee an observable Stark effect.

Quadratic rather than linear Stark effects have also been found, the zero-phonon line of the N_1 center (compare Fig. 4) being one example (Kaplyanskii et al. 1973, 1975). This observation in LiF proves nicely the inversion symmetry of the N_1 center. Using modulation techniques Stark measurements were recently extended to broad band transitions of several F-aggregates (Rhyner and Cameron, 1968; De Werd, 1972).

Compared to piezospectroscopic studies, from Stark measurements the symmetries and center orientations cannot be inferred as unambiguously. The strain-induced splitting is determined by a <u>second-rank tensor</u> (ellipsoid) being more "sensitive" to symmetry properties. The pseudo-Stark splitting is determined by the dipole moment <u>vector</u> that can be oriented in the same way for centers of different symmetry. Combined electric field and stress results in certain cases allow the unambiguous determination of the point group of a defect (Johannson et al. 1968, 1969).

The effects of an applied electric field on zero-phonon lines have been theoretically treated in detail by Kaplyanskii and Medvedev (1967), by Lanzl (1967) and others. (For further references see e.g. Johannson et al. (1969), Davis and Fitchen (1971), Kaplyanskii et al. (1973).) These treatments include electronic degeneracies that play an important role both in stress and electric field measurements.

3.2 Electronic Degeneracies

In defect centers with trigonal and higher symmetries orbitally degenerate electronic states may be involved in the zero-phonon transition. This type of degeneracy will be removed by components of the external stress or electric field <u>not</u> having the symmetry of the defect. Additional splitting components occur, showing pronounced polarization effects if the initial state for the transition is degenerate. The occupation of the split sublevels is determined by a Boltzmann distribution, reflected in a striking field and temperature dependence of the intensities of different transitions.

Several trigonal centers were studied both under uniaxial stress and in high electric fields. These experiments in various alkali halides have revealed the orbital degeneracy of certain states for the F_3^+, F_3 and F_3^- center. To calculate the expected effects degenerate perturbation theory has to be applied. Many results concerning stress effects in degenerate states are quoted by Fitchen (1968) and von der Osten (1968).

The example presented here will be the linear Stark effect for transitions between singlet (A) and doublet (E) states of centers with C_{3v} symmetry, in particular for the F_3^+ and F_3 center. The removal of the orientational and orbital degeneracy in these cases is described by two independent parameters (U_0 and U_1 in Table 1) differently defined in the literature. The shift and relative strength for each splitting component are listed in Table 1 for a [100] electric field. The splitting of the F_3^+ line in Fig. 5 is entirely consistent with these predictions for a C_{3v} $A_1 \rightarrow E$ transition, the expected intensities being schematically indicated. The lack of any temperature effect in the intensities proves the <u>excited state</u> to be degenerate (Johannson, 1969; Kaplyanskii and Medvedev, 1969). Contrary to this the occurrence of a temperature dependent dichroism for the R_2 zero-phonon line of the F_3 center suggests a degenerate (E) <u>ground state</u> for this center, which has been well established both experimentally and theoretically (Silsbee, 1965; Hughes, 1966).

VACANCY AGGREGATE CENTERS IN IONIC CRYSTALS

shift	$I_\parallel$	$I_\perp$
$\pm(U_0 + 2U_1)E$	0	2
$\pm(U_0 - 2U_1)E$	8/3	2/3

Table 1. Stark effect for A↔E transition in centers with C_{3v} symmetry (Lanzl, 1967). Note the symmetric splitting pattern. $E \parallel [100]$.

The small splittings common to most Stark measurements are usually masked by the linewidth even though narrow zero-phonon lines are studied. Instead of resolved splittings merely changes in lineshape are detected. These can be readily <u>interpreted</u> by applying a moment analysis. In particular, employing differential <u>modulation</u> techniques small changes in the optical spectra can be measured directly and analysed in terms of the parameters pertinent to the splitting. The absorption changes with and without field are represented for the F_3^+ ($A_1 \rightarrow E$) and F_3 ($E \rightarrow A$) zero-phonon line in Figs. 5 (right) and 6. They essentially correspond to a second and (at 1,7 and 4,2 K) first moment change, respectively, the latter reflecting the thermal redistribution in the split ground state of the F_3 center. From the moment changes as function of applied electric field the splitting parameters can be obtained (Fig. 7). The comparison with the

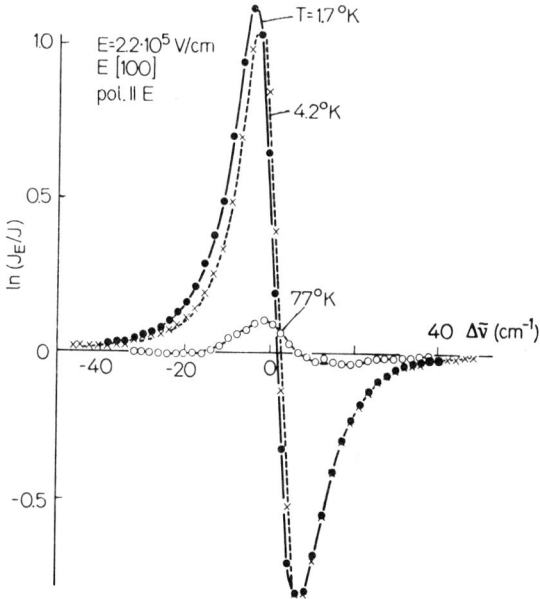

Fig. 6. Stark effect of the R_2 (3909 Å) absorption line in LiF: field-induced absorption change (Johannson et al. 1969)

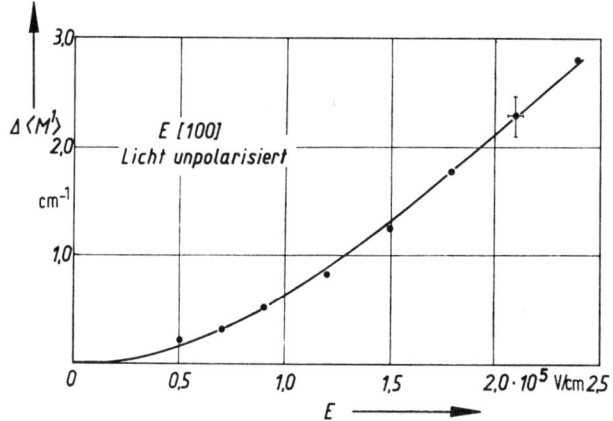

Fig. 7. Experimental and predicted first moment change of the R_2 line in LiF as function of electric field strength (Johannson, 1969)

theoretical prediction using these parameters show excellent agreement (full line). Details of this method are to be found in the literature (Johannson et al. 1967a, 1969; Johannson, 1969 and ref. therein).

Following Silsbee (1965) one can choose a convenient coordinate system for the F_3 center with the x-, y- and z-direction along $[2\bar{1}\bar{1}]$, $[0\bar{1}1]$ and $[111]$ for one certain orientation. Taking his vibronic wave functions ψ_x and ψ_y, which are even and odd under reflection at the (xz) plane, the splitting parameters U_0 and U_1 can be expressed in terms of the expectation values for the dipole moment perpendicular ($\|z$) and parallel ($\|x$) to the plane of the center. Transformation from the crystal to the center frame yields the non-zero matrix elements contained in Table 2 for the F_3 center in various alkali halides (Johannson et al. 1969; Davis and Fitchen, 1971). Presently no calculations of the dipole moment matrix elements exist with which these parameters can be compared. Any quantitative comparison should take the effect of Jahn-Teller distorsions in the E state into account that should appear as a reduction of U_1 (Davis and Fitchen, 1971).

	LiF[a]	NaF[b]	KCl[b]
$e \langle \psi_x|z|\psi_x \rangle = \sqrt{3}\, U_0$	1.14	0.21	0.12
$e \langle \psi_x|x|\psi_x \rangle = \sqrt{6}\, U_1$	1.66	1.25	1.55

Table 2. F_3 center: dipole moment expectation values (i.u. 10^{-5} cm^{-1} (V/cm)$^{-1}$) perpendicular and parallel to the plane of the center (a) Johannson, 1969; b) B. Jahnke, W. Büdenbender, priv. commun.)

Stark effect measurements to investigate the inversion symmetry of defects have already been proposed by Overhauser and Rüchardt (1958) but only recently accomplished by means of zero-phonon transitions.

4. VIBRATIONAL PROPERTIES

While zero-phonon lines are sensitive probes to study the symmetry properties of F-aggregate centers the vibrational sideband structure contains information about the vibrational modes that interact with the center. The shape of the optical absorption and emission of any defect depends on the spectrum of coupled modes and the electron-phonon coupling strength S (see e.g. Fitchen, 1968; von der Osten, 1968).

In cases of small and intermediate coupling ($1 \leq S \leq 6$) the more or less resolved sideband structure can be analyzed with regard to interacting modes. The usual way to obtain the information is to compare the various vibronic peaks to the vibrational spectrum of the host crystal. The one-phonon contribution to the sideband is intuitively correlated with van Hove singularities at or near Brillouin zone boundaries. The high density of states there should be reflected in the optical sideband (Fig. 8). There exists, however, a number of objections to this practice:

1. the correlation is based only on energetic considerations. Although efforts have been undertaken to derive selection rules (Loudon, 1964; Kennedy, 1966) for certain centres the use of symmetry arguments (Hughes, 1967) has not yet proved of great value in understanding the situation for aggregate centers. Computing selection rules is complicated by the fact that the center of symmetry of F-aggregates is not located at a lattice point.

2. The identification of lattice modes rely upon the knowledge of phonon dispersion curves. Only recently have unambiguous assignments been made using measured phonon spectra.

The work done along these lines in alkali halides and other substances has been reviewed by several authors (Hughes, 1967; Fitchen, 1968; Hughes and Henderson, 1970; Hayes and Stoneham, 1974 and others). As an example in Fig. 8 the sideband structure for the F_3 (D_{2d}) center in CaF_2 is compared to a single phonon density of states obtained from a shell model fit to neutron scattering measurements. The F_3 band arises from transitions between singlet orbital states allowing coupling only to symmetric (Γ_1) modes. This condition is not very restrictive for vibronic transitions and similarity to the full single phonon density of states is not surprising (Hughes, 1967). In the case of LiF the effect of isotopic substitution on the vibronic peaks helped to remove some ambiguities

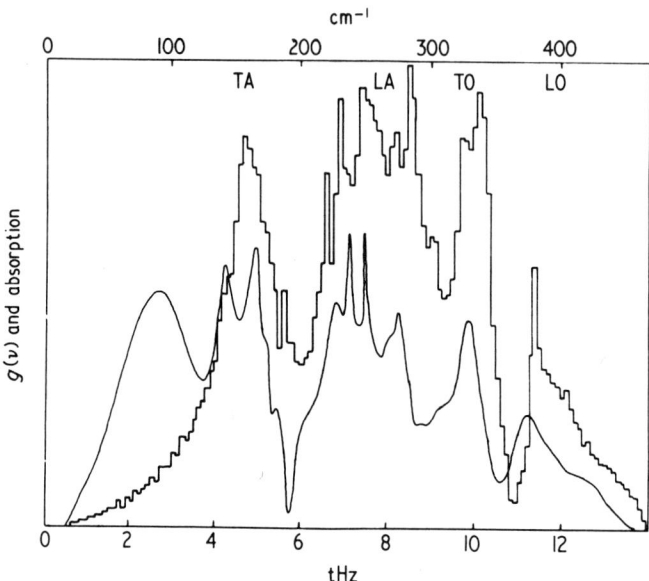

Fig. 8. Comparison of the phonon structure in the F_3 absorption band in CaF_2 at 1.5K with a single-phonon density of states (Beaumont et al. 1972)

in phonon-assignments (Hughes, 1966; Fetterman, 1968). In general, besides the lattice modes (eventually slightly modified close to the defect) two other types of modes can exist in crystals with defects:

i. pseudo-localized or resonant modes with frequencies within the continuum of lattice modes. With their high amplitude near the defect even small densities contribute appreciably to the sideband;

ii. localized modes with frequencies outside the lattice mode spectrum and amplitudes that decrease exponentially with distance from the defect.

In many aggregates relatively broad low frequency peaks are observed. The 85 cm^{-1} peak in Fig. 8 is not correlated with any critical point and apparently is due to a resonant mode associated with the defect. Such modes commonly occur also in the optical bands of F-aggregates in the alkali halides, the N_1 center being a prominent example. Localized modes so far were not observed in colour center sidebands.

For S>1 the observed sideband structure is the result of a superposition of zero-, one- and higher (n-) phonon contributions. The n-phonon parts, including all processes in which the number of

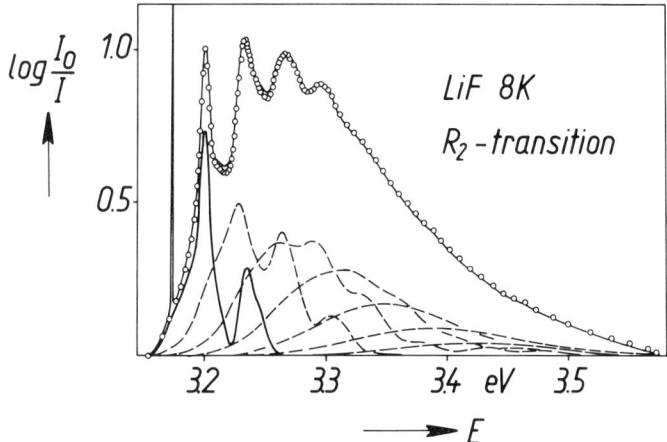

Fig. 9. R_2 absorption in LiF. Superposition of one-phonon (full-line) and higher phonon (dashed) processes gives the total sideband structure as shown by the full line (Giesecke et al. 1972)

phonons change by n, predominate the one-phonon contribution. In these cases the one-phonon absorption can be obtained using a method developed by Lanzl et al. (1968). It presumes the explicit knowledge of the shape of the zero-phonon line and the sideband measured with sufficiently high spectral resolution. It was applied to the N_1 and R_2 transition in NaF and to the R_2 absorption in LiF, KCl and KBr (Giesecke et al. 1972). The decomposed R_2 band in LiF is shown in Fig. 9. Note that both the first and part of the second peak (at 3.20 and 3.23 eV) arise from one-phonon processes. Although unexpected, this result agrees excellently with uniaxial stress and circular dichroism measurements (Piehl, 1974; Burke, 1968). Assuming a $1/\omega$-dependence of the coupling strength (Maradudin, 1966), the effectively coupled spectrum of modes in Fig. 10 (full line) is obtained. Comparison with the LiF phonon density of states (dashed) (Dolling et al. 1968) by no means show a one-to-one correspondence. These deviations are inherent in other examples (Giesecke et al. 1972) and are analogously found in LiF for the F_2^- and F_3^- transitions (Fetterman, 1968). They are presumably due to local interactions of the defect with the lattice changing the energies of a few normal modes and giving them a relative large amplitude at the defect.

The Fourier-transform $g_1(t)$ of the one-phonon absorption spectrum gives reliable values for the linear electron-phonon coupling strength S. They are compared in Table 3 with S-values obtained from the integrated absorption I_o in the zero-phonon line relative to the total absorption I_{tot} in the band (see Fitchen, 1968; von der Osten, 1968).

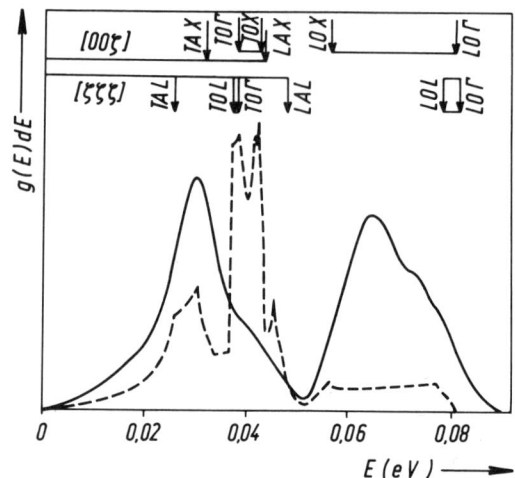

Fig. 10. Comparison of the spectrum of coupled modes (full line) for the R_2 transition with the phonon density of states (dashed). Critical point energies of unperturbed LiF are indicated (Giesecke et al. 1972)

	LiF T=7.4K	KCl T=6.1K	KBr T=8.4K
$S = g_1(t=0)$	3.38	3.40	4.56
$S = \ln(I_{tot}/I_o)$	3.3	3.4	4.8

Table 3. Linear electron-phonon coupling strength S for the R_2 transition in several alkali halides

While this method basically is a consistency check for the linear coupling theory several recent approaches exist to analyze or generate the vibrational sideband structure of F-aggregate centers (Ritter and Markham, 1969; Evans and Kemp, 1970; Mostoller et al. 1971).

The examples discussed in this section demonstrate the extent to which reliable interpretations of vibrational sideband structures can be made. The approach used by many workers in the field is certainly an oversimplification and it is hoped that exact calculations of the coupled spectrum for a F-aggregate center will be feasable in the future. Selection rules for vibronic transitions under uniaxial stress in context with degenerate electronic states will be briefly illustrated in section 5.

5. JAHN-TELLER DISTORTED SYSTEMS

If orbitally degenerate electronic states are involved in the optical transitions qualitative differences in the optical spectra will appear. F_3 type aggregate centers are of interest in this respect because their trigonal (C_{3v}) point symmetry allows for orbital degeneracies and implies the possibility of a Jahn-Teller effect. In a classic paper, Silsbee (1965) discussed the implication of a Jahn-Teller interaction with regard to the optical properties of the R center in KCl. The R_2 absorption of this center he studied under uniaxial stress corresponds to an E→A transition (see section 3). To interpret his results he used the model of Longuet-Higgins et al. (LHOPS 1958) who treated this problem numerically for the case of a doubly degenerate (E) electronic state interacting with a single pair of doubly degenerate vibrational modes of E symmetry causing the dynamic distortion. In particular Silsbee (1965) computed selection rules from the stress-split (E) ground state to higher vibronic levels of the excited state, that depend only on the number of E-mode quanta (for more details see Fitchen, 1968). From the stress-induced R_2 band dichroism the E mode was then tentatively

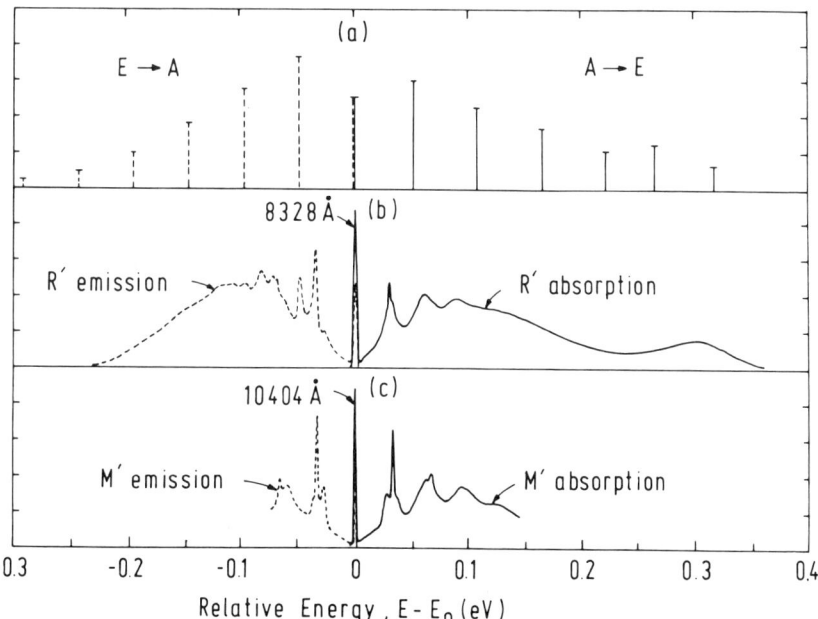

Fig. 11. Calculated Jahn-Teller vibronic spectra for an A↔E transition with $k^2 = 3.5$ and $\hbar\omega_E = 49$ meV (a). Optical spectra of the $F_3^-(R')$ and $F_2^-(M')$ center (b and c, respectively) in LiF at 4 K (Fetterman and Fitchen, 1968)

identified and its energy determined. Also the strength of the Jahn-Teller coupling was estimated from the broad band dichroism.

An example with still better resolved vibronic structure (Fig. 11b) to study the effect of Jahn-Teller distortions in optical spectra is the A↔E transition of the $F_3^-(R')$ center (Fetterman, 1968; Fetterman and Fitchen, 1968). New features predicted by the LHOPS model appear in the optical spectra: a double-humped band shape (Fig. 11b) expected for singlet (A) to doublet (E) transitions and a lack of "mirror" symmetry between absorption and emission. (Compare the "mirror" spectra of the $F_2^-(M')$ center in LiF with no orbital degeneracy in Fig. 11c).

Studying the emission under high [100] stress at low temperature (Fig. 12a) allows to establish the symmetries of the modes responsible for the various vibronic peaks. The selection rules for

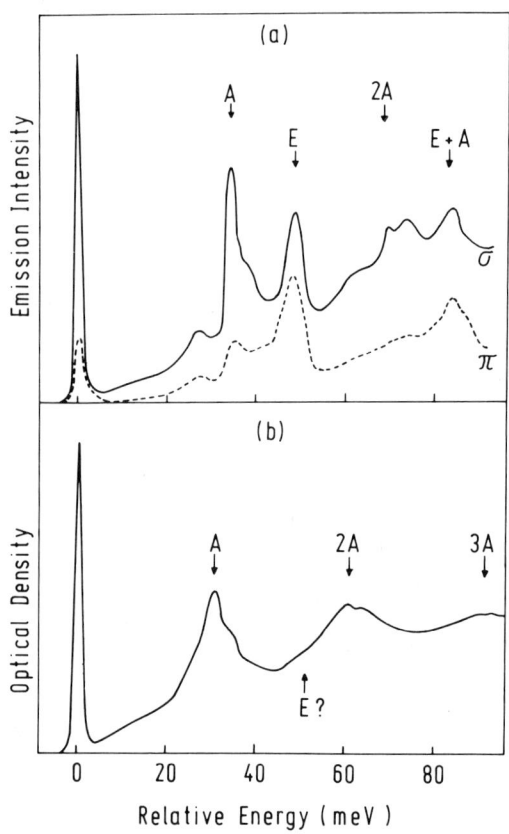

Fig. 12. a) Polarization of $F_3^-(R')$ emission in LiF under a [100] uniaxial stress of 17 kg/cm² at 1K. b) $F_3^-(R')$ absorption spectra and predicted position of the perturbed E mode peak (Fetterman and Fitchen, 1968)

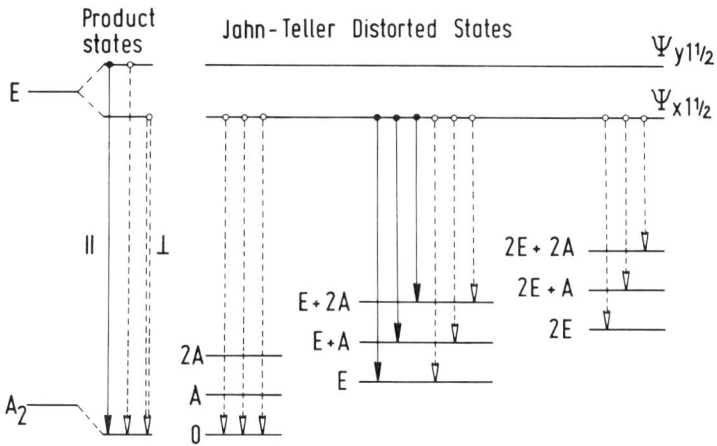

Fig. 13. Selection rules for polarized $F_3^-(R')$ emission under [100] stress involving A and E modes.

this case (Fig. 13) show, that only for vibronic transitions involving excitation of an <u>odd</u> number of E-mode quanta (and any number of A-mode quanta) will emission polarized parallel to the stress axis be allowed. In this way, the sideband peak at 49 meV in Fig. 12a was identified with the unperturbed E mode. From the intensity ratio of the E mode and the zero-phonon line a coupling strength $k^2 = 3\pm1$ is estimated.

For these values the LHOPS model predicts locations and intensities of the E mode transitions as shown in Fig. 11a. The double hump in the $F_3^-(R')$ absorption is nicely reproduced and indicate the presence of transitions to the perturbed E mode levels in the excited state. All resolved peaks can be assigned to A modes (Fig. 11 and 12b), the lack of resolved E mode absorption peaks suggests that the modes responsible for the dynamic distortions are strongly altered by the Jahn-Teller coupling. Similar effects in other centers indicate that coupling to many E modes might have to be taken into account rather than coupling to only a single pair of modes. Nevertheless, the gross features of the simple model are supported by the magnetic properties of the center (Davis, 1970; Davis and Fitchen, 1971).

Fig. 14 represents the R_2 absorption band in LiF (a) together with the [100] stress-induced dichroism ΔK (b) for light polarized parallel and perpendicular to the stress axis (R. Piehl, 1974). Transitions involving an <u>odd</u> number of E-mode quanta are not dichroic (Silsbee, 1965). Thus the deviation of the ΔK signal from the shape of the absorption curve K implies that the second vibronic peak of the spectrum is at least partly due to E mode excitation.

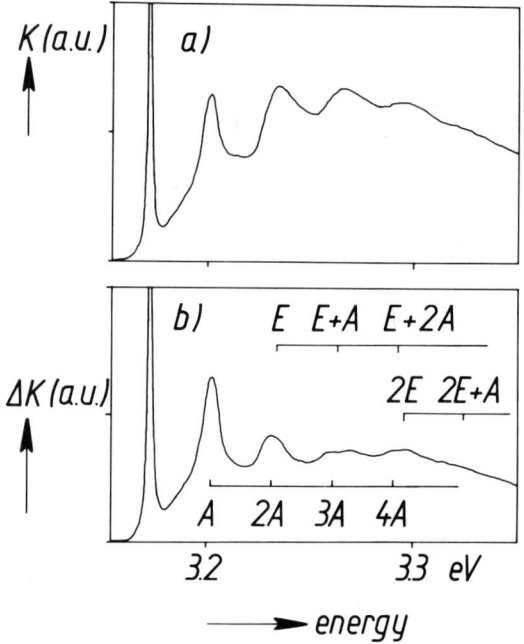

Fig. 14. R_2 absorption in LiF (a) and stress-induced dichroism (b). Positions of A and E mode peaks indicated (Piehl, 1974)

Comparing this results to the R_2 sideband analysis in section 4 the second peak of the one-phonon contribution at 68 meV (Fig. 10) must have E symmetry, in agreement with circular dichroism measurements by Burke (1968). Accidentally the E mode one-phonon process is superimposed by the 2 A mode process as indicated in Figs. 9 and 14b.

REFERENCES

Aegerter, M.A. and Lüty, F. (1970): J. Luminesc. <u>1,2</u>, 624

Aegerter, M.A. and Lüty, F. (1971): phys. stat. sol. (b) <u>43</u>, 227 and 245

Beaumont, J.H., Harmer, A.L. and Hayes, W. (1972): J. Phys. C - Sol. State Phys. <u>5</u>, 257

Burke, W. (1968): Phys. Rev. <u>172</u>, 886

Collins, W.C. (1973): phys. stat. sol. (b) <u>57</u>, 63

Compton, W.D. and Rabin, H. (1964) in: Sol. State Phys. <u>16</u>, 121

Davis, J.A. (1970): Ph.D. Thesis, Cornell University, Ithaca, unpublished

Davis, J.A. and Fitchen, D.B. (1971): phys. stat. sol. (b) 43, 327

De Werd, L.A. (1972): phys. stat. sol. (b) 52, 207

Dolling, G., Smith, H.G., Nicklow, R.M., Vijayaraghavan, P.R. and Wilkinson, M.K. (1968): Phys. Rev. 168, 970

Elsässer, K. and Seidel, H. (1971): phys. stat. sol. 43, 301

Engstrom, H. (1975): Phys. Rev. B 11, 1689

Evans, B.D. and Kemp, J.C. (1970): Phys. Rev. B 2, 4179

Fetterman, H.R. (1968): Ph.D. Thesis, Cornell University, Ithaca, unpublished

Fetterman, H.R. and Fitchen, D.B. (1968): Sol. State Commun. 6, 501

Fitchen, D.B. (1968) in: Physics of Color Centers, ed. W.B. Fowler, Academic Press, New York, p. 294

Fowler, W.B. (1968) in: Physics of Color Centers, ed. W.B. Fowler, Academic Press, New York, p. 54

Giesecke, P., von der Osten, W. and Röder, U. (1972): phys stat. sol. (b) 51, 723

Haarer, D. and Pick, H. (1967): Z. Physik 200, 213

Hayes, W. and Stoneham, A.M. (1974) in: Crystals with the Fluorite Structure, ed. W. Hayes, Clarendon Press, Oxford, p. 185

Henderson, B. and Wertz, J.E. (1968): Adv. Physics 17, 749

Herman, R., Wallis, M.C. and Wallis, R.F. (1956): Phys. Rev. 103, 87

Hughes, A.E. (1966): Ph.D. Thesis. Oxford University, Oxford, unpublished

Hughes, A.E. (1967): J. de Physique 28, Suppl. 8/9, C 4 - 55

Hughes, A.E. and Henderson, B. (1972) in: Point Defects in Solids, ed. J.H. Crawford jr. and L.M. Slifkin, Plenum Press, New York, p. 381

Inoue, M., Sati, R. and Wang, S. (1970): Can. J. Phys. 48, 1694

Jasperson, S.N and Schnatterly, S.E. (1969): Rev. Sci. Instr. 40, 761

Johannson, G. (1969): Z. Physik 228, 222

Johannson, G., Lanzl, F., von der Osten, W. and Waidelich, W. (1965): Phys. Letters 15, 110

Johannson, G., Lanzl, F., von der Osten, W. and Waidelich, W. (1967a): Phys. Letters 25 A, 598

Johannson, G., Lanzl, F., von der Osten, W. and Waidelich, W. (1967b): Z. Phys. 201, 430

Johannson, G., Lanzl, F., Mödl, H., von der Osten, W. and Waidelich, W. (1968): Z. Phys. 210, 1

Johannson, G., von der Osten, W., Piehl, R. and Waidelich, W. (1969): phys. stat. sol. 34, 699

Kaplyanskii, A.A. (1964): Opt. Spectry. 16, 329 and 557

Kaplyanskii, A.A. (1967): J. de Physique 28, Suppl. 8/9, C 4 - 39

Kaplyanskii, A.A. and Medvedev, V.N. (1967): Opt. Spectry 23, 404

Kaplyanskii, A.A. and Medvedev, V.N. (1969): Sovj. Phys. - Sol. State 11, 109

Kaplyanskii, A.A., Medvedev, V.N. and Skvortsov, A.P. (1973): Surf. Sci. 37, 650

Kaplyanskii, A.A., Medvedev, V.N. and Skvortsov, A.P. (1975): Opt. Spectry. 38, 227

Kemp, J.C. (1969): J. Opt. Soc. Am. 59, 950

Kennedy, L.Z. (1966): J. Chem. Phys. 44, 1746

Lanzl, F. (1967): Phys. Letters 25 A, 596

Lanzl, F., von der Osten, W. and Waidelich, W. (1966): Lectures. NATO Summer School, Ghent, Belgium, unpublished

Lanzl, F., von der Osten, W., Röder, U. and Waidelich, W. (1968) in: Localized Excitations in Solids, ed. R.F. Wallis, Plenum Press, New York, p. 575

Longuet-Higgins, H.C., Öpik, U., Pryce, M.H.L. and Sack, R.A. (1958): Proc. Roy. Soc. (London) 244, 1

Loudon, R. (1964): Proc. Phys. Soc. 84, 379

Maradudin, A.A. (1966): Sol. State Phys. 18, 273

Mostoller, M., Ganguly, B.N. and Wood, R.F. (1971): Phys. Rev. B 4, 2015

Von der Osten, W. (1968): Z. angew. Phys. 24, 365

Overhauser, A.W. and Rüchardt, H. (1958): Phys. Rev. 112, 722

Pick, H. (1965) in: Springer Tracts in Modern Physics 38, 1, ed. G. Höhler, Springer, Berlin

Piehl, R. (1974): TH Darmstadt, priv. commun.

Rhyner, C.R. and Cameron, J.R. (1968): Phys. Rev. 169, 710

Ritter, J.T. and Markham, J.J. (1969): Phys. Rev. 185, 1201

Schnatterly, S.E. (1965): Phys. Rev. 140, A 1364

Schneider, I. (1970): Phys. Rev. Letters 24, 1296

Schulman, J.H. and Compton, W.D. (1962): Color Centers in Solids. Pergamon Press, Oxford

Silsbee, R.H. (1965): Phys. Rev. 138, A 180

Stoneham, A.M. (1972): phys. stat. sol. (b) 52, 9
Ueta, M. (1952): J. Phys. Soc. Japan 7, 107

PERTURBATION SPECTROSCOPY AND OPTICAL DETECTION OF PARAMAGNETIC RESONANCE

Yves MERLE d'AUBIGNE

Laboratoire de Spectrométrie Physique

U.S.M.G., B.P. 53 - 38041 GRENOBLE CEDEX

   Zeeman and Stark effects have long been studied on the narrow spectral lines emitted by atoms in gases. Similar effects have been recently studied for impurities or defect in solids which have broad absorption or emission bands (1). This became possible due to the development of a new experimental technique, essentially lock-in detection, which allows one to measure very small variations of absorption or emission intensities produced when a perturbation is applied to a crystal. Stark effect studies are described by Spinolo and some of the consequences of stress have been outlined by Van der Osten. My discussion is restricted to measurements of circular dichroïsm in magnetic fields and linear dichroïsm under uniaxial stress. In this first lecture I outline the principle of these differential techniques and describe how they may uncover the structures of electronic levels much smaller than the width of an optical band. I then show that this perturbation spectroscopy is an excellent tool for the study of such phenomena as the electron-lattice coupling and the Jahn-Teller effect (2-4). Having outlined the general theory and experimental technique, a number of examples are discussed, in particular the use of perturbation spectroscopy in the study of zero phonon lines, broad bands and defects with orbitally degenerate ground states. In the second lecture a detailed account of the excited state resonance of a number of defects is given.

I - PERTURBATION SPECTROSCOPY

A - General Principles

Consider the example shown in Fig. 1 of optical transitions between a singlet ground state and an excited orbital triplet state ; the selection rules for the absorption of circularly and linearly polarized light are indicated in a) (by convention σ_+ and σ_- are left and right circularly polarized light, and π represents light polarized linearly parallel to the magnetic field). In the presence of a magnetic field the energies of the levels $M_L = \pm 1, 0$ of the triplet state are different so that in a gaseous specimen a clearly resolved splitting of the transition between the singlet and triplet would be resolved. In a solid each of these transitions is broadened by interaction of the centre with the vibrations of the crystal. If the width W of these transitions is now greater than the Zeeman structure $2g_L\beta H$ then the splitting pattern cannot be resolved with unpolarized light. However, if one measures separately the absorption coefficients k^+ and k^- for left and right circularly polarized light one obtains directly the Zeeman splitting $2g_L\beta H$. In practice, if the width W is much larger than the Zeeman structure this technique is not very precise, and it is preferable to measure directly the circular dichroïsm i.e. the difference $2\Delta k = k^+ - k^-$ of the absorption coefficients and the average absorption coefficient $k = \frac{1}{2}(k^+ + k^-)$. Then one recalculates the coefficient $k^\pm = k \pm \Delta k$ to obtain the Zeeman structure $2g_L\beta H$. From Fig. 1 one then deduces

$$\Delta k \simeq -g_L \beta H \frac{dk}{d\nu} \tag{1}$$

Note that although this particular example involves the application of a magnetic field as a perturbation, the degeneracy of levels may equally well be removed by other perturbations such as uniaxial stress or electric fields.

The dichroïsm can be measured with high accuracy, using apparatus similar to that of the single beam absorption spectrophotometer. The monochromator is set at a given wavelength to measure the transmitted light intensity through the sample $I^\pm$ when the incident light of intensity I_o is polarized σ^+ or σ^-. Then for a sample of thickness l

$$k^\pm l = -\log(I^\pm/I_o)$$

and for a small dichroïsm

$$(k^+ - k^-)l \simeq -(I^+ - I^-)/\bar{I} \text{ where } \bar{I} = \tfrac{1}{2}(I^+ + I^-) \tag{2}$$

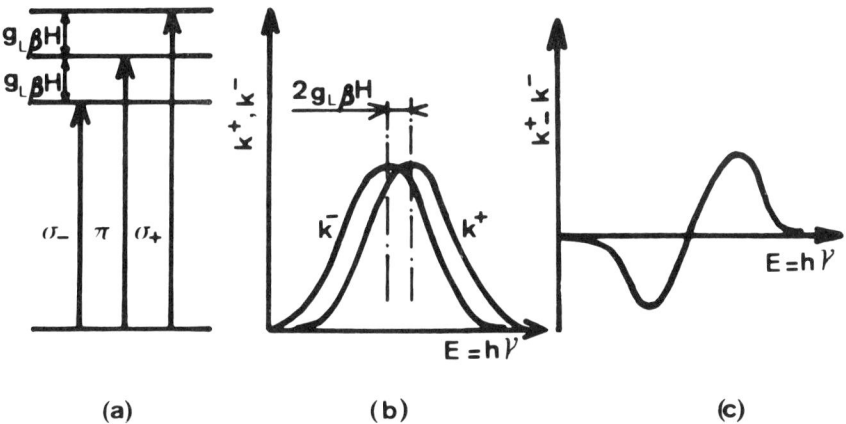

Fig. 1. Circular dichroïsm, a simple example.

In practice very small differences in the light intensities $I^+ - I^-$ are measured by lock-in techniques. One modulates the polarization of the incident light at some audiofrequency and detects the in-phase variation of the transmitted light. When the wavelength of the monochromator is swept through the absorption band the intensity of the transmitted light varies. It is practical to use a servocontrol of the power supply of the photomultiplier so that the dc component of the photo current is kept constant. Then, from equation (2) the dichroïsm is seen to be proportional to the ac component of the transmitted light.

In general, samples with optical densities of order unity are used so that at the absorption band maximum $k_m l \simeq 3$. In most practical cases the dichroïsm is small when the absorption bandwidth W is large, so that a low resolution monochromator may be used and shot noise is small. Relative variations of transmitted light as small as 10^{-4} are measured. For a symmetrical structureless band shape $\frac{dk}{d\nu} \simeq \frac{k_m}{W}$, so that from equation (1) and (2) one can resolve structures

$$g_L \beta H \simeq \Delta k l / \frac{dkl}{d\nu} \simeq \Delta k l \cdot W / k_m l \simeq 3.10^{-5} \text{ W},$$

much smaller than the bandwidth W.

Among the various kinds of polarization modulators the stress modulated quarter wave-plate (5) is the more widely used.

B - Electron-Phonon Coupling and the Method of Moments

The analysis made in Fig. 1 assumes that the three components of the absorption spectra are equally broadened by the interaction with the crystal vibrations and that the selection rules and the Zeeman interaction remain essentially the same as in the absence of electron-lattice coupling. Hence the bands are "rigidly shifted" by $\pm g_L \beta H$. We will now investigate the meaning of this "rigid shift" approximation. For orbitally degenerate states, the electronic and nuclear motions cannot be separated as in the usual Born-Oppenheimer approximation. Consequently the eigenstates are not simple products of an electronic state and a vibrational state. They are in fact linear combinations of such Born-Oppenheimer products which are called vibronic states. For instance, once the electron lattice interaction is taken into account, an electronic triplet $|\beta\rangle$, $\beta = x, y, z$ gives rise to a whole set of vibronic states ; since the lower energy states are still triply degenerate they can be written

$$|\beta\rangle\rangle = \langle x|\beta\rangle\rangle|x\rangle + \langle y|\beta\rangle\rangle|y\rangle + \langle z|\beta\rangle\rangle|z\rangle$$

where the double ket indicates a vibronic state, $\langle x|\beta\rangle\rangle$, $\langle y|\beta\rangle\rangle$ and $\langle z|\beta\rangle\rangle$ are vibrational wave functions.

The effect of an applied perturbation appears to be strongly quenched when these vibronic states are considered. A typical matrix element of the Zeeman interaction calculated inside the ground vibronic triplet is

$$\langle\langle y|\mathcal{H}_z|x\rangle\rangle = g_L\beta H_z \langle\langle y|L_z|x\rangle\rangle = g_L K(T_1)\beta H_z \langle y|L_z|x\rangle$$

where $K(T_1) = \langle\langle y|y\rangle\langle x|x\rangle\rangle - \langle\langle y|x\rangle\langle y|x\rangle\rangle$ is the difference of two overlap integrals and is smaller than or equal to one. Quite generally the matrix elements of an electronic operator transforming like the Γ representation of the point group will be reduced (2) by the factor $K(\Gamma)$ which depends only on the symmetry Γ of the operator. It is now clear that the simplified picture given in Fig. 1 has no real meaning.

Soon after the first circular dichroïsm measurements on the F centres Henry, Schnatterly and Slichter (3) (H.S.S.) proposed a method of interpretation of the experimental information based on the calculation of the moments of the bandshape : we shall briefly recall their results. The method of moments as used by H.S.S. was intended for application to F centres. It applies without modification to absorption taking place between an orbital singlet state and an orbitally degenerate excited level. Time dependent perturbation calculations give the absorption coefficient for light of polarization η and wave number E/h as

$$k^\eta = C E g_\eta(E) \tag{3}$$

where C is a constant and the "shape function" of the absorption band is

$$g_\eta(E) = Av_a \Sigma_b |<b|P_\eta|a>|^2 \delta(E_b - E_a - E) \tag{4}$$

Here one makes a thermal average on the initial states $|a>$ and a summation on the final states $|b>$. P_η is the dipole operator for light of polarization η. For centres which are not oriented the shape function before the application of the perturbation is the same for all the polarizations : $g_\eta(E) = f(E)$. The moments are defined by

$$A = \int f(E) dE \qquad \bar{E} = A^{-1} \int E f(E) dE$$
$$<E^n> = A^{-1} \int (E - \bar{E})^n f(E) dE \tag{5}$$

The moments change when the perturbation is applied, becoming

$$<\Delta E_\eta^n> = A^{-1} \int (E - \bar{E})^n \left[g_\eta(E) - f(E) \right] dE \qquad n \geqslant 1$$

Then from (4) and (5) one gets

$$A = Av_a \Sigma_b <a|P_\eta^+|b><b|P_\eta|a> \tag{6}$$

and similar expressions for the other moments. The advantage of this method is that, as it will be shown now, the moments can be calculated without having a detailed knowledge of the vibronic levels $|b>$.

The hamiltonian describing the F centre can be separated into two parts : H_O and H'. H_O contains the lattice hamiltonian H_L and H_E which represents the centre when the lattice is fixed in its equilibrium position. $H' = H_{EL} + H_{SO} + H_p$ is the sum of three terms representing the electron-lattice interaction, the spin-orbit coupling and the applied perturbation. The eigenstates of H_O are simple products of an orbital singlet ($|\alpha>$) or triplet ($|\beta>$) and a spin state $|M>$ and a lattice state $|X_L>$: i.e. $|\tau> = |\beta>|M>|X_L>$. The eigenstates $|b>$ of $H_O + H'$ are related to the states $|\tau>$ by a unitary transformation so that the principle of spectroscopic stability applies

and $\Sigma_b |b><b| = \Sigma_\tau |\tau><\tau| = \Sigma_{\beta, M, X_L} |\beta, M, X_L><\beta, M, X_L|$

Since the $|X_L>$ and the $|M>$ form a complete set :

$$\Sigma_M |M><M| = 1 \qquad \Sigma_{X_L} |X_L><X_L| = 1$$

and $\Sigma_b |b><b| = \Sigma_\beta |\beta><\beta|$

Substitution into equation (6) gives for A an expression which does not contain the eigenstates $|b>$. Similarly one obtains the variation of the first moment of the F centres when a magnetic field is applied. Hence we find

$$<\Delta E_{\pm}> = \pm |<y|L_z|x>|(g_L\beta H + \lambda <S_z>) \qquad (7)$$

where + and − stands for left and right circularly polarized light, g_L is the Landé factor, β the Bohr magneton, λ the spin-orbit coupling constant and $<S_z>$ the spin polarization. The coupling to the lattice does not appear explicitly in this expression, which is the same as that calculated in the rigid shift approximation.

The third moment variations are given by

$$<\Delta E_{\pm}^3> = 3<\Delta E_{\pm}> \left[<E^2>_C + \frac{1}{2} <E^2>_{NC} + \frac{1}{4} \lambda^2 \right]$$

where $<E^2>_C$ and $<E^2>_{NC}$ are the cubic and non cubic contributions to the second moment as given by

$$<E^2> = <E^2>_C + <E^2>_{NC} + \lambda^2/2$$

Measurements of linear dichroïsm under stress (6) are even more selective : analysis of the third moment variations allows one to distinguish between the contributions to $<E^2>_{NC}$ of the tetragonal and trigonal modes of vibration. When such studies are made over a wide temperature range, they allow a determination of the average frequency of the modes of a given symmetry (7).

C − Some Examples in Perturbation Spectroscopy

(i) <u>Zero phonon lines and broadbands</u>. Very direct information about the nature and strength of the Jahn-Teller coupling may be obtained when the coupling to the lattice vibrations is not too large so that the zero phonon line is observed. In absorption the transitions which result in this line end up in the ground vibronic state of the excited electronic level. Then from the discussion of I(B) it is clear that the Zeeman and spin-orbit structures are strongly reduced as a consequence of the coupling of the electronic and nuclear motions. More precisely one measures the spin-orbit coupling constant $\lambda K(T_1)$ and orbital Landé factor $g_L K(T_1)$ reduced by the Ham reduction factors $K(T_1)$. Comparison of the measurements made on the broad band (see equation (7)) gives a direct determination of $K(T_1)$ (8). Similarly, comparison of stress effect (9) on the broad band and on the zero phonon line give directly the reduction factors $K(E)$ and $K(T_2)$ for operators transforming like the representation E_g and T_{2g} respectively [the effect of stresses applied along [100] and [111] directions are represented by operators transforming like E_g and T_{2g} respectively].

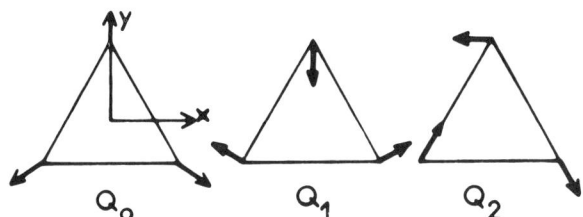

Fig. 2. Normal modes of a triatomic molecule. Q_0 is totally symmetrical, Q_1 and Q_2 are degenerate and transform like the E representation.

One may wonder how in the presence of strong quenching of the angular momentum a finite variation in the first moment of the band is observed (see equation (7)). In fact this variation is not due to Zeeman splittings but to small admixtures of the vibronic states induced by the magnetic field. Then for σ_+ light the oscillator strength of the transitions on the high and low energy side of the band is increased and decreased respectively. This produces an apparent shift of the band in the high energy direction.

(ii) <u>Orbitally degenerate ground state</u>. We shall shortly describe an example of application of these techniques to a system having an orbitally degenerate ground state level. Consider the F_3 (or R) centres formed by the association of three F centres in the configuration of an equilateral triangle. The three modes of vibrations are defined in Fig. 2 : they transform as the representations A_1 and E of the C_{3v} group respectively. The R_2 band is due to transitions taking place between the orbital doublet ground state E and the excited orbital singlet A_2 (Fig. 3). The ground vibrational state of the excited levels transform as A_2 and so does the vibrational state obtained by the excitation of one totally symmetrical phonon A_1. The vibrational state corresponding to the excitation of one E phonon transform like the representation E. Straightforward application of group theory leads to the selection rules shown in Fig. 3. One notices that the selection rules for absorption of σ_+ and σ_- light are reversed when the latter takes place with the simultaneous creation of an E phonon. One then expects a change of the sign of the dichroïsm for such an absorption. The circular dichroïsm observed in the R_2 band of the F_3 centres in KCl is shown in Fig. 4. A sudden decrease of the circular dichroïsm is observed for $E = 13570$ cm^{-1}. This shows that the peak of the vibrational structure observed at this energy is due to a non totally symmetrical E mode of vibration. Comparison of the dichroïsm observed in the zero phonon line and in the first phonon peak at 13540 cm^{-1} shows on the opposite that this peak is due to a totally symmetrical mode.

From Fig. 4 it is immediately seen that the integrated intensities $A_o^{\pm}$ of the zero phonon line are the same for σ_+ and σ_- light and indeed, the selection rules shown in Fig. 3 lead to :

$$\frac{A_o^+ - A_o^-}{A_o^+ + A_o^-} = \tanh \frac{g_L R \beta H}{kT} \tag{8}$$

where in the expression of the Zeeman splitting $g_L R \beta H$ the orbital Landé factor g_L is reduced by the Ham factor R. The situation for the F_3 centre is in fact more complicated since both the ground state and the excited level have an additional spin degeneracy of 2. It can be shown however that the expression (8) remains valid if one writes

$$H = H_o + \frac{\lambda}{g_L \beta} \langle S_z \rangle$$

where H_o is the applied field, λ is the spin-orbit coupling constant and $\langle S_z \rangle$ is the spin polarization. Since the sign of the dichroïsm changes for some of the vibrational peaks, one expects the dichroïsm of the whole band to be smaller than that observed in the zero phonon line. This is indeed the case and it can be shown (4) that if A_b^+ and A_b^- are the moment of order zero of the whole band one has

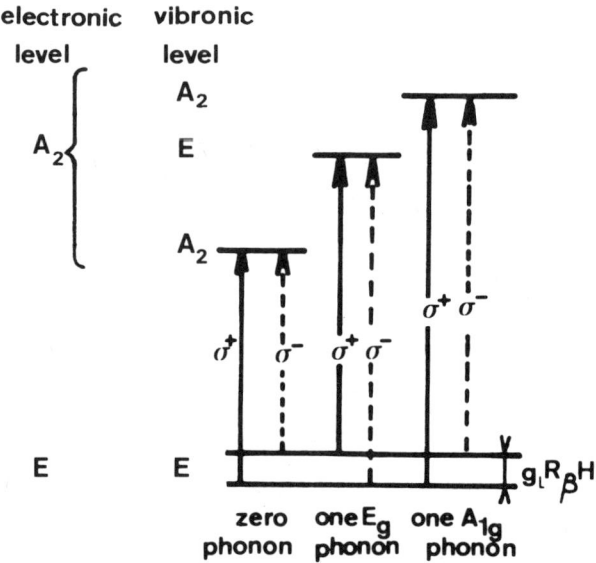

Fig. 3. Simplified model of the selection rules for F_3 centres.

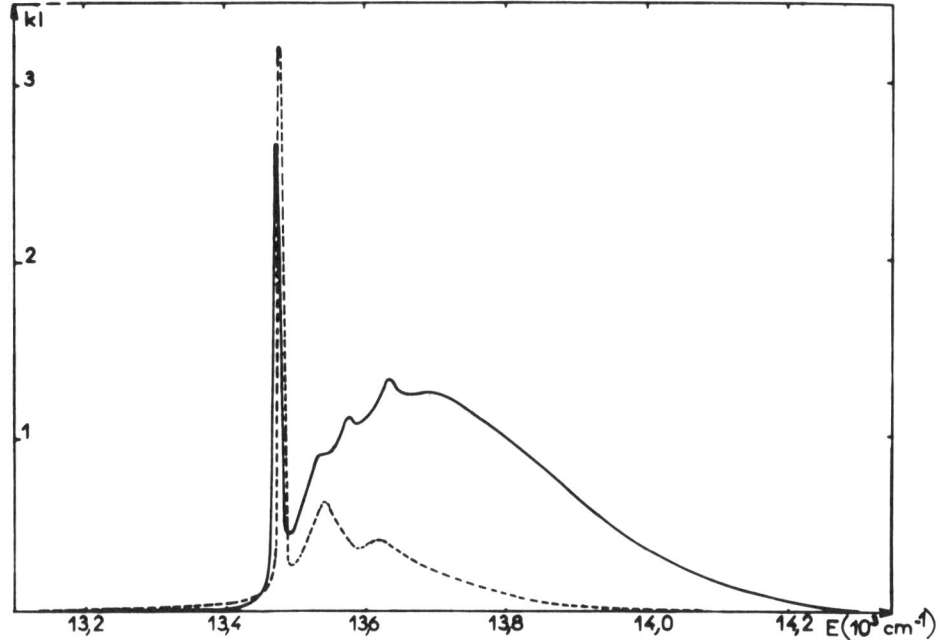

Fig. 4. F_3 centres in KCl. ——— absorption ;
---- circular dichroïsm.

$$\frac{A_b^+ - A_b^-}{A_b^+ + A_b^-} = R \tanh \frac{g_L R \beta H}{kT}$$

Comparing to the expression (8) giving the dichroïsm in the zero phonon line one notices that it is reduced by the Ham reduction factor. So again comparison of the dichroïsm measured in the zero phonon line and in the broad band allows a direct determination of the Ham reduction factors. Contrary to the case of a non degenerate orbital ground state the dichroïsm is smaller in the broad band than in the zero phonon line.

Similar studies can be done on the circular or linear polarization of emission bands (9). Then the emission intensity is related to the shape function $g_\eta(E)$ by

$$I_\eta(E) = C' E^4 g_\eta(E)$$

and in the expression (4) of $g_\eta(E)$ a and b are now the excited and ground state level respectively.

II - OPTICAL DETECTION OF PARAMAGNETIC RESONANCE IN EXCITED STATES

A - Introductory Remarks

An excellent review of optical techniques in EPR has been published recently by Geschwind (10). In this lecture I will restrict the discussion to measurements on the excited states of simple defects. In particular I describe in detail some results concerned with the excited triplet state of the F centre in CaO (11). These results give a very good illustration of the various aspects of one of the more important techniques for optical detection of magnetic resonance : the observation of the change in the polarization of light emitted by a centre when microwave transitions are induced between magnetic levels of the emitting state. Subsequently I show briefly that the same technique applied to the metastable triplet state of the self-trapped exciton in alkali halide crystals gives very similar effects (12,13). In outline I describe a technique which utilizes optical spin polarization memory in the pumping cycle : it is of importance to colour centre people since it allowed Mollenauer et al (14) to detect resonances in the excited state of F centres in the alkali halides. Due to lack of time little will be said of the technique used successfully by the Neuchatel group (15) in their studies of the excited levels of F and F aggregate centres in alkali halides.

B - Detection of Magnetic Resonance Using Polarized Luminescence

(i) <u>Experimental technique</u>. The basic principle of this method is illustrated in Fig. 5. A triplet state in a magnetic field emits light from the $M_S = 0, \pm 1$ levels according to the selection

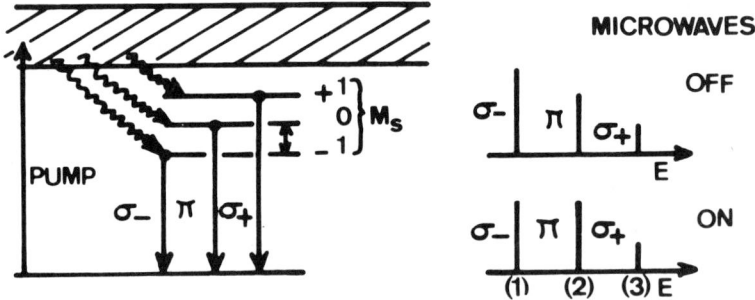

Fig. 5. Detection of EPR using polarized luminescence.

rules shown. If at low temperature the relaxation rates inside the triplet state are fast relative to the radiative lifetime then a Boltzman equilibrium is obtained between the populations of levels $M_S = 0, \pm 1$. Thus at sufficiently low temperature and high magnetic fields appreciable population differences exist between the three levels. This is shown schematically in Fig. 5 where the line (1) due to emission from the -1 level is more intense than the line (2) due to emission from the 0 level. Consequently when microwave transitions are excited between levels $M_S = -1$ and $M_S = 0$ the population difference is decreased, line (1) decreases in intensity and line (2) increases. In general the emission lines are broadened by interaction with lattice vibrations and/or internal strains so that the Zeeman effect is not resolved. The resonance can still be observed, however, since it gives rise to a change in the polarization of the emitted light. For example excitation of the $M_S = -1 \rightarrow 0$ transitions results in a decrease of σ- light and an increase of π polarized light.

A schematic diagram of the necessary equipment is shown in Fig. 6. The exciting radiation is filtered by passage through the filter 1 (in some cases this filter may be replaced by a monochromator). The sample is situated in a microwave cavity which is itself immersed in the helium bath of an optical cryostat. The latter is located in the gap of the magnet. The luminescence from the sample is passed through a polarization analyser and a filter (2) (which again may be a monochromator) before being focussed on the photomultiplier. A small mirror allows the observation of the light emitted in the direction of the magnetic field so that circular polarizations may be measured. Alternatively the light emitted in a direction perpendicular to the magnetic field may be monitored. The microwave part of the equipment is very simple. It includes a klystron and its power supply, an isolator, an attenuator and a 10 db coupler which allows one to look at the microwave power reflected by the cavity and check that it is correctly coupled. Many microwave cavity systems may be used depending upon the wavelength range of the microwaves and on the nature of the light sources used. Using a laser the light is very easily admitted into the cavity through a small hole. When the source is a conventional lamp the optical beam is wider and the light is usually passed through slits cut in the walls of the cavity. A cavity using the properties of guided waves above the cut off frequency has been designed in our laboratory (16).

Two methods are currently used to detect the resonance. In the first one (Fig. 6a) one monitors the circular (or sometimes the linear) polarization of the emitted light with a polarization modulator, irradiating with microwaves continuously. Then, as the field is swept through resonance a variation of the luminescence polarization is observed. In the second technique (Fig. 6b), light of a given polarization is monitored while chopping the microwave

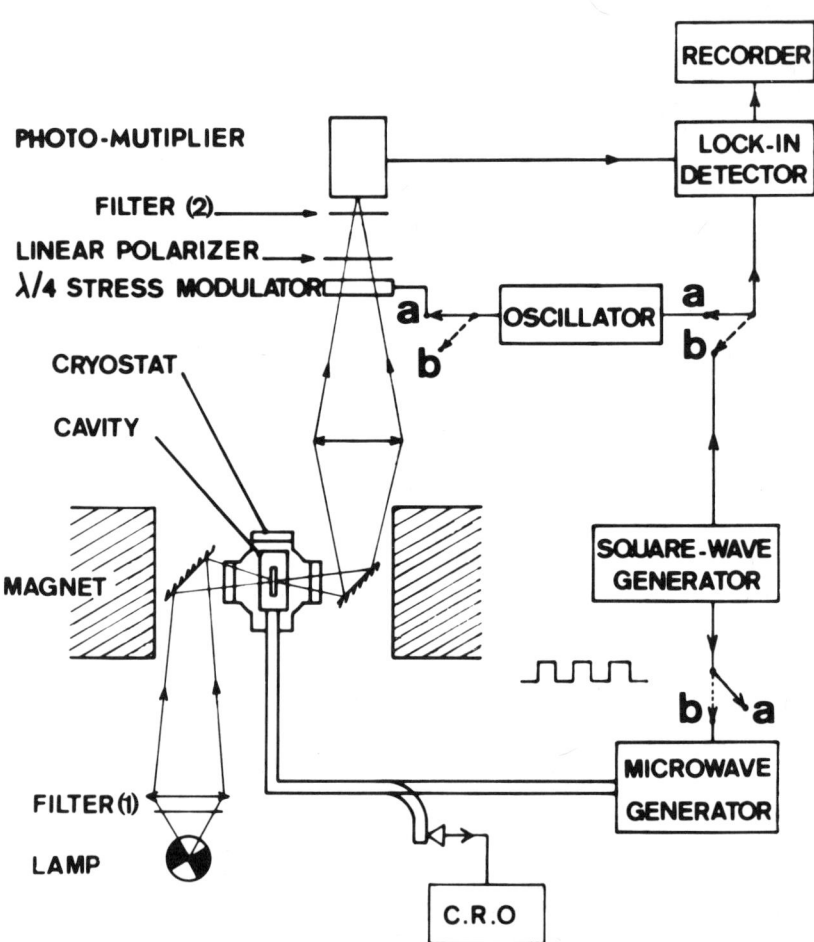

Fig. 6. Experimental arrangement for optical detection of magnetic resonance.

power at an audiofrequency. If this frequency is small with respect to the relaxation rates inside the excited level, the intensity of the light is modulated at the same frequency and may be detected with lock-in techniques. Examples of spectra obtained with this technique are shown in Fig. 7.

The same apparatus can be used for measurements of relaxation times ; one then looks for evolution of the light level when the microwaves are being chopped. Sometimes these variations can be directly seen on the oscilloscope, although more typically signal averaging techniques have to be used.

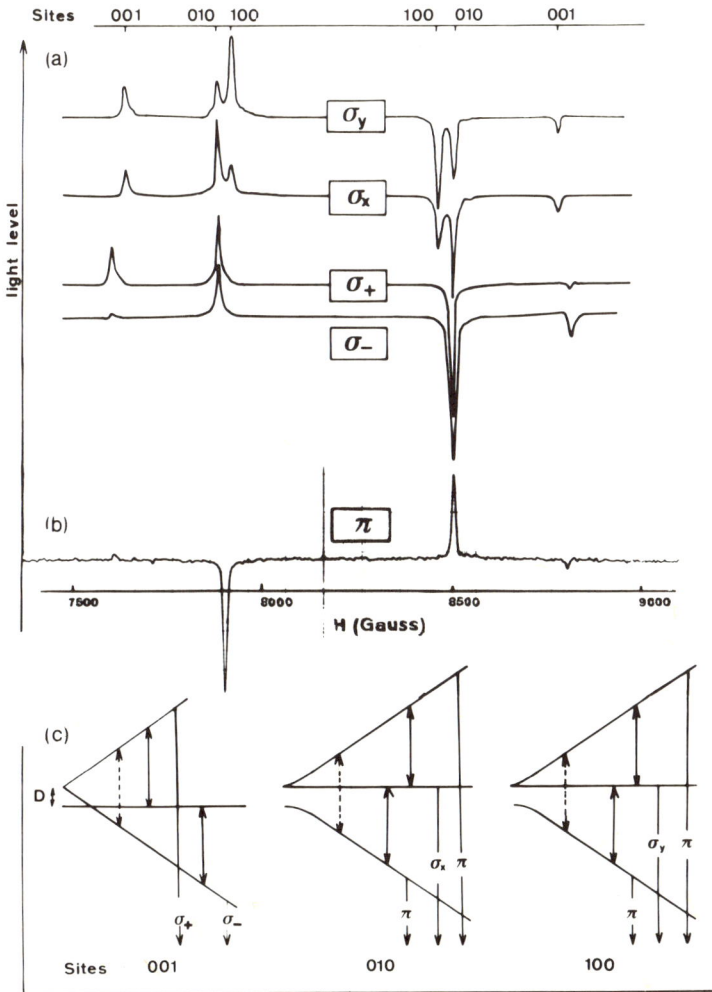

Fig. 7. Observation of the magnetic resonance using the variation in the intensities of the emitted light of various polarizations (a) σ_y, σ_x, σ_+, and σ_- (schematic) ; and (b) π (facsimile). (c) Predicted selection rules for the emission of polarized light. The magnetic field is along the [001] axis of the crystal, except for the σ_x and σ_y spectra where it is slightly misaligned. After Edel et al (11).

A very well known advantage of the optical detection of resonance is its high sensitivity. One makes transitions between states separated by approximatively 1 cm^{-1} but observe change in the number of photons of much higher energy (20 000 cm^{-1}). The minimum number of centres which can be detected depends very much on the system being studied. In favourable conditions (ruby) the sensitivity may be as high as 3×10^4 paramagnetic centres, a factor 10^7 better than the sensitivity obtained with a good conventional ESR spectrometer (10).

(ii) <u>The F centre in CaO - EPR spectra and optical selection rules</u>. In divalent compounds, the F centre is an anion vacancy having trapped two electrons. In CaO, which has the rocksalt structure, it is located at a site of cubic symmetry. The energy level scheme is analogous to that of the helium atom. The strong absorption band observed at 25 000 cm^{-1} is due to transitions from the non-magnetic $^1A_{1g}(1S)$ ground state level to a $^1T_{1u}(1P)$ level. The fluorescence band observed around 16 700 cm^{-1} by Henderson et al (17) has been attributed to the $^3T_{1u} \to {}^1A_{1g}$ transitions.

Using this fluorescence as the monitoring light, the EPR spectra shown in Fig. 7 were observed by Edel et al (11). The dependence of the spectrum on the field orientation shows the existence of three equivalent tetragonal sites. The spin hamiltonian is of the usual form for an $S = 1$ level and tetragonal symmetry. The existence of tetragonal spectra is interpreted as evidence of a strong Jahn-Teller coupling of the T_{1u} orbital states to E_g modes of vibration. Then a tetragonal deformation of the crystal around the centre takes place and the orbital degeneracy is lifted showing an orbital singlet as the ground state. As shown by Ham (2), second-order effects of the Zeeman orbital interaction and of the spin-orbit coupling give rise to a g shift and a zero-field splitting D. In the strong coupling approximation, the latter is given by

$$D = D_d + \lambda^2/3 \, E_{JT} - \lambda'^2/\Delta,$$

where D_d is the dipole-dipole interaction, λ the spin-orbit coupling coefficient inside the $^3T_{1u}$ level and λ' a matrix element of the spin-orbit interaction coupling $^1T_{1u}$ and $^3T_{1u}$ levels. Note that for Russel-Saunders coupling λ and λ' are both equal to half the spin-orbit coupling constant specified for a single electron. Analysis of the spectrum shown in Fig. 7 gives $D = +603$ G. From EPR measurements under uniaxial stress we found (18) λ to be of order 7 cm^{-1}. Definitions of the energy separation between $^1T_{1ux}$, $^1T_{1uy}$ and $^3T_{1uz}$, Δ and the Jahn-Teller energy E_{JT} are shown in Fig. 8. Order of magnitude estimates of these energies may be obtained from the optical measurements of Henderson et al (19).

The $^3T_{1u} \to {}^1A_{1g}$ transitions become partially allowed by the

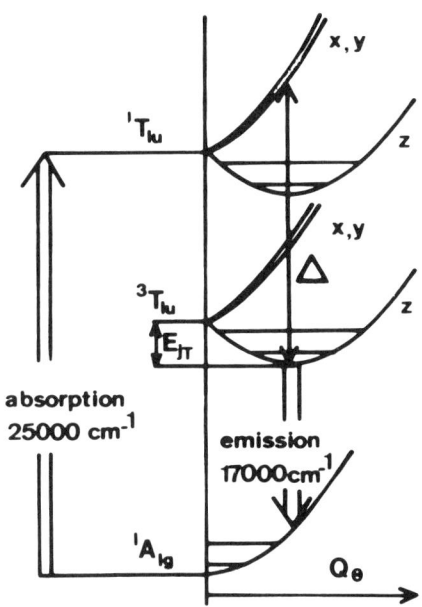

Fig. 8. Energy level diagramme of the F centres. Q_θ is a tetragonal deformation

mixing through spin-orbit interaction of the $^3T_{1u}$ and $^1T_{1u}$ states. Thus the selection rules shown in Fig. 7 are easily established. Consider first the [001] site, where the orbital part of the wave function behaves as a p_z function. Inside a p (or T_{1u}) level, the only matrix elements of the angular momentum different from zero are non-diagonal and of the form $<p_x|l_z|p_y>$. Consequently, the spin-orbit coupling mixes the ^3P_z states with ^1P_x or ^1P_y states only, and the emitted light is polarized along the x or y directions. This explains why, quite generally, a centre lying in a given direction emits light polarized in the plane perpendicular to this direction. When the field is applied along the axis of the centre as shown in Fig. 7 for the [001] site, the Zeeman effect is linear : the $|+1>$ and $|-1>$ states emit σ_+ and σ_- light respectively and the $|o>$ is non-radiative. Consideration of Fig. 7 shows that the selection rules are obeyed. Identification of the lines is shown at the top of the figure. One notes that the two extreme lines, which are due to transitions within the [00$\bar{1}$] site, are not observed in π polarization, as predicted by Fig. 7c. The low and high field lines are observed in σ₊ and σ₋ light respectively which shows that the zero field splitting D is positive. For D negative, the situation would be reversed. One also finds that the lines due to transitions within the [100] and [010] sites are mainly observed with light polarized along the y and x directions respectively. In that case, the selection rules are not perfectly obeyed, and we shall see that this is due to dynamical effects. From the sign of the variation of intensities in Fig. 7, one deduces that in a field of 8 000 G and at a temperature of 1.8 K, thermalization was approximately achieved in the excited level.

(iii) <u>Observation of level crossing</u>. Fig. 9 shows the circular and linear polarizations measured at 1,6° in the absence of microwave radiation. A narrow "resonance" is observed at 600 G which according to the value of D measured by resonance is just the field at which the $|-1>$ and $|o>$ levels of the [001] centres cross each other. Such level crossing, or zero frequency resonance effects have been observed recently in luminescence spectra of a variety of triplet states in inorganic and organic crystals. For

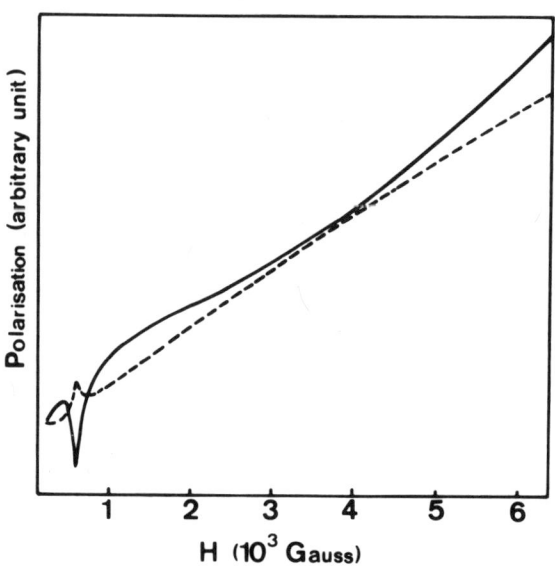

Fig. 9. Circular (dashed curve) and linear (solid curve) polarization of the light emitted by F centres in CaO after Edel et al (11).

levels having large zero field splitting the ordinary EPR has to
be done at very high frequency and is experimentally difficult.
The optical detection of level crossing is very simple. It has been used for the measurement of zero field splitting
of excitons in the alkali iodides (20).

(iv) <u>Observation of dynamical effects in the $^3T_{1u}$ states of
the F centre</u>. Fig. 10 shows spectra which are markedly different
from those of Fig. 7. They were obtained at 9 GHz instead of 24,
and with a well annealed sample to minimize internal strains. The
predicted selection rules are no longer obeyed, and a microwave
induced spin flip of the [001] centre has an effect on the intensity of the π light which is emitted by [100] and [010] centres
only. This effect does not originate in cross-relaxation since it
is concentration independent, rather it is an elegant demonstration that the three sites are actually different vibronic states
$|x>>$, $|y>>$ and $|z>>$ of the same centre between which tunneling
may take place in a time short with respect to the radiative lifetime.

(v) <u>Resonance of the self-trapped exciton</u>. It will now be
shown that the energy level scheme and selection rules described
above are also useful in the study of a very different type of
defect, the self-trapped exciton in alkali halides. When excited
by U.V. light, X-rays, or electrons, these crystals show a characteristic luminescence which is attributed to the radiative decay
of the self-trapped exciton. The hole is highly localized on two
adjacent covalently bonded halide ions, so that the self-trapped
exciton may be regarded as a V_K centre (self-trapped hole) having
trapped an electron. In most of the alkali halides, two emissions
bands are observed, the low energy one being assigned to transitions from a spin triplet (20). Magnetic resonance in this triplet
has been observed in a number of alkali halides using the technique described above (12,13).

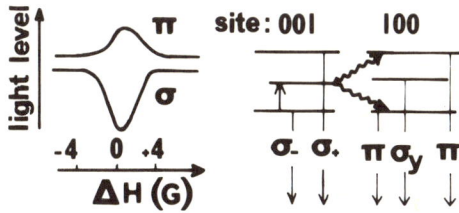

Fig. 10. Saturation of the high field line of the [001] centres.

Typical spectra from K Br are shown in Fig. 11. Using the selection rules shown in Fig. 7c, the spectra lines are easily identified. First recall that the symmetry of the defect is D_{2h} in K Br and that there are six equivalent sites. For a magnetic field along a [110] direction one expects to observe three groups of lines due to sites [110] and [1$\bar{1}$0], which are respectively parallel and perpendicular to the field, and due to the four remaining sites, which make an angle of 60° with the field. The low and high field lines of our spectra are observed in σ_+ and σ_- light respectively, so by reference to Fig. 7c, they are clearly due to [110] sites and the zero field splitting D is positive. The lines due to the [1$\bar{1}$0] sites overlap strongly with the parallel spectrum, so that they can be clearly observed only when the observation of the latter is forbidden by the optical selection rules, i.e. in π and σ_- lights for the low field line, and in π or σ_+ light for the high field line.

The study of this system would have been impossible by the conventional EPR. The lines of the various sites are very broad and overlap strongly, so that their identification by measuring the orientation dependence of the spectrum would be impossible. The identification of the lines and the precise measurement of their positions are only possible, because, by choosing the polarization of the monitoring light one can detect a definite tran-

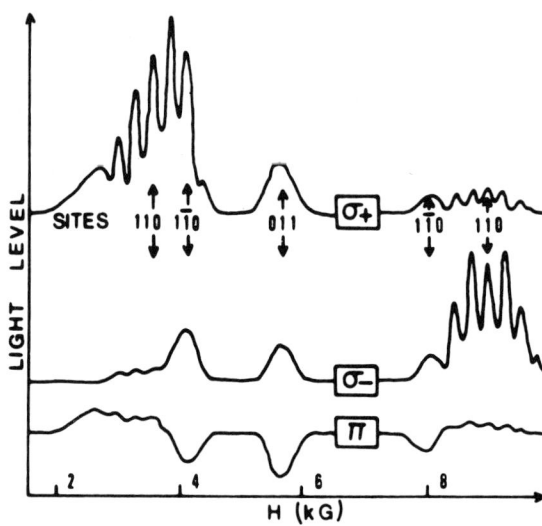

Fig. 11. EPR spectra of the self-trapped exciton in K Br. The magnetic field was along a [110] direction of the crystal. After Wasiela et al (12).

sition from a definite site.

Interpretation of these spectra is very similar to that used above for F centres. The first excited configuration of the bromine ion is $(4p)^5$ 5s. The "p" hole is shared between two adjacent bromine ions and the resulting molecular field splits the orbital triplet into $^3\pi_u$ and $^3\Sigma_u$ levels. If E is the value of this splitting, the parameter defining the spin hamiltonian of the $^3\Sigma_u$ level are calculated as above for the F centre, except that E replaces the Jahn-Teller splitting 3 E_{JT}, and that a straightforward modification has to be made to take into account the lower symmetry. The transitions from the $^3\Sigma_u$ states to the $^1\Sigma_g$ singlet ground state are made partially allowed by the mixing through spin-orbit interaction of the $^3\Sigma_u$ states with the more excited $^1\pi_u$ states. The situation is analogous to the F centre case, except that a p hole replaces the p electron : hence the selection rules are identical.

(vi) <u>Identification of defects</u>. The techniques we have described above are clearly of great value in elucidating many features in the electronic structure of defects. An extra benefit which they have is that they may actually be used to identify the new defects in crystals. Recently, in our laboratory, we have investigated the X-ray excited luminescence of $SrCl_2$ crystals. Two luminescence bands are observed situated respectively at 3100 Å and 3900 Å. Comparison with the luminescence and circular polarization spectra of the alkali halides suggests that 3100 Å and 3900 Å bands are due to singlet → singlet transitions and triplet - singlet transitions of the exciton. Measurements of the EPR in the excited state demonstrate unequivocally that the 3100 Å band is due to the triplet - singlet luminescence of the exciton whereas the 3900 Å band is due to some unidentified defect.

C - Less General Methods

(i) <u>The use of optical spin memory - F centres in alkali halides</u>. The method outlined in B cannot be applied to F centres in alkali halides since the circular polarization of the fluorescence light is very small. Mollenauer et al (14) devised an ingenious method which relies on spin memory. The changes in the excited states populations induced by resonance are reflected in changes in the ground states populations. The latter may be measured with great sensitivity by monitoring the magnetic circular dichroïsm of the absorption band. The optical pumping cycle of the F centres in alkali halides is shown in Fig. 12. The ground state populations are n_+ and n_-, u_+ and u_- are the rates per F centres for pumping in a given region of the absorption band with (say) σ_+ light. n_0^+ and n_0^- are the populations of the $\pm\frac{1}{2}$ states of the relaxed excited level. In the process of pumping from the ground state to the re-

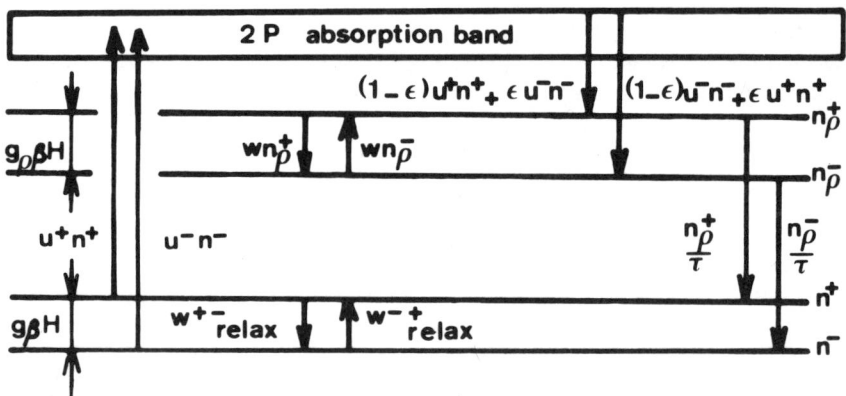

Fig. 12. Optical pumping cycle for F centres in alkali halides, after Mollenauer et al (14).

laxed excited state an electron has a probability ε of spin reversal (i.e. to lose its "spin memory"). Spin is assumed to be conserved in the transitions from the relaxed excited level to the ground state and by time reversal symmetry the lifetimes ($\simeq 10^{-6}$) of the two Zeeman sub-levels are identical. The relaxation times in both ground and excited state are assumed to be long compared with the lifetime. The pump power needs not be very large and in the first experiments of Mollenauer et al (14) the conditions $u^+ + u^- \ll 1/\tau$ was satisfied. Then the rate equations lead to the following expression for the steady state polarization of the ground state level

$$P_e = \frac{n_+ - n_-}{n_+ + n_-} = \frac{P_{es} - [T_P/T_1]\tanh(g\beta H/2kT)}{1 + T_P/T_1}$$

where $1/T_P = (\varepsilon + \omega\tau)/(u^+ + u^-)$ is the ground state spin-flipping rate due to optical pumping and resonance in the relaxed excited state ; $P_{es} = (u^- - u^+)/(u^- + u^+)$ is the value of P_e for saturated optical pumping $T_P \ll T_1$. One sees that the polarization P_e depends on the microwave transition rate ω through the pumping rate $1/T_P$. When the magnetic field is swept through resonance, the transition rate ω is large and a change of P_e is observed. Very small variation of P_e are easily detected when monitoring the circular dichroïsm. The optimum sensitivity is attained when the pump power is reduced to the point where $T_P \simeq T_1$. Then the response time of the system is of the order of T_1 (some seconds) which does not allow the use of lock-in detection. In order to detect Endor Mollenauer et al (21) developed a new detection scheme using saturated optical pumping $T_p \ll T_1$.

(ii) <u>Resonance of pairs of F centres</u>. EPR and Endor from close pairs of F centres, F_2 centres and F_3 centres have been observed by the Neuchatel group (15). They monitor the total intensity of the emitted light. At low concentration F centres optically excited in the F band decay radiatively with a quantum yield of unity. At concentrations larger than 10^{16} cm^{-3}, there is a high probability that some F centres have near neighbour F centres : for these centres the quantum yield is smaller due to non-radiative transitions which are attributed either to the formation of a transient F'+ vacancy pair (22) or to an instability in local vibrational modes leading to a radiationless disexcitation (23). In either case the intermediary state is a spin singlet and the non-radiative process is effective only for pairs of centres which have opposite spins. This has been clearly shown by Porret and Lüty (24) who observed that at low temperature and high magnetic field where the spins are highly polarized the quantum yield of a concentrated sample is dramatically enhanced. At lower field, when the applied field is of the same order of magnitude as the hyperfine interaction a similar but smaller effect is observed at temperature as high as 50° K. Because of the hyperfine interaction the eigenstates are not pure singlets of triplet spin states. The probability of non-radiative decay is larger for the state which contains the larger proportion of singlet state. A microwave induced spin flip is then observed as a variation of the total intensity of the emitted light (and not of its polarization).

BIBLIOGRAPHY

(1) J. Margerie, J. de Physique $\underline{28}$ C4 - 103 (1967) ; N.Y. Karlov, J. Margerie and Y. Merle d'Aubigné, J. Phys. Rad. $\underline{24}$ 717 (1963) ; F. Lüty and J. Mort, Phys. Rev. Lett. $\underline{12}$, 45 (1964).
(2) F.S. Ham, Phys. Rev. $\underline{138}$, A 1727 (1965).
(3) C.H. Henry, S.E. Schnatterly and C.P. Slichter, Phys. Rev. $\underline{137}$, A 583 (1965).
(4) Y. Merle d'Aubigné and P. Duval, J. Phys. $\underline{29}$, 896 (1965).
(5) M. Billardou and J. Badoz, Compt. Rend. $\underline{262}$, 1672 (1966) ; J.C. Kemp, J. Opt. Soc. Am. $\underline{59}$, 915 (1960) ; S.N. Jasperson and S.E. Schnatterly, Rev. Sci. Instrum. $\underline{40}$, 761 (1969).
(6) S.E. Schnatterly, Phys. Rev. $\underline{140}$, A 1364 (1965).
(7) R.E. Hetrick, Phys. Rev. $\underline{188}$, 1392 (1969).
(8) Y. Merle d'Aubigné and A. Roussel, Phys. Rev. $\underline{B3}$, 1421 (1971).
(9) J. Duran, Y. Merle d'Aubigné and R. Romestain, J. Phys. C : Solid State Phys. $\underline{5}$, 2225 (1972).
(10) S. Geschwind in Electron Paramagnetic Resonance, edited by S. Geschwind (Plenum, New York 1972).
(11) P. Edel, C. Hennics, Y. Merle d'Aubigné, R. Romestain and Y. Twarowski, Phys. Rev. Lett. $\underline{28}$, 1268 (1972).
(12) A. Wasiela, G. Ascarelli and Y. Merle d'Aubigné, J. Phys. suppl. $\underline{34\ C9}$, 123 (1973) and Phys. Rev. Lett. $\underline{31}$, 993 (1973).
(13) M.J. Marrone, F.W. Pattern and M.N. Kabler, Phys. Rev. Lett. $\underline{31}$, 467 (1973).
(14) L.F. Mollenauer, S. Pan and S. Yngvesson, Phys. Rev. Lett. $\underline{23}$, 683 (1969).
(15) Y. Ruedin, P.A. Schnegg, C. Jaccard and M.A. Aegerter, Phys. Stat. Sol. (b) $\underline{54}$, 565 (1972) ; P.A. Schnegg, C. Jaccard and M. Aegerter, Phys. Stat. Sol. (b) $\underline{63}$, 587 (1974).
(16) J. Chamel, R. Chicault and Y. Merle d'Aubigné, to be published.
(17) B. Henderson, S.E. Stokowski and T.C. Ensign, Phys. Rev. B $\underline{183}$, 826 (1969).
(18) Le Si Dang, Y. Merle d'Aubigné and Y. Rasoloarison, to be published.
(19) B. Henderson, Y. Chen and W.A. Sibley, Phys. Rev. $\underline{B6}$, 4060 (1972).
(20) W.B. Fowler, M.J. Marrone and M.N. Kabler, Phys. Rev. $\underline{B8}$, 5909 (1973).
(21) L.F. Mollenauer, S. Pan and A. Winnacker, Phys. Rev. Lett. $\underline{26}$, 1643 (1971).
(22) F. Lüty, Halbleiterphysik, Vol. VI, 238 (1961) (Vichweg, Braunschweig).
(23) C. Jaccard and M. Aegerter, Phys. Lett. $\underline{44A}$, 391 (1973).
(24) F. Porret and F. Lüty, Phys. Rev. Lett. $\underline{26}$, 843 (1971).

EXCITED STATE PROPERTIES OF LOCALIZED CENTRES

Giorgio M. Spinolo

Istituto di Fisica, Università degli Studi di Milano

Via Celoria 16, Milano, Italy

The aim of the present lectures is to give a brief general outline of the excited state problem in localized centres either point defects or foreign impurity atoms in ionic crystals, in practice, alkali halides. Attention will be focussed on phenomenology and we will go through the main techniques discussing how the results of the experiments help to construct an increasingly detailed picture of the excited state.

In real crystals, as it is well known, there are always present their number depending on temperature, positive and negative ion vacancies. A negative ion vacancy, with respect to the local charge distribution, acts as a positive charge, captures an electron and becomes an F centre, the basic colour centre. Much in the same way behave positive ion vacancies.
In the same crystals there are also divacancies and more complex clusters of vacancies and so we have, after each one has captured an electron, F_2, F_3, F_4,.... (the traditional names are M, R, N,... respectively) centres.
If one of the nearest neighbours of the F centre (an alkali ion) is substituted by a different alkali ion we have an F_A centre; if one of the next nearest neighbours of the F centre (an halide ion) is substituted by a different halide ion we have an F_H centre (until now observed only in the body centred cubic caesium halides).

The information I have just given together with a number of other more specialized facts, are now rather well established but

it took about twenty years, from the middle thirties to the late
fifties,to prove them. The most important instrument to study the
colour centre models were at the beginning the photochemical reactions; later on, optically induced dichroism and electron spin
resonance have also been extremely helpful. Few of the arguments
I have just mentioned receive greater attention in parallel lectures.

In our crystals we can also have impurities either added in
the melt during growth or introduced by diffusion at high temperatures. The problem of identification of the ionic model is simple
when the impurity is not associated with defects, as occurs for divalent ions like Ca^{++}, Sr^{++}, Ba^{++} and others. The electronic problem
is,in many ways,similar to that of the colour centres.

THE BASIC ASPECTS OF THE ELECTRONIC STRUCTURE OF LOCALIZED CENTRES

To describe the electronic structure of a localized centre is
still useful to consider the configurational coordinate diagram of
a two level system. One important fact is that,for a centre   in
ionic crystals,we cannot describe always the cycle absorption-emis-
sion with only one c.c. diagram. After absorption, being the charge
distribution of the excited state different from that of the ground
state, there is a rearrangement of the ions around the centre, a
relaxation due to the anharmonic part of the potential. In absorption the transition is between the ground state and the unrelaxed
excited state (u.e.s.);in emission the transition is between the
relaxed excited state (r.e.s.) and the relaxed ground state. This
is certainly the case for the F centre; the F_2 centre, due to its
rather localized nature, can be sufficiently well described in the
frame of a single diagram.
With the c.c. diagram one can also give an elementary description
of electron-phonon interaction, the Stokes shift and configuration
crossing and mixing. However, the minimum sofistication of each
problem of interaction,(electron phonon, Jahn-Teller effect, electronic configuration) or of external field perturbation (Stark,
Zeeman, uniaxial stress) puts in evidence many limitations of the
model which must be regarded as a zero order description. Even
fundamental aspects of the electronic structure of the centre,as
the distance of the relaxed and unrelaxed excited states from the
conduction band,can hardly be represented in a c.c. diagram.

Let us now proceed to define in some detail the physical
quantities of interest with particular regard to those having greater connections with the phenomenology of the excited state. What

one wants to know about the u.e.s. is the energy position of the
levels respect to the ground state and respect to the conduction
band and the symmetry of the associated wave functions, the magnitude of the electron-phonon interaction, the magnitude of the spin-orbit splitting; the relative probability of the various transitions,
the effect of different perturbing fields. Further, one wants to
know which is the energy position, symmetry and electronic structure
of the r.e.s. and how is changed the situation respect to the u.e.s.
Other problems are the lifetime of the r.e.s., the ioniziation
energy from the r.e.s. to the conduction band, the energy position
in the relaxed situation of the higher excited states and so on.

Many other interesting details and particular cases can be
studied: among them we will consider the off-center effects, referred to the Cu^+ impurity. Cu^+ (and also Ag^+) substitutional in alkali
halides, in some host matrices does not occupy the regular lattice
position of the ion it substitutes but it goes off-center in one of
the eight 111 or of the six 100 directions. This fact has consequencies on most of the physical quantities we have considered and appears to be of great didactical value.

THE UNRELAXED EXCITED STATE

The Oscillator Strength

One of the simplest methods to have information on the u.e.s.
is to study the absolute magnitude and the temperature dependence
of the oscillator strength through optical absorption methods.
The Smakula's equation:

$$Nf = 1.29 \cdot 10^{17} \frac{n}{(n^2+2)^2} \alpha_{max} W \qquad (1)$$

where N is the number of centres, f is the oscillator strength of
the transition considered, n the refractive index, α the absorption
coefficient (in cm^{-1}) at the absorption peak and W the width at half
height (in eV), gives the possibility to evaluate if the transition
responsible of the absorption band is allowed or forbidden. This
is possible when one has measured N, mainly through chemical methods.

To make the question simple, we can remember that if a transition is allowed by the symmetry properties of the initial and final
wave functions its oscillator strength is temperature independent.
In other words, the area under the absorption band is constant,
even if the peak shifts and the width changes as a consequence of
the electron-phonon interaction.

If a transition is forbidden we should not be able to observe it and so this should be the end of all. In practice, we do not always have "pure" wave functions, from the point of view of symmetry, but quite often local fields due to static distortions of the lattice or odd parity vibrations mix together wave functions of different parity and this fact has relevant effects on the matrix elements of the transition probability. Going into details of the mixing mechanism, the odd parity vibration puts an oscillating electric field on the centre. The strength of the field depends on the amplitude of the vibration and the latter on the average number of phonons, say, temperature; the oscillating field produces mixing of wave functions and so the matrix elements and the oscillator strength of the transition will increase with increasing temperature. Ag^+ and Cu^+ in NaCl (1) offer the possibility to check that f is proportional to the average number of phonons, $\bar{n}$:

$$f \propto \bar{n} \propto \coth \frac{\hbar\omega_{eff}}{2kT} \qquad (2)$$

ω_{eff} being the effective frequency of the interacting odd modes.

We have also forbidden transitions with f temperature independent. To distinguish them from the allowed transitions, it is important to have an evaluation of the absolute magnitude of f. If there is a static distortion of the lattice around the centre, there will be a static mixing and if this mixing is much greater than that introduced by vibrations f will be of the order of 10^{-2} but essentially temperature independent. A typical example is Cu^+ in KCl, where Cu^+ goes off-centre. (1)
In general we can say that f for an allowed transition goes from 1 to 10^{-2} and is temperature independent and for a forbidden transition goes from 10^{-2} to 10^{-4} and below and may be temperature dependent.

Just to have a wider outlook on the subject, we remember that in transition metal compounds we have d-d forbidden transitions bound to the metal. These can be doubly forbidden both for the selection rule on ℓ and for the selection rule on spins according to which, transitions between, say, the doublet and the quartet levels in a d^3 system are not allowed. For what regards the possibilities to bypass the selection rule on ℓ, the situation is essentially similar to that described before for alkali halides. For the spin-forbidden transitions the oscillator strength is given by the spin-orbit coupling (and correspondent mixing) with wave functions of different spin, and is temperature independent. The other possibility is that spin-waves (magnons) produce a spin inversion in an ion neighbour to that in which the spin forbidden transition occurs, so to preserve the local total spin. These processes are more likely in a nearly or completely magnetically ordered situation (you need spin waves) and so near or below the magnetic transition temperature (2) (Néel or Curie, according to the cases): higher oscillator strength at lower temperatures, just the opposite of what occurs for the phonon allowed transitions!

The Energy Level Scheme

We have just commented on one of the most simple methods to have information on the u.e.s. if one has, through different techniques, typically ESR and ENDOR, some knowledge on the ground state. Other fundamental information that can be obtained on the u.e.s. with optical absorption, is that regarding the energy level distribution and the shape of the absorption peaks.

We can simply consider again the F center in alkali halides: a detailed absorption study of the K band (3) has clarified that the higher unrelaxed excited states of the F centre are arranged in a hydrogenic series. The limit of the series gives, obviously, information on the effective dielectric constant which operates in that region of the potential and an evaluation of the radius of the electronic orbits of the u.e.s. Other u.e.s. have also been found at higher energies (the L bands) (4) and attributed to resonant states of the continuum, in other words, of the conduction band. To strengthen this interpretation of absorption data, it has been very helpful to perform some photoconductivity measurements. The results show that at a temperature near 4.2°K in the F band (corresponding to the 1s → 2p transition of the F centre electron) there is essentially zero photoconductivity, while in the K band (corresponding to the envelope of 1s → np transitions of the F centre electron) the photoconductivity yield increases as one approaches the extreme tail of the band. In the L band region the photoconductivity yield is nearly unity.(3) With the coupled information of absorption and photoconductivity, we have established that all the observed bands are due to allowed transitions and have an hydrogenic series disposition; further, we have located the u.e.s. with respect to the conduction band. To study other centres, as for example, F_2 and F_A which have an axial symmetry, one has a tool to check the orientation of the dipoles corresponding to the optical transitions, the optically induced dichroism. (5)

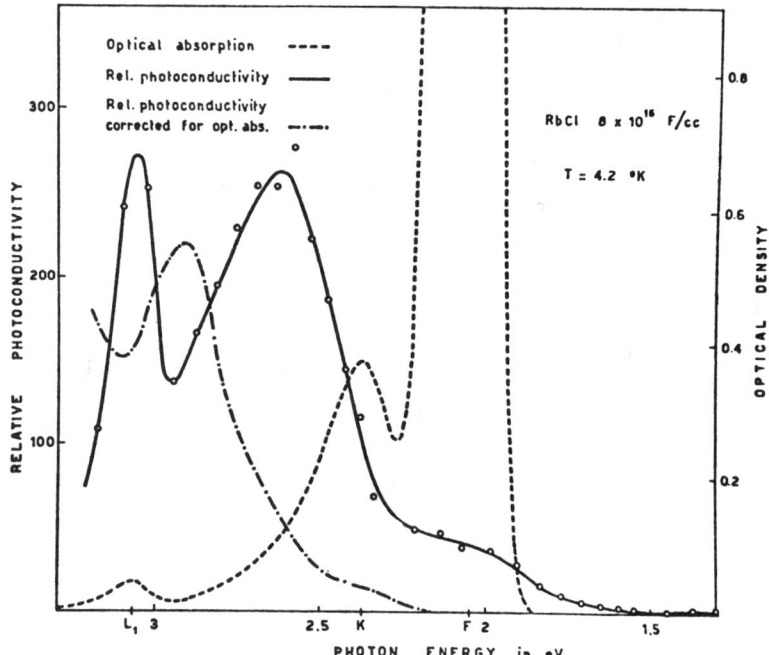

Fig.1 - A comparison of the optical absorption (scale on the right), relative photoconductivity corrected for variations in exciting light intensity, and relative photoconductivity corrected for optical absorption for coloured RbCl at liquid helium temperature.

Let us now consider briefly the studies on the shape of the absorption bands. These measurements are mainly directed to study the electron phonon interaction in the weak (zero and multiphonon structures) and strong (wide unstructured bands) limits. Many works have been made on the shape of the F band, trying to interpret its somewhat puzzling "double gaussian" shape.(6) The effort in this direction, however, stopped when it was realized that the unsymmetrical shape of the F band was mainly related to a purely electronic effect as the spin-orbit interaction. In practice, the F band turns out to be the superposition of the $1s \rightarrow 2p_{\frac{1}{2}}$ and $1s \rightarrow 2p_{3/2}$ transitions.

External Field Perturbation

After it was realized that a complex structure was present in the F band, several studies were performed with various perturbing fields and a deeper insight was obtained on the u.e.s. of the F centre. Analogous studies were performed on other colour centres and impurity bands with modulation techniques. Through these efforts it has been reached a satisfactory understanding on many unrelaxed excited states of localized centres.

As an example of what can be learned on the u.e.s. with the application of external fields we will briefly comment the results obtained with the Faraday and Stark effects (the effect of a stress field will be discussed in parallel lectures).
The magnetic splitting of the ground state can be studied with electron spin resonance and electron nuclear double resonance. To have a corresponding knowledge on the u.e.s. one must think of Zeeman experiments and since the expected splittings are of the order of $10^{-3} - 10^{-4}$ eV and the F band is 0.2 eV wide, some trick to by-pass the difficulty appears necessary: the Faraday rotation is a sensitive tool to detect small changes in the dispersive properties of crystals in a magnetic field.

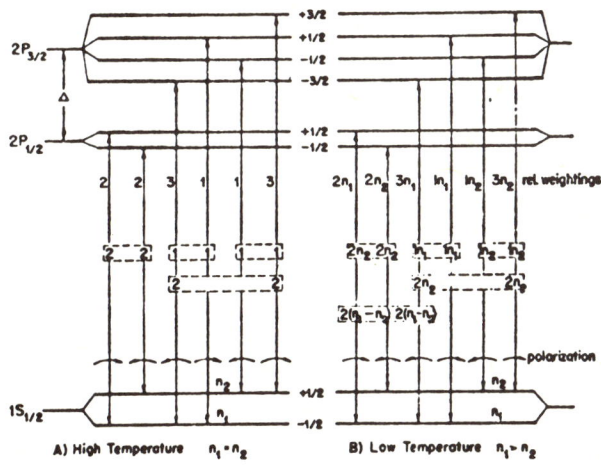

Fig.2 - Energy level diagram used to interpret the results, showing the allowed transitions with their weightings taken from the atomic case. The number of electrons in the spin levels $-\frac{1}{2}$ and $+\frac{1}{2}$ are indicated by n_1 and n_2, respectively. In Fig.2A $n_1=n_2$ and represents the diamagnetic case while in Fig.2B $n_1>n_2$ corresponding to the paramagnetic case. Note that the energy separation between the S and P states is about 2.3 eV whereas Δ is only about 10^{-2} eV. In the case of the F center Δ is negative so that the $P_{\frac{1}{2}}$ levels lie above the $P_{3/2}$ levels.

A study of the temperature dependence of the rotation angle clearly showed that there are two distinct contributions: the "diamagnetic" and the "paramagnetic" one. (7) The diamagnetic effect arises from the Zeeman splitting when equal electron populations exist in the two spin levels of the ground state, and is temperature independent. At low temperatures, below 15°K the Faraday rotation has an additional component proportional to 1/T, the paramagnetic contribution, which is directly related in magnitude to the population difference in the spin levels of the ground state. Both contributions are linear in the magnetic field so that a Verdet constant can be easily evaluated. θ_{dia} depends only on g splittings while θ_{par} depends also on Δ, the spin orbit splitting. Their ratio is equal to:

$$\frac{\theta_{par}}{\theta_{dia}} = \frac{(\Delta E)_g}{(\Delta E)_e} \cdot \frac{\Delta}{2kT} \qquad (3)$$

where $(\Delta E)_e$ and $(\Delta E)_g$ are the mean effective splitting of the u.e.s. and the splitting of the ground state, respectively.
Δ, turns out to be of the order of $3 \cdot 10^{-3}$ eV in KCl and is found to be negative, in other words, the $P_{\frac{1}{2}}$ level lies above the $P_{3/2}$ level: the fact has found subsequently a theoretical explanation.(8)

Until now we have considered only the allowed transitions of the F centre, essentially the $1s \rightarrow np(n=2,3,...)$ transitions. With the electronic field perturbation, the Stark effect, it has been possible to prove the existence of $1s \rightarrow 2s$ and $1s \rightarrow 3d$ transitions (9,11); the energy position, the width at half height and the oscil-

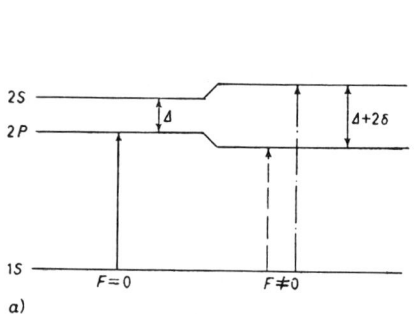
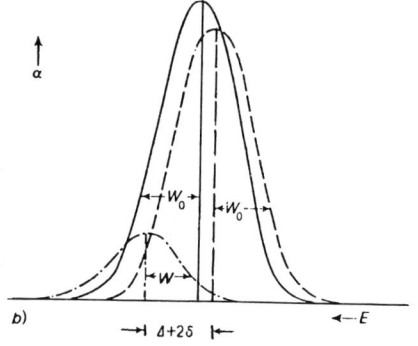

Fig.3 - a) Energy level diagram and b) model explanation of the Stark effect broadening.

lator strength of the corresponding absorption bands has also been measured.

In practice, with the application of the electric field the two nearly degenerate 2p and 2s levels are mixed together so that the 1s → 2s transition becomes partially allowed and correspondently the 1s → 2p transition becomes weaker.

According to the moment theory, (10) the area (zeroth moment) and the centre of gravity (first moment) of the F band does not change with the application of the electric field but the second moment does. This variation, (ΔE_z^2), in the direction of the field, is equal to:

$$\Delta E_z^2 = (eF)^2 |<2s|z|2p>|^2 \tag{4}$$

F, being the intensity of the local field.

It has been observed a broadening of the absorption band due to an increase of the separation 2s-2p induced by the field and to the presence of the partially allowed transition 1s → 2s at about 0.11 eV above the 1s → 2p. As foreseen, the displacement of the levels upon application of the electric field depends quadratically on field strength.

With the same technique but with light perpendicular to the electric field it has been proved also the presence of a 3d state at 0.21 eV above the 2p level (11) but below the 3p level: again an inversion like in the 2p, 2s case.

THE RELAXED EXCITED STATE

The Lifetime Problem in the F-Centre and in Cu^+ in Alkali Halides

The F-centre. After that the electron has been excited to some bound energy level of the unrelaxed excited state or, in other words, if it has not gone directly to the conduction band, it will lose a certain amount of energy emitting about 20 phonons. Part of the charge will now be outside of the box of the nearest neighbours and consequently the surrounding lattice will relax. In general no matter to which of the u.e.s. the electron has been excited, after a time of the order of 10^{-12}s the electron will be in the lowest r.e.s. An instructive case to be considered is that of the F_A centre. It is well known that the u.e. states are different according that the direction of the electric dipole is parallel or perpendicular to the axis centre of the vacancy-foreign alkali ion. Now, no matter if one excites in the F_{A1} or F_{A2} absorption band there will be only one emission band and only one lifetime: the electron will always reach the lowest r.e.s.

For what regards the reciprocal position of the various r.e.s. of the F centre we must pay attention to the potential: the absorption spectrum is sufficiently well explained by using a Simpson potential (3) but certainly in the r.e.s. the Coulombic part will gain importance and so we should be prepared to see the s, p and d levels nearer to each other and maybe in a different order (12). With relaxation the energy distance from the lowest r.e.s. to the conduction band will become smaller and it will be possible to measure it with photoconductivity experiments as a function of temperature.

The electron, after sitting in the r.e.s. for some time, it decays to the relaxed ground state emitting a photon of energy lower than that of the exciting photon (Stokes shift). The time spent in the r.e.s. at low temperatures is strictly related to the downwards transition probability and so from the study of τ_R, the radiative lifetime of the state, we can have information on the symmetry of the r.e.s. When the temperature is increased, the ionization processes take place and so τ becomes shorter because of this new channel of decay. The formula that gives $\frac{1}{\tau}$, the total decay probability, versus $\frac{1}{\tau_R}$, the radiative transition probability and $(1/\tau_0) \cdot \exp(-\Delta E/KT)$, the ionization transition probability is the following:

$$\frac{1}{\tau} = \frac{1}{\tau_R} + \frac{1}{\tau_0} e^{-\Delta E/KT} \tag{5}$$

where $\frac{1}{\tau_0}$ is the escape trials frequency and ΔE the ionization energy, the energy distance from the r.e.s. to the conduction band. The relation with which one studies τ_R has been proposed by Fowler and Dexter (13). In substance it says that:

$$\tau_{eg} \propto \frac{E_{ge}}{E_{eg}^3} \frac{|<z>|^2 \text{ abs}}{|<z>|^2 \text{ emiss}} \tag{6}$$

where e and g stand for excited and ground, E_{ge} and E_{eg} are the absorption and emission peak energies, $|<z>|^2$ are the matrix elements for absorption and emission.

Swank and Brown (14) were the first to measure τ vs T for the F centre and found that τ_R was of the order of the μs for several alkali halides. (The radiative lifetime for an allowed transition should be $\simeq 10^{-8}$ s). The temperature region in which the ionization

takes place was easily understood and ΔE evaluated. The problems that remained were the following: i) why the lifetime is so long and which is the symmetry of the r.e.s.; ii) why there is at low temperatures (10-60 K) a change in τ_R; iii) how τ_R depends on the host crystal fundamental parameters. To give an answer to these questions a good number of experimental and theoretical studies were performed. These range from studies on τ of the F_A centre (15) to electric field effects (16), and include magnetic circular dichroic effects (17) and ENDOR spectra of the r.e.s. (18).
We cannot say that everything is now clear but certainly it has been proved experimentally (18) that the r.e.s. extends over several shells of neighbours and so the early "large orbit" (19) proposal is substantially correct. One of the reason for the lifetime to be long is that the overlap integral between the excited state and the ground state wave function is smaller in emission than in absorption. Furhter, the 2p and 2s states are mixed and the 2s may be predominant, in other words the transition to the 1s is partially forbidden. The mixing should be dependent on the amplitude of the odd parity modes coupled to the r.e.s. and so should change with temperature. Up to now, though, it has not been reached a satisfactory quantitative explanation of the temperature behaviour of τ vs T below the ionization temperature range (20). On the grounds of an hydrogenic model it has been possible (20) to show that $\tau_R(0°K)$ depends on the ε_∞ of the host crystal. It appears now obvious that if the orbit of the r.e.s. is large and the potential is

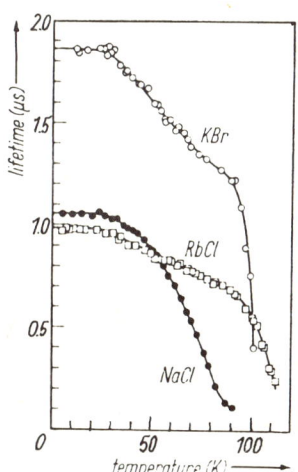

Fig.4 - Experimental plots of the F centre lifetime in KBr, RbCl and NaCl, measured as a function of temperature by the SP technique.

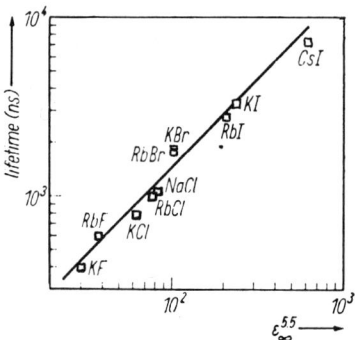

Fig.5 - Plot of the experimental lifetime values at very low temperature, $\tau_R(0)$ for F centres in different alkali halides versus $\varepsilon_\infty^{5.5}$ (ε_∞ is the high-frequency dielectric constant). The interpolating curve $\tau_R(0)=K\varepsilon_\infty^{5.5}$ is also drawn.

essentially coulombic, the radius of the orbit and therefore the overlap integral with the ground state and the lifetime are controlled by the dielectric constant. The presence of a large orbit in the r.e.s. helps also to understand the reason why the lifetime of the F_A centre is in general (let us not examine the few special cases like the F_A:Li in KCl) rather similar to that of the F centre. If the orbit overlaps hundreds of ions it is clear that the substitution of one, even if it is a nearest neighbour, should not change substantially the symmetry and shape of the wave function. In the case of the F_H centre (21) there is a lower Stokes shift respect to the F, clear sign of a smaller relaxation, the lifetime gets considerably shorter than the $\tau_R(F)$ and there are apparently two different relaxed excited states with two different lifetimes.

It may be interesting to consider the F_2 centre since its difference respect to the F. In absorption there are several bands partially overlapping with the F and with well defined dichroic properties: to understand the energy level distribution one should make references to the hydrogen molecule. In the F_2 centre the Stokes Shift from absorption to emission is rather small respect to the F centre and the bands both in absorption and emission are narrower: small relaxation, small electron phonon interaction. The emission, again at variance with the F centre, is polarized, the ionization energy of the r.e.s. is rather high so that around 300°K the luminescence efficiency is still good: the r.e.s. is deep, the electron does not escape easily from it.

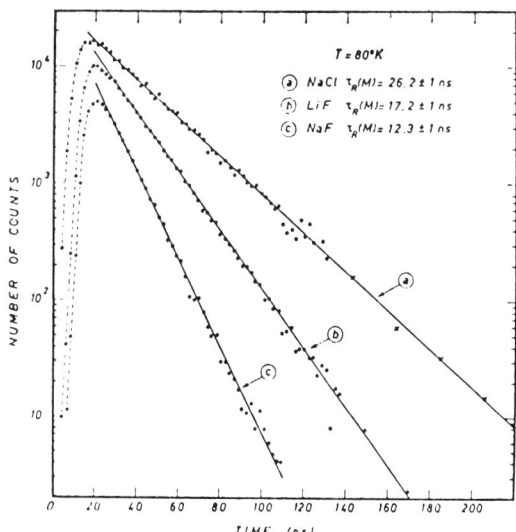

Fig.6 - Time decay of the M-centre luminescence at 80°K in NaCl (excitation filter: 725 mµ interference, Zeiss; detection: 87 A, Wratten Kodak), NaF (excitation: 500 mµ interference, Zeiss; detection: 650 mµ interference, Zeiss) and LiF (excitation: 450 mµ interference, Zeiss; detection: 675 mµ interference, Zeiss).

All these characteristics are clearly not contradictory with the lifetime result that yields a τ_R of the order of 10-30 ns for the different alkali halides studied (22).

The problem of the ionization of the r.e.s. is rather simple to describe and has a clear phenomenology. At temperatures around 100°K, in several alkali halides, the luminescence yield decreases rapidly and correspondently photoconductivity is observed. It is sufficiently well proved that as the temperature raises the electrons may be excited to the conduction band and so a second decay process is offered: either decay radiatively or jump to the conduction band. At higher temperatures the second process is favoured and the total lifetime gets considerably shorter. The rate of change of $1/\tau(T)$ versus $1/T$ gives the possibility to obtain ΔE, the thermal ionization energy of the r.e.s. Efforts to relate ΔE to the fundamental lattice parameters of the crystal have not been successful.

The Cu^+ substitutional ion. In alkali halides it is possible to introduce into the lattice, either by diffusion at high temperatures or by doping the melt, a certain number of Cu^+ ions. These ions occupy the position of the alkali they substitute but in most cases they go off-centre, as it has already been pointed out. The ion has a size similar to Na but smaller than K and Rb and so it is rather free to move: the repulsive forces with the nearest neighbours are stabilizing, they would keep the ion in the central position; the dipolar forces are destablishing in the sense that the dipoles induced electrostatically on the Cu^+ ion during its vibrations inside the vacancy are such to favour an off-centre position. From the balance between these forces depends the in-centre or off-centre position of Cu^+.

Now, in the ground state, Cu^+ is in the $3d^{10}$ electronic configuration and the absorption bands at ~5 eV correspond to the various sub-bands of the crystal field split $3d^{10} \to 3d^9 4s$ transition. The transition is forbidden and may be made allowed by odd parity vibrations in NaCl. The same transition in KCl, KBr and other alkali halides, where the mixing is realized by the static field due to the off-centre position of the ion, has a comparatively large temperature independent oscillator strength (10^{-2}).

After reaching the r.e.s. the electron can again either decay to the ground state and emit a photon around 3.3 eV or be thermally ionized to the conduction band. The r.e.s. is rather deep ($\Delta E \simeq 1$ eV)

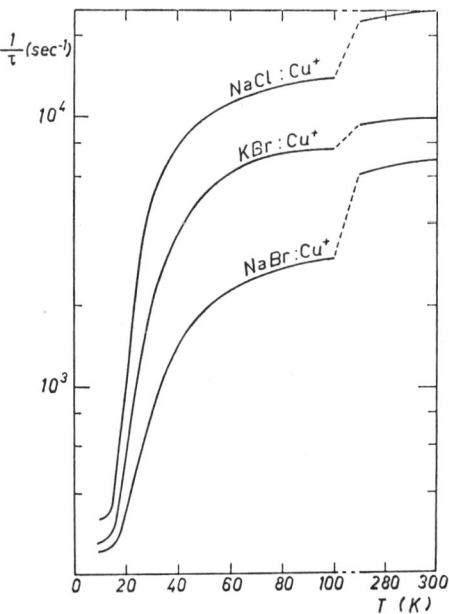

Fig. 7 - Lifetime as a function of temperature for $NaCl:Cu^+$, $KBr:Cu^+$, $NaBr:Cu^+$.

and so we have luminescence up to 100-200 centigrades (23).
The lifetime in KI and RbI is of the order of few µs, is temperature independent down to 4.2°K and does not give problems of interpretation. The ion in the r.e.s. is still off-centre (temperature independence) and the transition is rather forbidden (the mixing promoted by the off-centre position is not very high): for comparison we can remember that $\tau(Cu^+)$ is rather longer than τ for the F centre. In the bromides and in NaCl though, τ, even at room temperature is of the order from several tens to one hundred µs. Very interesting is the fact that starting from 30-40 K and going to lower temperatures τ gets longer and longer and reaches values around 3-4 ms at about 4.2°K: that is a really forbidden transition! The exceptionally long lifetime should be due to the absence both of dynamic (we are at low temperatures) and of static mixing: there is no effect that helps the transition to become allowed. We have therefore a strong indication that in these matrices, when Cu^+ is in its r.e.s. it reaches an in-centre position even if in the ground state was off-centre (the bromides). It clearly means that the increased size of the 4s orbit has shifted the balance in favour of the repulsive forces. In the low temperature region (<30-40 K) the τ vs T function can be analyzed and it exhibits a well defined exponential behaviour $\tau \propto \exp(-\Delta E/KT)$. The new ΔE it is not an ionization energy but rather the energy barrier that the ion has to overcome for going in an off-centre position. We think in fact that as n_{ph} increases, the dipolar forces bound also to the amplitude of the vibrations become predominant respect to the repulsive ones and the ion goes off-centre. From this transition region up to about 100 centigrades the lifetime changes slowly as a function of temperature with an hyperbolic cotangent law: in the off-centre position there is the possibility that both the static and the dynamic mixing are at work.

EXCITED STATES SPECTROSCOPY

The F-centre. The long lifetime of the r.e.s. offered the possibility to obtain a further insight into the nature of the excited states. It has been possible to keep through a laser or a conventional source, a non zero population in the r.e.s. and to measure the absorption from the first r.e.s. to the higher ones below the conduction band. The early work of Park and Faust (25) showed that there was some absorption induced by a pulsed ruby laser (5 Joule; 500 µs duration) in the 0.1-0.5 eV region, normally transparent. More recently Kondo and Kanzaki (26) repeated with a considerably improved sensitivity that type of measurements in KCl, KBr and KI.

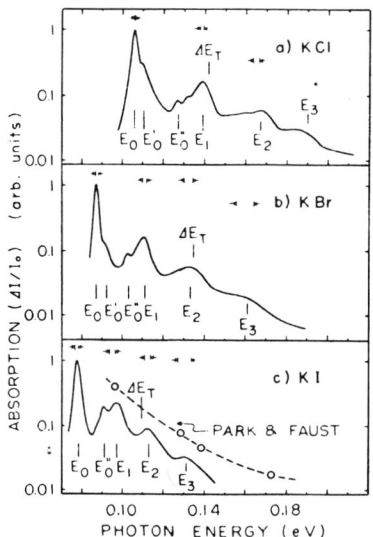

Fig.8 - Transient-absorption spectra of additively coloured crystals induced by F-band excitation at 10 K. The induced absorption has been normalized at the maximum value. ΔE_T represents thermal ionization energy of the F centre. The previous results of Park and Faust on KI are shown in (c) for comparison. Spectral bandwidths in the experiments are indicated at the top of each spectrum.

They obtained a transient absorption bound to the r.e.s. population produced by a pumping laser light. The results were interpreted as being due to 2s → 3p, 2s → 3d, 2s → 4p transitions in the r.e.s.; the analysis was performed in the frame of a Simpson potential by a variational procedure and using hydrogenic wave functions. All these transitions turn out to be at energies below the thermal ionization and so are undoubtably related to bound states. Higher energy absorption peaks are observed as well and it is proposed that those correspond to one, two and three phonon structures. Further work is highly desirable in this field because it could lead to a much clear understanding of the relaxed F centre: the potential now appears to be more and more coulombic outside the vacancy or, if you prefer, for the higher excited states: the symmetry of the first r.e.s., though, remains still matter of debate and consequently also that of the higher ones.

The Cu^+ substitutional ion. The Cu^+ ion is in a situation even better than the F centre for excited states spectroscopy studies

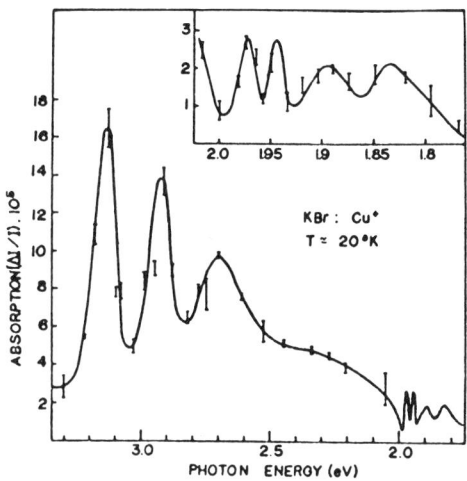

Fig.9 - Optical absorption from the $3d^94s$ in KBr. In the insert we see, enlarged, the absorption in the 1.8-2.8 eV region.

since its exceptionally long lifetime, few ms, as it has been said. Due to this fact, Bussolati et al.(2)) performed this type of measurements pumping the $3d^{10} \to 3d^94s$ transition of Cu^+ which is found around 4.7 eV in KBr and NaBr. An absorption induced by the pumping light has been observed; four narrow and small intensity bands are observed at energies slightly lower than 2 eV; a broad threshold type absorption was observed from 2.0 to 2.4 eV and also three bands peaking 2.7, 2.92 and 3.13 eV: the oscillator strength of the most intense band was evaluated to be of the order of $10^{-3} - 10^{-4}$. The attribution of these bands is not straightforward: among the possibilities we might include $3d^94s \to 3d^94s$ crystal field split forbidden transitions; $3d^94s \to 3d^94p$ allowed transitions, and $3d^94s \to$ conduction band, transitions. Since the rather high energy at which the transitions are observed the interpretation that, at the moment, has better chances to be the right one is the latter one. Photoconductivity measurements, now in progress, should give a quite clear answer in this regard. It seems interesting to point out that in the case these transitions have their final state in the conduction band, we would have direct information on the conduction band density of states and this would be a very useful information complementary to that on the joint density of states obtained with reflectivity and absorption techniques above the band-to-band edge.

BIBLIOGRAPHY

1. K. Füssgaenger, Phys. Stat. Sol. $\underline{34}$, 157 (1969).
   K. Füssgaenger, Phys. Stat. Sol. $\underline{36}$, 645 (1969).
2. M. Kozielski, I. Pollini and G. Spinolo, Phys. Rev. Lett. $\underline{27}$ 1223 (1971).
3. G. Spinolo and D.Y. Smith, Phys. Rev. $\underline{140}$, A2117 (1965).
   D.Y. Smith and G. Spinolo, Phys. Rev. $\underline{140}$, A2121 (1965).
4. F. Lüty, Halbleiterprobleme, Band VI, Verlag F. Vieweg and Sohn, Braunschweig (1961) pp 238.
5. K. Kojima, N. Nishimaki and T. Kojima, J. Phys. Soc. Japan $\underline{16}$, 2033 (1961).
   F. Okamoto, Phys. Rev. $\underline{124}$, 1090 (1961).
6. C.C. Klick, D.A. Patterson and R.S. Knox, Phys. Rev. $\underline{133}$, A1717, (1964).
7. F. Lüty and J. Mort, Phys. Rev. Lett. $\underline{12}$, 45 (1964).
   J. Mort, F. Lüty and F. Brown, Phys. Rev. $\underline{137}$, A566 (1965).
8. D.Y. Smith, Phys. Rev. $\underline{137}$, A547 (1965).
9. G. Chiarotti, U.M. Grassano, G. Margaritondo and R. Rosei, Nuovo Cimento $\underline{64}$, 159 (1969).
10. C.H. Henry, S.E. Schnatterly and C.P. Slichter, Phys. Rev. $\underline{137}$, A583 (1965).
11. M. Bonciani, U.M. Grassano and R. Rosei, Phys. Rev. B$\underline{8}$, 5855 (1973).
12. W.B. Fowler, E. Calabrese and D.Y. Smith, Sol. State Comm. $\underline{5}$, 569 (1967).
13. W.B. Fowler and D.L. Dexter, Phys. Rev. $\underline{128}$, 2154 (1962).
14. R.K. Swank and F.C. Brown, Phys. Rev. $\underline{135}$, A450 (1964).
15. G. Spinolo and F.C. Brown, Phys. Rev. $\underline{135}$, A450 (1964).
16. L.F. Stiles, M.P. Fontana and D.B. Fitchen, Phys. Rev. $\underline{2}$, B2077 (1970).
17. M.P. Fontana and D.B. Fitchen, Phys. Rev. Lett. $\underline{23}$, 1497 (1969).
18. G. Baldacchini and L.F. Mollenauer, Phys. Rev. Lett. $\underline{29}$, 465 (1972).
19. W.B. Fowler, Phys. Rev. $\underline{135}$, A1725 (1964).
20. L. Bosi, S. Cova and G. Spinolo, Phys. Stat. Sol. (b) $\underline{8}$, 2317 (1975).
21. L. Bosi, S. Cova and G. Spinolo, Phys. Rev. B $\underline{15}$, 4542 (1974).
    L. Pelsers and G. Jacobs, Phys. Rev. $\underline{188}$, 1322 (1969).
22. L. Bosi, C. Bussolati and G. Spinolo, Phys. Rev. $\underline{1B}$, 890 (1970).
23. G. Baldini, A. Jean and G. Spinolo, Phys. Stat. Sol. $\underline{25}$, 557 (1968).
24. M. Bertolaccini, G. Padovini, P. Gagliardelli and G. Spinolo, to

be published.
25. K. Park and W.L. Faust, Phys. Rev. Letters 17, 137 (1966).
26. Y. Kondo and H. Kanzaki, Phys. Rev. Lett. 34, 664 (1975).
27. C. Bussolati, P. Gagliardelli and G. Spinolo, Sol. State Comm. 12, 1253 (1973).

Part II

Defects in Low Symmetry Environments

INTERNAL FRICTION AND DEFECTS NEAR DISLOCATIONS

R. DE BATIST

S.C.K./C.E.N., MOL (BELGIUM)

also : RUCA, ANTWERP (BELGIUM)

1. INTRODUCTION

Although de La Palice did not know about dislocations, it is at present a lapalissade (1) to state that the mechanical properties of crystalline materials are determined by the behaviour of the crystal dislocations. Dislocations disturb the crystalline periodicity of the lattice and thereby generate a stress and strain field in their surroundings. This dislocation stress field interacts with the strain fields of other crystalline defects such as other dislocations and also point defects. The result of these interaction effects is that a dislocation, moving through the crystal under the influence of say an external shear stress, will be hindered in its motion by these other defects. An understanding of dislocation-point defect interaction (DPDI) is therefore very important if one wishes to explain the mechanical behaviour of a crystal.

Various ways have been followed, and are still being pursued, to investigate the influence of point defects on the plasticity of crystals. The most direct approach is of course to study the yield stress of the material in a number of different structural states characterized by : dislocation density, varied by plastic predeformation or a thermal recovery treatment; point defect concentration, varied by means of changes in composition (for chemical, or extrinsic point defects) or by means of irradiation, plastic deformation or stoichiometry (for physical, or intrinsic point defects); strength of interaction between dislocation and point defect, varied by changing the temperature. In spite of numerous efforts during the past decades (see e.g. 2, 3, 4) it appears that there are usually too many uncontrolled (and often uncontrollable) variables in the experiments for the analysis in terms of the thermodynamics of dislocation mobility to be able to yield trustworthy results. Major difficulties appear to be the changes in internal structure occurring during temperature- or strain rate changes, required to derive the thermodynamic quantities such as activation enthalpy, activation volume, ..., which describe the DPDI and the fact that different interaction processes occur simultaneously and can not be separated very well. Very often, however, the interactions can be broken down in two broad classes : rather steeply temperature dependent and almost temperature independent ones. Long-range dislocation - dislocation interactions result in a fairly constant yield stress, whereas short-range interactions can be thermally activated and are therefore only noticeable below some critical temperature. Stress relaxation experiments have frequently been used in order to determine the long-range internal stress, very often, though, without due attention to some of the inherent restrictions of this experimental technique (5).

Since dislocations cause very pronounced anelastic effects, it is clear that DPDI may also be investigated by means of internal friction studies. Dislocation relaxation effects (the Bordoni peak in fcc metals) are not very useful in this respect, since the changes in dislocation loop length, resulting from DPDI, cause only minor changes in the peak shape of the dislocation relaxation peak and especially since these secondary aspects of the Bordoni peak are not very well understood yet. Dislocation resonance and dislocation hysteresis effects, on the other hand, have been very widely used for studying DPDI. From the loop length dependence of the modulus and the internal friction, one may obtain information about the kinetics of point defect recovery processes; from the dislocation break-away effects leading to hysteretic damping, the strength of the dislocation point defect binding force may be estimated.

The fact that we are interested here in non-metallic crystals implies that in addition to the purely mechanical aspects alluded to above, also interaction processes involving coupling effects between mechanical quantities and electrical or optical quantities may occur in these materials. Electromechanical coupling is observed e.g. in alkalihalides, where dislocations carry an electrical charge resulting in an effective piezoelectricity, although of course a perfect alkalihalide crystal with its centro-symmetric lattice cannot be piezoelectric. Optomechanical coupling arises when dislocations interact with colour centres whereby the binding energy changes with excitation of the centre, thus possibly leading to photoplasticity. Such interaction processes can also be studied very profitably by means of dislocation internal friction experiments.

This then will be the plan of these lectures : first we will discuss briefly the various anelastic effects caused by dislocations; this will then be applied to studying DPDI effects in non-metallic crystals; finally, coupling processes will be discussed.

2. ANELASTICITY DUE TO DISLOCATIONS

2.1. Dislocation Relaxation

As mentioned in the introduction, dislocation relaxation is not very useful for the study of DPDI effects. It is an intrinsic dislocation effect in which the dislocation loop length enters only through the pre-exponential factor in the Arrhenius expression for the relaxation time, $\tau = \tau_Q \exp(H/kT)$. It will not be further discussed here; interested readers are referred to (6, 7).

2.2. Dislocation Resonance

A dislocation line in a crystal possesses potential energy proportional to its length and hence it will tend to be as short as possible : it has the equivalent of a line tension. Therefore Koehler (8) suggested to use for the equation of motion of a dislocation line (parallel to the x-axis) in its glide plane in the crystal (x, y plane) the vibrating string analogue, assuming the dislocation to be fixed at its extremities (x = o and x = L). Thus :

$$A \ddot{y} + B \dot{y} - C y'' = b R \sigma \quad (2.1.)$$

Here, A represents the effective mass per unit length of the dislocation, B its damping constant, C its line tension and $b R \sigma$ is the force per unit length exerted by the shear stress $R \sigma$ (acting in the x - y plane in the y direction). In an internal friction experiment, the specimen is set into vibration in one of its eigenmodes, and the resulting shear stress will be an harmonic function of the time. From eqn (2.1.), one may then calculate the dislocation strain ε_d resulting from the dislocation displacement, and from ε_d, obtain the internal friction

$$\phi = \frac{\Delta W}{2W} = \frac{\int_{period} \varepsilon_d \, d\sigma}{2W}$$

and the sound velocity, which then gives the modulus defect, $\Delta M/M$.

Granato and Lücke (9) have shown how ϕ and $\Delta M/M$ can be expressed by means of a rapidly converging series, of which usually only the first term is retained :

$$\phi = C_1 \frac{\omega d}{(\omega_o^2 - \omega^2)^2 + (\omega d)^2} \qquad (2.2.)$$

$$\frac{\Delta M}{M} = C_1 \frac{\omega_o^2 - \omega^2}{(\omega_o^2 - \omega^2)^2 + (\omega d)^2} \qquad (2.3.)$$

$$(d = \frac{B}{A} \; ; \; \omega_o = \frac{\pi}{L} \sqrt{\frac{C}{A}} \; ; \; C_1 = \frac{8Gb^2 \Lambda}{\pi^2 A})$$

Dislocation resonance experiments are mostly carried out in either one of two frequency ranges. In the first one, one selects frequencies in the vicinity of those yielding maximum damping. This usually requires pulse-echo techniques, using travelling waves rather than standing waves, with frequencies in the range of 1 to 100 MHz. One of the aims of the experiments is then to trace out $\phi(\omega)$ at constant temperature and to determine the damping constant B from a knowledge of the density of mobile dislocations, Λ. Several difficulties arise in the interpretation. First, eqn. (2.2.) has been derived assuming a delta function distribution for the dislocation loop length, L. For the random distribution (8), the numerical factors in front of eqn. (2.2.) and (2.3.) are changed. Second, not only is the dislocation distribution uncertain, also the dislocation density cannot be determined very accurately. The main problem, however, appears to be the existence of dislocations with widely differing properties (usually assumed to be edge-type and screw-type, resp.) leading to two separate, and different contributions to ϕ and $\Delta M/M$.

In a second type of experiments, the frequency is maintained well below the one for which ϕ becomes maximum and the expressions (2.2.) and (2.3.) reduce to

$$\phi = C_2 \omega \Lambda L^4 \qquad (2.4.)$$

$$\Delta M/M = C_3 \Lambda L^2 \qquad (2.5.)$$

$$\left[C_2 = 8Gb^2 B (\pi^6 C^2)^{-1} \; ; C_3 = 8Gb^2 (\pi^4 C)^{-1} \right]$$

Hence, the damping varies as the fourth power of the loop length L; the modulus defect as L^2. This type of experiment is of course particularly useful for studying DPDI, since it is expected that the presence of point defects near a dislocation line will restrict its mobility and hence will reduce the dislocation loop length. From the temperature dependence of the rate of pinning, deduced from the changes in damping and modulus defect, one obtains information about the migration enthalpies of the point defects involved. A study of the time law governing the rate of arrival of point defects at dislocation sinks sometimes allows conclusions to be drawn concerning the geometry of the diffusion process and the nature of the point defects. Here too, difficulties may arise as a result of the existence of various kinds of point defects interacting with various kinds of dislocations.

Equation (2.1.) has been derived by assuming the solid to be linearly elastic. When this restriction is relaxed, additional effects arise due to the anharmonicity of the lattice and to the higher order terms in the dislocation line tension, C (10). From the equation of motion in one-dimensional form, for a displacement u in the z direction,

$$\rho \ddot{u} = \frac{\partial \sigma}{\partial z}, \text{ or } \rho \frac{\partial^2}{\partial t^2}(\frac{\partial u}{\partial z}) = \frac{\partial^2 \sigma}{\partial z^2}$$

(ρ is the density of the undeformed material, σ the normal stress) it is clear that when a sinusoidal wave of frequency ω is introduced into the solid (at z = o) it will distort as it propagates and the stress will contain harmonics of the fundamental wave for z>o :

$$\sigma = \sum_{i=0}^{\infty} \sigma_i \cos i(\omega t - kz - \delta_i) \qquad (2.6.)$$

(k is the wave vector and i$\delta_i - \delta_1$ the phase angle relative to the fundamental wave). The displacement, u, contains a contribution due to the lattice and one due to the dislocation : $u = u_l + u_d$. Neglecting higher order terms, the lattice displacement gradient $\partial u_l / \partial z$ is related to the stress by :

$$\sigma = a_1 \frac{\partial u_l}{\partial z} + a_2 (\frac{\partial u_l}{\partial z})^2 \qquad (2.7.)$$

The dislocation displacement gradient can be calculated from the deviation, y, of the dislocation from its equilibrium position (along the x axis) in the absence of stress :

$$\frac{\partial u_d}{\partial z} = \frac{Nbq}{L} \int_0^L y \, dx$$

N is the dislocation density, b the burgers vector, L the dislocation loop length and q converts the shear strain into longitudinal strain. y is the solution of the equation replacing (2.1.) :

$$A \ddot{y} + B \dot{y} - C (y'' - C'y'^2 y'') = bR\sigma \qquad (2.1.a.)$$

in which C takes account of the dislocation character (edge or screw) and C' expresses the effect of non-linearity upon the dislocation line tension.

From equations (2.1.a.) and (2.6.), expressions are obtained by Hikata and Elbaum (10) for the amplitudes of the various harmonics;

$$\sigma_i = K_i f_i(L) \, F_i(\alpha,z) A_{10}^i \qquad (2.8.)$$

K_i contain numerical factors and a number of material and experimental constants such as density, sound velocity and frequency of the imposed fundamental wave ;

$$F_1(\alpha,z) = \exp(-\alpha_1 z)$$
$$F_2(\alpha,z) = [\exp(-2\alpha_1 z) - \exp(-\alpha_2 z)] \, (\alpha_2 - 2\alpha_1)^{-1}$$
$$F_3(\alpha,z) = [\exp(-3\alpha_1 z) - \exp(-\alpha_3 z)] \, (\alpha_3 - 3\alpha_1)^{-1}$$
$$f_2(L) = (f_1 + f_d + f_{1d})^{1/2}$$

where f_1 gives the lattice contribution, given by a_2 of eqn. (2.7.), f_d gives the dislocation contribution, proportional to the static bias stress σ_o^2 and containing the quantities $S_o = (\omega_o^2 - \omega^2)^2 + (\omega d)^2$ and $T_o = (\omega_o^2 - 9\omega^2)^2 + (3\omega d)^2$ [ω_o and d as defined for equation (2.2.) and (2.3.)], and f_{1d} is a cross term proportional to $f_1^{1/2}$ and $f_d^{1/2}$;

$$f_3(L) = L^{-4} S_o^{-3/2} T_o^{-1/2}.$$

When one considers the magnitude of the various terms in (2.8.), it becomes clear that the third harmonic is much more useful for dislocation studies than the second harmonic where the contribution of the lattice anharmonicity cannot be neglected. It should also be noted that a dislocation contribution to the second harmonic can only be observed when the dislocation line has been displaced from it y = o position as the result of a static (internal or external) bias stress, so that the displacement of the dislocation becomes a non-symmetric function of the stress.

2.3. Dislocation Hysteresis

In a dislocation resonance experiment, the effective length of the dislocation string is considered to remain constant, except for the arrival (or disappearance) of pinning points resulting from some kind of mobile point defect. However, the efficiency of a point defect to act as a pinning point is determined by the strength of the DPDI. In an internal friction experiment, the vibrating dislocation string exerts a force on the point defect pin which is proportional to the shear stress acting in the dislocation glide plane, and the pin remains a pin only when this force does not exceed the maximum interaction force possible between dislocation and point defect. This interaction force is determined by the dislocation point defect binding energy. Of course, thermal fluctuations may help the dislocation overcome the pin barrier. Therefore, the pinning point density may be temperature dependent, even in the absence of temperature dependent point defect mobility effects. It may also be stress dependent, i.e. the dislocation loop length may change as a function of the vibration amplitude applied in an internal friction experiment. This leads to thermomechanical breakaway effects and to internal friction which is strongly amplitude dependent but frequency independent. This hysteretic dislocation damping was also first treated by Granato and Lücke (9) in the zero temperature limit (i.e. neglecting thermal breakaway) with the following result:

$$\pi \phi = \frac{\Delta M}{M} = \frac{C_4}{\varepsilon_o} \exp\left(-\frac{C_5}{\varepsilon_o}\right) \qquad (2.9.)$$

$$\text{with } C_4 = \frac{2\Omega\Lambda L_N^3 b^2 F_M}{\pi^3 a C L_p^2} \quad , \quad C_5 = \frac{\pi F_M}{4 a G L_p}$$

and where one has introduced two kinds of loop length, L_N, the network length, determined by the strong pins remaining unaffected during the internal friction experiment, and L_p, the weak pin length, determined by the distance between point defect pins from which the dislocation can be torn loose during its vibration; F_M is the binding force between dislocation and point defect.

Incorporating the possibility for thermal breakaway into the Granato-Lücke description of hysteretic dislocation damping allows one to study DPDI effects through either the amplitude dependence or the temperature dependence of the damping and modulus defect. However, the damping peaks which are expected to be caused by thermal breakaway processes cannot unambiguously be interpreted as such. It is very difficult indeed to distinguish between the various models which have been suggested for a number of relaxation peaks attributed to DPDI processes (7). Therefore, the best way to make use of the thermomechanical breakaway effect still seems to be the study of the amplitude dependence of damping and modulus defect. Measuring this amplitude dependence at various temperatures then yields information about the temperature dependence of the various parameters appearing in the constants C_4 and C_5 in eqn. (2.9.).

3. DISLOCATION POINT DEFECT INTERACTION EFFECTS IN NON METALLIC CRYSTALS

3.1. Recovery Kinetics

The internal structural equilibrium of a crystalline solid may be disturbed as a result of thermal or mechanical excitation or by irradiation with energetic particles or photons. A study of the way in which the material reestablishes equilibrium allows one to obtain fundamental information about the various structural defects generated in the solid and about the interaction effects between such defects. We will not discuss here recovery processes involving mere annihilation of either point defects or dislocations, but will focus our attention on the use of internal friction experiments for the study of recovery processes in which point defects migrate towards dislocations, resulting in a lowering of the dislocation mobility.

This means that we will assume here that point defects will be able to act as pinning points, thus changing the dislocation loop length and hence the dislocation damping. The general problem of the kinetics of migration of a point defect to a dislocation has been discussed e.g. by Bullough (11) in terms of the second order partial differential equation :

$$\frac{\partial C}{\partial t} = D\nabla^2 C + \frac{D}{kT}\nabla.(C\nabla E) \qquad (3.1.)$$

Here, D is the diffusion coefficient of the point defect with concentration $C(x_i, t)$ (i = 1, 2, 3) and E is the interaction energy between a dislocation (lying along the x_3 axis) and a point defect situated at (x_1, x_2, x_3). The DPDI energy consists of a number of contributions and depends upon the distance, r, between the point defect and the dislocation line axis :

$$E = A_1 r^{-1} + A_2 r^{-2} + \ldots \qquad (3.2.)$$

The first order term arises from the difference in size between the point defect and the lattice hole in which it is forced to reside. The second order term is caused partly by the non-linearity of the elastic properties of the host material, partly by the inhomogeneity interaction which occurs as a result of the difference between the elastic constants of the matrix and of the inclusion. For impurities and self-interstitials, the term in r^{-1} is usually dominant. For vacancies, on the other hand, A_1 may vanish and the second order term will be necessary.

Equation (3.1.) cannot be solved in its most general form, one of the major difficulties being the definition of appropiate boundary conditions at the dislocation core (one usually assumes

INTERNAL FRICTION AND DEFECTS NEAR DISLOCATIONS 313

zero flow far from the dislocation, i.e. somewhere midway between the dislocations in an idealized crystal containing a number of parallel dislocations). Supposing first that the point defects retain their identity near the dislocation, a Maxwell atmosphere will be built up as soon as the drift and diffusion parts of the flow appearing in eqn. (3.1.) balance each other and

$$C = C(r) = C_o \exp\left(-\frac{E}{kT}\right) \qquad (3.3.)$$

The well-known Cottrell-Bilby (12) solution is obtained by neglecting the diffusion contribution in eqn. (3.1.) and results for $E = A_1 r^{-1}$ in :

$$C(r,t) = rC_o (r^3 + 3\ LDt)^{-1/3} \qquad (3.4.)$$

This gives for the number of point defects removed from solid solution per unit length of dislocation :

$$N(t) = C_o \pi (3LDt)^{2/3} \qquad (3.5.)$$

It is obvious from (3.4.) that the diffusion flow can be neglected only at early times, when the concentration gradient near the dislocations as well as the over-all reduction in point defect concentration implied by (3.4.) remain both small. Therefore, one expects a gradual decrease in the point defect accumulation rate, so that eqn. (3.5.) is generalized for all times by an expression of the form

$$N(t) = N(\infty)\left[1 - \exp(-\alpha\ t^n)\right] \qquad (3.6.)$$

Bullough and Newman (13) have shown that kinetics of the form of eqn. (3.6.), with $n = 2/3$ as frequently observed experimentally [the so-called Harper kinetics (14)] cannot be obtained with the simple atmosphere type model. Their numerical solution very nearly approximates the Harper behaviour, however, when precipitation effects are introduced in such a way that the point defects become tightly bound to the dislocation, forming a rod-like precipitate. Recently, Polak (45) has been able to derive from eqn. (3.1.) kinetics very close to the Harper form by assuming a random distribution of sinks for the migrating point defects. He finds $n = 0.4$ to 0.5 for random planar sinks and $n \simeq 2/3$ for random linear sinks.

For point defects for which the second order term in (3.2.) becomes dominant, the exponent n in (3.6.) takes the value $1/2$. A different value for n is also expected when the diffusion geometry deviates from the cylindrical symmetry assumed in the preceding discussion. Examples of this situation are diffusion of point defects towards small dislocation loops [leading to $n = 0.6$ and to $n = 1.5$ for the pure drift kinetics of the impurity-loop and the self-interstitial-loop interaction, resp. (13)] and the plane diffusion geometry resulting from dispersion of a point defect atmosphere in the glide plane of an oscillating dislocation [leading to n = 1/3 (15, 16)].

Several investigators have studied recovery effects near room temperature in a number of ionic crystals. Table I shows there is little consistency in the results of both the recovery kinetics and the activation enthalpy governing the recovery process. The value of the activation enthalpy derived from such experiments depends upon the manner in which the data are analysed (value of the kinetic exponent, separation of the measured damping in hysteretic and resonance contributions, neglect of non-dislocation contributions to the damping, ...) Hence, it is not obvious how the calculated enthalpies should be related to actual physical processes. When all these considerations are taken into account, one must conclude that such recovery experiments are valuable only when sufficient precautions are taken to fully characterise the experimental conditions and to ensure oneself about the analysis to be applied to the experimental results.

Table I

Material	Temperature range	Kinetics	Activation enthalpy	Reference
LiF	275 - 325 K	$t^{1/3}$	0.3 eV	17
LiF	300 - 340 K	$t^{2/3}$	0.85 eV	18
NaCl	273 - 326 K	$t^{2/3}$	0.7 eV	19
NaCl	300 - 330 K	$t^{1/3}$	0.32 eV	20
AgCl	273 - 323 K	$t^{2/3}$	0.23 eV	21
AgCl	225 - 340 K	$t^{1/3}$	0.42 eV	16

3.2. Equilibrium Effects

The interaction energy between a dislocation and the various obstacles to its motion is an important parameter for understanding the mechanical behaviour of a crystal. It may be determined experimentally by means of plastic deformation or stress relaxation experiments (e.g. 2, 3, 5), although the experimental conditions can not always be sufficiently well specified to allow unequivocal interpretation. The strong loop length dependence of the dislocation resonance damping provides a way of using internal friction to study the small-displacement behaviour of DPDI by measuring e.g. the temperature dependence of the pinning point density (eqn.3.3.). Kim et al (16) e.g. have thus estimated the binding energy between dislocations and the (unspecified) pinning points in AgCl to be about 0.15 eV. These authors also used the Granato-Lücke (9) analysis of the amplitude dependence to obtain a binding energy of 0.2 eV. It is clear from eqn. (2.9.) that the distance between weak pins, L_p can be derived from the slope as well as from the intercept of the Granato-Lücke plots of $\ln(\varepsilon_o \phi)$ vs $(1/\varepsilon_o)$. In some cases (e.g. (22)) this yields values for the binding energy which agree within experimental uncertainty. Phillips and Pratt (23) on the other hand find a value of - 0.26 eV from the slope and of - 0.68 eV from the intercept of their Granato-Lücke plots for NaCl crystals. We will come back to these results in § 4 where we will discuss some effects related to the fact that dislocations can carry an electrical charge in an insulating crystal. Other difficulties may arise when comparing the number of point defect pins derived from the resonance and the hysteretic damping. French and Harris (24) e.g. find that the number of additional point defect pins observed during neutron irradiation of NaCl single crystals at 22 °C increases more rapidly with irradiation time when it is derived from amplitude dependent damping data than when calculated from amplitude independent results. French (25) has further shown that the additional pinning points responsible for the amplitude dependence decay away in a few minutes time when the neutron beam is interrupted, whereas the number of pins derived from dislocation resonance damping remained steady. These authors interpret their results in terms of point defects created continuously during irradiation in the close vicinity of the dislocation line but at the same time also disappearing gradually through a recombination process. Only a fraction of the point defects is assumed to be formed close enough to the dislocation core to act as a pinning point at low amplitudes; the other point defects are therefore only observed at larger strain amplitudes. Only those will continue recombining when the neutron beam is interrupted and will thus lead to a decrease in the observed number of "amplitude dependent" pinning points.

The typical Granato-Lücke type behaviour expressed by means of a linear Granato-Lücke plot is only expected to occur under certain well specified experimental conditions where thermal effects upon the distribution of the weak pins along the dislocations line and upon the mechanical breakaway process can be neglected. Ritchie and Sprungmann (26) have investigated the effect of temperature on the amplitude dependent internal friction of a single crystal of MgO containing long loop lengths of fresh dislocations. They are able to interpret their results in terms of the Blair, Hutchison and Rogers (27) development of the Granato-Lücke theory for hysteretic dislocation damping. The results of the Blair, Hutchison, Rogers treatment can be summarized in a temperature-strain diagram as given in figure 1 for the special case of very long and equal dislocation loop lengths L_p between the pins. Different unpinning behaviour occurs in the five regions in the $T - \varepsilon$ domain :
- In A, unpinning is activated over a group of pins but only few of the dislocations are liberated in a given cycle.
- In B, a single pin activates unpinning of the whole row, again only for a few of the dislocations in a given cycle.

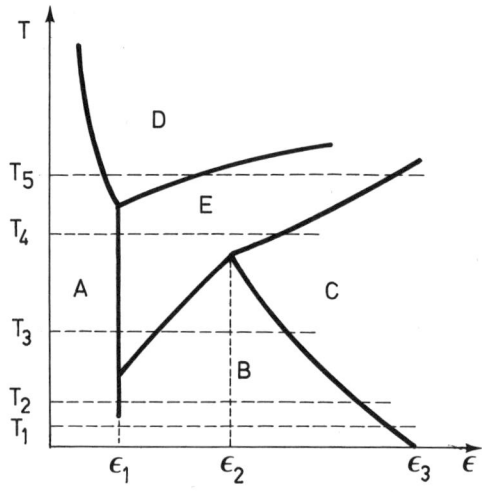

Figure 1 :

Temperature-strain diagram used in the theory of Blair, Hutchison and Rogers

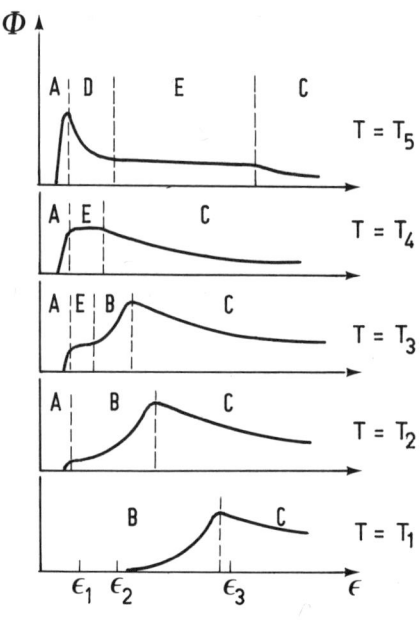

Figure 2 :

Amplitude dependent damping to be expected at various temperatures corresponding to different fields in the temperature-strain diagram of Figure 1

- In C, nearly all the dislocations unpin following activation of a single pin.
- In D, nearly all the dislocations unpin following activation of a group of pins.
- In E, relaxation type behaviour occurs, the activation energy for unpinning being effectively constant throughout the strain cycle.

The amplitude dependence of the internal friction is determined by the $T - \varepsilon$ domains which are sampled during the experiment, and hence will be a function of the temperature. A few representative $\phi - \varepsilon$ curves to be expected at various temperatures are shown in fig. 2. Ritchie and Sprungmann have succeeded in constructing experimentally the $T - \varepsilon$ diagram for MgO as shown in fig. 1 and from this have determined the critical parameters $T_0, \varepsilon_1, \varepsilon_2$ and ε_3. This allows them to estimate the binding energy between dislocation and pin to be approximately 1.1 eV. They also conclude that the interaction energy is much more short ranged than expected for elastic interaction; it varies as r^{-9}. This is explained as suggesting that the pinning force is mostly electrical in nature, and that it might arise from the overlap energy which contributes predominantly to the pinning between a dislocation and a positive ion vacancy (28).

4. CHARGED DISLOCATIONS.

Non-conducting crystals offer a unique way of determining interaction effects between dislocations and point defects, and hence of studying the behaviour of point defects within the dislocation core sites as well as the dislocation core structure itself, because of the additional complication that edge dislocations may maintain an effective electrical line charge. This results in macroscopically observable electrical effects, in addition to the mechanical effects usually associated with dislocations. Brantley and Bauer (29) have used a Nye-like (30) loop to picture the relations that exist in a centro-symmetric (intrinsically non-piezoelectric) crystal due to the presence of electrically charged mobile edge dislocations (fig. 3). The coupling between the electric and elastic properties caused by the mobility of the charged dislocations also induces changes in the "direct" material constants (dielectric and elastic constants). Furthermore, the transport properties may be changed too; e.g. the relationship between the stress and the strain rate will depend upon the interaction between charged dislocations and charged point defects, and this interaction may be changed by irradiation of the material with photons capable of exciting point defects, thereby changing their charge state.

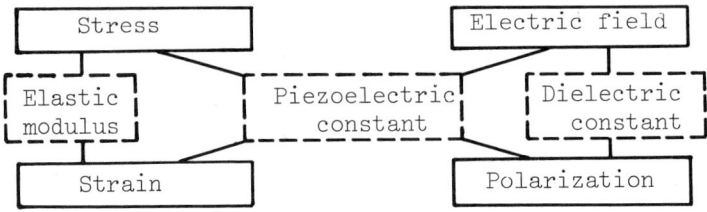

Figure 3 :

Dislocation induced piezoelectricity

An edge dislocation in an ionic crystal will acquire an electrical charge if the numbers of cations and anions along its core are different. This difference may be caused by the presence of either a different number of cation- or anion vacancies, or an excess of jogs of either sign. A cation vacancy may be absorbed or emitted at a positive jog, thereby changing the jog charge from $+ e/2$ to $- e/2$. Eshelby et al (31) have shown how dislocations will become charged when the free energy for formation of a cation vacancy differs from that for an anion vacancy and how this dislocation line charge will be compensated by the build-up of an atmosphere of oppositely charged point defects. It is precisely this atmosphere formation which makes internal friction experiments a useful tool for the study of charged dislocations. Indeed, this charge cloud may be expected to change the dislocation damping constant B and the distance L_p between weak pinners. Furthermore, fatigue effects can be generated if diffusion processes lead to changes in the cloud configuration. These kinetic aspects, however, have been covered already in § 3.1.

Hysteretic dislocation damping was first extensively used for the study of charged dislocations in NaCl, in combination with direct measurements of the vibration induced surface polarisation, by Strumane et al (22); analogous experiments have been carried out on CsI by De Batist et al (32). Recently, Kardashev et al (33) have extended the temperature range down to liquid helium temperature for measurements on LiF, NaF and NaCl. In order to obtain a measurable polarisation resulting from the oscillation of the dislocation with respect to the compensating charge cloud (assumed to remain stationary), it is necessary that there is a surplus of dislocations moving

simultaneously in the same direction. This is accomplished by bending the specimen, thus introducing an excess of edge dislocations of the same mechanical sign. The main conclusions to be derived from these experiments are :
- the density of weak pinners is smaller than the dislocation line charge density;
- the dislocation charge is an equilibrium property;
- breakaway of screw dislocations is easier than that of edge dislocations.

The Granato-Lücke theory has been used in (22) to obtain estimates for L_p and for the effective binding energy between weak pins and dislocations. Strumane et al. as well as Phillips and Pratt (23) obtain negative values for the binding energy; the magnitude, however, is different and this remains hard to explain. Kardashev et al. (33) on the other hand use the theory of Indenbom and Chernov (34) in order to derive the temperature dependence of the interaction force between dislocation and weak pins and, using the proportionality between activation energy for thermal breakaway and temperature, they calculate the activation distance d. They find that at low temperatures d becomes independent of the amplitude of oscillation and is given very nearly by the quantummechanical zero-point fluctuations.

Next, let us consider the effect of the dislocation line charge on the damping constant, B. This influence will be most clearly observed in the neighbourhood of an isoelectric temperature, where the line charge goes through zero (31). Robinson and Birnbaum (35) and Robinson (36) have solved eqn (2.1.) using the expression for B derived by Brown (37). It is found that the experimental data can be better interpreted in terms of charge cloud damping than in terms of phonon damping. In particular the effect of irradiation with X rays (38) or gamma rays (39), resulting in an increase of the damping constant, finds a very natural explanation in the charge cloud damping theory, since B is expected to be proportional to the number of point defects in the charge cloud and to the square of the dislocation line charge, whereas for phonon damping B is expected to remain constant. The disappearance of the charge contribution to the damping in the vicinity of the isoelectric temperature, T_i, has been reported for KCl by Kliewer and Koehler (40). This is not a very easy experiment, however, since the actual behaviour of the damping near T_i depends very sensitively upon the precise structural conditions (loop length distribution, line charge, ...) of the specimen. From the dependence of T_i upon the concentration of divalent impurities, one may then estimate the free energies of formation for cation and anion vacancies, as well as their temperature dependence. Unfortunately, the available results imply rather large uncertainties (7).

Since non-metallic crystals are not only dielectrics, but very often are transparent for visible or near-visible light, optomechanical coupling effects may also profitably be used for the study of DPDI. Such experiments can obviously be particularly relevant for the identification of the point defect involved in the interaction process. Gordon (41) has reviewed the information available in this field a decade ago, especially the relationship between dislocation damping effects and the production and excitation of F centres in alkali halides. Since then, following the work of Nadeau (42), also the effect of illumination upon the plastic deformation behaviour has received a fair amount of attention. Simultaneous measurements of yield stress and ultrasonic attenuation in coloured NaCl crystals by Ermakov et al (43) are interpreted in terms of the transformation by irradiation with F light of F centres into F' and α centres, on the assumption that F' and α centres are much more strongly bound to dislocations than F centres. An analogous interpretation has been suggested for the effect of illumination with UV light of oxygen doped NaCl crystals (44), where it is assumed that excitation of the O^{2-} ion liberates the associated anion vacancy an thus creates an α centre. It is clear that this approach to the study of DPDI is a very profitable one and may be expected to yield valuable information about the structure of point defects in the neighbourhood of dislocations and about the detailed force-distance relationship describing this interaction.

REFERENCES

(1) See e.g. "Petit Larousse" (1964)
(2) Second International Conference on Strength of Metals and Alloys, ASM (1970)
(3) Dislocation Dynamics ed A.R. Rosenfield et al, Mc Graw Hill, New York (1968)
(4) The Interactions between Dislocations and Point Defects, ed. B.L. Eyre - AERE - R 5944 (1968)
(5) R. De Batist, Reviews on the Deformation Behaviour of Materials (to be published)
(6) A.S. Nowick & B.S. Berry, Anelastic Relaxation in Crystalline Solids, Academic Press, New York (1972)
(7) R. De Batist, Internal Friction of Structural Defects in Crystalline Solids, N. Holland, Amsterdam (1972)
(8) J.S. Koehler in Imperfections in Nearly Perfect Crystals J. Wiley, New York (1952)
(9) A.V. Granato & K. Lücke, J. Appl. Phys. $\underline{27}$, 583, 789 (1956)
(10) A. Hikata & C. Elbaum, Phys. Rev. $\underline{144}$, 469 (1966)
(11) R. Bullough in (4)
(12) A.H. Cottrell & B.A. Bilby, Proc. Phys. Soc. $\underline{62}$, 49 (1949)
(13) R. Bullough & R.C. Newman, Proc. Roy. Soc. $\underline{A\ 266}$, 198 and 209 (1962) - Rep. Progr. Phys. $\underline{33}$, 101 (1970)
(14) S. Harper, Phys. Rev. $\underline{83}$, 709 (1951)
(15) R.H. Chambers & R. Smoluchowski, Phys. Rev. $\underline{117}$, 725 (1960)
(16) J.S. Kim, L. Slifkin & A. Fukai, J. Phys. Chem. Solids $\underline{35}$, 741 (1974)
(17) S.H. Carpenter, Acta Met. $\underline{16}$, 73 (1968)
(18) G. Guénin, J. Perez & P.F. Gobin, Cryst. Latt. Defects, $\underline{3}$, 199 (1972)
(19) D.C. Phillips, P.L. Pratt, Phil. Mag. $\underline{21}$, 217 (1970)
(20) C.R. Scorey, Phil. Mag. $\underline{21}$, 723 (1970)
(21) A.A. Blistanov, G.V. Malakhov & M.P. Shaskol'skava, Sov. Phys. S.S. $\underline{6}$, 1935 (1965)
(22) R. Strumane, R. De Batist & S. Amelinckx, Phys. stat. sol. $\underline{3}$, 1379 (1963)
(23) D.C. Phillips & P.L. Pratt, Phil. Mag. $\underline{22}$, 809 (1970)
(24) I.E. French & R.W. Harris, J. Phys. C $\underline{4}$, 331 (1971)
(25) I.E. French, J. Phys. C $\underline{4}$, 1725 (1971)
(26) I.G. Ritchie & K.W. Sprungmann, Scripta Met. $\underline{7}$, 323 (1973)
(27) D.G. Blair, T.S. Hutchison & D.H. Rogers, Can. J. Phys. $\underline{49}$, 633 (1971)
(28) F. Bassani & R. Thomson, Phys. Rev. $\underline{102}$, 1264 (1956)
(29) W.A. Brantley & C.L. Bauer, Phil. Mag. $\underline{20}$, 441 (1969)
(30) J.F. Nye, Physical Properties of Crystals, Clarendon Press, Oxford (1957)
(31) J.D. Eshelby, C.W.A. Newey, P.L. Pratt & A.B. Lidiard, Phil. Mag. $\underline{3}$, 75 (1958)
(32) R. De Batist, E. Van Dingenen, Yu. N. Martyshev,

I.M. Silvestrova & A.A. Urusovskaya, Sov. Phys. Crystallography 12, 881 (1968)
(33) B.K. Kardashev & S.P. Nikanorov, Sov. Phys. Solid State 13, 128 (1971); 16, 690 (1974)
B.K. Kardashev, S.P. Nikanorov & O.A. Voinova, Sov. Phys. Solid State 16, 687 (1974); Phys. stat. solidi 256, 359 (1974)
(34) V.L. Indenbom & V.N. Chernov, Phys. stat. solidi 14a, 347 (1972)
(35) W.H. Robinson & H.K. Birnbaum, J. Appl. Phys. 37, 3754 (1966)
(36) W.H. Robinson, J. Mater. Sci. 7, 115 (1972); also J.L. Tallon & W.H. Robinson, Phil. Mag., 27, 985 (1973)
(37) L.M. Brown, Phys. stat. sol. 1, 585 (1961)
(38) T. Suzuki, A. Ikushima & M. Aoki, Acta Met. 12, 1231 (1964)
(39) G.A. Bielig, J. Appl. Phys. 42, 4758 (1971)
(40) K.L. Kliewer & J.S. Koehler, Phys. Rev. 157, 685 (1967)
(41) R.B. Gordon in Physical Acoustics 3 B, ed. W.P. Mason, Academic Press. New York (1965)
(42) J.S. Nadeau, J. Appl. Phys. 35, 669 (1964)
(43) G.A. Ermakov, E.V. Korovkin & Ya. M. Soifer, Sov. Phys. Solid State 16, 457, 1139 (1974)
(44) I.M. Spitkovskii & N.A. Tsal', Sov. Phys. Solid State 15, 233 (1973)
(45) J. Polak, Crystal Latt. Defects 5, 155 (1974)

DEFECTS IN NON-CRYSTALLINE OXIDES

David L. Griscom

Naval Research Laboratory

Washington, D.C. 20375

I. INTRODUCTION

The study of defects in crystals traces its beginnings to the early 19th century and has been intensively pursued for the past 50 years, leading to a tremendously detailed body of knowledge concerning "color centers" in materials with long range order.[1,2] On the other hand, non-crystalline oxides have been actively developed for their aesthetic and utilitarian qualities since at least the second millennium B.C.,[3,4] and they have now reached a new peak of technological importance with the recent advent of fiber optics and metal-oxide-semiconductor (MOS) structures to name a few. Yet full application of the most modern techniques of solid state science to the problem of defects in oxide glasses has been confined mostly to the last two decades, and but a few review articles have been written on the subject.[5-7] Understandably, the broadening and overlapping of both optical and electron spin resonance (ESR) bands due to random local distortions, coupled with the intrinsic inability to exploit crystal symmetry, has discouraged most color center scientists from dealing with glasses. However, the ESR method in particular is now proving to be a surprisingly powerful tool for unraveling the electronic structures of magnetic defects in non-crystalline oxides[7] and other glassy solids.

II. NON-CRYSTALLINE OXIDES AND THEIR DEFECT STRUCTURES

In order to begin understanding the natures of defects in non-crystalline oxides, it is first necessary to know something about the structure of a "perfect" non-crystalline oxide. For the present purposes the latter may be taken to be an oxide _glass_. One definition of a glass is a material which has been cooled from the melt to a rigid condition without crystallizing.[4] Therefore, any oxide which can be melted and quenched in a large enough batch ($\gtrsim$ 20 mg) to determine the absence of crystallinity (e.g., by X-ray diffraction) is considered to be a glass-forming composition. Not all oxides form glasses. A number of theories and semiempirical guide-lines have been proposed for predicting whether or not a given substance can be vitrified.[4] While it is beyond the scope of this paper to discuss all of these rules, it is perhaps instructive to mention one rule which generally holds for the oxides, namely, that those systems which form glasses have a high proportion of "mixed" chemical bonds, i.e., bonds which are roughly 50% covalent and 50% ionic. Thus SiO_2, which meets this criterion, can be readily quenched to a glass but MgO, which is $\sim$ 75% ionic, cannot. Since glass formation is dependent on cooling rate as well as composition, the production of extremely small masses of vitreous MgO cannot be ruled out, for example, in the vicinity of "thermal spikes" resulting from implantation of energetic particles. However, the present paper will deal only with oxide glasses which can be routinely produced in substantial quantities.

Much is known about the structures of oxide glasses, e.g., from X-ray diffraction and nuclear magnetic resonance (NMR) studies, but only a few essential aspects of glass structure will be covered here. The reader is referred to Rawson's book[4] and the recent glass literature for more detailed discussions. The cationic species in oxide glasses fall into two broad catagories called "network formers" and "network modifiers". Network-forming oxides are those which form partially covalent bonds, the directional natures of which are responsible for the cations being either tetrahedrally or trigonally coordinated by oxygens; each oxygen in turn joins just two metal ions. Examples of network-forming oxides are: SiO_2, GeO_2, B_2O_3, P_2O_5, and As_2O_3. Network-modifying oxides are ones which do not form glasses by themselves but which contribute their oxygens to the glass network when melted together with substantial amounts of one of the network formers just listed. Network-modifying cations take up interstitial positions. Almost any metal oxide which is not itself a glass former can serve as a network modifier; the most common modifiers in transparent glasses include the alkali and alkaline earth oxides, lead oxide, and zinc oxide. The extra oxygens contributed by the modifier may enter the network in two possible ways, namely as nonbridging (or singly-bonded) oxygens or as bridging oxygens; the latter effect the four-coordination of cations which otherwise would be either three-coordinated (e.g., boron) or occupying an interstitial positions (e.g., aluminum). The four-coordination of boron upon addition of modifier oxides to B_2O_3 glass has been demonstrated and quantitatively determined by NMR.[8]

Figure 1 presents a schematic view of an oxide glass, wherein a three-dimensional random network of RO_4 tetrahedra is

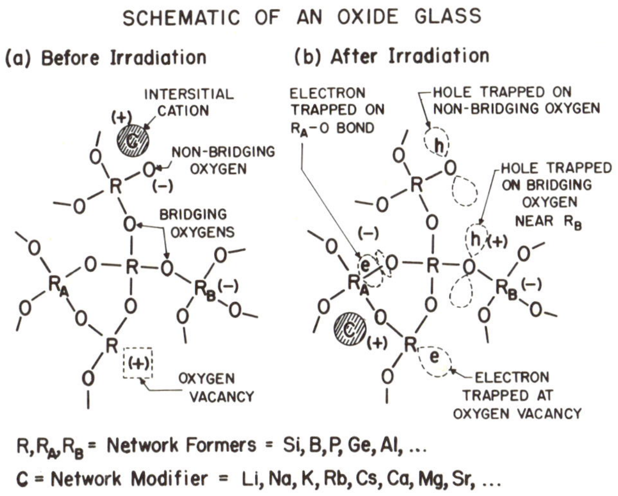

DEFECTS IN NON-CRYSTALLINE OXIDES 325

Fig. 1. Schematic diagram of a complex oxide glass showing several types of radiation-induced paramagnetic centers (b) in their relationships with preexisting structural features (a). Dashed 'balloons' enclose regions of high probability density for trapped electrons (e) or holes (h). (Figure taken from Ref. 7.)

portrayed in a stylized two-dimensional projection. Part (a) of this figure illustrates some of the imperfections which might exist in the as-quenched (unirradiated) glass. As indicated, a nonbridging oxygen represents a negatively charged point defect, which on the average must be charge compensated by a nearby interstitial modifier cation (C). Likewise, a normally-trivalent ion such as boron or aluminum (R_B) which enters the network isomorphously for a silicon requires an addition electron to complete the fourth bond and hence may also be looked upon as a negatively charged point defect. The extra half $O^=$ ion which effects the four-coordination of R_B may have been supplied by a modifier oxide or it may have been scavenged from an oxygen vacancy-interstitial pair (i.e., a Frenkel defect) quenched into the glass. Each half $O^=$-ion vacancy is, of course, a positively charged point defect. As in crystalline solids, charged point defects in glasses are capable of trapping electrons or holes upon irradiation by high energy photons.

Figure 1b schematically indicates the predicted orbitals (dashed curves) for electrons and holes trapped in the generalized oxide glass described above. The existence in irradiated oxide glasses of most of these paramagnetic structures will be confirmed and further elaborated by examples in Section IV below. However, it is encouraging to note that essentially the same defects have been previously identified in single crystals of irradiated α-

quartz and other silicates. Perhaps the best known defect in α-quartz is the E' center[9], which in the schema of Fig. 1b would be an electron trapped in a dangling sp^3 silicon orbital at the site of an oxygen vacancy. Similarly, the defect responsible for the coloration of smoky quartz has been shown[10] to be a hole trapped on an oxygen next to a substitutional aluminum, i.e., the bridging-oxygen defect of Fig. 1b with R_B=Aℓ. A non-bridging-oxygen trapped-hole defect has been elucidated in the mineral phenacite.[11] Another electron trap in quartz occurs when a substitutional metal ion has the same valency as silicon but a greater electron affinity, e.g., R_A=Ge. In this case one supposes the electron to occupy a Ge—O antibonding orbital as illustrated; there is evidence that such defect structures are stabilized by a nearby interstitial alkali ion which diffuses to that site following irradiation.[12,13] This alkali is released from its charge-compensating position near a substitutional aluminum upon hole trapping at the aluminum site.[13]

III. ESR SPECTRAL ANALYSIS

The spin Hamiltonian appropriate to virtually all of the defect centers to be discussed in this paper comprises simply the electron Zeeman and hyperfine terms:

$$\mathcal{H} = \beta \vec{H} \cdot \overleftrightarrow{g} \cdot \vec{S} + \vec{S} \cdot \overleftrightarrow{A} \cdot \vec{I} \quad , \tag{1}$$

where the electronic spin is S=1/2 and the other quantities have their usual meanings. Moreover, in all of these cases the spectra could be adequately fitted by treating the hyperfine interaction to second order in perturbation theory. Assuming that the principal axes of $\overleftrightarrow{g}$ and $\overleftrightarrow{A}$ are coparallel and that the orientation of the applied magnetic field $\vec{H}$ with respect to these axes is given by the Euler angles θ and φ, the resonance condition due to J.F. Baugher[14] is

$$H(m,\theta,\varphi) = \frac{h\nu}{g\beta} - \frac{A}{g\beta} m$$
$$- \frac{m^2 \sin^2\theta}{2g\beta h\nu} \left\{ \frac{g_1^2 g_2^2 (A_1^2 - A_2^2)}{g_a^2 g^2} \cdot \frac{1}{A^2} \sin^2\varphi \cos^2\varphi + \frac{g_a^2 g_3^2 (A_3^2 - B^2)^2}{A^2} \cos^2\theta \right\}$$
$$- \frac{[I(I+1) - m^2]}{4g\beta h\nu} \left\{ \frac{A_1^2 A_2^2}{B^2} + \frac{A_3^2 B^2}{A^2} + \frac{g_1^2 g_2^2 g_3^2}{4 g_a^4 g^2} \cdot \frac{A_3^2 (A_2^2 - A_1^2)^2}{B^2 A^2} \cos^2\theta \sin^2 2\varphi \right\} \tag{2}$$

where

$$g = \sqrt{g_a^2 \sin^2\theta + g_3^2 \cos^2\theta} \quad ,$$

$$g_a = \sqrt{g_1^2\sin^2\varphi + g_2^2\cos^2\varphi} \quad ,$$

$$A = \frac{1}{g_a}\sqrt{g_a^2 B^2 \sin^2\theta + g_3^2 A_3^2 \cos^2\theta} \quad ,$$

$$B = \frac{1}{g_a}\sqrt{g_1^2 A_1^2 \sin^2\varphi + g_2^2 A_2^2 \cos^2\varphi} \quad ,$$

A_1, A_2, and A_3, g_1, g_2, and g_3 are the principal components of $\overleftrightarrow{A}$ and $\overleftrightarrow{g}$, respectively, and m is the magnetic quantum number of the (single) nucleus of spin I with which the unpaired spin is undergoing the hyperfine interaction.

To interpret the ESR spectra of powders or glasses, Eq. (2) must be averaged over all angles, leading to the calculation of the so-called "powder pattern". Powder patterns were originally discussed in the ESR context by Sands[15] and have been recently reviewed by Taylor, Baugher, and Kriz;[14] in essence, they represent the absorption spectra which would be observed in ideal powdered samples in the limit of zero single-crystal linewidth. An actual spectrum may be computer simulated by first calculating the powder pattern and then convoluting it with the suitable single crystal broadening function(s) (Gaussian and/or Lorentzian). Usually the first derivative is taken, since this is the form in which the experimental spectrum is conventionally obtained. Powder patterns can be cast into closed-form algebraic expressions in such simple cases as those illustrated in Fig. 2, but in general it is necessary to evaluate them numerically. In principal, this should be done by calculating the resonance field from Eq. (2) for each of a very large number of random solid angle orientations and histogramming the results on a magnetic field scale.[16] In practice, however, a less-noisy powder pattern is obtained for a given number of computed orientations by considering a uniform grid in $\cos\theta - \varphi$ space (corresponding to equal elements of solid angle). This is what has been done in the present case by means of a program developed by Taylor and Bray.[16]

ESR spectra of glasses differ from those of powdered crystals in a fundamental way. A single set of spin Hamiltonian parameters may be adequate for simulating the spectrum of a single class of defects in a powder, but in a true glass there always exist random variations in 'crystal' fields, bond angles, etc. which lead to statistical variations in the spin Hamiltonian parameters for any given defect type. One unique aspect of the Taylor-Bray program[16] is that it provides an economical means of including such statistical distributions in the simulation process. A simple example of its application involves an analysis of the O_2^- molecular ion in powdered-crystal and amorphous hosts. The experimental spectrum in Fig. 2a is that of O_2^- in polycrystalline sodium peroxide;[17] the superimposed dotted curve is a computer simulation

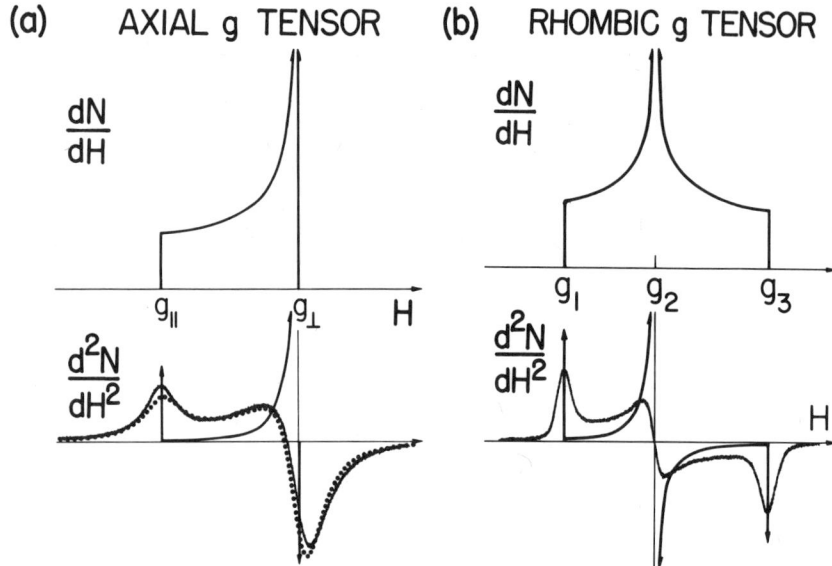

Fig. 2. ESR powder patterns (top) and their first derivatives (bottom) for (a) axial and (b) orthorhombic g tensors and no hyperfine interactions. Heavy curves, at bottom are experimental spectra of (a) O_2^- and (b) O_3^- in polycrystalline hosts (after Ref. 17). Dotted curve is a computer simulation.

assuming an axially symmetric g tensor, no hyperfine interaction, and a Lorentzian convolution function. On the other hand, the spectrum of O_2^- in the <u>amorphous</u> peroxyborates[17,18] is dramatically different in shape (Fig. 3). However, that shape has been reasonably well simulated (dotted curve in Fig. 3) by assuming a broad skewed distribution in $g_\parallel$ values. While this distribution was empirically determined, its physical origin is easily understood in terms of the g value theory as developed by Känzig and Cohen[19] and illustrated in Fig. 4. It is clear that the g shifts are primarily determined by the splitting Δ between the two 2p π_g orbitals. In sodium peroxide Δ is fixed by the crystal structure, but in the amorphous peroxyborates it is reasonable to suppose that Δ varies from site to site in a manner well approximated by a Gaussian distribution function, $P(\Delta)$. To convert this to a distribution in g values, one calculates

$$P(g) = P(\Delta)(\partial g/\partial \Delta)^{-1} \quad , \tag{3}$$

Since g is a non-linear function of Δ, $P(g)$ is in general a skewed distribution [7,20,21] such as the one employed in Fig. 3. In carrying out the simulation of Fig. 3, a relatively narrow distribution

DEFECTS IN NON-CRYSTALLINE OXIDES

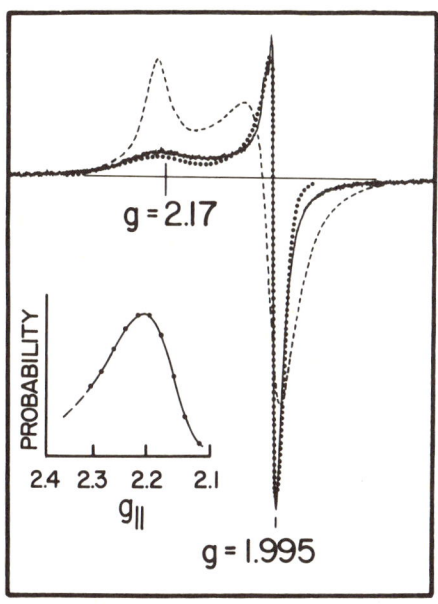

Fig. 3. ESR spectrum of O_2^- molecular ions in the amorphous peroxyborates (noisy trace) and in polycrystalline Na_2O_2 (dashed curve). Dotted curve is a computer simulation of the former assuming the distribution of $g_\parallel$ values given in the inset.

$$g_\parallel = g_e + 2\left[\frac{\lambda^2}{\lambda^2 + \Delta^2}\right]^{\frac{1}{2}}$$

$$g_\perp = \left(g_e + \frac{\lambda}{E}\right)\left[\frac{\Delta^2}{\lambda^2 + \Delta^2}\right]^{\frac{1}{2}}$$

$$\frac{\lambda}{\Delta} \sim 0.1$$

$$\frac{\lambda}{E} \sim 0.002$$

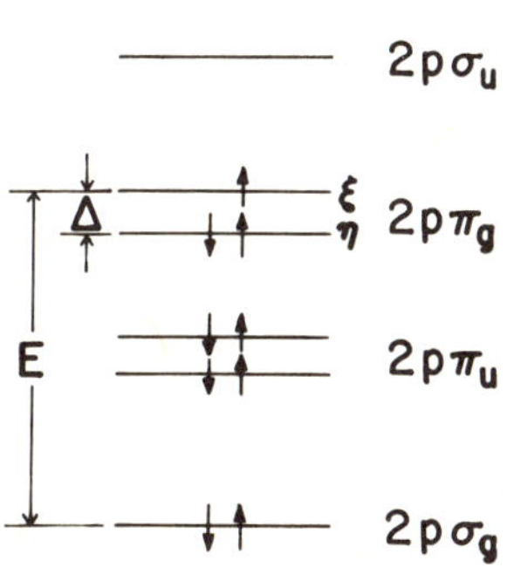

Fig. 4. g value formulae and electronic energy level scheme for O_2^-. g_e is the free electron g value (2.0023) and λ is the spin orbit coupling constant for the O^- ion (~ 0.014 eV).

in $g_\perp$ was also included in the program; this $g_\perp$ distribution was determined from the distribution in $g_\parallel$ via the equations of Fig. 4. Thus, the present example demonstrates how chemistry and physics can place rigid constraints on the type and degree of randomness introduced into the spin Hamiltonian parameters of defects in non-crystalline solids. Because of the manifest existence of such constraints, glass spectra frequently contain more extractable information than do simple powder spectra!

IV. SOME FUNDAMENTAL DEFECT CENTERS IN COMMON OXIDE GLASSES

1. The E' Center in Glassy SiO_2

Probably the best known and most extensively studied intrinsic defect in a glass-forming oxide is the E' center in SiO_2. In α-quartz, a crystalline polymorph of SiO_2, the E' center has been shown to comprise an electron trapped in a nonbonding sp^3 hybrid orbital on a silicon at the site of an oxygen vacancy.[9] Although details of the structure of this center continue to be the focus of theoretical analysis,[22,23] the basic model has remained fixed by the primary hyperfine splittings with ^{29}Si (4.7 percent abundant, I=1/2) measured by ESR: $A_{iso} \approx 410$ G, $A_{aniso} \approx 22$ G.[9] The existence of the E' center in irradiated glassy fused silica has long been inferred from the similarity of induced UV absorption bands[9,24] and of ESR powder-pattern line shapes[25] (for nonmagnetic ^{28}Si and ^{30}Si nuclei) to those observed in α-quartz. Indeed, Fig. 5 shows that there is excellent agreement between the ESR spectrum of a γ-irradiated commercial fused silica and the computed[26a] powder line shape using the principal g values experimentally determined for α-quartz.[9]

More recently, Griscom, Friebele, and Sigel[27] have observed and analyzed the primary ^{29}Si hyperfine structure (hfs) of the E' center in glassy fused silica and have found the splittings to be essentially identical to those measured in crystalline quartz. These hyperfine lines are somewhat more difficult to observe in the vitreous material because of their relatively greater breadth ($\sim$ 40 G vs $\sim$ 1 G), but their identification was ultimately confirmed by studying a sample of non-crystalline SiO_2 enriched to 95% in ^{29}Si.[27] Figure 6a illustrates the ^{29}Si hfs of the E' center (420 G doublet) in a heavily irradiated sample of fused silica of normal isotopic abundance; (a 70-G inner doublet is tentatively assigned to a center involving a nearby hydrogen). The dotted curve is a computer simulation of the ^{29}Si hfs only, based on Eq. (2) and employing the same principal g values as were determined for the central, $^{28}Si + ^{30}Si$ part of the spectrum (Fig. 5). The fit was optimized by varying the parameters $A_\parallel$ and $A_\perp$; however, the initial trial fits failed to reproduce the differences in

DEFECTS IN NON-CRYSTALLINE OXIDES 331

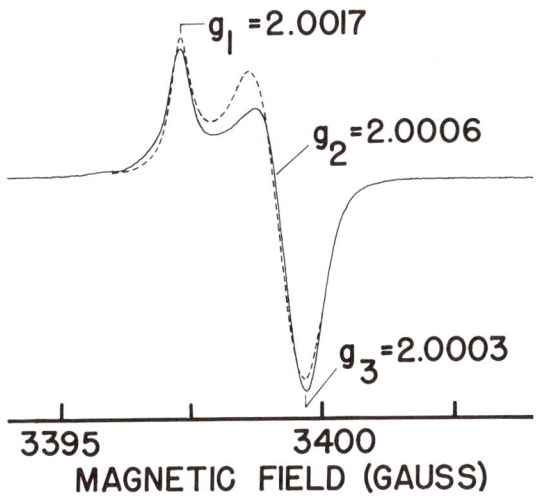

Fig. 5. ESR spectrum of the E' center (for non-magnetic silicon isotopes) in irradiated Corning 7943 fused silica (unbroken curve). Dashed curve is a computer simulation based on measured g values for the same defect in α-quartz. (Figure taken from Ref. 26a).

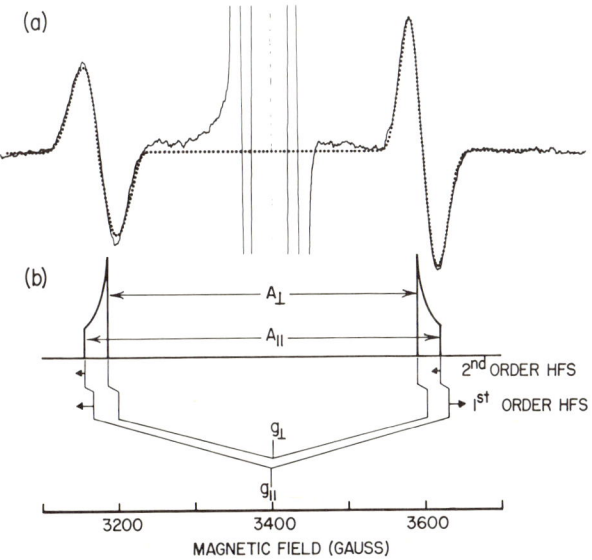

Fig. 6. X-band ESR spectrum at high gain of irradiated Corning 7943 fused silica (a). ^{29}Si hyperfine structure has been computer simulated (dotted curve) by a superposition of powder patterns of the type diagrammed in (b). (Figure taken from Ref. 27.)

height and width observed between the high- and low-field members of the doublet. The final, successful simulation was achieved by introducing a Gaussian distribution in $A_{iso} = (A_\| + 2A_\perp)/3$. Thirty one individual spectra were computed separately and added together according to a Gaussian weighting function of halfwidth $\Delta A_{iso} = 30$ G. It is thereupon recognized that the low-field peak appears relatively squat because here the first- and second-order hyperfine shifts act in consonance under the distribution in A values, while the high-field peak is sharpened because in this case the first- and second-order shifts are negatively correlated and thus tend to mitigate the effect of the distribution (see Fig. 6b).[27]

In principal, the hyperfine parameters derived in the computer simulation process could be used to calculate the exact bond angles at the defect site, but imprecise knowledge of the atomic s-state and p-state coupling constants A_s and A_p for silicon, coupled with the absence of full-blown theoretical calculations, effectively precluded this. However, the measured distribution of A_{iso} values implies that there exists a distribution of bond angles at the E' sites, and the mean site-to-site <u>variation</u> in the (defect)—Si—O bond angle ρ can be reasonably estimated from the expression:[28]

$$\tan \frac{\rho}{2} = - \left[2(1+\xi^2) \right]^{1/2} , \qquad (4)$$

where $\xi = (A_{aniso}/A_p)/(A_{iso}/A_s)$ is the hybridization ratio and $A_{aniso} = (A_\| - A_\perp)/3$. This is done by means of Eq. (3) upon making the substitutions $g \rightarrow \rho$ and $\Delta \rightarrow A_{iso}$. In this way the assumed Gaussian distribution in A_{iso} transforms to a very slightly skewed distribution in ρ with a halfwidth $\Delta \rho \approx 0.7$ deg. This relatively tiny statistical distribution in bond angles is the most apparent effect of vitreous disorder upon the structure of this fundamental defect center.

2. A Nonbridging-Oxygen Hole Center in Silicate Glasses

The most common oxide glasses in commerce or nature are complex silicates, i.e., SiO_2 plus modifier oxides. Since these occur in infinite variety, a logical starting point to begin investigating radiation-induced defects in such materials has been a few simple binary silicate systems. Among these, the alkali silicates have been the most widely studied (see Ref. 29 for a literature review). Schreurs[29] carried out an exceedingly comprehensive ESR study of defects in these glasses as functions of both type and quantity of alkali oxide present; he also made some tentative correlations of the ESR spectra with certain radiation-induced optical bands in the visible. A weak ESR singlet at g=2.000 was reasonably ascribed to the E' center (see above). However, the ESR spectra of irradiated alkali silicate glasses also comprise two far-more-intense components centered near g=2.01, which Schreurs

DEFECTS IN NON-CRYSTALLINE OXIDES

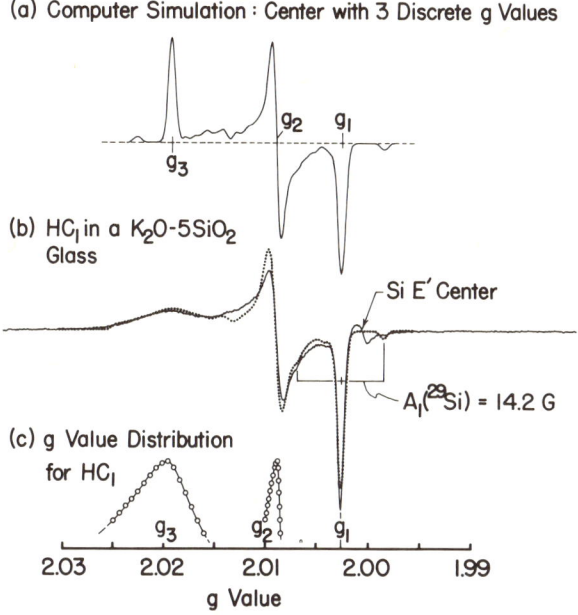

Fig. 7. X-band (9 GHz) ESR spectrum of an irradiated potassium silicate glass (unbroken curve in b). Curve of (a) and dotted curve of (b) are computer line shape simulations of a hole-center component; the latter is based on the g value distributions of (c).

has nicely separated and designated HC_1 and HC_2. HC_1 is characterized by an orthorhombic g tensor and dominates for glasses of low alkali oxide contents ($\lesssim$ 10 mole %), while HC_2, which exhibits an axial g tensor, becomes increasingly apparent when the alkali oxide exceeds ~ 30 mole %.[29] By means of cerium doping and studying the intensity of these bands as a function of Ce^{3+} and Ce^{4+} concentration, it was earlier concluded that all of the g=2.01 spectra are due to trapped holes.[30] It is readily predictable on the basis of the discussion of Sec. II, above, that these trapped-hole centers most likely involve nonbridging oxygens (which are created on a one-for-one basis as molecules of alkali oxide are added to the melt). Schreurs assigned the HC_1 to silicon-oxygen tetrahedra having two nonbridging oxygens and HC_2 to tetrahedra for which three oxygens were nonbridging, but, as will be described below, these detailed assignments are probably not correct.

The basic X-band (9 GHz) ESR spectrum of HC_1 in $K_2O \cdot 5SiO_2$ glass is illustrated in Fig. 7b (unbroken curve). Weak features identified by Schreurs as the E' center or as [29]Si hfs of HC_1 are indicated. Figure 7a is a computer simulation of the HC_1 component assuming a discrete orthorhombic g tensor and an essentially isotropic hyperfine interaction with (4.7%-abundant)^{29}Si. The simulation is clearly much improved (dotted curve in Fig. 7b) by introducing a broad skewed distribution in g_3 and a small distribution in g_2 (Fig. 7c). An equally satisfactory fit to a spectrum obtained at Ka-band (35 GHz) was generated under exactly the same

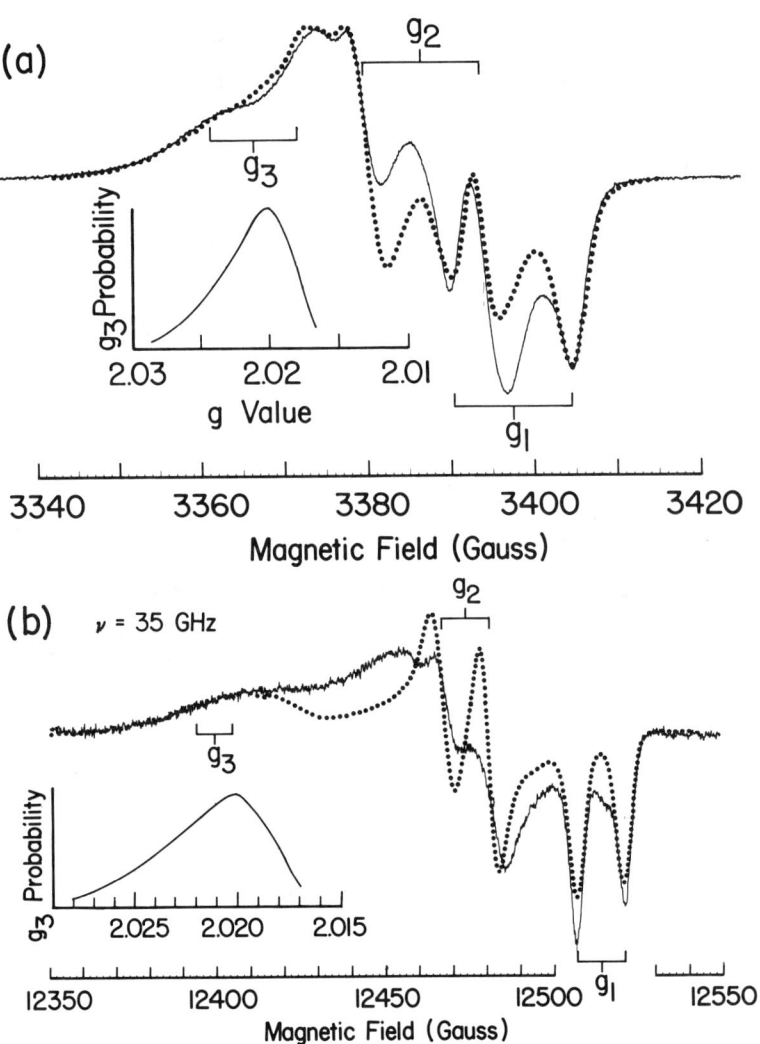

Fig. 8. ESR spectra of an irradiated $K_2O \cdot 5SiO_2$ glass enriched to 95% in ^{29}Si, as obtained at 9 GHz (a) and 35 GHz (b). Dotted curves are mutually-consistent computer simulations of the HC_1 hole-center component, employing the g value distribution shown in the insets

assumptions. Thus the basic g tensor of HC_1 and its approximate statistical variation is well established by Fig. 7. However, the same cannot be said for the ^{29}Si hfs, which is too poorly resolved to decide whether the HC_1 has one or two near-neighbor silicons.

Sidorov and Tyul'kin[31] studied several γ-irradiated sodium and potassium silicate glasses which had been enriched to 68% in ^{29}Si. Although the glass compositions and isotropic enrichment ratio employed were not ideal for isolating the hfs of either HC_1 or HC_2, these authors nevertheless concluded that the observed spectra could be ascribed to sites with a single neighboring silicon. This conclusion has now been substantiated by means of an as-yet-incomplete study[32] of an irradiated $K_2O \cdot 5SiO_2$ glass enriched to 95% in ^{29}Si. The results at the current state of progress are illustrated in Fig. 8, where the X-band and Ka-band spectra of this glass are displayed together with computer simulations (dotted curves) incorporating the assumptions that only HC_1 is important, that the hyperfine interaction is with a single ^{29}Si nucleus, and that the distribution in g_3 values $P(g_3)$ is that given by the insets. The functional form of $P(g_3)$ was constrained to be that derived via Eq. (3) by assuming that

$$g_3 - g_e = 2\lambda/\Delta \tag{5}$$

and that $P(\Delta)$ is a Gaussian. Here g_e is the free electron g value (2.0023) and λ is the spin orbit coupling constant for the O^- ion (taken to be ~ 0.014 eV). The simulations further incorporate a distribution of g_2 values considerably narrower than that shown in Fig. 7c, as well as Gaussian distributions in all components of the hyperfine tensor, each with a halfwidth $\Delta A \sim 3G$. All distributed parameters were taken to be positively correlated. In the course of carrying out more than a dozen trial simulations, it was determined that the fit was more-or-less insensitive to the assumed degree of correlation, except in the g_3 region of the X - band spectrum (Fig. 8a) where a positive correlation of A_3 and g_3 seemed to be essential to a successful outcome. The most probable (peak) values of the spin Hamiltonian parameters used to compute the spectra of Fig. 8 are listed in Table I.

Certain obvious discrepancies between the experimental an computed spectra of Fig. 8 can be ascribed to the presence of an HC_2 component in the spectrum. The alkali oxide content is sufficiently great in this glass (16.7%) that some HC_2 is to be expected[29] and, by operating at microwave power levels high enough to selectively saturate HC_1,[29] the discrepancy is indeed exacerbated. Finally, by following an annealing schedule outlined by Schreurs (1/2 h at $150°C$)[29] to preferentially bleach HC_2, it has recently been found that the experimental spectrum can be brought into closer agreement with the present simulations. In any event, the

Table I. Spin Hamiltonian parameters for the HC_1 trapped-hole center in irradiated $K_2O \cdot 5SiO_2$ glass, as determined by optimization of computer line-shape simulations.

	Principal axis designation		
Parameter	1	2	3
g value [a]	2.0026	2.0090	2.0210
^{29}Si hyperfine coupling constant (G) [b]	14.15	14.15	10.5

[a] Statistical distributions in these parameters are shown in Figs. 7 and 8. Most probable values are tabulated here.

[b] These parameters were assumed to be distributed about the mean values shown according to Gaussian distribution functions of halfwidths $\Delta A \sim 3G$.

simulation of the 35-GHz spectrum (Fig. 8b) remains unsatisfactory due apparently to the assumption of an overly narrow g_2 distribution.

Despite the incomplete status of the HC_1 problem, the present results provide a basis for discussing possible structural models for this defect center. The simulations are certainly adequate to demonstrate that the hyperfine interaction is with a solitary silicon,[32] thereby eliminating a bridging-oxygen model. Moreover, this interaction is small enough ($A_{iso}/A_s \approx 0.008$) to conclude that the unpaired spin is essentially localized <u>away</u> from the silicon. If one hypothesizes that in its ground state the hole is localized in the $2p_z$ orbital of an oxygen bonding to a single silicon, the g tensor of Table I can be analyzed by means of the well known relation

$$g_i - g_e = 2\lambda \sum_{n \neq 0} \left[\langle 0|L_i|n\rangle\langle n|L_i|0\rangle / (E_o - E_n) \right] , \quad (6)$$

where the L_is are the operators of orbital angular momentum and

$$|0\rangle = |O2p_z\rangle$$
$$|1\rangle = |O2p_y\rangle$$
$$|2\rangle = \eta|O2p_x\rangle + \zeta|Si\ \underline{sp}^3\ hybrid\rangle . \quad (7)$$

In principle at least, $|0\rangle$ and $|1\rangle$ should be nearly degenerate and the matrix element of $|L_x|$ coupling them should be unity. For this reason the largest g shift, g_3-g_e, is to be associated with the x direction (i.e., for the applied magnetic field parallel to the bond axis) and Eq. (6) reduces to Eq. (5). From this result and Table I one can make two important observations: (1) the unique axis of the ^{29}Si hyperfine tensor is parallel to the <u>bond</u> direction— as expected for the nonbridging oxygen model—and (2) the distribution in g_3 values can be related to the assumed Gaussian distribution in energy splittings between the 'nearly degenerate' $O2p_z$ and $O2p_y$ orbitals. However, the mean value of Δ calculated by Eq. (5) is ~ 1.5 eV, implying that the degeneracy has been emphatically lifted. The most cogent explanation of this large splitting would seem to be the presence of a nearby modifier cation in the x-y plane. This hypothesis is supported by the data of Schreurs[29] which can be interpreted as showing that the shift g_3-g_e varies inversely as the field strength of the modifier ion (K^+, Na^+, Rb^+, or Ca^{++}), defined as Ze/r^2 where Z is the valency and r is the ionic radius. The intermediate g shift, g_2-g_e, which is independent of modifier ion, is then surmised to be inversely proportional to the energy of the Si-O bond, although this energy cannot be readily calculated without some apriori knowledge of the degree of covalency, i.e., the molecular orbital coefficients η and ζ in Eqs. (7).

3. A Bridging-Oxygen Hole Center

A number of commercially important glasses (e.g., Pyrex and Vycor) contain boron oxide as a major constituent. When these are irradiated, they display a distinctively-structured ESR spectrum similar to that observed in alkali borate glasses of low alkali oxide content.[33] By means of isotopic enrichment experiments, Lee and Bray[33] showed that the observed structure is due to a relatively weak hyperfine interaction with a single boron. Their hypothesis that the defect consisted of a hole trapped on an oxygen neighboring the boron nucleus was later confirmed by Griscom, Taylor, Ware and Bray[34], who also computer simulated the spectrum under the assumptions of orthohombic g and hyperfine tensors with a skewed distribution of g_3 values (Fig. 9). Notwithstanding the successful spectral analysis, it has remained difficult to decide whether this boron-oxygen hole center (BOHC) comprised a bridging or nonbridging oxygen.[7]

One very recent application of high purity borosilicate glass has been as the cladding material for a pure-silica-core optical fiber having the desirable property of an extremely low intrinsic attenuation for visible light.[35] Since a similar high level of performance is sometimes required in radiation environments, studies of radiation-induced defects in fiber glasses in general[26]

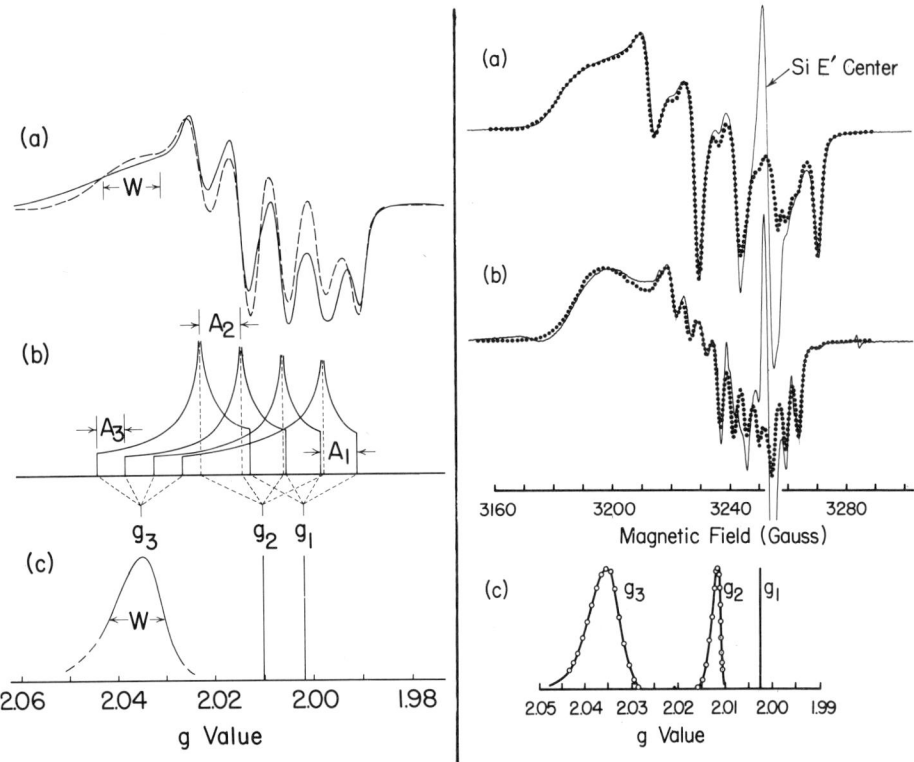

Fig. 9. (left) An analysis of the X-band (9 GHz) ESR spectrum of the boron-oxygen hole center (BOHC) in irradiated alkali borate glasses. Unbroken curve in (a) is an experimental spectrum for a 20% K_2O-80% B_2O_3 glass. Dashed curve in (a) is a computer simulation based on the powder patterns shown in (b) and the distribution of g values shown in (c). Because the width W of the g_3 distribution is greater than the hyperfine splitting A_3, no structure is observed on the low-field shoulder in (a). (After Ref. 34; figure taken from Ref. 7.)

Fig. 10. (right) X-band (9 GHz) ESR spectra of irradiated glasses of composition $B_2O_3 \cdot 3SiO_2$ with (a) natural boron isotopic abundance (81% ^{11}B, 19% ^{10}B) and (b) 95% encirhment in ^{10}B. Dotted curves are computer simulations of the BOHC component, involving the distributions of g values shown in (c). (Figure taken from Ref. 36.)

and borosilicate glasses in particular[36] have been undertaken. The $B_2O_3 \cdot 3SiO_2$ cladding glass composition has proved to be an excellent medium for studying the BOHC, since it is the only type of hole center produced here and its spectrum is not seriously broadened by secondary hyperfine interactions with either borons or alkalis. Mutually consistent computer simulations of the BOHC spectra in $B_2O_3 \cdot 3SiO_2$ glasses of two widely differing boron isotopic compositions are shown in Fig. 10, together with the g value distribution that optimized both fits.[36]

The overall paramagnetic defect concentration in $B_2O_3 \cdot 3SiO_2$ glass was found to exceed that of pure fused SiO_2 by a factor of $\sim 10^2$ for equal γ-ray doses $\gtrsim 10^6$ Rad,[36] and it is instructive to to inquire into the possible sources of this greater radiation sensitivity. Most complex oxide glasses color readily under irradiation due to the presence of <u>modifier oxides</u> (e.g., Refs. 37, 38), but the pure $B_2O_3 \cdot 3SiO_2$ glass studied here is simply a mixture of two network-<u>forming</u> oxides. One clue to what might be going on in the present case is the fact that the borosilicate glass exhibits up to 35 times more <u>Si E' centers</u> than pure silica. Since the E' center is associated with oxygen vacancies, it could be suggested that borosilicates contain more Frenkel defects. It is in fact true that all glasses contain a great many Frenkel defects when they are melted, and the number of these which would be retained upon quenching may well depend on details of the glass structure. Since boron can exist in both three and four coordination,[8] it seems quite reasonable that a few normally-three-coordinated borons may become four coordinated by gettering interstitial oxygens during the melting process, thereby giving rise to an abundance of quenched-in oxygen vacancies. As mentioned in Sec. II, above, any four-coordinated boron should serve as a hole trapping site. Thus, this somewhat heuristic approach gives strong reason to believe that the BOHC may comprise a hole trapped on a (bridging) oxygen at the site of a four-coordinated boron.

The spin Hamiltonian parameters for the BOHC (Table II) can be shown to be consistent with this model. First, the isotropic part of the hyperfine interaction is sufficiently small (A_{iso}/A_s (^{11}B) ≈ 0.015) as to support a model wherein the hole is primarily localized away from the boron. Moreover, the observed g tensor can be analyzed under the constraints of the model by means of Eq. (6). Again, the ground state is taken to be $|0\rangle = |O2p_z\rangle$. It is then assumed that the atomic compositions of the nearest lying excited states are essentially the same as those calculated by Yip and Fowler[39] for the Si_2O cluster in quartz, i.e.,

$$|1\rangle = |O2p_y\rangle$$

$$|2\rangle = .64\,|O2p_x\rangle + .36\,\{\underline{sp^3}\text{ B and Si orbitals}\} \quad . \quad (8)$$

Table II. Spin Hamiltonian parameters for the boron-oxygen hole center (BOHC) in irradiated $B_2O_3 \cdot 3SiO_2$ glass, as determined by optimization of computer line-shape simulations (after Ref. 36).

Parameter	Principal axis designation		
	1	2	3
g value[a]	2.0025	2.0115	2.0355
^{11}B hyperfine coupling constant (G)	13.6	15.3	8.7

[a] Statistical distributions in these parameters are shown in Fig. 10c. Most probable values are tabulated here.

(A notable feature of the Yip-Fowler result is the absence of sp hybridization on the oxygen — an eventuality which was not anticipated in previous attempts to analyze the g tensor of the BOHC.[7,34]) Since state $|1\rangle$ is nonbonding and state $|2\rangle$ is bonding, it is normal to associate the largest g shift, $g_3 - g_e$, with the x direction in Fig. 11a. In this way, Eqs. (6) and (8) lead to the calculated energy separations

$$E_0 - E_1 = 0.85 \text{ eV}$$
$$E_0 - E_2 = 1.99 \text{ eV}.$$

In Fig. 11b, these separations are compared with the experimental[40] and calculated[39] valence band structures for 'perfect' SiO_2 by arbitrarily setting the uppermost energy level in each manifold equal to zero. It can be seen that the splitting between the two nonbonding orbitals at the BOHC site is in fairly good agreement with the photoemission data[40], whereas the Si-O-B bond strength at that site appears to be substantially weaker than the average Si-O-Si bond energy for 'perfect' SiO_2.

It is interesting to note the strong similarity between the spin Hamiltonian parameters for the HC_1 (Table I) and those for the BOHC (Table II) — a similarity which exists despite the identification of the former as a nonbridging-oxygen defect and of the latter as a bridging-oxygen center. Thus, a host of centers having similar g and hyperfine tensors in various oxide glasses (see Refs. 7 or 41 for a tabulation) are probably either bridging- or nonbridging- oxygen defects depending on details of the structures of the respective glasses.

DEFECTS IN NON-CRYSTALLINE OXIDES 341

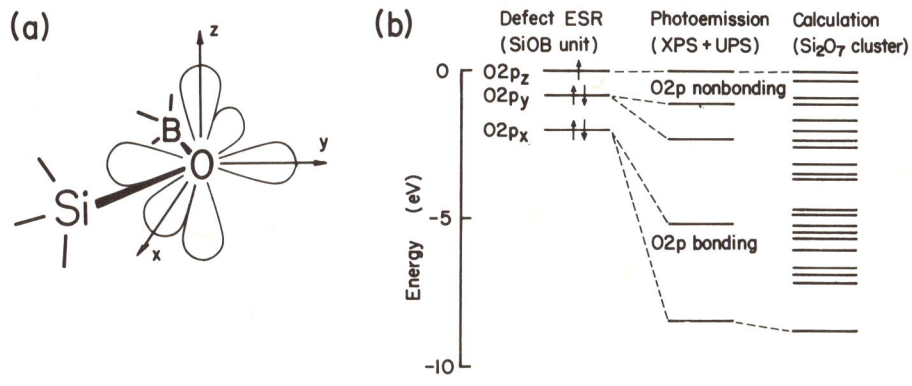

Fig. 11. Model for the boron-oxygen hole center (BOHC). (a) Steric picture. (b) Energy level scheme with comparison to photoemission data[40] and band structure calculations[39] for SiO_2. Dotted lines connect levels presumed to be of similar orbital compositions. (Figure taken from Ref. 36.)

4. The Boron E' Center

In addition to the BOHC and the Si E' center (both discussed above), one other defect type was isolated in irradiated glasses of the $B_2O_3 \cdot 3SiO_2$ composition. This center is apparently the boron analog of the Si E' center, i.e., an electron trapped in the dangling sp^3 orbital of a boron pyramidally bonded to three network oxygens.[36] Figure 12a shows the X-band ESR spectrum of this center in a glass of normal isotopic abundance, as obtained with a broader field scan and higher spectrometer gain than required to observe the BOHC. Figure 12b shows the corresponding spectrum of a glass enriched to 95% in the ^{10}B isotope. Resolution of ^{11}B and ^{10}B hyperfine splittings is seen to be excellent, and achievement of the dotted computer simulations provided measurements of the relevant parameters (i.e., $A_\parallel$, $A_\perp$, and their distributions)[36] needed to infer the defect geometry. Table III presents the results of a spin-population analysis of the B E' center, showing that the electronic structure at the B E' site is nearly identical to that of the Si E' center in both glassy and crystalline forms of SiO_2. Within the accuracy to which the various parameters of Eq. (4) are known, the bond angle ρ at the defect site is in each case close to the tetrahedral angle of 109.5°.

The discovery of the B E' center is of particular interest because of the support it lends to the theory of Feigl, Fowler, and Yip[23] for the mechanism of E' center formation in α-quartz (see

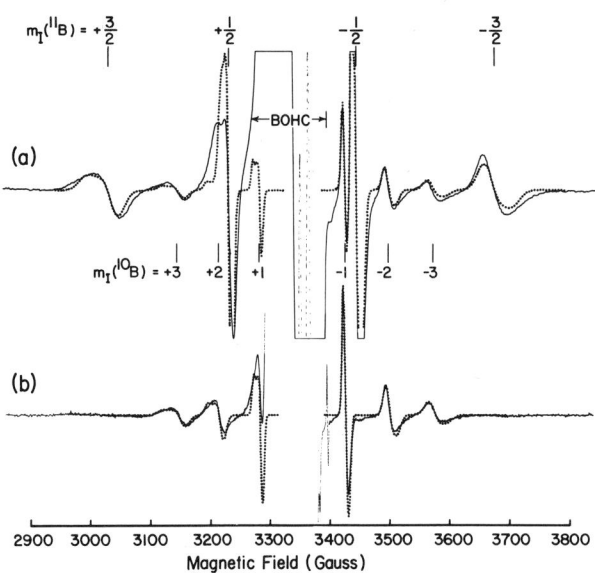

Fig. 12. X-band (9 GHz) ESR spectra irradiated $B_2O_3 \cdot 3SiO_2$ glasses (a) for boron in its natural isotopic abundance and (b) for boron enriched to 95% in ^{10}B. Dotted curves are computer simulations of the boron E' center contributions to the spectra (neglecting the $m_I=0$ component which is obscured by the BOHC). (Figure taken from Ref. 36.)

Fig. 13). Qualitatively speaking, the theory considers two SiO_3 units facing each other across an oxygen vacancy and shows that the geometries of each of these 'half defects' depend on the number of electrons in the structure. If there are no electrons in the dangling Si orbital perpendicular to the plane defined by the three oxygens, the theory says that the SiO_3 unit will relax to planarity. If, on the other hand, one of the 'half defects' should trap an electron in this orbital, the energy of the structure is minimized by $\underline{sp^3}$ hybridization of the silicon orbitals, thus leading to a pyramidal SiO_3 unit with nearly tetrahedral O-Si-O bond angles. While the existence of planar SiO_3 units has not yet been demonstrated directly, it is well known[8] that the BO_3 unit is a planar structure. Therefore the present study[36] has shown that if one adds an extra electron to a <u>bonafide</u> planar unit the structure indeed distorts into the pyramidal configuration with the unpaired electron residing in the dangling $\underline{sp^3}$ orbital of the apex atom (in this case a boron), as predicted.[23]

DEFECTS IN NON-CRYSTALLINE OXIDES

Table III. Spin population analyses of the silicon and boron E' centers in glass with comparison to the silicon E' center in α-quartz.

Center	$\|\langle\psi\|ns\rangle\|^{2\,a}$	$\|\langle\psi\|np\rangle\|^{2\,b}$	$\Delta\rho^{\,c}$
B E' in glass[36]	0.26	0.46	± 0.5 deg
Si E' in glass[27]	0.25	0.65	± 0.7 deg
Si E' in α-quartz[9]	0.24	0.63	---

^a Spin density in boron 2s or silicon 3s atomic orbitals: $\|\langle\psi\|ns\rangle\|^2$ = A_{iso}/A_s, where $A_s(^{29}Si)=1710$ G; $A_s(^{11}B)=835$ G.

^b Spin density in boron 2p or silicon 3p atomic orbitals: $\|\langle\psi\|np\rangle\|^2$ = A_{aniso}/A_p, where $A_p(^{29}Si)=34$ G; $A_p(^{11}B)=19.6$ G.

^c Half-width of slightly-skewed distribution in bond angles derived via Eq. (4) from observed Gaussian distributions in A_{iso}.

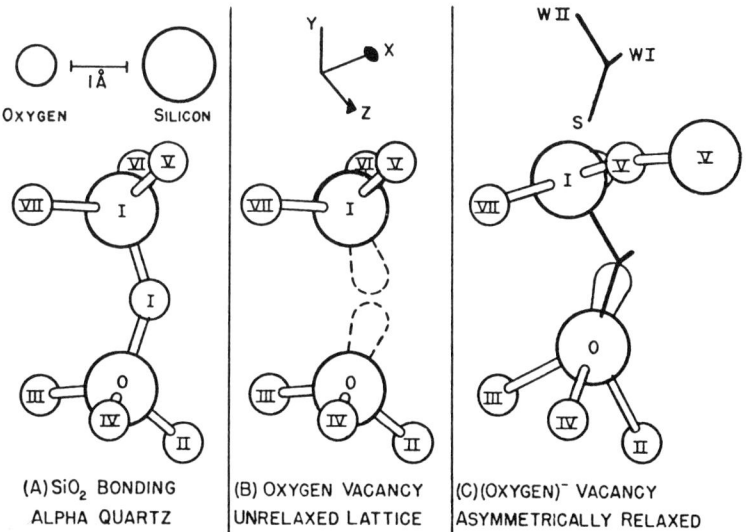

Fig.13. Atomic clusters in quartz, presented as projections onto the plane defined by Si_O, O_I, and Si_I. (a) Normal quartz. (b) Unrelaxed oxygen vacancy in α-quartz. (c) Asymmetrically relaxed O^- vacancy in α-quartz, according to Feigl, Fowler, and Yip.[23] Essentially identical SiO_3 'half defects' evidently occur in glassy SiO_2,[27] although not necessarily in pairs with well-defined separations and orientations. (Figure taken from Ref. 23.)

5. Other Radiation-Induced Defects

Due to space restrictions, the present paper has limited its emphasis to only a few of the most fundamental and best characterized defect types in common oxide glasses. It is appropriate, however, to give at least brief mention to a number of other centers which are either equally fundamental, particularly well characterized, or otherwise interesting.

It is possible that one of the most common defects in multicomponent oxide glasses consists of one or more electrons trapped on or near a 'cluster' of modifier cations. ESR evidence has been given for the formation of such clusters under irradiation in alkali borate glasses[38] and in doped α-quartz.[42] While this type of defect must certainly be fundamental, it has been very difficult to detect and analyze because of its relatively broad hyperfine envelope[38] as well as the fact that centers involving even numbers of electrons are nonparamagnetic. Thus, complex borate or silicate glasses irradiated at room temperature usually display hole center resonances of intensities proportional to the modifier oxide contents (up to ~ 20 mole %)[37] but no detectable (or distinguishable) electron center resonances. Alkali-associated trapped electron centers in non-crystalline oxides may be in some ways analogous to F-aggregate centers[1,2] in the alkali halides.

At least one cation-associated defect is apparently a hole trap. Friebele[43] has shown recently that a familiar ESR spectrum[44] in irradiated lead-bearing oxide glasses is due to Pb^{2+} ions which have captured holes to become Pb^{3+}. Because of the large ^{207}Pb hyperfine splitting ($A_{iso} \approx 13,000$ G) the spectrum could not be analyzed by means of Eq. (2), and an exact diagonalization of Eq. (1) was performed[43] in order to computer simulate the spectrum.

An additional E' center analog, the Ge E' center, has been elucidated[45] in a germanium-doped silica composition employed as the core material in a low-loss optical fiber. Here it was shown that the g tensor is extremely sensitive to random strains normally associated with the vitreous state, as well as to the number of germanium ions immediately neighboring the GeO_3 unit trapping the electron. The basic defect has also been reported in glassy and crystalline GeO_2[46] and it has been extensively characterized in Ge-doped quartz by Feigl and Anderson.[9]

The Si E' center is always observed in irradiated pure fused silica.[26] Since it is interpreted to be an electron-type defect in bulk glasses, one infers that barring charge build-up there exists an equal population of hole-type centers in these materials. Weeks[47] made the first observation of an ESR spectrum attributable to oxygen-associated hole centers (OHCs) in heavily-irradiated silica. However, the spectral line shape was quite different from those of the HC_1 and BOHC discussed above and has until recently defied any form of analysis. Computer simulation studies now in progress[48] suggest that the OHC spectrum in pure fused silica can be resolved into a minimum of 3 discrete overlapping components, one of which bears some similarities to the HC_1.

Radiation-induced defect centers in alkali borate glasses with alkali halide additions have been studied by ESR,[26,37] optical,[49] and Raman[50] spectrometry. Among the species observed have been atomic halogens[20] (at cryogenic temperatures) and hal_2^- molecular ions which remained stable at room temperature.[37] The X-band ESR spectrum of Cl_2^- ions in glass is illustrated in Fig. 14 together with a computer simulation based on a resonance condition[51] somewhat more complicated than Eq. (2).

6. Ferromagnetic Precipitates

Oxide glasses containing Fe^{2+} or Fe^{3+} ions can be made to precipitate spherical ferromagnetic particles, roughly in the 50-1000Å size range. Particles of metallic iron may occur in glasses which were sufficiently reduced in the melt[52] or were subjected to subsolidus reduction in the range 700-1000°C.[53] On the other hand, glasses which are relatively oxidized (or reduced glasses

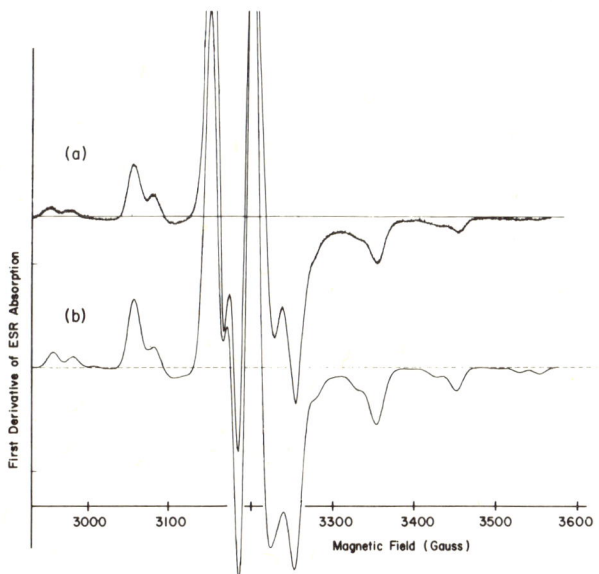

Fig. 14. X-band ESR spectrum of an irradiated glass prepared from NaCl and B_2O_3 (a) Experimental spectrum. (b) Computed spectrum for Cl_2^- molecular ions. (Figure taken from Ref. 37.)

exposed to an oxidizing environment) precipitate ferrimagnetic particles of ferric iron spinel[54,55] (e.g., magnetite) upon annealing in the range 600-750°C. Depending on one's point of view, these various precipitates may or may not be regarded as 'defects', although there is some evidence that the precipitation process can be enhanced by irradiation.[56]

In any event, ferromagnetic precipitates in glass share one common aspect with the defect centers discussed above, namely, that their ESR spectra can be analyzed by computer simulation techniques. Providing that the particles are small with respect to the microwave skin depth, sufficiently dilute, and highly equidimensional (within ~ 5% of perfect sphericity), and further that they are single domain* and characterized by cubic magnetocrystalline anisotropy, the appropriate resonance condition to be used in place of Eq. (2) is[57-59]

$$H(\theta,\varphi) = \frac{h\nu}{g\beta} - \frac{2K_1}{M_s}\left[1-5(\cos^2\theta\sin^2\theta+\sin^4\theta\sin^2\varphi\cos^2\varphi)\right], \quad (9)$$

where K_1 is the first order magnetocrystalline anisotropy constant and M_s is the saturation magnetization. A powder pattern calculated from Eq. (9) for positive K_1 is shown as the unbroken curve in Fig. 15b where $H_a = 2K_1/M_s$. For negative K_1, the corresponding powder pattern is obtained by reflecting about the center of gravity (c.g.) of the spectrum. It is thus clear that both the sign and magnitude of the anisotropy constant can be determined for an ensemble of spherical particles with randomly oriented crystallographic axes.

Soils returned from the moon have been found to comprise as much as 50% glass and this glass is laden with microscopic and submicroscopic ferromagnetic inclusions, much of which is iron.[60] The heavy curve in Fig. 15a is an experimental X-band spectrum[55] obtained at room temperature for a sample of lunar soil which was annealed at 650°C in an evacuated, sealed silica tube. Lighter-weight curves in Fig. 15a are computed spectra based on the powder patterns of Fig. 15b. The dotted curve is the difference between the experimental and computed spectra and represents the contribution of nonspherical and/or multidomain iron particles.[52,61] The natures of the (spherical) particles giving rise to the spectral component which was successfully simulated have been the subject of debate since it is now known[59] that both metallic iron and certain ferric iron spinel precipitates can exhibit positive anisotropy at 300°K. One means which has been proposed to discriminate between these two candidate phases involves studying the temperature dependence of the integrated area under the ESR

---

*Multidomain effects in the resonance spectra of iron are discussed in Ref. 52.

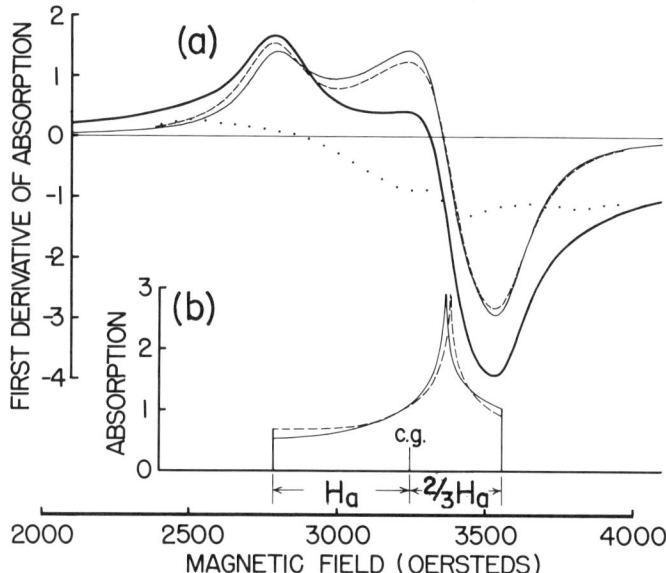

Fig. 15. X-band ferromagnetic resonance spectrum of a typical lunar soil following vacuum heat treatment at 650°C for 880h (heavy curve in a). Lighter weight curves in (a) are computer simulations based on the powder patterns of (b). Unbroken theoretical curves are calculated from Eq. (9), while dashed curves represent a more exact treatment. The point labeled "c.g." locates the center of gravity of the spectrum in the limit of vanishing H_a; it corresponds here to a g value of 2.083. $H_a = 2K_1/M_s$ is the anisotropy field. (After Ref. 55.)

absorption curve.[61] Application of this method to the lunar soils leads to the conclusion[55,61] that both metallic iron and ferric iron spinel phases are present—a finding which might at first seem surprising in view of the overall reduced natures of lunar materials but which is nevertheless consistent with known phase-equilibrium data.[62]

V. ELECTRONIC STRUCTURE OF SiO_2 WITH DEFECTS

Because of its central relevance to fiber optics, MOS devices, and other applications, silicon dioxide and its structure have been the subject of numerous experimental and theoretical investigations (see Ref. 39 for a review of the recent literature). Keeping pace with these studies have been a number of theoretical papers dealing with intrinsic defects (mainly the E' center)[22,23]

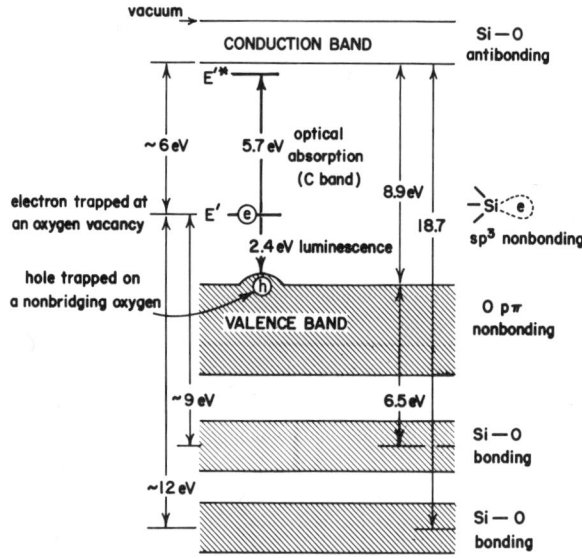

Fig. 16. Electronic structure of SiO_2 with defects. Data are drawn from a variety of sources cited in the text.

in both crystalline and non-crystalline forms of the material. Figure 16 summarizes much that is known or inferred about the electronic structure of SiO_2 with defects. DiStefano and Eastman[63] have determined the band gap of 8.9 eV from photoconductivity measurements and the position of the valence band relative to the vacuum from the photoemission threshold. The structure and width of the valence band is known from photoelectron spectroscopy[40] and can be understood theoretically.[39] It can be qualitatively reasoned that the energy of a dangling sp^3 silicon hybrid orbital (i.e., the E' center) must lie somewhere within the band gap and that oxygen-associated trapped-hole states must lie near or just above the top of the valence band.

The actual position of the E' level in Fig. 16 has been induced from rather diverse considerations. First, it has been inferred that the first excited state of the E' center (E'*) must lie below the conduction band edge, since excitation into the familiar E' center absorption band (5.7 eV)[9] produces neither bleaching, photocurrent, nor photoluminescence.[64] However, under pulsed-electron irradiation, a transient luminescence is observed at ~ 500 nm (2.4 eV) in both pure fused silica and α-quartz which decays at the same rate as an induced transient absorption at 5.7 eV.[64,65] Various details of the data are best explained by assuming the formation of transient E' centers, most of which decay nonradiatively (possibly by rehealing of broken Si-O bonds).[64] The 2.4 eV emission is then ascribed to a

DEFECTS IN NON-CRYSTALLINE OXIDES 349

relatively small number of holes which are released into the valence band and recombine with trapped electrons at E' sites. If this interpretation is correct, the E' level is rather precisely fixed, since the combined energies of the absorption and emission bands total 90% of the band gap.

This energy-level information, coupled with the results of very recent band structure calculations,[39] provide an excellent opportunity to apply Eq. (6) to the calculation of the g tensor for the E' center in SiO_2. The present calculation will nevertheless involve a number of approximations, the first of which is to neglect Si 3d orbitals, as was done by Yip and Fowler.[39] Secondly, it will be assumed that all Si-O bonding orbitals involving O 2p states have the orbital composition calculated[39] for the Si_2O molecule (Table IV) and that their energies relative to the E' level are all equal to the average for the two corresponding bands of Fig. 16, i.e.,$-(1/2) \cdot (9 \text{ eV} + 12 \text{ eV}) = -10.5$ eV. It will be further assumed that the make-ups of the Si-O <u>antibonding</u> orbitals are likewise given by Table IV and that their average position in energy is about halfway between the conduction band edge and the vacuum, i.e., ~ 7 eV above the E' level (see Fig. 16). To proceed with the calculation, one only requires expressions for four tetrahedrally directed silicon hybrid orbitals which will comprise the basis set. Choosing the z axis to lie along the direction of the dangling orbital containing the unpaired electron, the (normalized) Si <u>3p-state parts</u> of these orbitals can be given as

$$\chi_0 = P_z$$
$$\chi_1 = \frac{1}{3} P_z - \frac{\sqrt{2}}{3} P_y + \sqrt{\frac{2}{3}} P_x$$
$$\chi_2 = \frac{1}{3} P_z + \frac{2\sqrt{2}}{3} P_y \qquad (10)$$
$$\chi_3 = \frac{1}{3} P_z - \frac{\sqrt{2}}{3} P_y - \sqrt{\frac{2}{3}} P_x$$

Thus, according to Eq. (6) $g_\| = g_z \approx g_e$; (the measured value of $g_\| = 2.0017^9$ compares with $g_e = 2.00235$). It follows that $g_x = g_y = g_\perp$. For simplicity, only $\Delta g_y (= g_\perp - g_\|)$ need be calculated, and for this purpose it is easily shown that $|\langle \chi_1|L_y|\chi_0\rangle|^2 = |\langle \chi_3|L_y|\chi_0\rangle|^2 = 2/3$ and that $|\langle \chi_2|L_y|\chi_0\rangle|^2 = 0$. Since the ground state $|0\rangle$ of the E' center is well characterized by ESR (Table III), it is known that $|\langle 0|\chi_0\rangle|^2 \approx 0.65$. As mentioned above, the individual Si 2p-state compositions of the relevant excited states $|\langle n|\chi_{i \neq 0}\rangle|^2$ are to be taken from Table IV; they are $0.56 \div 2 = 0.28$ for antibonding states and $0.21 \div 2 = 0.105$ for bonding states. Taking the spin orbit coupling constant for silicon to be $\lambda = 0.0185$ eV[66], the calculation of Eq. (6) for the E' center reduces to

Table IV. Energy levels and orbital compositions of a hypothetical Si_2O molecular cluster according to Yip and Fowler.[39]

Bonding Nature	Energy (eV)	Orbital Composition[a]
Antibonding Si 3p Orbitals	six levels −12 to −18	36% O 2p, 8% Si 3s 56% Si 3p (see note b)
Nonbonding O 2p Orbitals	−26.46 −26.56	100% O 2p
Bonding O 2p Orbital	−30.20	64% O 2p, 15% Si 3s 21% Si 3p
Bonding O 2s Orbital	−44.15	93% O 2s, 3% Si 3s 4% Si 3p

[a] All orbitals listed include contributions of <u>both</u> silicons. To obtain probability density on a single given silicon, given numbers should be divided by 2.

[b] Compositions of the six antibonding orbitals were not given by Yip and Fowler. 'Average' values shown here are 'ball-park' estimates based on simple molecular orbital considerations.

$$(g_\perp - g_\parallel)_{calc} = -2 \times 0.0185 \left\{ \frac{0.65 \times 0.28 \times (4/3)}{7} - \frac{0.65 \times 0.105 \times (4/3)}{10.5} \right\}$$

$$= -0.00096$$

This compares with[9]

$$(g_\perp - g_\parallel)_{exp} = 2.00045 - 2.0017$$

$$= -0.00125.$$

The agreement seems reasonable, given the assumptions which were made and the uncertainties in the various parameters entering into the calculation.

In summary, it appears that the structural picture of defects in non-crystalline oxides now emerging from ESR and optical studies is readily reconciled with recent experimental and theoretical advances in understanding the band structures of these materials. This is particularly encouraging in view of the high technological importance of such non-crystalline oxides as SiO_2. Indeed, formation of E' centers in the SiO_2 layers of irradiated MOS devices has now been shown to be a substantial source of the space charge build-up responsible for device degradation under irradiation [67].

REFERENCES

1. J.H. Schulman and W.D. Compton, Color Centers in Solids, (MacMillan, New York, 1962).
2. W.B. Fowler (ed). Physics of Color Centers (Academic Press, New York, 1968).
3. D.P. Barag, IX Congrès International du Verre, Artistic and Historical Communications, p. 183, 1972.
4. H. Rawson, Inorganic Glass-Forming Systems (Academic Press, London, 1967).
5. E. Lell, N.J. Kreidl and J.R. Hensler, in: Progress in Ceramic Science, Vol. 4, Ed. J. Burke (Pergamon, Oxford, 1966).
6. A. Bishay, J. Non-Crystalline Solids $\underline{3}$, 54 (1970).
7. D.L. Griscom, J. Non-Cryst. Solids $\underline{13}$, 251 (1973/74).
8. P.J. Bray and J.G. O'Keefe, Phys. Chem. Glasses $\underline{4}$, 37 (1963); P.J. Bray, in: Interaction of Radiation with Solids, Ed. A. Bishay, (Plenum Press, New York, 1967).
9. R.A. Weeks, J. Appl. Phys. $\underline{27}$, 1376 (1956); C.M. Nelson and R.A. Weeks, J. Am. Ceram. Soc. $\underline{43}$, 396 (1960); ibid $\underline{43}$, 399 (1960); R.H. Silsbee, J. Appl. Phys. $\underline{32}$, 1459 (1961); F.J. Feigl and J.H. Anderson, J. Phys. Chem. Solids $\underline{31}$, 575 (1970).
10. M.C.M. O'Brien, Proc. Roy. Soc. (London) A$\underline{231}$, 404 (1955).
11. H. Lozykowski, R.G. Wilson, and F. Holuj, J. Chem. Phys. $\underline{51}$, 2309 (1969).
12. J.H. Anderson and J.A. Weil, J. Chem. Phys. $\underline{31}$, 427 (1959).
13. J.H. Mackey, Jr., J. Chem. Phys. $\underline{39}$, 74 (1963).
14. P.C. Taylor, J.F. Baugher, and H.M. Kriz, Chem. Rev. $\underline{75}$, 205 (1975).
15. R.H. Sands, Phys. Rev. $\underline{99}$, 1222 (1955).
16. P.C. Taylor and P.J. Bray, J. Mag. Res. $\underline{2}$, 305 (1970).
17. D.L. Griscom, Ph.D. Thesis, Brown University, Providence, R.I., 1966.
18. J.O. Edwards, D.L. Griscom, R.B. Jones, K.L. Watters, and R.A. Weeks, J. Am. Chem. Soc. $\underline{91}$, 1095 (1969).
19. W. Känzig and M.H. Cohen, Phys. Rev. Lettr. $\underline{3}$, 509 (1959).
20. D.L. Griscom, Solid State Comm. $\underline{11}$, 899 (1972).
21. G.E. Peterson, C.R. Kurkjian, and A. Carnevale, Phys. Chem. Glasses $\underline{15}$, 52 (1974).
22. A.R. Ruffa, Phys. Rev. Lett. $\underline{25}$, 650 (1970); A.J. Bennett and L.M. Roth, J. Phys. Chem Solids $\underline{32}$, 1251 (1971); A.R. Ruffa, J. Non-Cryst. Solids $\underline{13}$, 37 (1973).
23. F.J. Feigl, W.B. Fowler, and K.L. Yip, Sol. State Comm. $\underline{14}$, 225 (1974); K.L. Yip and W.B. Fowler, Phys. Rev. B$\underline{11}$, 2337 (1975).
24. See, for example G.W. Arnold and W.D. Compton, Phys. Rev. $\underline{116}$, 802 (1959); P.W. Levy, Phys. Chem. Solids $\underline{13}$, 287 (1960); R.A. Weeks and E. Lell, J. Appl. Phys. $\underline{35}$, 1932 (1964).
25. R.A. Weeks and C.M. Nelson, J. Appl. Phys. $\underline{31}$, 1555 (1960).
26. (a) E.J. Friebele, D.L. Griscom, R.J. Ginther, and G.H. Sigel, Jr., Proc. Tenth International Cong. on Glass, Kyoto,

Japan, p. 6-16 (1974); (b) E.J. Friebele, R.J. Ginther, and G.H. Sigel, Jr., Appl. Phys. Lett. <u>24</u>, 412 (1974); G.H. Sigel, Jr. and B.D. Evans, Appl. Phys. Lett. <u>24</u>, 410 (1974); B.D. Evans and G.H. Sigel, Jr., IEEE Trans. on Nuc. Sci. <u>NS-21</u>, 113 (1974).
27. D.L. Griscom, E.J. Friebele, and G.H. Sigel, Jr., Sol. State Comm. <u>15</u>, 479 (1974).
28. A.R. Reinberg, J. Chem. Phys. <u>41</u>, 850 (1964).
29. J.W.H. Schreurs, J. Chem. Phys. <u>47</u>, 818 (1967).
30. R.F. Tucker, in *Advances in Glass Technology*, Proc. Sixth International Cong. on Glass, Washington, 1962 (Plenum, New York, 1963), Vol. I, p. 121.
31. T.A. Sidorov and B.A. Tyul'kin, Dokl. Akad, Nauk SSSR <u>175</u>, 872 (1967).
32. D.L. Griscom, Bull. Am. Ceram. Soc. <u>54</u>, 436 (1975).
33. S. Lee and P.J. Bray, J. Chem. Phys. <u>39</u>, 2863 (1963).
34. D.L. Griscom, P.C. Taylor, D.A. Ware, and P.J. Bray, J. Chem. Phys, <u>48</u>, 5158 (1968).
35. F.W. Dabby, D.A. Pinnow, F.W. Ostermayer, L.G. Van Uitert, M.A. Saifi, and I. Camibel, Appl. Phys. Lett. <u>25</u>, 714 (1974).
36. D.L. Griscom, G.H. Sigel, Jr., and R.J. Ginther, submitted to J. Appl. Phys. (June, 1975).
37. D.L. Griscom, P.C. Taylor and P.J. Bray, J. Chem. Phys. <u>50</u>, 977 (1969).
38. D.L. Griscom, J. Non-Crystalline Solids <u>6</u>, 275 (1971).
39. K.L. Yip and W.B. Fowler, Phys. Rev. B<u>10</u>, 1400 (1974).
40. T.H. DiStefano and D.E. Eastman, Phys. Rev. Lett. <u>27</u>, 1560 (1971); H. Ibach and J.E. Rowe, Phys. Rev. B<u>10</u>, 710 (1974).
41. D.L. Griscom, P.C. Taylor, and P.J. Bray, J. Chem. Phys. <u>53</u>, 469 (1970).
42. R.V. Lorenze and F.J. Feigl, Phys. Rev. B<u>8</u>, 4833 (1973).
43. E.J. Friebele, Bull. Am. Ceram. Soc. <u>54</u>, 436 (1975); manuscript in preparation.
44. Y.M. Kim and P.J. Bray, J. Chem. Phys. <u>49</u>, 1298 (1968).
45. E.J. Friebele, D.L. Griscom, and G.H. Sigel, Jr., J. Appl. Phys. <u>45</u>, 3424 (1974).
46. T.A. Purcell and R.A. Weeks, J. Phys. Chem. Glasses <u>10</u>, 198 (1969); R.A. Weeks and T.A. Purcell, J. Chem. Phys. <u>43</u>, 483 (1965); G.F.J. Garlick, J.E. Nicholls, and A.M. Ozer, J. Phys. C<u>4</u>, 2230 (1971).
47. R.A. Weeks, J. Appl. Phys. <u>27</u>, 1376 (1956).
48. D.L. Griscom, unpublished work.
49. D.L. Griscom, J. Chem. Phys. <u>51</u>, 5186 (1969).
50. M. Hass and D.L. Griscom, J. Chem. Phys. <u>51</u>, 5185 (1969).
51. D. Schoemaker, Phys. Rev. <u>149</u>, 693 (1966); W. Dreybrodt and D. Silber, Phys. Status Solidi <u>16</u>, 215 (1966).
52. D.L. Griscom, C.L. Marquardt, E.J. Friebele, and D.J. Dunlop, Earth Planet. Sci. Lett. <u>24</u>, 78 (1974).
53. G.W. Pearce, R.J. Williams, and D.S. McKay, Earth Planet. Sci. Lett. <u>17</u>, 95 (1972).

54. M.P. O'Horo, Bull. Amer. Ceram. Soc. 53, 324, 356 (1974); manuscript in preparation.
55. D.L. Griscom, E.J. Friebele, and C.L. Marquardt, Proc. Lunar Sci. Conf. 4th, Geochim. Cosmochim. Acta, Suppl. 4, Vol. 3, 2709 (1973).
56. R.A. Weeks, D.L. Griscom, and P.M. Bell, Proc. 18th Ampere Congress, Nottingham, 1974, p. 167.
57. E. Schlömann, J. Phys. Chem. Solids 6, 257 (1958).
58. F.-D. Tsay, S.I. Chan, and S.L. Manatt, Geochim. Cosmochim. Acta 35, 865 (1971).
59. D.L. Griscom, Geochim. Cosmochim. Acta 38, 1509 (1974).
60. R.M. Housley, R.W. Grant, and M. Abdel-Gawad, Proc. Lunar Sci. Conf. 3rd, Geochim. Cosmochim. Acta, Suppl. 3, Vol. 1, 1065 (1972).
61. D.L. Griscom, C.L. Marquardt, and E.J. Friebele, J. Geophys. Res. 80, 2935 (1975).
62. R.J. Williams and E.K. Gibson, Earth Planet. Sci. Lett. 17, 84 (1972).
63. T.H. DiStefano and D.E. Eastman, Solid State Comm. 9, 2259 (1971).
64. G.H. Sigel, Jr., J. Non-Cryst. Solids 13, 372 (1973/74).
65. D.L. Griscom and G.H. Sigel, Jr., Bull. Am. Phys. Soc. 13, 1474 (1968); G.H. Sigel, Jr. and D.L. Griscom, Bull. Am. Ceram. Soc. 48, 447 (1969).
66. P.W. Atkins and M.C.R. Symons, The Structure of Inorganic Radicals, Elsevier, London, 1967.
67. C.L. Marquardt and G.H. Sigel, Jr., submitted to IEEE Trans. on Nucl. Sci. NS-22, No. 6, (1975).

INSULATOR AND SEMICONDUCTOR SURFACES

A. M. Stoneham

Theoretical Physics Division

Harwell, United Kingdom

1. INTRODUCTION

Whilst both the science and technology of surfaces are long-established, it is the practical application of surface phenomena which has dominated the subject. Catalysis, corrosion, lubrication, and even many semiconductor devices are often surface-controlled, and their importance is great. Until recently, the understanding of surface behaviour was based on empirical rules and phenomenology. The move towards a microscopic and fundamental understanding has only just begun, stimulated both by fashion and the development of an impressive range of new experimental tools.[†] In this paper, I attempt to identify what has been done and put the consequences into a scientific and technological perspective.

Surfaces lead to three broad classes of phenomena. First, even perfect surfaces have direct observable effects, like the production of surface states and surface phonon modes, and the modification of crystal properties within a few atomic layers of the surface. Secondly, surfaces, whether perfect or imperfect, have indirect effects. For example the relative concentrations of defects can be modified, possibly as far as several microns from the surface, without appreciable effects on the properties of the defects themselves. In this case, the details of the outer few atomic planes may be unimportant. Thirdly, there may be surface defects - either intrinsic or impurity - which have important properties.

[†] Experimental methods are surveyed in "Characterisation of Solid Surfaces" (edited P.F. Kane and G.R. Larrabee; Plenum Press 1974) and will not be discussed further here.

This survey is primarily concerned with the solid-vacuum interface. But analogous phenomena occur for other interfaces, for example between two solids (e.g. the metal-semiconductor interface) or between a solid and a liquid electrolyte. Some aspects of these cases will be mentioned in passing.

2. SURFACES FREE FROM IMPERFECTIONS

It is said that there are three classes of surface: Industrial (= dirty), cleaned (= dirty) and vacuum-cleaved (= clean). Theorists alone know the further class, Ideal, in which the atoms near the surface are undisplaced from their positions in an infinite bulk crystal.

2.1 Surface Geometry

<u>Ionic Crystals</u> In ionic crystals, the main changes in atomic configuration near a surface displace the anions and cations differently. For the (001) face of an alkali halide, the smaller and more polarising ions (the cations except for CsF and possibly RbF) move inwards relative to the larger, more polarisable ions. The sign of the displacements of each sublattice relative to the perfect lattice is predicted to alternate from plane to plane as one moves into the crystal. The amplitude of the displacements falls rapidly to zero over a distance determined mainly by the difference in ionic radii, $|r_+ - r_-|$.

Benson & Claxton[1] have made detailed shell-model calculations for all 17 NaCl-structure alkali halides. These can be compared with experiment in one case, LiF. Here Low Energy Electron Diffraction experiments of McRae & Caldwell[2] have been analysed by Laramore & Switendick[3]. There are some minor problems, for the analysis of the experiments assumed only a single lattice plane relaxes; further, the experiments were carried out at fairly high temperatures. Setting these aside, the surface cations are deduced to move in by about 12% of the nearest-neighbour distance, compared with predictions of about 5%. The distance between the surface cations and their neighbouring anions in the second plane is predicted to reduce by about 7%, compared with the LEED estimate of 5%.

Data for other crystals are less detailed. Similar LEED analyses for ZnO[4] suggest that the lattice contracts by 0.1 to 0.2Å at the (0001) surface, but has parameters similar to the bulk for the (1010) surface. For MgO, where it is possible to obtain very small cubic crystallites, both theory[5] and experiment[6,7] agree that the average ionic separation is less for smaller crystallites.

INSULATOR AND SEMICONDUCTOR SURFACES

<u>Semiconductors</u> In semiconductors, the atomic displacements at the surface can be more profound, and usually merit the name <u>reconstruction</u>. The symmetry and the number of atoms in the surface unit cell may be changed. If an unreconstructed surface has a structure defined by basis vectors $\underline{\ell}_1$ and $\underline{\ell}_2$, and if the new reconstructed surface has a structure corresponding to $(N\underline{\ell}_1, M\underline{\ell}_2)$ instead, then the surface is described as an (NXM) surface[8]. Clearly, the symmetry of the surface can be lowered. Further, there can be many different reconstructed forms of a given surface. The (111) surface of Si exists in (1x1), (1x2), (2x2) and (7x7) forms, depending on temperature; possibly other forms exist too, although impurities are necessary for some. Reconstruction has been observed for different surfaces (eg (100) and (111) for Si, Ge and InSb) of tetrahedral (Si, Ge, GaAs, GaSb, InSb) and trigonal (Te, CdS, ZnO) semiconductors, both polar and non-polar, as well as certain ionic systems like UO_2.

The semiconductor case can be contrasted with the ionic crystal case, where (1x1) surfaces[†] alone are the rule. Several explanations have been offered. The earliest[9] (Lander & Morrison) regarded the structure as consisting of warped "benzene-ring" units, as in the perfect structure, but with a number of vacancies in the outer layer. Phillips[10] extended these ideas, relating the favoured vacancy distributions in different crystals to the different relative magnitudes of the metallic and directional contributions to the bond energies. Other explanations invoke excitonic instabilities or soft phonon modes. The soft-phonon modes suggestion in any simple form is hard to reconcile with the large number of surface phases seen.

2.2 Surface Energies and Related Quantities

Consider a finite crystal, sufficiently large that its edges and corners are a negligible fraction of the whole. Then a part of the cohesive energy can be attributed to the surface, namely

[Cohesive energy of finite crystal] - [Number of atoms] x

x[Cohesive energy per atom in infinite crystal].

[†] Technically one could argue that the basis vector <u>normal</u> to the surface had doubled because of the alternating signs of displacements in successive planes. But each plane parallel to the surface retains the unreconstructed vectors.

If the total surface area is A, then this energy is σA, where σ is the <u>surface energy</u>, quoted per unit area. Basically, it is the energy per unit area required to cut the infinite crystal and separate it into two halves. Typical values in ergs/cm^2 are around 1000 for metals, 100 for alkali halides, and 20 for polymers.

Three simple but important points should be made. First, surface energies are anisotropic: <111> and <100> faces may have very different energies. Secondly, unlike simple liquids, σ is not equal to the <u>surface tension</u>, γ, in general. The surface tension refers to the reversible stretching of a surface:

$$\gamma = \frac{d}{dA} (\sigma A)$$

and is relevant, for example, in considering the equilibrium of a bubble as the gas pressure varies. Thirdly, lattice relaxation near a surface <u>reduces</u> the surface energy.

General features of surface energies are discussed in refs. (1, 11-15). Experimental data are prone to all sorts of problems, and are often mutually inconsistent; data given here should be considered representative rather than reliable.

<u>Surface Energies of Ionic Crystals</u> In the NaCl structure, the <100> faces have lowest energy. The simplest models of the surface show why this is and give sensible energies. Born & Stern[11] showed that there is a Madelung contribution $+ 0.26(Ze)^2/a^3$ which is reduced by a repulsive term $- 0.14(Ze)^2/a^3$ if an r^{-9} approximation to the repulsive forces is used. The sum for an unrelaxed surface, $0.1173(Ze)^2/a^3$ is in quite tolerable agreement with experiment. The analogous value for <110> faces is almost three times larger, $0.315(Ze)^2/a^3$; for <111> faces, which consist of cations only or anions only, the value is higher still.

The effects of lattice relaxation have been discussed by Shuttleworth[14], Tosi[13], and Benson & Claxton[1]. Roughly, in cases where the anions and cations have similar radii (eg NaF or RbI), the relaxation is confined to very near the surface; the reduction in surface energy is small, around 10-15%. If the anions and cations have widely different radii (eg CsF, NaI) the distortion is significant well into the crystal, and the surface energy can be reduced by around 40%.

Some figures are given below in ergs/cm^2

	LiF	NaCl	KCl	MgO
Surface energy (cleavage expt)	340	300	110	1200
Surface enthalpy (solution expt)	-	276	252	1000, 1150
Surface tension (expt)	-1000	370	-	-
Born & Stern (theory)	414	151	108	1448
Shuttleworth (theory)	363	214	171	-
Benson & Claxton (theory)	222	167	152	-

Extensive measurements [15] have been made for UO_{2+x} and UC_{1+y}, albeit with the added difficulties of non-stoichiometry. Two significant trends appear, however. First, the surface energies are very sensitive to stoichiometry: in UO_{2+x}, a change of x from 0.001 or less to 0.1 increases σ by a factor of about 5. Secondly, the surface energies decrease with temperature at constant x.

For low-symmetry surfaces, there are extra energy terms associated with surface steps. Yamada's calculations are reviewed by Tosi[13].

Surface Energies of Semiconductors Suprisingly little reliable information is available here, considering the importance of these systems. The anisotropic cleavage properties of diamond and graphite are widely known, yet surface energies are hard to find. There are technical problems, of course, plus some ill-defined complications. For example, graphite needs about 1000 ergs/cm^2 to cleave, where the planes are pulled apart, whereas friction implies much smaller energies, 50-100 erg/cm^2 when sliding is involved.

For the Group IV valence crystals, data are mainly theoretical; some uncertainties still remain concerning the role of the s^2p^2-sp^3 promotion energy. Values for diamond are large: 5000 ergs/cm^2 for {111} and 8000 ergs/cm^2 for {100} are cited by Bowden & Taylor,[16] and these are in line with estimates for {111} of 5350 ergs/cm^2 by Sinclair & Lawn [17]. There seems no reason to accept van Vechten's guess [18] of 1750 ergs/cm^2. The theory of ref (17) seems to agree with experiment [19] quite well for Si(1460 ergs/cm^2 compared with the experimental [19] 1230) and Ge(1070 ergs/cm^2 compared with 1060), although they differ greatly from the surface tensions quoted by ref (18) from ref (20). β-SiC has a high surface energy, 2223 ergs/cm^2 [15], intermediate between those of diamond and Si. α and β-Sn follow the trend of smaller surface energies for higher atomic numbers, values of 520 ergs/cm^2 being cited [21].

The theoretical energies ignore surface relaxation and reconstruction. Related energies are those associated with stacking-faults, where the sequence of crystal planes is altered. These energies are (21a) known from experiment as 55 ergs/cm^2 (Si), 38-45 ergs/cm^2 (Ge) and 43 ergs/cm^2 (GaP$_{0.3}$As$_{0.7}$); predictions give 1860 ergs/cm^2 for diamond, 55 ergs/cm^2 for Si, 26 ergs/cm^2 for Ge, and only 3 ergs/cm^2 for grey tin.

<u>Importance of Surface Energies</u> (a) <u>Crystal habit</u> In cases where crystals are formed by an equilibrium process (as opposed to growth based on dislocations where kinetic factors matter), low-energy surfaces should dominate. Thus in MgO smokes, [100] cubes are found. Natural diamonds are frequently octahedral ([111] faces), although dodecahedral forms are common too, whereas cubic forms are rare.

(b) <u>Ease of Cleavage</u> Cleavage occurs easily along low energy surfaces. The diamond industry exploits the easy [111] cleavage, and the associated quality that [111] faces are hard. By contrast, the [100] faces are soft, and must be sawn to machine - a process which means about 5% of a diamond's volume may be lost. The III-V semiconductors cleave preferentially along [110]. Graphite and mica have a spectacular anisotropy for cleavage.

(c) <u>Mechanism of Fracture</u> The relative importance of intergrannular and transgrannular fracture depends on the ratio f of the grain boundary energy to the cleavage energy. When f is small, fracture tends to take place between grains, rather than through them.

	LiF	KCl	NaCl	UO$_{2+x}$	Al$_2$O$_3$	ZrO$_2$
f	0.41	0.48	0.53	0.6-0.7	0.71	0.78

The effective surface energy for the initiation of fracture, σ_I, may differ from the mean surface energy σ_F over all fracture processes(22). If there are many crack sources in initiation, $\sigma_I > \sigma_F$, as for glasses and perspex. If subsidiary cracking occurs as fracture proceeds, $\sigma_F > \sigma_I$, as in graphite. In other cases, eg Al$_2$O$_3$, $\sigma_I \sim \sigma_F$.

(d) <u>Growth from the Melt</u> One fairly general result [25,26] is that the way in which a crystal grows from the melt is determined by a parameter

$$\alpha = \text{(Entropy of melting per mole) (Crystallographic factor)}$$
in units R

where the first factor varies with the material and the second depends on the structure of the surface, being high for close-packed surfaces. When α is small (≤ 2, typically metals), rounded surfaces are formed, without nucleation problems. But for larger α (> 2, including most compounds and insulators), certain facets are favoured, and the rates of growth are primarily determined by surface steps, etc.

2.3 Surface States of Perfect Surfaces

The surface gives rise to a discontinuity in the otherwise periodic potential of a large crystal. This discontinuity leads to two broad classes of states with large amplitudes near the surface: Localised Surface States and Surface Resonances. For the localised state, the wavefunction falls to zero as one moves away from the surface. The resonance, however, is best regarded in terms of bulk states whose amplitude is enhanced near the surface. We shall not discuss here work on polaron effects near a surface,[25] relevant for the motion of slow electrons parallel to the surface.

A surfeit of simple models exist which show the way surface states emerge. Fortunately, Davison & Levine[26] have reviewed them in detail, and we can concentrate on the more realistic descriptions of surfaces developed recently. But it is probably useful to mention first a common distinction made between Tamm states, which appear when the surface perturbation is large in some sense, and Shockley states, which appear when the perturbation is small but where the bands are crossed. This distinction is not very useful, particularly since ambiguities can arise and, in anycase, both types can be regarded as special limits of more general models.

Ionic Crystals   In these cases, the surface has remarkably little influence. The point is that the ions can be regarded roughly as free ions whose levels are shifted by the Madelung potential. But the surface and bulk Madelung constants are remarkably similar (for NaCl (100) α_M is 1.681, cf. the bulk 1.7476). The energy shifts involved are thus of order $0.067 e^2/a$ or, if the nearest neighbour distance a is in Å, [0.96 eV]/a, typically 1/3 eV. Various authors have argued small effects occur in optical spectra. Questions of band-bending have been discussed in RbI[27], and in MgO[28] a significant reduction in effective bandgap has been suggested. In bulk MgO, the gap is around 8.7 eV, whereas for microcrystals a 5 eV threshold with a 5.7 eV peak is attributed to the effects of the surface. Similar effects are seen in CaO and SrO.

Semiconductor Surfaces   Two broad classes of methods have been used here: the various approximate molecular-orbital methods have been tried [26,29], and there have been pseudopotential approaches, in which core pseudopotential parameters are deduced from the bulk band data (eg the self-consistent calculations of refs 30, 31, 32). Both types of method are approximate, differing mainly in the way in which information is put in - either in deducing coulomb and resonance integrals, or in the choice of pseudopotential (which is not uniquely defined by a finite set of band energies) - but, at least for Si, both give similar results.

One system on which much work has been done is the (111) surface of Si. Rowe & Ibach[33] observed three surface states on the reconstructed (7 x 7) surface by a differential electron energy-loss method. One (S_1) lay in the gap, just above the valence band. Another (S_2) lay in the valence band, in a minimum in the density of states; this resonance lies just below a p-like peak. The third (S_3), was just below the valence band. Both Applebaum & Hamann[30] and Pandey & Phillips [29] have reproduced these peaks satisfactorily by their different methods for <u>relaxed</u> but <u>unreconstructed</u> (1x1) surfaces. Roughly, the S_2 state is built of $p_{||}$ orbitals oriented parallel to the surface, whereas the S_1 and S_3 states are constructed from bonding and antibonding combinations of S and $p_\perp$ orbitals. The relaxation needed to give satisfactory predictions appears to be an inward motion of the surface plane by about 1/3 Å; if this is reduced significantly the S_2 and S_3 states are absent. More recent work [34] suggests a slightly more complicated picture, including some effects of the hybridization of the S_1 and S_2 states. Some similar results for Ge have been obtained [31,35] which indicate, amongst other things, modified dielectric properties close to the surface.

Studies of surface states of Si date back long before the work just cited, and many aspects were sorted out by Allen & Gobelli [36]. Electrical, optical and spin resonance results are reviewed by Davison & Levine [26]. The conclusions are interpreted in terms of two peaks in the surface density of states which lie in the bulk band gap. The states in the lower peak are occupied, and the others empty. Evidence for this structure comes mainly from the variation of the surface charge state with Fermi level and the early observation that the rectification properties of Si-metal junctions depended little on the metal work function. Properties deduced from this and other work (including conductivity [37], optical absorption [38] and spin resonance [39] indicate that the surface-state density usually exceeds $10^{14}(cm^2$ (cf. 3.10^{15} sites per unit area in a monolayer); spin resonance on fresh surfaces gives about $0.8.10^{14}$ spins/cm^2, although without hyperfine structure or g-tensor anisotropy. The surface states are not electrically-conducting, even for degenerate n-type Si, where there may be a surface charge for roughly every tenth atom.

2.4 Phonon Modes near Surfaces

Just as the surface introduces surface electronic states, so it leads to phonon modes near the surface. Indeed, there are many formal similarities between the two cases [40], (although these can be overstressed), including the appearance of both localised states and resonances (occasionally called pseudo surface waves).

The surface waves and analogous modes which appear within elasticity theory are well-known, [41,42,43] and are related to the acoustic bulk modes. The most important are Rayleigh waves, in which the individual atoms move in an elliptical motion in the plane of the propogation direction and the surface normal. The associated amplitudes fall off exponentially within the solid, and the waves move with a velocity slower than that of bulk waves. Related are the Stoneley waves, which occur when there are two media in contact, rather than a bulk-vacuum interface. Lamb waves occur in plates of finite thickness, but reduce to Rayleigh waves in appropriate circumstances. Love waves are the horizontally-polarised shear waves which can be trapped in a surface layer.

The simplest elastic solutions omit two important practical features: elastic anisotropy (which is trivial in principle but complex in practice) and piezo-electric effects. The piezo effects, which couple atomic displacements and a macroscopic electric field, usually stiffen the effective elastic constants; in $LiNbO_3$, for instance, they enhance the surface sound velocity by about 15%. But the piezo terms are especially important because they can be influenced by applied electric fields. This has lead to a new area of device technology. The advantages of "acoustoelectronics" are that the surface modes are accessible so that they can be manipulated easily, and they have a velocity much less than light. Processing in the time-(rather than frequency-) domain is possible, with low attenuation and good control and signal/noise ratios. The manipulation is handled in many ways: generation by using microwaves with spaced metal electrodes, guiding and modifying velocities by adding extra films (Au on Si slows surface waves, whereas Al on glass speeds them), and a whole range of techniques to filter, amplify, store and compare signals. The slow velocities are of immense importance in producing light, cheap delay lines. In addition to surface waves, much interest recently has been in acoustic waves in thin insulating films on metals.

3. DEFECTS ON SURFACES

We discuss here simple surface defects - mainly point defects, possibly involving impurities - excluding dislocations, grain boundaries, shear planes and more complex defects. Macroscopic defects - surface roughness and asperities - will not be discussed although their importance in friction, growth from solution, and catalysis is considerable.

3.1 Ionic Crystals

<u>Vacancy Centres</u>   Probably the best documented vacancy centres are those in the alkaline earth oxides[44]. Here spin resonance, diffuse reflectance and adsorption work all point to a model in which an electron is trapped at an oxygen vacancy on a (001) face. These F_s^+ centres differ from the bulk F centres mainly in their axial symmetry. The electronic charge is centred not at the vacant site, but further into the crystal, giving larger hyperfine constants on the axial cation.

F_s^+ centre	MgO(theory)[45]	MgO(expt)[44]	CaO(expt)[44]
Isotropic hfs (Gauss)	−11.7	± 25.5	
Optical transition (eV)	∼ 5	2.05	1.90
	(these two energies probably refer to different transitions)		

A related centre has been seen associated with hydrogen, possibly with a hydrogen ion on an adjacent surface site [44F]. For this F_s^+(H) centre a novel temperature dependence of the hyperfine constants is seen [44h].

F_s^+(H) centre	MgO	CaO	SrO
$g_{\parallel}$	2.0016	1.9992	1.9846
$g_{\perp}$	2.0003	1.9969	1.9792
A (gauss)	9.4-12 gauss(T dep.)	−	−
B (gauss)	≤ 0.5 gauss	−	−
Opitcal bands (eV)	1.77, 1.0	1.65, 0.91	1.60, 0.76

Analogous centres have been proposed for other systems, as in connection with the fast adsorption of oxygen on neutron irradiated NiO[46]. Likewise, Smart & Jennings[47] have made point-ion calculations which suggest surface F centres should be detectable in NaCl and NaF. Attempts to see surface states by EPR following electron bombardment have been made for LiF, NaCl, KCl, CaF_2 and BaF_2; in LiF bulk F centres were seen, and in all cases [48] conduction electron resonances from metal particles were observed. The antimorph of the F_s^+ centre, the V_s^- centre (roughly an O^- ion next to a surface cation vacancy) has also been detected in MgO[49] by its characteristic hyperfine constants for ^{17}O ($A_{\parallel}$=302 MLz, $A_{\perp}$=29 MLz).

Radiation damage by neutrons has an additional effect in increasing the apparent surface area, (44e,50) probably by the fragmentation of aggregates at grain boundaries.

<u>Intrinsic Adatoms</u>   Again, the alkaline earth oxides seem to be favoured hosts. The systems can be identified by using adsorbed ^{17}O those detected include $(O^-)_s$(51,52,53), $(O_2^-)_s$(51,52,54) and $(O_3^-)_s$(51,49b,55). These species have been seen on many other hosts, including TiO_2 and ZnO. The $(O_2^-)_s$ species is particularly interesting, both because it occurs so widely (52,56) and because its EPR allows the detailed mechanism of oxygen isotope exchange on MgO to be unravelled(57). The isotope-exchange work suggests that another species (possibly O_4^-. with square rather than pyramidal structure) acts as an intermediate. The study of intermediates is of great potential importance in catalysis.

3.2 Valence Crystals

<u>Intrinsic Defects</u>   I know of no observations of isolated intrinsic defects on valence crystal surfaces. To some extent this is a question of definition; reconstructed surfaces can perhaps be regarded as regular arrays of vacancies. But intrinsic defects in bulk Si, for example, can be notoriously mobile, and it may be that any intrinsic isolated defects move rapidly to sinks. Surface states dependent on steps on the (111) Si surface have been detected (58), giving a new state some 0.4 eV above a main surface peak. The closest analogues of the defects defected by EPR in ionic crystals give rise to the g=2.0055 line observed by Haneman and co-workers (59a,b) in various forms of crushed and cleared silicon. This has been interpreted in terms of delocalised electrons at a reconstructed surface. Graphite is (as always) a special case, and it is customary to assume there is little difference between surface and bulk substitutional or vacancy defects.

Much work has been done on crushed semiconductor crystals (carbon, Si, Ge, Si-Ge alloys and II-VI and III-V semiconductors). Most of the signals seen appear to be associated with oxygen impurity or with carbon from hydrocarbons in the vacuum system. The data are reviewed in ref (59b).

3.3 Impurity Adatoms

Most of the enormous amount of work on impurity adatoms has been for metal substrates, where the heavy metal catalysts are so important. But even for insulators and semiconductors, there are enough examples that a complete survey is impossible.

Broadly, adsorbed impurities are classified as <u>physically adsorbed</u> if they are bound with an energy less than about 0.5eV, and as <u>chemisorbed</u> if the binding is stronger. An example of physical adsorbtion is LiF: He, where the binding energy is less than 0.01 eV. In this case, the He-LiF potential produces four weakly-bound states [60] (0.0058 eV, 0.0022 eV, 0.0005 eV and 0.0001 eV). Examples of chemisorption include oxygen on graphite, Si and Ge, where the effects have great technological importance. The heats of adsorption are around 9.5 eV (Si) and 5.3 eV (Ge)[61]. For graphite, it is the oxygen which is partly responsible for the lubricating properties, a result which was very important in the design of aero-engines for high altitude flying. The oxidation of graphite moderators in reactors is also important. For Si, the effects are on the electronic properties of devices. The coverage tends to stabilize at 1.5 monolayers, a result neatly explained [62] by the assumptions that (a) O_2 adsorbs to give two O atoms bonded to adjacent dangling bonds on a (111) surface, (b) interchange of one O and its Si neighbour occurs, which, after rebonding, allows (c) the addition of one further O atom. Another case of chemisorption, Si:H, has received much theoretical attention [63].

As the concentration of adatoms increases, their mutual interaction becomes important. There are several forms of interaction: direct adatom-adatom interactions, and those through the substrate. The substrate interactions involve three main elements:

(i) Interactions of the strain or polarisation fields of the individual adatoms;

(ii) Effects of the lattice structure of the substrate. These are especially important in questions of epitaxy where the substrate geometry helps to determine the known ordering of surface adatoms. Reconstruction of adsorbed layers may also occur (eg for oxygen on silicon); indeed, some reconstructions of nominally perfect surfaces may be stabilized by impurities;

(iii) Effects on charged defects or free carriers near the surface

Very little work has been done on either the study of these interactions for realistic atomic models or the detailed experimental analysis of contributions.

In principle, certain impurities in semiconductors should lead to spread-out, weakly-bound donor and acceptor states. The theory of these states had been given [64] in the effective-mass limit. Whilst there are some hints from experiment (notably from Na near Si-SiO interfaces[64,65], or from surface plasmon effects at high densities) I know of no detailed spectroscopic work on these shallow defects.

3.4 Surface Diffusion and Related Phenomena

Surface diffusion is important in processes like corrosion, creep, sintering, and crystal growth. It has been studied by tracer methods, by sintering and creep experiments, and by a variety of experiments in which surface structures are monitored with time. No atomistic theory has been done, and the results are disappointingly inconsistent.

Three mechanisms have been discussed for surface diffusion (we shall not discuss the fascinating motions on metal surfaces at low temperatures seen by Ehlich and co-workers). First, a vacancy mechanism could be involved[66], where one expects an activation energy somewhat smaller than the bulk value. Secondly, motion of a surface adatom may be involved [67], when it has been suggested that the activation energy would be typically 2/3 of the heat of vapourization. Thirdly, at higher temperatures, mass transfer via the vapour phase dominates [68]. This leads to very high pre-exponential factors. Experimental data are often too poor to decide which mechanism operates, and it is rare to find consistency between the results of even careful and reliable workers. There are special reasons. For example, surfaces may have a different stoichiometry from the bulk, and they may involve different proportions of anion and cation motion. Further, in an effective "surface depth" is used in analysing tracer measurements, and it leads to some uncertainty. We now give some examples, including in addition work on <u>dislocation diffusion</u> and diffusion along <u>grain boundaries</u>.

<u>Alkali Halides</u> I know of only a few tracer measurements [69, 70], all giving low activation energies relative to bulk values. Impurity examples are included in the Table only to indicate the degree of consistency.

<u>Oxides</u> Much work has been done on MgO,[72] Al_2O_3[72] and UO_2[15]. An extra problem appears here, notably for MgO and Al_2O_3, where the diffusion may be extrinsic, i.e. partly impurity-controlled. Results in UO_2[15] are clearer, at least when reviewed collectively. As indicated in the figure, surface diffusion of both U and O proceeds at a rate somewhat faster than that for O motion in bulk UO_2, where oxygen moves much faster. Above about 1700°C, vapour-phase diffusion dominates. Stoichiometry problems cause difficulties, for even modest diviations cause large changes in rates. However, there seems no reason to believe the high values often cited for the activation energy and pre-exponential factor.

The enhancement of diffusion by dislocations [72] has been seen for $MgO:Ba$[73] and $ThO_2:Th$[74].

Table 1

Tracer measurements of surface diffusion on Alkali Halides

System	Temperature °C	Activation energy in eV			Bulk[71] Diffusion
		Adatom[69]	Adjacent Surface Layer[69]	"Surface"[70] (unspecified)	
NaCl: ^{22}Na	200–550	0.47	0.46		~ 2
NaCl: ^{36}Cl	250–600	0.35	0.62		~ 2
KBr : ^{82}Br	250–550	0.41	0.59		~ 2.6
NaCl: ^{110}Ag	250–500	0.70	0.70	0.17 (360–800°C)	
NaCl:Au	360–800			0.17	
KCl :Au	360–800			0.12	

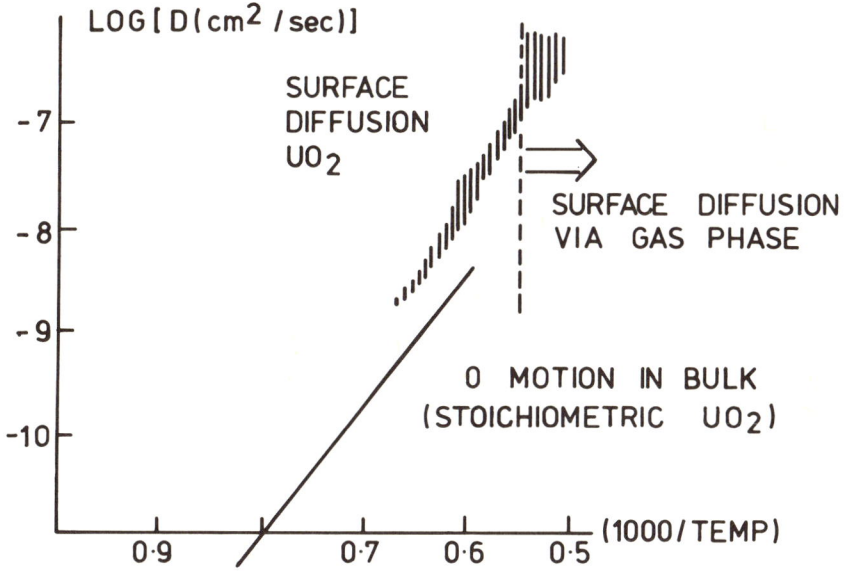

Fig. 1

Carbides   Analogous results to oxides are found. For UC, the grain boundary activation energy has been measured for C[75], giving a value around 55% of that for bulk diffusion.

Silicon   Various estimates of activation energies for surface self-diffusion are available, including (a) the recovery of the (7x7) structure on a (111) surface after a laser pulse[76], giving 1.1 eV, (b) silane studies [77] giving 1.81 eV, (c) field emitter studies [78] indicating 2 eV, and (d) a study of growth processes[79] suggested about 0.2 eV. It is not trivial relating these to each other or to bulk studies. In the bulk, the diffusion of intrinsic defects depends on their charge state and the presence of ionisation. On the surface, these effects will be additionally complicated by the range of possible orientations and sensitivity to impurities.

4. NEAR-SURFACE EFFECTS

The effects discussed here all arise because a crystal need not be electrically neutral near a surface [80]. In the interior of a crystal, any process in which charges (electronic or ionic) are produced gives no systematic separation of the charges, nor is it possible to avoid charge balance. But if a surface (or grain boundary or dislocation) is present, the different species of charged defect remain separately in equilibrium with these extended "sinks" or "sources" of defects. If a charged point defect of one sign needs less energy to produce from the extended defect, this charge will dominate in the bulk and the extended defect will have the opposite sign. Clearly, the charges in the bulk will tend to cluster near to the surface. The region of charge imbalance extends over a distance of the order of the screening length L from the surface:

$$L = \{\frac{kT}{e^2/\varepsilon} (\sum_i c_i |z_i|)^{-1}\}^{\frac{1}{2}}$$

where ε is the dielectric constant, and c_i, Z_i are the concentrations and charges of the charged species present. In some systems the sign of the surface charges with temperature: at low temperatures, extrinsic (impurity) effects may be important, whereas intrinsic defects dominate at high temperatures. The "isoelectric temperature" is that at which the impurity-induced vacancy concentration is equal to the intrinsic vacancy concentration in the absence of electrostatic effects.

<u>Silver Halides and the Photographic Process</u> In AgCl[81], there are large fields (10^4–10^5 volts/cm) in the space charge layer. The surface is negative with respect to the bulk by about 0.1 volts at low temperatures where divalent cations are important, changing sign at high temperatures when an excess of silver vacancies in the bulk occurs. There is no direct evidence of surface states. In Ag Br[82], the (111) surface is also negative, by $-0.14 \pm .05$ volts, with fields of order $(3^{+4}_{-1}).10^4$ volts/cm. These correspond to $(2^{+3}_{-1}).10^{11}$ charges/cm^2; for the (200) surface, the densities are a factor of 20 down.

The fields are important in photography for three reasons: they reduce recombination by separating the electrons and holes produced optically; they repel electrons from the surface where they may be trapped; and they lead to an excess of cation interstitials in the bulk where processes like these may occur:

$$e^- + Ag^+_{int} \rightarrow Ag^0_{int}; \quad Ag^0 + e^- + Ag^+ \rightarrow Ag^0_2$$

Slifkin[83] has commented that it is easier to produce an Ag^+ interstitial than a vacancy at a surface jog, whereas the converse seems true for dislocations in AgCl and AgBr (see also ref 84).

INSULATOR AND SEMICONDUCTOR SURFACES

Segregation and Phase Separation Near Grain Boundaries Just as surfaces and dislocations can be charged, so can low-angle and other grain boundaries in ceramics. Kingery's[85] survey notes that the potentials involved are again a few tenths of a volt, with the boundaries being negative for NaCl and MgO, and positive for Al_2O_3; it is not clear whether extrinsic or intrinsic effects dominate.

Adsorption on microcrystals of MgO When oxygen is adsorbed on MgO microcrystals containing bulk F^+ centres (electrons at anion vacancies), the oxygen coverage follows "Elovich" kinetics[44d,e]. Thermal activation is involved, either through electron or oxygen migration, with an activation energy related to the F^+ and adsorbed molecule concentrations in a way expected from the Coulomb interactions involved.

Carrier concentrations near semiconductor surfaces It is too far from our theme to discuss the device applications of these phenomena, but it is useful to give some vocabulary. An __inversion layer__ occurs near a surface when the minority carriers in the bulk dominate in the space charge. In an __accumulation layer__ it is the majority carriers which dominate near the surface, and few majority or minority carriers remain in a __depletion layer__. Since there are inhomogeneous space charges near the surface, the energy of an electron at the bottom of the conduction band or a hole at the top of the valence band differs from the corresponding bulk value. This can be described by __band bending__ near the surface. One important extra feature when the space charge is electronic is that it can be modulated optically. We also note that an applied electric field in a device may push carriers towards the surface, where they are localised (albeit in a different sense to para 2.3).

Electrodes and Electrolytes Two simple models of electrolytes near electrodes are common, both useful whilst only partly true[86,87]. In the Helmholz-Perrin model, a simple double-layer of charge results, and is treated as a capacitor- there is no adsorption of electrolyte ions on the electrode, and screening is complete at very short distances. In the Gouy-Chapman model, all the screening is the longer-range Debye-Hückel screening. More complicated models recognise that chemisorption (usually of the larger anions) occurs, complicated by questions of hydration and dipole moments of water molecules and interactions between ions at the electrode surface. Thus the usual compromise between the simple models of an anode have anions chemisorbed at a distance σ_I from the surface, excluding cations; the closest approach of the cations is the somewhat larger distance σ_O, beyond which Debye-Hückel screening occurs. In this Stern-layer picture σ_I and σ_O define the inner- and outer-Helmholz planes. The electrode/electrolyte systems are particularly interesting because they show species-dependent and other deviations from the pure Debye-Hückel approach rather clearly.

5. DYNAMIC PHENOMENA AT INTERFACES

Here we touch on some aspects of surfaces in which complex phenomena of great technological importance arise. Whilst only the briefest comments are possible, the basic ideas are relevant in corrosion, catalysis, crystal growth and electrochemistry.

<u>Oxidation of Metals</u> The mechanism of oxidation passes through several stages [88]. After the initial chemisorption, oxygen is incorporated rapidly into the substrate by metal and oxygen ions exchanging places. This occurs even at the lowest temperatures, and may involve several atomic layers; the amount incorporated depends on the chemical and electrical image forces. A slow logarithmic growth follows, usually limited by cation diffusion under the electric field (the Cabrera-Mott voltage) which is incorporated. Reordering of the film and the growth of islands of oxide can lead to faster logarithmic growth, assisted by fast diffusion paths. Finally, at long times the electric field may become irrelevant, with bulk diffusion in the oxide determining the well-known $\sqrt{t}$ growth law.

This description misses much out. First, when is it sensible to change from a "chemisorbed oxygen" picture to a "metal oxide" picture? Auger spectroscopic work suggests the changeover occurs after a mere one or two monolayers[89], although transitions associated with the interface can be seen beyond this. Secondly, when a thick layer of good oxide is formed, it acts as a diffusion barrier to subsequent growth. The stability of the oxide against flaking off ("spalling") is important. Here questions of epitaxy and oxide-metal mismatch occur, and the way in which mismatch strains can be resolved by misfit-didocations[90].

<u>Epitaxial Metal on Surfaces</u> Epitaxy and mismatch also play a roll in the growth of metal films from the vapour onto a non-metallic substrate. Here there are further questions[91]. First, how does nucleation into islands of metal begin? Is it nucleation on some critically-sized group of metal atoms, or are ledges, etc involved? And is the growth rate determined primarily by surface diffusion rates, or is the driving force of different surface energies for different configurations the critical one?

Two unexpected features are found. First, some fairly substantial groups of metal atoms can rotate and migrate[92], although they appear to become immobile once certain epitaxial orientations are acquired. Second, there appears to be much less coalescence than expected[93], although whether from electrostatic charging, contamination or misfit stresses is not known.

Catalysis   Catalysts operate in four main ways:

(i) They may merely adsorb reactants inertly, thereby increasing the probability of the reactants coming together; zeolites and other large-surface area systems are often in this class;

(ii) They may provide special "active" sites which encourage one of the critical steps in a reaction; surface ledges are often cited in this class;

(iii) They may supply one of the reactants - for example, it can be verified by isotope-exchange work that oxides often act as sources of oxygen in certain oxidations, presumably replenishing their own oxygen in independent reactions;

(iv) They may act as a source of electrons or holes. This shows up clearly in the light-sensitivity of certain reactions- for example $H_2+D_2 \leftrightharpoons 2HD$ on MgO:Fe is strongly affected by light with energy appropriate to photoionising the Fe[94].

Even the simplest-sounding catalysed reactions become exceedingly complex when looked at in detail, and at present it seems only easy to make unjustifiable statements of negligible value in almost all cases.

1. G.C. Benson & T.A. Claxton, J.Chem.Phys. __48__ 1358 (1968)

2. E.G. McRae & C.W. Caldwell, Surf.Sci. __2__ 509 (1964)

3. G.E. Laramore & A.C. Switendick, Phys.Rev. __B7__ 3615 (1973)

4. A.R. Lubinsky & G.B. Duke, Bull.Am.Phys.Soc. __20__ 388 (1975)

5. P.J. Anderson & A. Scholz, Trans.Farad.Soc. __64__ 2973 (1968)

6. A. Cimino, P. Porta & M. Valiga, J.Am.Ceram.Soc. __49__ 152 (1966)

7. I.F. Guilliatt & N.H. Brett, Trans.Farad.Soc. __65__ 3328 (1969)

8. e.g. G.A. Somorjaj & H.H. Farrell, Adv.Chem.Phys. __20__ 215 (1971)

9. J.J. Lander & J. Morrison, J.Appl.Phys. __34__ 1403 (1963) and Surf.Sci. __4__ 241 (1966)

10. J.C. Phillips, Surf.Sci. __40__ 459 (1973)

11. M. Born & O. Stern, Sitzber Preuss.Akad.Wiss.Berlin __48__ 901 (1919)

12. C. Herring, in "Structure & Properties of Solid Surfaces" (Chicago University Press 1953)

13. M.P. Tosi, Sol.St.Phys. __16__ 92 (1964)

14. R. Shuttleworth, Proc.Phys.Soc. __A62__ 167 (1949), __A63__ 444 (1950)

15. D.A. Mortimer, AERE Reports R7751 (1974) and R7941 (1975)

16. F.P. Bowden & D. Tabor, "Friction & Lubrication of Solids II" p 160-1 (Oxford 1964)

17. J.E. Sinclair & B.R. Lawn, Proc.Roy.Soc. __A329__ 83 (1972)

18. J.C. Phillips & J. van Vechten, Phys.Rev.Lett. __30__ 220 (1973)

19. R.J. Jaccodine, J.Electrochem.Soc. __110__ 524(1963); J.J. Gilman, J.Appl.Phys. __31__ 2208 (1960)

20. B.N. Oshcherin, Bull.Acad.Sci.USSR,Inorg.Mat. __10__ 14 (1970)

21. American Institute of Physics Handbook

21a L. Chen & L.M. Falicov, Phil.Mag. __29__ 1, 1133 (1974)
    S. Mader & A.E. Blakesee, IBM J.Res.Dev. __19__ 151 (1975)

22. R.W. Davidge & G. Tappin, J.Mat.Sci. 3 165 (1968)

23. D.P. Woodruff "The Solid-Liquid Interface" (Cambridge 1973)

24. K.A. Jackson, D.R. Uhlmann & J.D. Hunt, J.Cryst.Growth 1 1 (1967)

25. T.D. Clark, Sol.St.Comm. 16 861 (1975)

26. S.G. Davison & J.D. Levine, Sol.St.Phys. 25 1 (1970)

27. M. Sydor, Phys.Rev.Lett. 27 1286 (1971) and Phys.Rev. B7 4012 (1973)

28. R.L. Nelson & J.W. Hale, Disc Farad.Soc. 52 77 (1971)

29. Pandey & J.C. Phillips, Sol.St.Comm. 14 439 (1974); Phys.Rev.Lett. 32 1433 (1974)

30. J. Applebaum & D. Hamann, Phys.Rev.Lett. 31 106 (1973), 32 225 (1974); Phys.Rev. B8 1777 (1973)

31. D.J. Chadi & M.L. Cohen, Phys.Rev. B11 732 (1975)

32. M. Schlüter, J.R. Chelikowsky, S.G. Lovie & M.L. Cohen, Phys.Rev.Lett. 34 1385 (1975)

33. J.E. Rowe & H. Ibach, Phys.Rev.Lett. 31 102 (1973), 32 421 (1974)

34. J.E. Rowe & M.M. Traum & N.V. Smith, Phys.Rev.Lett. 33 1333 (1974)

35. J.E. Rowe, Phys.Rev.Lett. 34 398 (1975)

36. F.G. Allen & G.W. Gobeli, J.Appl.Phys. 35 597 (1964), Phys. Rev. 137 245 (1965)

37. D.E. Aspnes & P. Handler, Surface Science 4 353 (1966)

38. G. Chiarotti, S. Nannarone, R. Pastore & P. Chiaradia, Phys.Rev. B4 3399

39. D. Haneman, Phys.Rev. 170 705 (1968)

40. A.M. Stoneham, "Theory of Defects in Solids" (Oxford 1975)

41. I. Tolstoy "Wave Propagation" (Mc Graw-Hill 1973)

42. G.W. Farnell, Phys.Acoust. 6 109 (1970)

43. K. Dransfield & E. Salzmann, Phys.Acoust. 7 219 (1970)

44. (a) R.L. Nelson & A.J. Tench, J.Chem.Phys. 40 2736 (1964)

    (b) R.L. Nelson, A.J. Tench & B.J. Harmsworth, Trans.Farad. Soc. 63 1427 (1967)

    (c) A.J. Tench & R.L. Nelson, Trans.Farad.Soc. 63 2254 (1967)

    (d) R.L. Nelson & A.J. Tench, Trans.Farad.Soc. 63 3039 (1967)

    (e) R.L. Nelson, J.W. Hale, B.J. Harmsworth & A.J. Tench, Trans.Farad.Soc. 64 2521 (1968)

    (f) D.R. Smith & A.J. Tench, Chem.Comm. 1968, p 1113

    (g) R.L. Nelson & J.W. Hale, Disc. Farad.Soc. 52 77 (1971)

    (h) A.J. Tench, Surf.Sci. 25 625 (1971)

    (i) A.J. Tench & G.T. Pott, Chem.Phys.Lett. 26 590 (1974)

45. R.R. Sharma & A.M. Stoneham, to be published

46. H.B. Charman & R.M. Dell, Trans.Farad.Soc. 59 470 (1963)

47. R.T. St.C. Smart & P.J. Jennings, Trans.Farad.Soc. 67 1193 (1971)

48. G.C. Fryburg & R.A. Lad, Surf.Sic. 48 353 (1975)

49. (a) D.D. Eley & M.A. Zammitt, J.Catal. 21 377 (1971)

    (b) A.J. Tench, J.Chem.Soc.Faraday I 68 1181 (1972)

50. S.E. Ermentov, Izv.Akad.Nauk.Kaz.SSR (Ser Fiz Mat) 2 60 (1971)
    S.E. Ermentov & S.K. Kusainov, ibid 2 56 (1971)

51. A.J. Tench & J.F.J. Kibblewhite, J.Chem.Soc.Comm. 1973 p955

52. J.H. Lunsford, Catal.Rev. 8 135 (1973)

53. A.J. Tench & T. Lawson, Chem.Phys.Letz. 7 459 (1970)
    N.B. Wong & J.H. Lunsford, J.Chem.Phys. 55 8007 (1971)

54. A.J. Tench & P. Holroyd, Chem.Comm. 1968 p471

55. N.B. Wong & J.H. Lunsford, J.Chem.Phys. 56 2664 (1972)

56. M. Che, A.J. Tench & C. Naccache, J.Chem.Soc.Faraday Trans I, 70 263 (1974)

57. M. Che, B. Shelimar, J.F.J. Kibblewhite & A.J. Tench, Chem.Phys.Lett. $\underline{28}$ 387 (1974)

58. J.E. Rowe, S.B. Christman & H. Ibach, Phys.Rev.Lett. $\underline{34}$ 875 (1975)

59. (a) M. Ching & D. Haneman, J.Appl.Phys. $\underline{37}$ 1879 (1966)
    D. Haneman, Phys.Rev. $\underline{170}$ 705 (1968)
    D. Haneman, M. Ching & A. Taloni, Phys.Rev. $\underline{170}$ 719 (1968)
    M. Ching, Bull.Am.Phys.Soc. $\underline{14}$ 787 (1969)

    (b) D. Haneman in "Characterization of Solid Surfaces" (edited P.F. Kare & G.R. Larrabee, Plenum Press 1974)

60. J.A. Meyers & D.R. Frankel, Bull.Am.Phys.Soc. $\underline{20}$ 305 (1975)
    D.E. Housten & D.R. Frankel, Phys.Rev.Lett. $\underline{31}$ 298 (1975)

61. (a) A. Many, Y. Goldstein & N.B. Grover, "Semiconductor Surfaces" (North Holland 1965)

    (b) L. Brewer, Chem.Rev. $\underline{52}$ 1 (1953)

62. M. Green & K.H. Maxwell, J.Phys.Chem.Sol. $\underline{13}$ 145 (1960)

63. e.g. J.A. Appelbaum & D.R. Hamann, Phys.Rev.Lett. $\underline{34}$ 806 (1975) and Bull.Am.Phys.Soc. $\underline{20}$ 304 (1975)

64. J.D. Levine, Phys.Rev. $\underline{140A}$ 586 (1965)
    F. Stern & R.E. Howard, Phys.Rev. $\underline{163}$ 816 (1967)

65. e.g. R.J. Tidey & R.A. Stradling, J.Phys. $\underline{C7}$ L356 (1974)

66. C.E. Birchenall, Trans.Met.Soc. AIME $\underline{227}$ 784 (1963)

67. J.Y. Choi & P.G. Shewman, Trans. AIME $\underline{224}$ 589 (1962)

68. P.S. Maiya, J.Nucl.Mat. $\underline{40}$ S7 (1971)

69. Ya.E.Geguzin & Y.U.S. Kagano V S Kii (1967); see Diffusion Data 1972 p611

70. I.V. Zolorukhin & V.M. Ievlev (1969); see Diffusion Data 1973 p437

71. R.J. Friauf, A.I.P. Handbook (McGraw-Hill 1972)

72. J. Mimkes, Thin Sol.Films $\underline{25}$ 221 (1975)
    K. Aihara & A.C.D. Chaklader, Acta.Met. $\underline{23}$ 855 (1975)

73. B.C. Harding, Phil.Mag. $\underline{16}$ 1039 (1967)

74. A.D. King, J.Nucl.Mat. __38__ 347 (1971)

75. J.L. Routhbort & H. Matzke, J.Am.Ceram.Soc. __58__ 81 (7S)

76. S.M. Bedair, Surf.Sci. __42__ 595 (1974)

77. B.A. Joya, R.R. Bradley & G.R. Booker, Phil.Mag. __15__ 1167 (1967)

78. F.G. Allen, J.Phys.Chem.Sol. __10__ 87 (1961)

79. H.C. Abbink, R.M. Broudy, & G.P. McGarthy, J.Appl.Phys. __39__ 4673 (1968)

80. J. Frenkel, "Kinetic Theory of Liquids" (Oxford 1946)
    K. Lehovec, J.Chem.Phys. __21__ 1123 (1953)
    J. Eshelby, C. Newey, P. Pratt & A.B. Lidiard, Phil.Mag. __3__ 75 (1958)

81. L. Slifkin, W. McGowan, A. Fukai & J-S Kern, Phot.Sci. & Eng. __11__ 79 (1962)

82. J.F. Hamilton, Prog.Sol.St.Chem. __8__ 167 (1973)

83. L. Slifkin, J. de Phys. __34__ C9 247 (1973)

84. R.W. Whitworth, Adv.Phys. __24__ 203 (1975)

85. W.D. Kingery, J.Am.Ceram.Soc. __57__ 1, 74 (1974)

86. J. O'M. Bockris & D.M. Drazic, "Electrochemical Science" Taylor & Francis 1972

87. N.F. Mott & R.J. Watts-Tobin, Electrochimica Acta __4__ 79 (1961)

88. F.P. Fehlner & N.F. Mott, Oxid Met. __2__ 59 (1970)

89. A.P. Janssen, R. Schoonmaker, J.A.D. Matthew & A. Chambers Sol.St.Comm. __14__ 1263 (1974)

90. e.g. J.H. van der Merwe, Treat.Mat.Sci.Tech __2__ 1 (1973)

91. J.P. Hirth, J. Crysr Growth __17__ 63 (1972)
    R.W. Cahn, Nature __250__ 702 (1974)

92. A. Masson, J.J. Metois & R.N. Kern, Surf.Sci. __27__ 463, 483 (1971)

93. D. Robertson, J.Appl.Phys. __44__ 3924 (1973)

94. C.G. Harkins, W.W. Chang & T.W. Leland, J.Phys.Chem. __73__ 130 (1969)

Part III

Diffraction Techniques for Point and Extended Defects

STRUCTURAL INFORMATION AND DEFECT ENERGIES STUDIED BY

X-RAY METHODS

H. Peisl

Sektion Physik der Ludwig-Maximilians-Universität

8 München 22, Geschwister-Scholl-Platz 1

INTRODUCTION

X-ray scattering has been enormous succesful in clearing up arrangements of atoms in periodic crystal lattices during the past 60 years (1). Even for very complicated biological systems structure determination has become almost routine. In many cases the most interesting physical properties are due to deviations from a perfect order. Thermal properties of solids are due to dynamic displacements of the lattice atoms from their regular sites. Scattering of X-rays and - more successfull - of neutrons gave detailed information on these lattice vibrations. Order and disorder phenomena in alloys are to be mentioned as an other field where X-ray scattering methods delivered information on deviations from perfect atomic arrangements.

Defects in a crystal lattice cannot be avoided. Often they are introduced deliberately, as in semiconductors, to cause certain properties of the material or they occur as damage of the material in a radiative environment. Typical defect concentrations in these cases are quite small ($<10^{-3}$) although the property changes may be large.

It has been only recently that X-ray scattering has been used to study the small but effective deviations from a perfect crystal lattice caused by point defects. Crystal lattice defects, as well extrinsic (impurities) as intrinsic (vacancies and interstitials), in general need more or less volume in the crystal. The defect volume can be determined from the average expansion or contraction of the lattice volume by X-ray measurements of the lattice parameters. As the defects are localized, the lattice distortions are not homogeneous throughout the lattice. The displacements of the lattice atoms vary with distance from the defect. The magnitude and symmetry of the long ranging distortion-field can be determined from diffuse X-ray scattering in the vicinity of the Bragg peaks. Diffuse scattering intensity distributions all over the space gives information on the local environment of the defect.

Only in the last decade high precision measuring methods have been developed and used to study lattice parameter changes due to defects. Advances in detector and electronique data evaluating techniques as well as the development of high power X-ray sources made it possible to deduce reliable information from the extremly small diffuse scattering intensity. For a review see e.g. (2-7).

THEORETICAL CONSIDERATIONS

First we shall summarize the important theoretical concepts to describe defects in a crystal lattice and describe the fundamental aspects of studying defects by X-ray methods.

Defects as Elastic Dipoles

The action of a defect in a crystal lattice may be described by a force array $f_j^{(m)}$ which would cause the same displacement field in a defect free lattice. If forces $f_j^{(m)}$ are applied to the atoms (m) at positions $X_i^{(m)}$ in the undistorted lattice the elastic dipole tensor or double force tensor is the dipole moment of the force array

$$P_{ij} = \sum_m X_i^{(m)} f_j^{(m)}. \tag{1}$$

Defects (concentration c) in a crystal lead to an average stress $\sigma_{ij} = c/\Omega \cdot P_{ij}$. The average strain is given by $\varepsilon_{ij} = (c/\Omega) S_{ijkl} P_{kl}$. S_{ijkl} is the tensor of the elastic stiffness constants and summation is to be carried out over equal indices.

STRUCTURAL INFORMATION AND DEFECT ENERGIES

The resulting crystal volume change is

$$\Delta V/V = \sum_i \varepsilon_{ii} = (c/\Omega)(S_{11}+2S_{12}) \text{ Trace } P_{ij} \qquad (2)$$

This can be measured as a change of the average atomic distance i.e. the lattice parameters. For a cubic crystal and small changes

$$\Delta V/V = 3\Delta a/a = c(\Delta v/\Omega). \qquad (3)$$

$\Delta v = (S_{11}+2S_{12})$ Trace P_{ij} is the volume change due to a single defect. Measuring $\Delta a/a$ as a function of the defect concentration c yields the relative defect volume $\Delta v/\Omega$ in atomic volumes.

For defects with an anisotropic tensor P_{ij} alignment of the defects leads to different lattice parameter changes in different directions. If different orientations ν are occupied to an amount ϱ^ν the lattice parameter change in a direction given by the unit vector e_i is

$$(\Delta a/a)_e = (c/\Omega)\varrho^\nu S_{ijkl} P^\nu_{kl} e_i e_j.. \qquad (4)$$

The individual displacements $\vec{u}_k$ of the lattice atoms can be obtained from the elastic equilibrium equation

$$C_{ijkl} \partial_j \partial_l u_k = P_{ij} \partial_j \delta(r-r_j). \qquad (5)$$

C_{ijkl} is the tensor of the elastic compliance constants.

In the following the main discussion will be how to determine the elastic dipole tensor by X-ray scattering methods.

X-Ray Scattering from a Defect Free Crystal Lattice

In the kinematic approximation the elastic X-ray scattering intensity is obtained by summing up the scattered amplitudes from the individual atoms or unit cells with the right phase relation.

Let $\vec{k}_o$ and $\vec{k}$ be the wave vectors of the incoming and the scattered X-ray wave, respectively; for elastic scattering $|\vec{k}_o| = |\vec{k}| = 2\pi/\lambda$, λ = X-ray wavelength, the scattering vector $\vec{K} = \vec{k}_o - \vec{k}$, $|\vec{K}| = 4\pi/\lambda \cdot \sin\vartheta$.

The X-ray wave scattered from an individual atom has the amplitude $f_m(\vec{K})$. A wave scattered from an atom at $\vec{r}_m$ is shifted in phase by $\exp(i\vec{K}\vec{r}_m)$ compared with a wave scattered from an atom at $\vec{r}_m = 0$.

$$I(\vec{K}) = \left| \sum_m f_m(\vec{K}) \exp(i\vec{K} \cdot \vec{r}_m) \right|^2 \qquad (6)$$

Summation exceeds over all m atoms in the crystal. For an ideal periodic arrangement of the atoms in a crystal lattice the scattering intensity is zero except for $\vec{K} \cdot \vec{r}_m = 2\pi n$, $n = 1, 2, \ldots$ This occurs whenever the scattering vector $\vec{K}$ is equal to a reciprocal lattice vector $\vec{G}$, defined by $\vec{G} \cdot \vec{r}_m = 2\pi n$. The Bragg law $n\lambda = 2d \cdot \sin\vartheta$ in vector notation is then given by

$$\vec{G} = \vec{K}. \qquad (7)$$

Scattering maxima can be predicted easily by geometrical considerations given in Fig. 1. For a given incoming wave vector k_o in the reciprocal lattice all possible scattered wave vectors must end on a sphere with the radius $|\vec{k}_o| = |\vec{k}| = 2\pi/\lambda$, the "Ewald-sphere". Whereever a reciprocal lattice point lies on the Ewald sphere scattering maxima occur.

Scattering from a Crystal Lattice with Defects

Due to the introduction of defects the lattice atoms are shifted to new sites $\vec{r} = \vec{r}_m + \vec{u}_m$. All defects, located at sites $\vec{r}_D$ contribute to $\vec{u}_m$ by their individual displacement field $\vec{u}_m^D$. The scattered intensity is now given by

$$I_D(\vec{K}) = \left| \sum_m f_m(\vec{K}) \exp\left[i\vec{K} \cdot (\vec{r}_m + \vec{u}_m)\right] \right.$$
$$\left. + \sum_D f_D(\vec{K}) \exp(i\vec{K} \cdot \vec{r}_D) \right|^2. \qquad (8)$$

The introduction of defects gives rise to the following changes

1. The <u>Bragg peaks are shifted</u> due to scattering from an average new lattice. This can be measured as a

STRUCTURAL INFORMATION AND DEFECT ENERGIES

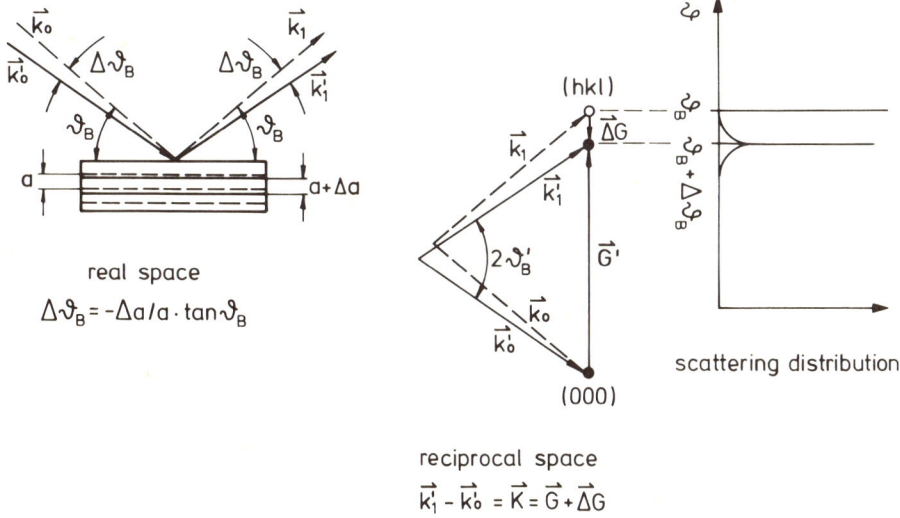

Figure 1. X-ray scattering geometry

lattice parameter change.

2. Deviations from this average lattice cause, as in the case of lattice vibratins, a <u>Debye-Waller factor T</u>.

3. Furthermore, these deviations from a perfect periodicity in the average defective lattice, give rise to a <u>diffuse scattering</u> distributed over the whole reciprocal space.

$$I_{Bragg} = \exp(-2T) \left| \sum_m f_m(\vec{K}) \exp\left[i\vec{K}(1+\tilde{\mathcal{E}})\cdot\vec{r}_m\right] \right|^2 \quad (9)$$

$\tilde{\mathcal{E}}$ is the average strain in the lattice with defects.

Bragg-peaks occur at new reciprocal lattice points

$$(G+\Delta G)(1+\tilde{\mathcal{E}})\vec{r}_m = 2\pi n, \quad (10)$$

$$\Delta G/G = \Delta K/K = -\Delta a/a = \text{ctg}\,\vartheta_B\,\Delta\vartheta_B. \quad (11)$$

Diffuse Scattering

For a low defect concentration and a random distribution of the defects we obtain the diffuse scattering intensity by incoherent summation of the scattering from individual defects: "single defect approximation".

The diffuse scattering from a single defect is obtained by subtracting from the scattering amplitude A_1 of a crystal with one defect the scattering amplitude A_o of a crystal with no defect.

$$I_D(\vec{K}) = n \left| A_1(\vec{K}) - A_o(\vec{K}) \right|^2 \qquad (12)$$

$A_1(\vec{K})$ is obtained in a similar way like $A_o(\vec{K})$ of a defect free crystal.

$\vec{r}_m$ is now the position of the lattice atoms in the average distorted lattice. At a site $\vec{r}_D$ a defect with a scattering amplitude f_D is situated. If we take the defect site as the origin, we have to add to the defect scattering amplitude the scattering amplitudes from the displaced lattice atoms with the proper phase factor.

$$A_1(\vec{K}) = f_D(\vec{K}) + f(\vec{K}) \sum_m \exp\left[i\vec{K} \cdot (\vec{r}_m + \vec{u}_m^D - \vec{r}_D)\right] \qquad (13)$$

In the defect coordinates we obtain also

$$A_o(\vec{K}) = f(\vec{K}) \sum_m \exp\left[i\vec{K} \cdot (\vec{r}_m - \vec{r}_D)\right] \qquad (14)$$

In most cases e.g. point defects and substitutional or interstitial impurities the displacements $\vec{u}_m$ are small, so that $\vec{u}_m^D \cdot \vec{K} \ll 1$ and the phase factor can be expanded.

$$\exp(i\vec{K} \cdot \vec{u}_m^D) = 1 + i\vec{K} \cdot \vec{u}_m^D + \ldots$$

$$I_D(\vec{K}) = n \left| f_D(\vec{K}) + f_{NN}(\vec{K}) + if(\vec{K})\vec{K} \cdot \right. \qquad (15)$$

$$\left. \cdot \sum_m \vec{u}_m^D \exp i\vec{K} \cdot (\vec{r}_n - \vec{r}_D) \right|^2 \qquad (16)$$

The sum in eq (16) is the Fourier transform $\tilde{u}$ of the displacement field $\vec{u}_m^D$

STRUCTURAL INFORMATION AND DEFECT ENERGIES

$$\tilde{u}(\vec{K}) = \sum_m \vec{u}_m^D \exp\left[i\vec{K}\cdot(\vec{r}_m - \vec{r}_D)\right] \qquad (17)$$

$f_{NN}(\vec{K})$ are contributions from the next nearest neighbours of the defects, where the expansion may not have been allowed. We may add this to the defect scattering term f_D. Thus the diffuse scattering is given by

$$I_D(\vec{K}) = n\left|f_D(\vec{K}) + if(\vec{K})\cdot\vec{K}\cdot\tilde{u}(\vec{K})\right|^2 \qquad (18)$$

The square of the first term is the "Laue scattering" due to the defect and eventually its immediate neighbourhood $I_L(\vec{K}) = nf_D^2(\vec{K})$. For a self interstitial $f_D = f$, for a vacancy $f_D = -f$ and for impurities $f_D = f_I - f$. This intensity is a structureless diffuse background, varying like $f_D(\vec{K})$.

The square of the second term is the "distortion-scattering".

$$I_D(\vec{K}) = n\, f^2(\vec{K})\left|\vec{K}\cdot\tilde{u}\right|^2 \qquad (19)$$

Close to the Bragg peaks $\vec{K}\approx\vec{G}$ the scattering intensity is given by the Fourier transform of the distortion field. Let $\vec{q}$ be a vector from any point in reciprocal space to the next nearest reciprocal lattice vector $\vec{G}$.

$$\vec{q} = \vec{K} - \vec{G}. \qquad (20)$$

We may rewrite equation (17)

$$u(\vec{K}-\vec{G}) = \sum_m \vec{u}_m^D \exp\left[i(\vec{K}-\vec{G})\cdot(\vec{r}_m - \vec{r}_D)\right]$$

$$= \tilde{u}(\vec{K})\exp(i\vec{G}\cdot\vec{r}_D), \qquad (21)$$

$$\tilde{u}(\vec{K}) = \tilde{u}(\vec{q})\exp(-i\vec{G}\cdot\vec{r}_D). \qquad (22)$$

The scattering for small $\vec{q}$ has been named Huang diffuse scattering (HDS) (12). It corresponds to the one-phonon scattering from thermal vibrations (TDS).

Huang scattering due to an isotropic defect in an isotropic medium. In order to gain a rough idea of what one expects experimentally, let us consider the most simple case: an isotropically distorting defect in an isotropic medium. This shall show all the important features.

The displacement field of an isotropic defect in an isotropic elastic continuum is

$$\vec{u}(\vec{r}) = \frac{P}{4\pi C_{11} r^2} \cdot \frac{\vec{r}}{r} \,. \tag{23}$$

C_{11} is an elastic constant, $\vec{r}$ denotes the distance from the defect. The Fourier transform is

$$\tilde{u}(\vec{g}) = i \frac{P}{\Omega C_{11} g} \cdot \frac{\vec{g}}{g} \,. \tag{24}$$

Ω is the mean atomic volume.

The Fourier transformation gives correlations of $\vec{u}(r \to \infty)$, the displacements at large distances from the defect, and $\tilde{u}(\vec{g} \to 0)$, which governs the scattering close to a reciprocal lattice point.

Equations (19) and (24) lead to an expected Huang scattering intensity.

$$I_{HDS} = \frac{n\, f^2(\vec{K})\, P^2}{\Omega^2\, C_{11}^2} \left| \frac{\vec{K} \cdot \vec{g}}{g^2} \right|^2$$

$$= n\, \frac{f^2(\vec{K})\, P^2}{\Omega^2\, C_{11}^2}\, \frac{K^2}{g^2}\, \cos^2(\vec{K},\vec{g}) \,. \tag{25}$$

We obtain isointensity contours which are spheres touching the reciprocal-lattice point (Huang spheres). Maximum diffuse scattering intensity is to be expected in the direction of $\vec{G}$ and no intensity perpendicular to $\vec{G}$. The intensity falls off like $1/g^2$ with the distance from the Bragg peak.

Huang scattering from an anisotropic defect in a cubic medium. In this case we get the Fourier transform of the displacement field by Fourier transformation of the elastic equilibrium condition [equation (5)].

$$\tilde{u}^\nu = i\, P^\nu_{ij}\, D^{-1}_{ik}\, g_j \qquad (26)$$

D^{-1}_{ik} is the inverted dynamical matrix. For $\vec{g} \to 0$ it is given by the elastic constants of the crystal
$D_{ik} = -C_{ijkl}\, g_j g_l$.

The Huang scattering intensity is

$$I_{HDS} \propto n \frac{|f|^2}{\Omega^2} \sum_\nu \left| -K_i D^{-1}_{ij} P^\nu_{jk} g_k \right|^2. \qquad (27)$$

The elastic dipole tensor can thus be determined by measuring I_{HDS}.

In general anisotropic defects have various equivalent orientations ν in a lattice. If the defects are randomly distributed amongst these orientations one obtains an averaged scattering intensity. This averaging and rearranging gives for cubic crystals (4,13)

$$I(\vec{K}) \propto (n/\Omega^2)|f(K)|^2 (G/g)^2 (\gamma_1 \Pi_1 + \gamma_2 \Pi_2 + \gamma_3 \Pi_3) \qquad (28)$$

Π_1, Π_2, Π_3 are quadratic dipole tensor parameters

$$\begin{aligned}
\Pi_1 &= 1/3 \left(\sum_i P_{ii}\right)^2 = 1/3\,(\text{Trace } P_{ij})^2, \\
\Pi_2 &= 1/6 \sum_{i>j} (P_{ii} - P_{jj})^2, \qquad (29) \\
\Pi_3 &= 2/3 \sum_{i>j} P^2_{ij}.
\end{aligned}$$

$\gamma_1, \gamma_2, \gamma_3$ are constants which contain the elastic constants of the crystal and the relative orientation of $\vec{G}$ and $\vec{g}$. In certain high symmetry directions one or two of the γ are zero. For these directions Π_1, Π_2, Π_3 can be determined very easily. Π_1 determines the strength of the defect in the same way as obtained by lattice parameter measurements [equation (3)]. Π_2 and Π_3 give the deviation from a cubic symmetry of the displacement field. Table 1 gives the information on can obtain about the symmetry of the dipole tensor (13).

Defect symmetry Prefered orientation	π_2	π_3	h00	hh0
Cubic (isotropic)	0	0	$P \perp 100$	$P \perp 110$
Tetragonal 100	+	0	$P \perp 100$	$L \parallel 001$
Trigonal 111	0	+	−	$L \parallel 110$
Orthorhombic 110	+	+	−	−

Table 1. π_2 and π_3 for defects of different symmetry and planes P and lines L of zero scattering intensity.

Absolute Defect Concentration and Defect Pairs

The Huang scattering intensity I_{HDS} and the relative lattice parameter change $\Delta a/a$ depend in different ways on the dipole tensor,

$$I_{HDS} \propto c\, P^2, \quad \Delta a/a \propto c\, P. \qquad (30)$$

Hence measuring I_{HDS} and $\Delta a/a$ gives a possibility to determine the defect concentration c. In cases where other methods exist to determine the concentration c of defect pairs (e.g. Frenkel pairs) with different dipol tensors for the interstitial P_I and the vacany P_V one can determine P_I and P_V separately.

$$I_{HDS} \propto c(P_I^2 + P_V^2), \quad \Delta a/a \propto c(P_I + P_V), \qquad (31)$$

gives two equations for P_I and P_V.

Effect of Clustering

Huang scattering is very sensitive to cluster formation. For a random defect distribution the scattered intensity from individual defects has been added incoherently, $I_{HDS} \propto n\, p^2$. If we take as an individual defect a cluster, we have again $I_{HDS} \propto n_{Cl}\, p_{Cl}^2$. Here c defects have formed n_{Cl} cluster of an average size $z = c/n_{Cl}$. For the approximation of linear superposition of the defect strength, $P_{Cl} = zP$, we obtain

$$I_{HDS}^{Cl} = z \, I_{HDS} \, . \tag{32}$$

The scattered intensity is enhanced by a factor z if a given defect concentration forms clusters. The reason for this enhancement is that the scattered intensity from the defects in a cluster has to be added coherently. More detailed considerations allow to get informations about the shape of clusters and small dislocation loops (14).

Asymptotic Distortion Scattering

Strongly distorted regions of the lattice determine the diffuse X-ray scattering farther away from Bragg reflections. Here the phase factor equation (15) can no longer be expanded, $\vec{K} \cdot \vec{u}^p \gg 1$. This case has been treated by several authors (2, 9 - 11): In the case of isotropic defects in an isotropic medium the scattering intensity in the asymptotic approximation is

$$I_A \propto c \, \frac{f^2}{\Omega^2} \, \frac{|P|}{C_{11}} \, \frac{|\vec{K}|}{q^4} \, . \tag{33}$$

The scattering intensity depends linearly on the defect strength and falls off like $1/q^4$. The physical interpretation of the asymptotic scattering is that scattering takes place locally from small strongly distorted regions. Interference between scattering amplitudes from different regions having equal distortion may give rise to intensity oscillations which give further information about the distortion field (15,16). The phase of these oscillations is given by

$$\varphi - \varphi_0 \propto (P \cdot |\vec{K}| \cdot q^2)^{1/3} \tag{34}$$

Diffuse Scattering far away from Bragg Peaks

Whereas close to the Bragg peaks Huang scattering gives the Fernfeld of the displacements, the scattered intensity farther away and between two Bragg peaks gives information about the immediate vicinity of the defect (Nahfeld). The scattering intensity is mainly given by the periodicity of $\tilde{u}(\vec{K})$ given in equation (22). Defects with the same $\tilde{u}(\vec{q})$ would give different scattering intensity distributions if they are on different lattice sites. Measuring diffuse scattering between Bragg peaks can thus give information on the defect site in the lattice (6,17).

Small Angle Scattering

Small angle scattering is centered around $\vec{G} = 0$, the origin of the reciprocal lattice. $\vec{G} \to 0$ also means $\vec{K} \to 0$ and therefore $\vec{K} \cdot \vec{u}_m^p \to 0$ and the expansion in equation (15) is sufficient accurate.

$$I_D(\vec{K} \to 0) = n \left| f_D(\vec{K}) + i\vec{K} \cdot \tilde{u}(\vec{K} \to 0) \cdot f(K) \right|^2 \tag{35}$$

In the Fourier transform $\tilde{u}(K \to 0)$ is only determined by the displacements far away from the defect. The same information as from Huang scattering ($\vec{K} \parallel \vec{G}$) can be obtained. However $\vec{K} \to 0$ makes the contribution small. It is more advantageous to measure at a higher $\vec{K}$ value, where the decrease of $f(\vec{K})$ is not yet noticeable, as $\vec{K}$ enters quadratically.

For an isotropic defect in an isotropic medium $iK \tilde{u}(\vec{K} \to 0) = -\Delta v/v \gamma$. $\Delta v/v$ is the volume change due to the defect, γ is the Eshelby constant. In unfavourable cases (e.g. an interstitial atom) the scattering by the defect may just be chancelled by the second term. An interstitial atom scatters like one lattice atoms. If $\Delta v/v \gamma$ is about one atomic volume no small angle scattering occurs. In the case of a vacancy both terms have the negative sign and add. The important fact that small angle scattering can be observed is a change in electron density. Vacancies, large vacancy clusters and large interstitial clusters in diatomic solids are the most favourable cases where small angle scattering can be applied. One big advantage is that it gives also results when the defects cause no lattice distortions but only a change in electron density.

EXPERIMENTS

Typical defect concentrations ($c \lesssim 10^{-3}$) and defect strength ($\Delta v = -1\Omega \ldots +3\Omega$) are of such an order of magnitude that relative lattice parameter changes are quite small and the diffuse scattering intensity has to compete with various types of background scattering. Evaluated techniques have to be used to obtain the experimental results with high enough accuracy.

Lattice Parameter Change

Measurements of the relative lattice parameter change $\Delta a/a$ have to be carried out with an accuracy of at least $d(\Delta a/a) \approx 10^{-5}$ or better. According to equation (11) one has to measure the shift of a Bragg peak in order to determine $\Delta a/a$.

$$\Delta \vartheta_B = - \tan \vartheta_B \cdot \Delta a/a \qquad (36)$$

Without to big an effort it is possible to measure $\Delta \vartheta_B \approx 10'' \triangleq 5 \cdot 10^{-5}$ rad. The smallest $\Delta a/a$ which can be measured with this angle resolution depends on the angle ϑ_B of the Bragg peak under consideration. Backscattering experiments ($\vartheta_B \to 90°$) yield most accurate results. Only in very favourable cases the use of characteristic X-radiation (e.g. CuK_α) gives Bragg peaks close to $90°$. $\vartheta_B = 85°$ gives $\tan \vartheta_B \approx 10$, thus allowing to detect $d(\Delta a/a) \approx 5 \cdot 10^{-6}$.

The highest accuracy in lattice parameter measurements has been achieved by using Bremsstrahlung instead of characteristic radiation in a double crystal backscattering diffractometer (19,20). The Bragg angle ϑ_B is almost $90°$ ($\vartheta_B = 89,6°$, $\tan \vartheta_B \approx 200$) and is not determined by the wavelength but only by geometrical conditions. The intensity distribution of a Bragg peak was not scanned by varying the angle ϑ, but by varying the wavelength used. For this the temperature of one crystal was varied ($\Delta T = 1$ deg $\triangleq \Delta\lambda/\lambda \approx 10^{-5}$).

Fig. 2 shows the relative lattice parameter change of LiF as a function of the defect concentration induced by γ-irradiation at room temperature (20). An average defect volume of $\Delta v = 2,5 \Omega$ for a Frenkel pair in LiF was determined. The accuracy seems now high enough to study low defect concentrations, details of annealing stages, defect orientation effects (8), effect of clustering on defect volumes, change of defect volume due to exitation or change of charge stage.

Lattice Parameter Versus Length Change

Measuring relative lattice parameter change gives relative volume changes of the crystal without knowing

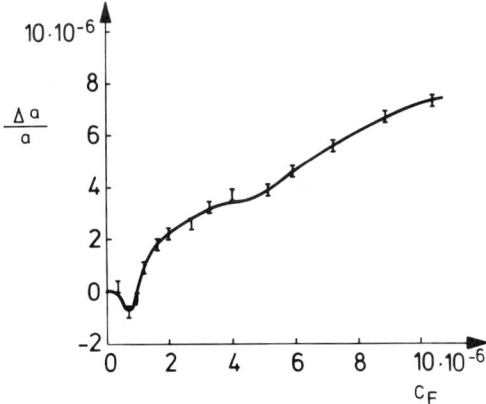

Fig. 2. Relative lattice parameter change versus defect concentration in γ-irradiated LiF (20).

the crystal volume. Measuring the relative volume change of a crystal macroscopically gives the identical information only in cases where the number of lattice sites is not changed. This is the case if substitutional impurities are introduced or Frenkel pairs are created during irradiation. On the other hand the experimental observation of lattice parameter and macroscopic volume (in cubic crystals, length change) can be used to obtain information on the type of the defects. For Frenkel pairs, substitutional or interstitial impurities one expects $3\Delta a/a = \Delta V/V = 3\Delta L/L$. On the other hand, if the number of lattice sites is changed during defect formation there is a difference between $\Delta a/a$ and $\Delta L/L$. For Schottky type defects for every vacancy created an extra lattice site has to be formed. $\Delta a/a$ measures only the lattice distortion due to the vacancy, whereas $\Delta L/L$ measures in addition the extra atomic volume due to a newly formed lattice site. Simultaneous measurements of $\Delta L/L$ and $\Delta a/a$ can be used to decide about the defect type as well as to determine the concentration of the new lattice sites and thus the vacancy concentration.

$$3(\Delta L/L - \Delta a/a) = C_v . \qquad (37)$$

<u>Defect formation energy.</u> The $(\Delta L/L - \Delta a/a)$ method can be used to determine the formation energy of Schottky defects in thermal equilibrium [see e.g. v. Guérard et al.

(21)]. Normal thermal expansion gives an equal contribution to $\Delta L/L$ and $\Delta a/a$, respectively. Fig. 3 shows values for KCl of $\Delta L/L$ and $\Delta a/a$ as function of temperature from room temperature to 750° C (22). For temperatures higher than 500° C $\Delta L/L$ becomes greater than $\Delta a/a$, that means that vacancies are being created in thermal equilibrium. The positive value of c_v indicates that the predominant defects are of Schottky type. The concentration of thermally created lattice sites at the melting point (770° C) is $c_v(T_m) \approx 2 \cdot 10^{-3}$. The vacancy concentration in thermal equilibrium is $c_v = \exp(S/k) \exp(-H/kT)$, where S and H are the effective entropy and enthalpy of formation, respectively. Thus plotting $3(\Delta L/L - \Delta a/a)$ versus 1/T in the usual semilog plot, one can determine S and H.

Defect type in alkali halides with colour centres.
F-centres in alkali halides can be formed by various

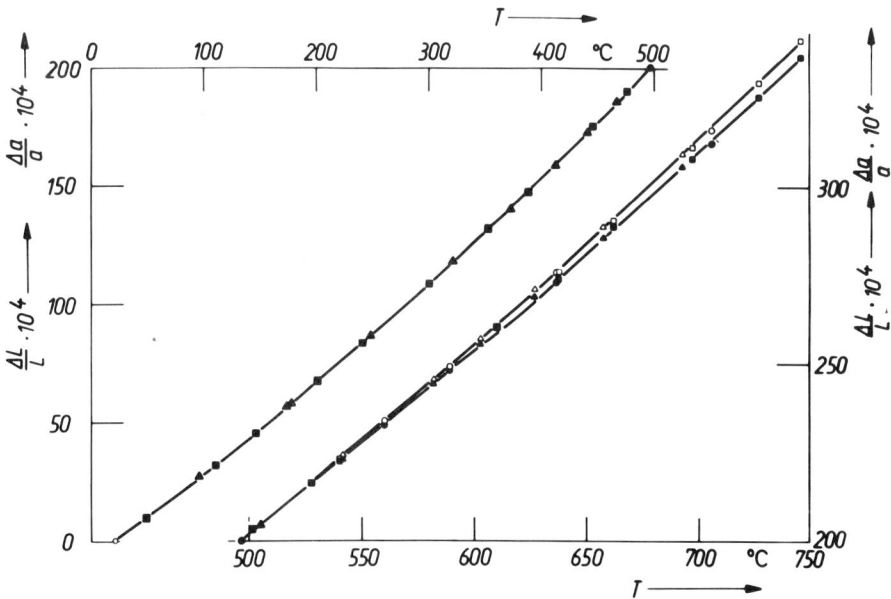

Fig. 3. Relative length change $\Delta L/L$ and relative lattice parameter change $\Delta a/a$ of KCl versus temperature (22).

methods. Additive and electrolytic colouration as well
as irradiation with ionizing radiation creates F-centres which give the crystals a characteristic colour.
It has been an object of long discussions (23) whether
defects in irradiated alkali halides are formed as
Schottky defects, eventually with the help of impurities or dislocations, or as Frenkel defects by an intrinsic displacement process. Measurements of relative
volume change and lattice parameter changes versus defect concentration gave an unambiguous decision.

Fig. 4 shows in the left hand part what one expects for the various types of defects. For Frenkel defects one expects for varying defect concentration
$\Delta L/L(c) = \Delta a/a(c)$ or $\Delta V/V(c) = 3\Delta a/a(c)$. For Schottky
defects $3\Delta L/L(c) - c_v = 3\Delta a/a(c)$ is expected. Whether
$\Delta a/a(c)$ is positive or negative depends on the fact

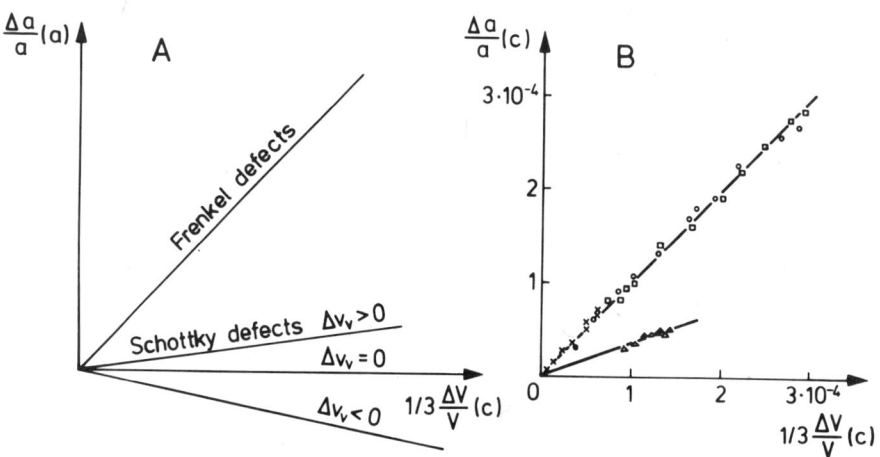

Fig. 4. Defect type determined by the $\Delta a/a$ versus
$\Delta L/L$ method.
A: Predictions for Frenkel and Schottky defects (various
   volume changes due to the vacancy are assumed).
B: Experimental results for defects in KCl
   Δ additively coloured (24),
   x X-irradiated at room temperature (24),
   o X-irradiated at 8 K (26),
   ▫ annealed after X-irradiation at 8 K (26).

whether the vacancies expand or contract the lattice. In the right hand part of Fig. 4 experimental results for KCl are given. Triangles are results for additively coloured KCl crystals (24). In this case direct length or volume change measurement were not possible, therefore the density change was measured, from which the volume change could be deduced. Fig. 4 shows clearly that Schottky defects are present in additively coloured KCl. The same result was obtained on KBr (25).

Crosses are measurements on crystals X-irradiated at room temperature (24), circles are results from X-irradiation at liquid helium temperature and squares are data taken during an isochronal annealing run after irradiation at 8 K. (26) In all cases only Frenkel defects are present in the crystals. A series of alkali halides was investigated by this technique and leads to the suggestion that Frenkel defect are created in irradiated alkali halides. (27)

<u>Defect volumes and elastic dipole tensor</u>. According to equation (3) the defect volume and from this the elastic dipole tensor can be determined by measuring the lattice parameter change versus defect concentration. In additively coloured alkali halides the defect concentration can be determined from optical absorption measurements. Measurements on KCl (24) already displayed in Fig. 4 were used to determine the defect concentration by means of equation (37). This gave excellent agreement with the optically determined defect concentration (28). Lattice parameter versus defect concentration gave Δv(F-centre) = + 0,6 Ω for KCl. The lattice is expanded by an F-centre.

Fig. 5 shows changes of lattice parameter of KCl during irradiation at 6 K versus irradiation time and during isochronal annealing versus annealing temperature (26). Measuring of the optical absorption spectra under identical conditions allowed to seperate the contributions of the two types of Frenkel pairs created. The charged pair constitudes of an anion vacancy (α-centre) and an interstitial ion and needs an additional defect volume Δv (vacancy-interstitial ion pair) = 3,2 Ω. The uncharged pair is an F-centre and an interstitial atom (H-centre). The defect volume of this pair is Δv(F, H pair) = 1,2 Ω. As we know Δv(F) we can obtain Δv(H) = 0,6 Ω.

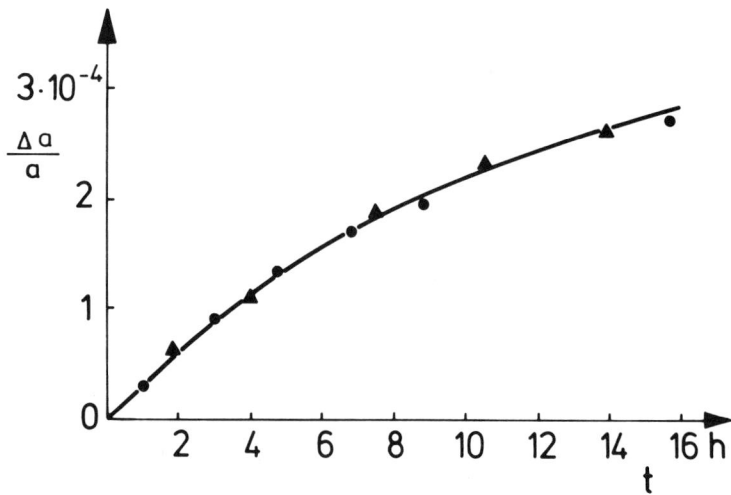

Fig. 5. Relative lattice parameter change of KCl after X-irradiation at 6 K versus irradiation time (26)

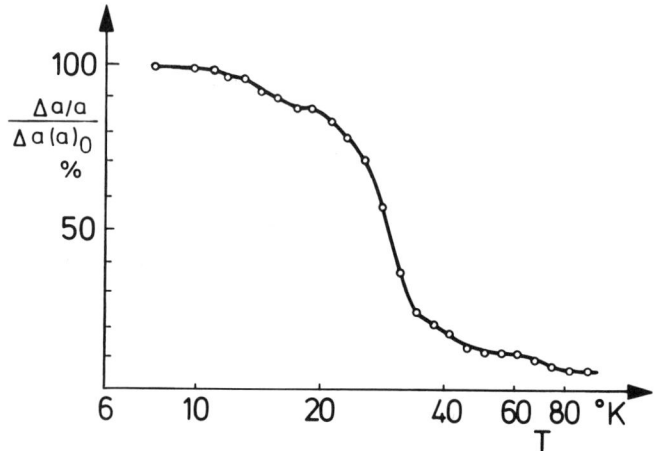

Fig. 6. Relative lattice parameter change of KCl during isochronal annealing. The remaining percentage of change after annealing at the temperature T is plotted. (26)

Measurements of the Huang Scattering from Point Defects

Point defects in KBr. LOHSTÖTER et al (29,30) have investigated KBr after X-irradiation at 6 K. Fig. 7 shows a typical result. Diffuse scattering intensity close to 600 reciprocal lattice point is given for the [100] and [010] directions. After the background scattering (crosses) due to thermal diffuse scattering, Compton scattering and stray radiation had been measured the crystals were irradiated and Frenkel defects formed. This gives additional scattering intensity in the [100] direction. In the [010] direction the scattering intensity stays constant, which tells immediately that $\pi_3 \approx 0$. Additional measurements for some other reciprocal lattice points and directions are given in Fig. 8. The additional scattering intensity is plotted on a log-log scale versus g/G. The experimental points are clearly

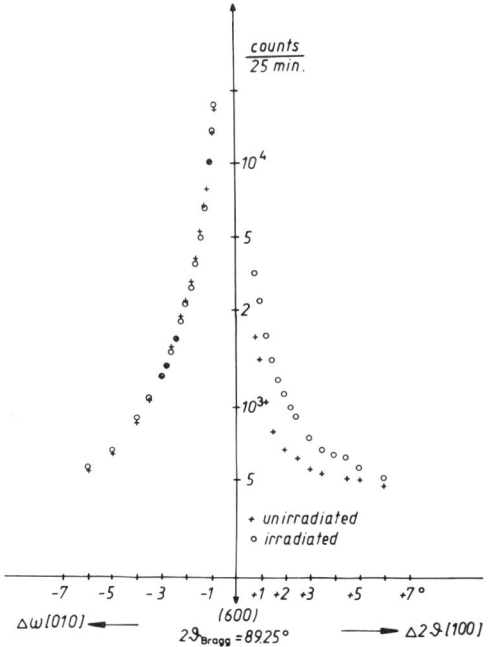

Fig. 7. Huang diffuse scattering X-ray intensity distribution of KBr at 6 K close to a 600 reflection in the [100] and [010] directions before and after X-irradiation at 6 K (30).

on straight lines with slope -2, demonstrating the predicted $1/g^2$ relation [see equation (29)] expected for Huang scattering. From these results the authors were able to determine Π_1 and Π_2. They were able to separate the contribution of the various defects. The elastic dipole tensors of the F and the H centres are known from lattice parameter measurements and paraelastic properties (31). Thus the contribution of the neutral defect pair could be subtracted. The defect concentration is known from optical absorption measurements and so equations (32) can be used to separate the contribution of the vacancy and the interstitial ion: $\Delta v_V = 0,7 \Omega$ and $\Delta v_I = 2,5 \Omega$.

The corresponding elastic dipole tensors are

$$P_{ij}(F) = \delta_{ij}\, 2 \text{ eV}, \quad P_{ij}(\alpha) = \delta_{ij}\, 2,8 \text{ eV}$$

$$P_{ij}(H) = \begin{pmatrix} 4,9 & -0,4 & 0 \\ 0 & 4,9 & 0 \\ -0,4 & 0 & -2,7 \end{pmatrix} \text{ eV} \quad P_{ij}(I) = \begin{pmatrix} 6,9 & & 0 \\ & 6,9 & \\ 0 & & 16,3 \end{pmatrix} \text{ eV}$$

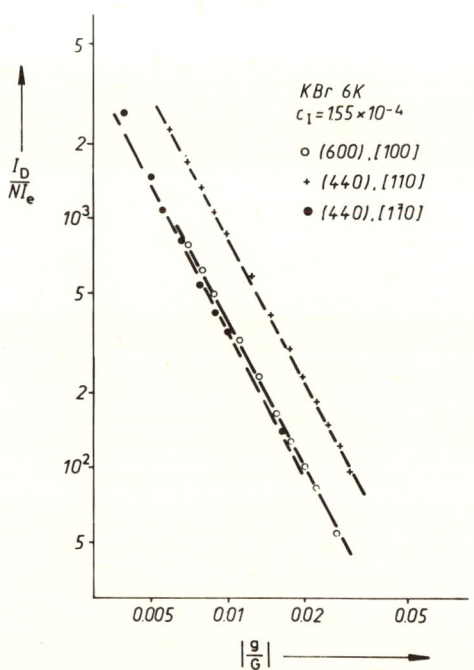

Fig. 8. Huang diffuse-scattering intensity in electron units as a function of the relative distance from reciprocal-lattice points (30).

Defects in neutron irradiated MgO were studied by GRASSE (32) using Huang scattering. Fig. 9 shows as a typical result the diffuse scattering intensity close to the (400) reciprocal lattice point in [100] and [010] direction. These and additional measurements close to other reciprocal lattice points and directions gave $\pi_2 = \pi_3 = 0$, $\pi_1 = 867$ eV^2 and an elastic dipole tensor $P_{ij}^2 = \delta_{ij}$ (51 $\mp$ 10) eV. The defects cause a cubic distortion field. Until now it was not possible to seperate the contributions of the various defects. Thus P_{ij} must be attributed to an "average defect". The defect concentration was obtained from optical absorption measurements as well as from additional lattice parameter change [equations (31)]. All the results are consistent if one assumes that isolated point defects are present. This remains also valid during an annealing programm where defects get mobile and annihilate completely. There is no evidence for cluster formation.

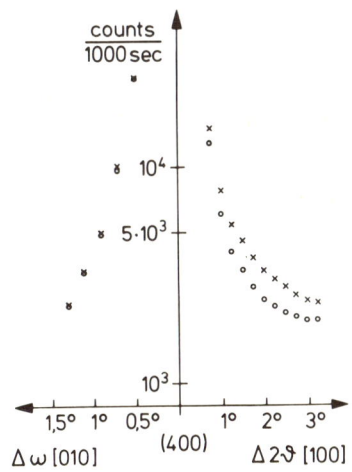

Fig. 9. Huang diffuse scattering X-ray intensity distribution of MgO at 20 K close to the (400) reciprocal lattice point in the [100] and [010] directions before and after neutron irradiation at reactor temperature (32).

Diffuse Scattering from Defect Clusters

Clusters in γ-irradiated LiF were observed by SPALT (33) after room temperature irradiation. Fig. 10 shows the diffuse scattering intensity close to the (400) reflection in [100] and [-100] direction. The cluster defect strength is high enough that one observes also asymptotic scattering with the expected $1/q^4$ dependence. Closer to the Bragg peak the $1/q^2$ dependence can also be observed. The defect strength and concentration were known in this case. The Huang and the asymptotic scattering intensity were calculated under the assumption of a random defect distribution (line A and H in Fig.10). Good agreement exists in the asymptotic region, complete disagreement in the Huang region. This can be easily ex-

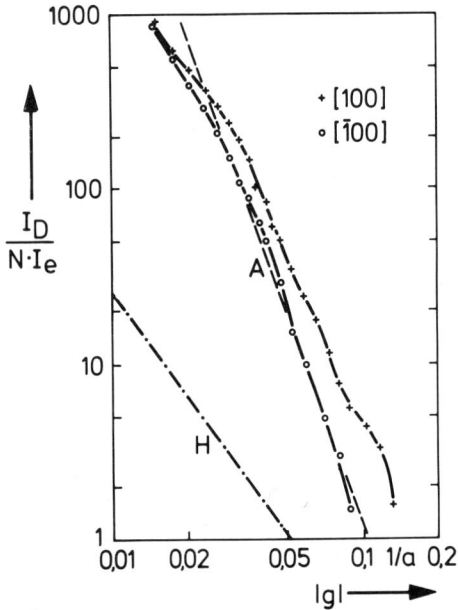

Fig. 10. Diffuse scattering from γ-irradiated LiF.
A: asymptotic distortion scattering in [100]-directions.
H: Huang scattering expected for a random distribution of point defects. (33)

plained by the enhancement of the Huang scattering intensity due to cluster formation [equation (32)]. An average of z = 200 - 300 atoms in a cluster was determined. From the q-value where $1/q^2$ goes over to $1/q^4$ an average cluster diameter of about 7 lattice constants was derived. In all cases where both types of scattering can be observed the different dependence on P [equation (25) and (33)] gives a further possibility to determine the concentration of the clusters.

These results also show the intensity oscillations in the asymptotic scattering region predicted by TRINKAUS et al (15). These oscillations can be seen more clearly if one plots I_q^4 versus q (15). The phase of these oscillations allows to determine z and P and gives quite good agreement.

Cluster formation during annealing of KBr. SPALT et al (30) studied also cluster growth and change of the defect symmetry during thermal annealing. After annealing to 30 K, close to 600 reflection scattering in [010] direction was detected which increased with increasing annealing temperature. Obviously the defect symmetry is changed by clustering. At temperatures above 60 K a $1/q^2$ dependence is observed only close to the reflection. The main intensity distribution shows now $1/q^4$ dependence as expected for asymptotic scattering from strongly distorting defect clusters.

Small Angle Scattering

LiF has recently been investigated by X-ray small angle scattering after γ-irradiation at room temperature by WAGNER (34). They interprete their results by the existence of two types of clusters with different average sizes. One cluster size agrees with the one observed by SPALT (33) and is ascribed to interstitial fluorine clusters. The electron density in a fluorine cluster is higher as in the surrounding LiF crystal so that small angle scattering can be observed. The same is true for Li clusters which seem to be the second type of clusters present in irradiated LiF. These clusters have only been detected in small angle scattering. The reason why they could not been observed in diffuse Huang or asymptotic scattering is most probably the fact that Li cluster are associated with no or very small distortions. Li exists in a fcc phase which has nearly the same lattice parameter as LiF. If the Li clusters are formed in this fcc phase in LiF almost no lattice distortion is expected.

After prolonged irradiation an intensity maximum occured in the scattering distribution. The authors speculate that this might be due to the formation of a periodic structure of the clusters.

REFERENCES

(1) For an introduction see e.g.
WARREN, B.E.: X-ray Diffraction, Reading, Mass.: Addison-Wesley, 1969.
(2) KRIVOGLAZ, M.A.: The Theory of X-Ray and Thermal-Neutron Scattering by Real Crystals. New York: Plenum 1969.
(3) SCHMATZ, W.: In "Treatise on Mat. Sci. and Techn." 2, 105-229 (1973)
(4) DEDERICHS, P.: J. Phys. F: Metal Phys. 3, 471 (1973).
(5) PEISL, H. and H. TRINKAUS: Comments Solid State Phys. 5, 167 (1973).
(6) EHRHARD, P., W. SCHILLING and H.-G. HAUBOLD: Adv. Solid State Phys. 14, 87 (1974).
(7) Proceedings of the International Discussion Meeting on Studies of Lattice Distortions and Local Atomic Arrangements by X-Ray, Neutron and Electron Diffraction in: J. Appl. Cryst. 8, 79 (1975).
(8) PEISL, H., R. BALZER, and H. PETERS: Phys. Letters 46A, 263 (1974).
(9) TRINKAUS, H.: Z. Angew. Phys. 31, 229 (1971).
(10) DEDERICHS, P.H.: Phys. Rev. B4, 1041 (1971).
(11) TRINKAUS, H.: Z. Naturforsch. 28a, 980 (1973).
(12) HUANG, K.: Proc.Roy.Soc. A 190, 102 (1947).
(13) TRINKAUS, H.: Phys. Stat. Sol. (b) 51, 307 (1972).
(14) TRINKAUS, H.: Phys. Stat. Sol. (b) 54, 209 (1972).
(15) TRINKAUS, H., H. SPALT, and H. PEISL: Phys. Stat. Sol. (a) 2, K97 (1970)
(16) SPALT, H.: Z. Angew. Phys. 29, 269 (1970).
(17) HAUBOLD, H.-G.: Rep. KFA Jülich, JÜL-1099-FF.
(18) GUINIER, A. and G. FOURNET: Small Angle Scattering of X-rays, New York: J. Wiley and Sons, 1955.
(19) SYKORA, B. and H. PEISL: Z. Angew. Phys. 30, 320 (1970).
(20) STIER, W.: Diplomarbeit, TH Darmstadt, 1973.

(21) VON GUERARD, B., H. PEISL, and R. ZITZMANN:
 Appl. Phys. 3, 37 (1974).
(22) VON GUERARD, B., H. PEISL, and W. WAIDELICH:
 Phys. Stat. Sol. 29, K59 (1968).
(23) CRAWFORD, Jr., J.H.: Advances Phys. 17, 93
 (1968).
(24) PEISL, H., R. BALZER, and W. WAIDELICH: Phys.
 Rev. Letters 17, 1129 (1966).
(25) VON GUERARD, B., H. PEISL, and W. WAIDELICH:
 Z. Physik 220, 473 (1969).
(26) BALZER, R.: Z. Physik 234, 242 (1970).
(27) BALZER, R., H. PEISL, and W. WAIDELICH: Phys.
 Stat. Sol. 28, 207 (1968).
(28) BALZER, R., H. PEISL, and W. WAIDELICH:
 Z. Physik 204, 405 (1967)
(29) LOHSTÖTER, H., H. SPALT and H. PEISL: Phys. Rev.
 Letters 29, 224 (1972).
(30) SPALT, H., H. LOHSTÖTER, and H. PEISL: Phys.
 Stat. Sol. (b) 56, 469 (1973).
(31) BACHMANN, K., and H. PEISL: J. Phys. Chem.
 Solids 31, 1525 (1970).
(32) GRASSE, D.: Diplomarbeit, TU München 1974.
(33) SPALT, H.: Z. Angew. Phys. 29, 269 (1970).
(34) WAGNER, W.: Diplomarbeit, TH Darmstadt 1972.

X-RAYS AND ELECTRON MICROSCOPY

K. H. G. Ashbee

H. H. Wills Physics Laboratory, University of Bristol, Tyndall Avenue, Bristol. BS8 1TL

WAVES IN PERIODIC STRUCTURES

The equation of scalar wave propagation in a simple periodic medium, described by

$$\nabla^2 u = \frac{1}{c^2} \frac{\partial^2 u}{\partial t^2} \quad ; \quad \frac{1}{c^2} = \alpha + 2\beta \cos(\underline{g} \cdot \underline{r})$$

can, by the Bloch theorem, be satisfied by a function of the form,

$$u = \sum_{n=-\infty}^{n=+\infty} a_n \exp i \left\{ (\underline{k} + n\underline{g}) \cdot \underline{r} - \omega t \right\}$$

For this to be a solution, however, we must set some restriction on the values of the coefficients a_n. The permitted values are deduced by substitution into the wave equation and are

$$\frac{a_{n+1} + a_{n-1}}{a_n} = \frac{1}{\beta} \left\{ \frac{(\underline{k} + n\underline{g})^2}{\omega^2} - \alpha \right\}$$

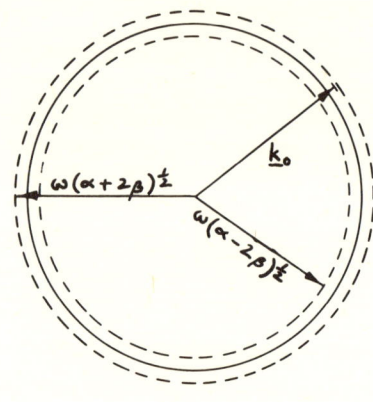

Fig. 1

Assuming that the wave frequency is fixed and that one amplitude is larger than any other, $|a_o| > |a_n|$, then

$$\omega(\alpha-2\beta)^{\frac{1}{2}} < |\underline{k}_o| < \omega(\alpha+2\beta)^{\frac{1}{2}}$$

β is usually small, so $|\underline{k}_o| \simeq \omega\alpha^{\frac{1}{2}}$. That is, the locus of the tips of allowed $\underline{k}_o$ is close to a sphere, the free wave sphere. This is represented graphically in Figure 1. The surface so defined in k-space is a constant frequency surface, otherwise known as a dispersion surface. (A dispersive medium is one in which the velocity of a wave propagating through it depends on the wavelength. In light optics, the dispersion surface is usually called the indicator surface.) Since the solution is periodic in $\underline{g}$, it is evident that the true dispersion surface can be drawn with a multiplicity of origins, at $n\underline{g}$. Examples of dispersion surfaces are sketched in Figure 2. (a) $|\underline{k}| \ll |\underline{g}|$. The dispersion surface takes the form of a very small sphere around each reciprocal lattice point. The conduction of electrons in crystals is represented by this situation. (b) $|\underline{k}| \lesssim \frac{1}{2}|\underline{g}|$. The surface is locally distorted towards the Brillouin planes but is still mainly comprised of spheres around each relp. (c) $|\underline{k}| \gtrsim \frac{1}{2}|\underline{g}|$. Adjacent spheres become linked at their distortions so as to form a continuous surface, periodic in $\underline{g}$, with necks intersecting the Brillouin planes normally. (d) $|\underline{k}| \simeq |\underline{g}|$. X-ray diffraction, neutron diffraction and low energy electron diffraction all approximate to this condition. Each sphere overlaps with its nearest and next nearest neighbours. (e) $|\underline{k}| > |\underline{g}|$. Bubbles have now appeared around each reciprocal lattice point. (f) $|\underline{k}| \gg |\underline{g}|$ is the condition prevailing in transmission electron microscopy. It corresponds to the case of a relatively flat sphere of

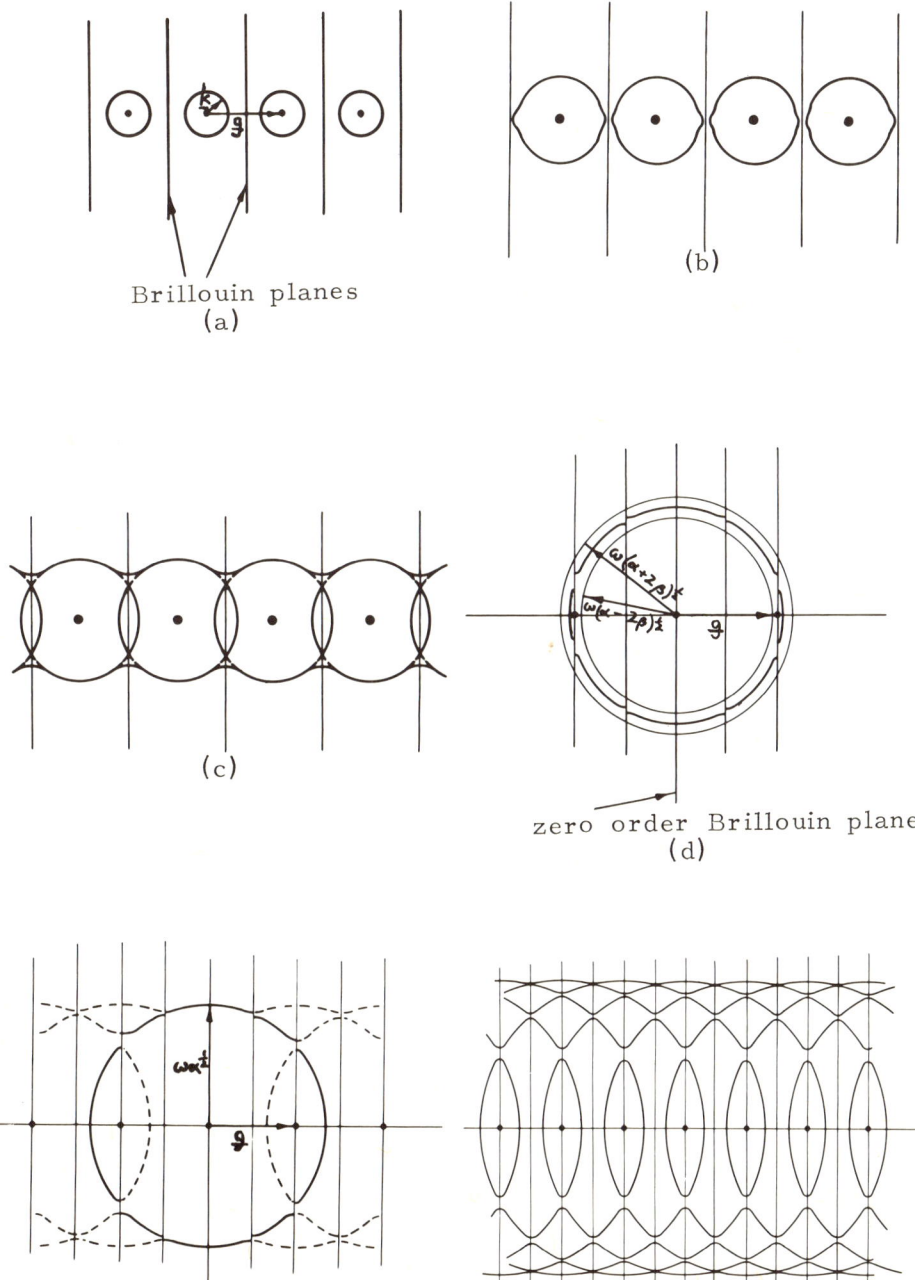

Fig. 2

reflection in Ewald's construction; a large number of reciprocal lattice points are simultaneously intersected.

Figure 2 is a summary of the extended zone scheme. Since each $\underline{k}$ is equivalent to $\underline{k} + n\underline{g}$, all $\underline{k}$ may be translated to the first Brillouin zone, as illustrated in Figure 3, to give the so-called reduced zone scheme.

Two Wave Approximation

Suppose that, in figure 2(c) for example, amplitudes a_o and a_{-1} are very much greater than all other amplitudes. The dispersion surface is re-drawn in Figure 4(a). Remote from the zero order Brillouin plane, the surface coincides with the free wave sphere but near the plane there are distortions associated with overlap. Consider the overlap region.

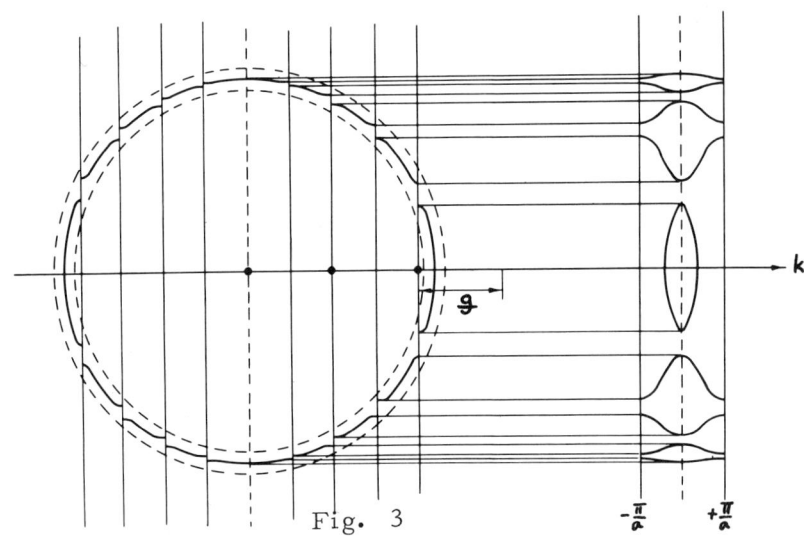

Fig. 3

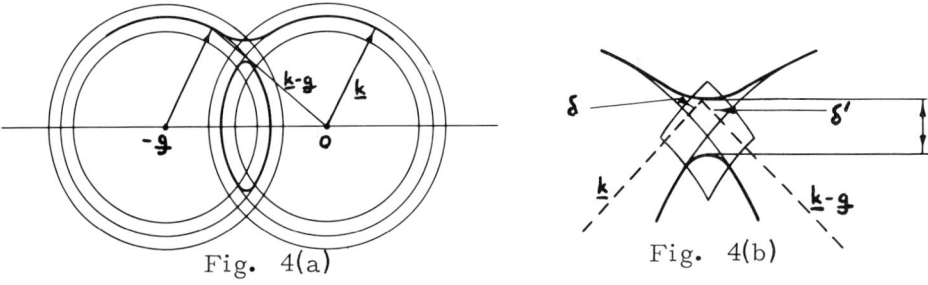

Fig. 4(a) Fig. 4(b)

X-RAYS AND ELECTRON MICROSCOPY

$$|\underline{k}| = \omega\alpha^{\frac{1}{2}} + \delta$$

$$|\underline{k} - \underline{g}| = \omega\alpha^{\frac{1}{2}} + \delta'$$

Substitution into the amplitude equations reveals that

$$\delta\delta' = \frac{\omega^2 \beta^2}{4\alpha}$$

This is the equation to the dispersion surface in the neighbourhood of the Brillouin plane. It is the equation to a hyperbola relative to the two intersecting free wave spheres and is sketched in Figure 4(b).

Figure 5(a) illustrates how the dispersion surface construction might be used to determine the wave vectors inside a crystal. A simpler problem to analyse, however, is the two-wave problem shown in Figure 5(b). The total amplitude is

$$A = a_0 \exp 2\pi i (\omega t - \underline{k}\cdot\underline{r}) + a_{-1} \exp 2\pi i (\omega t - (\underline{k}-\underline{g})\cdot\underline{r})$$

Consider the point X which lies on the Brillouin zone boundary. Here $a_0 = \pm a_{-1}$ and

$$A = 2(i)\, a\, \underbrace{\exp 2\pi i (\omega t - (\underline{k}-\tfrac{1}{2}\underline{g}))}\, \begin{array}{l}\cos(\pi \underline{g}\cdot\underline{r}) \text{ if } a_0 = +a_{-1} \\ \sin(\pi \underline{g}\cdot\underline{r}) \text{ if } a_0 = -a_{-1}\end{array}$$

The part enclosed by the bracket represents the travelling part of the wave and the cosine (or sine) term represents the standing (interference) part. The intensity

$$I = AA^* = a_0^2 \left\{ 1 + \left(\frac{a_{-1}}{a_0}\right)^2 + 2\frac{a_{-1}}{a_0}\cos(2\pi \underline{g}\cdot\underline{r}) \right\}$$

is sketched in figure 6(a) and reveals that $a_{-1}/a_0 < 0$ corresponds to far less absorption than $a_{-1}/a_0 > 0$

Returning to the problem shown in 5(a), elimination of the variables from the relationships

$$a_{\text{incident}} = a_0^I + a_0^{\overline{II}}, \quad 0 = a_{-1}^I + a_{-1}^{\overline{II}} \quad \text{(boundary conditions)}$$

and $\frac{a_{-1}^I}{a_0^I} = \frac{\delta_I}{c}$, $\frac{a_{-1}^{\overline{II}}}{a_0^{\overline{II}}} = \frac{\delta_{\overline{II}}}{c}$, where $c = \frac{\beta\omega}{2\alpha^{\frac{1}{2}}}$ (dispersion equation)

leads to the intensity profiles shown in Figure 6(b).

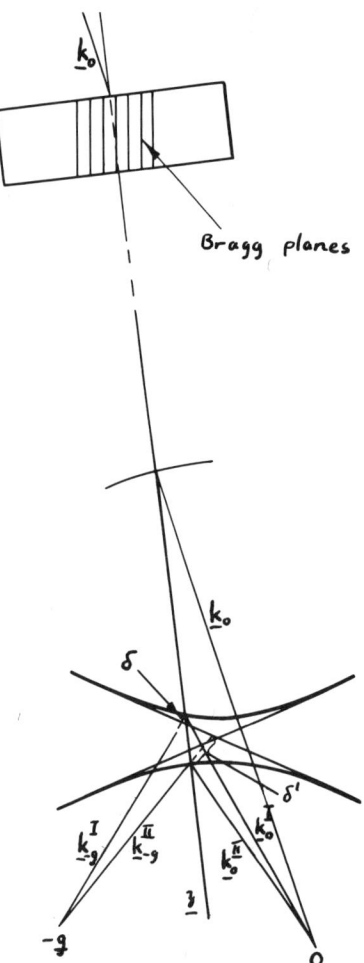
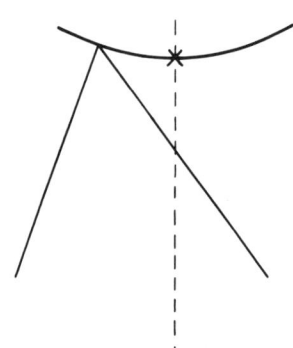

Fig. 5(a) Geometrical relationships between the allowed k during Bragg reflection. Note that $\underline{z}$, the normal to the crystal surface ties these vectors to the incident wave vector $\underline{k}_o$. This is simply Snell's law.

Fig. 5(b)

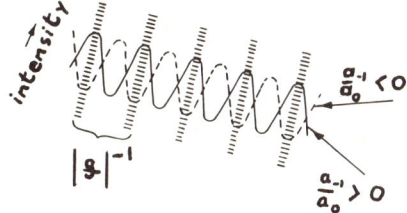 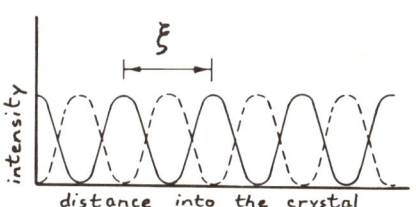

Fig. 6(a) Showing the relationship of intensity profiles to lattice planes (shaded bands).

Fig. 6(b)

These profiles are the origin of Pendellösung effects, such as thickness fringes produced by wedge-shaped crystals and extinction contours produced by elastically bent crystals. The extinction distance ξ_g is given by

$$\frac{1}{\xi_g} = \frac{\beta \omega}{\alpha^{\frac{1}{2}} \cos \theta_g}$$

where θ_g is the Bragg angle. Typical values for ξ_g are

    500Å   - 100 keV electrons
    25 μm   - 20 keV X-rays
    100 μm   - 1/40 eV neutrons
    1/5 μm   - $\frac{1}{2}$ μm light

MAPS OF BRILLOUIN ZONE BOUNDARIES

The Bragg condition is satisfied for a range of θ and gives rise to a contour of contrast which has finite width. For 100keV electrons propagating through a typical non-metallic crystal, $|\underline{k}/\underline{g}|$ is large, of order 100, many beams are excited and many Bragg reflections contribute to the bright field image. With a flat foil, the area illuminated is too small to encompass the whole of each Bragg contour but with an elastically buckled foil extensive patterns of Bragg contours are seen. Examples are shown in Figure 7(a), close inspection of which reveals that the contours occur as $\pm \underline{g}$ pairs, this being a consequence of elastic bending, see Figure 7(b).

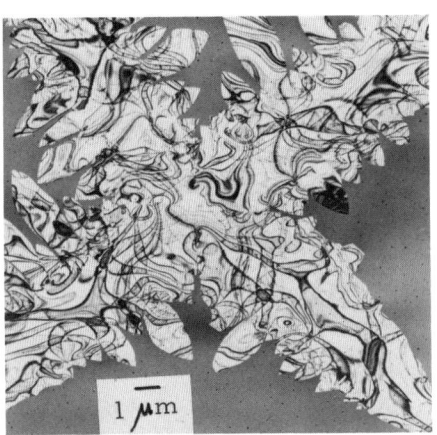

Fig. 7(a) Bragg contours in a polycrystalline foil of tantalum pentoxide (courtesy of R. E. Pawel)

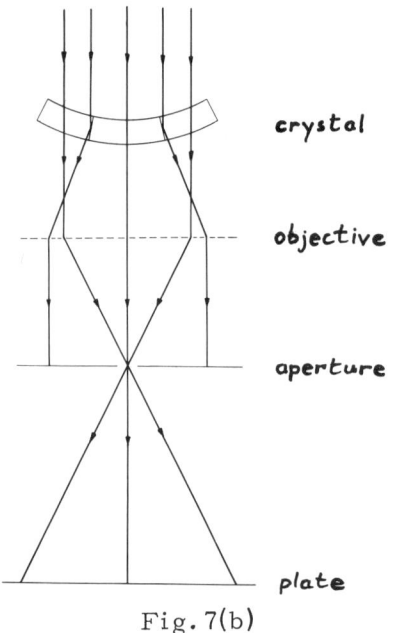

Fig. 7(b)

Bragg contours are real space images of the gnomic projections of the intersections of Brillouin zone boundaries with the Ewald sphere and since they are produced by regions of perfect crystal, they offer the possibility of crystallographic analysis that is not subject to the complications of grain boundaries, lattice defects and impurity segregations that are sampled in the much larger specimens required by conventional techniques. Many non-metals, corrosion products for example, crystallise with a large number of closely related structures of low symmetry and appear to be ideally suited to structural analysis by real space crystallography.

Kikuchi and Kossel lines are also images of the gnomic projections of Brillouin zone boundary-Ewald sphere intersections. Kikuchi lines occur as pairs of lines which exhibit different contrast to the background contrast of electron diffraction patterns from thick foils prepared from near perfect crystals, see Figure 8. The deficiency line of each pair has lower intensity than the background and occurs nearer the origin of the pattern than the excess line which has a higher intensity than the background. The formation of pairs of Kikuchi lines may be understood by the following argument. Thermal scattering of electrons is characterised by wide angles of scatter without appreciable changes of energy $(\Delta \varepsilon = kT \sim \frac{1}{40} eV)$. Some of these diffusely scattered electrons will be oriented at exactly the Bragg angle for each set of planes and will be diffracted as shown in Figure 9(a). Scattered electrons moving along the path OQ' are diffracted to P' and those

X-RAYS AND ELECTRON MICROSCOPY

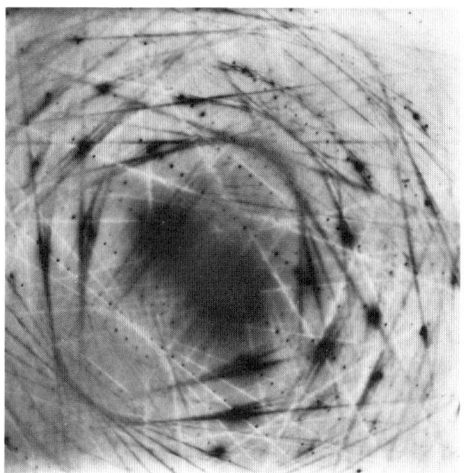

Fig. 8   200 keV electron diffraction pattern from beryl. Identification of the spot pattern provides an approximate orientation determination. Identification of the Kikuchi pattern permits an exact orientation determination. The separation between the diffraction spot and the Kikuchi line for a given reflection is a measure of the deviation from the exact Bragg condition. Note that, if the foil orientation is changed, the Kikuchi lines sweep across the pattern whereas the diffraction spots merely change in intensity.

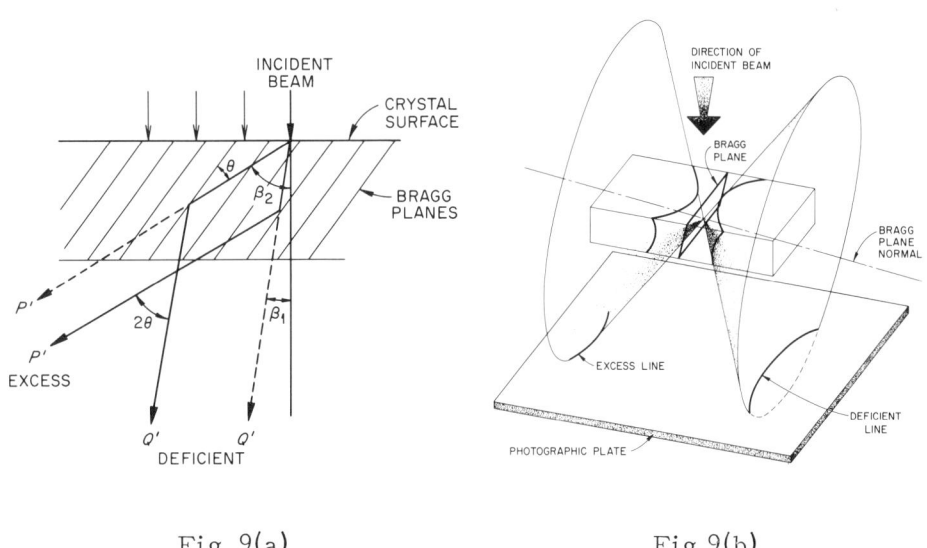

Fig 9(a)   Fig 9(b)

moving along the path OP' to Q'. Since the angular
distribution of diffusely scattered electrons is peaked about
the direction of the incident beam, i.e. since the intensity of
scattered electrons falls rapidly with β, the intensity diffracted
to Q' is less than that towards P'. Hence the intensity at P'
exceeds that of the background intensity whereas that at Q' is
below it. In perspective, the diffracted electrons form two
cones coaxial with the plane normal as shown in Figure 9(b).

Patterns analogous to Kikuchi patterns produced by
diffraction of a divergent beam of X-rays, are called Kossel
patterns. The divergent X-ray beam plays the same role as
the diffusely scattered electrons in electron diffraction. The
only difference is one of scale; since the Ewald sphere is
very much smaller in the X-ray case, the curvature of Kossel
lines (see Figure 10) is more obvious than that of Kikuchi
lines.

Bragg contours, Kikuchi lines and Kossel lines may be
indexed by inspection by comparing with computed maps. To
see how such maps are constructed, consider the geometry
of Kikuchi lines. In Figure 11(a) OO' is the incident beam
direction, ON is the normal to plane MO, L is the camera
length and P' and Q' are a pair of Kikuchi lines formed by
Bragg diffraction of diffusely scattered electrons incident along
PO and QO respectively. In k-space, set $L = 1/\lambda$ and
construct the Ewald sphere as shown in Figure 11(b). The
sphere intersects the two cones at small circles SS' and TT'.
SS' and TT' are the intersections of the sphere with the $\pm \underline{g}$

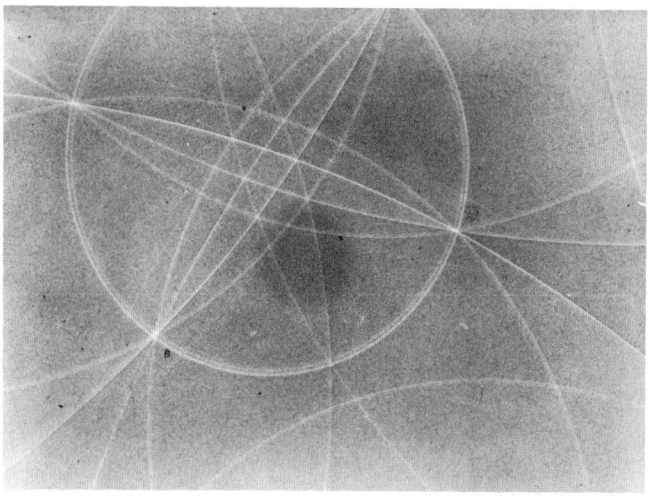

Fig. 10 Kossel pattern from a cubic crystal (courtesy of
H. J. Neuhäuser)

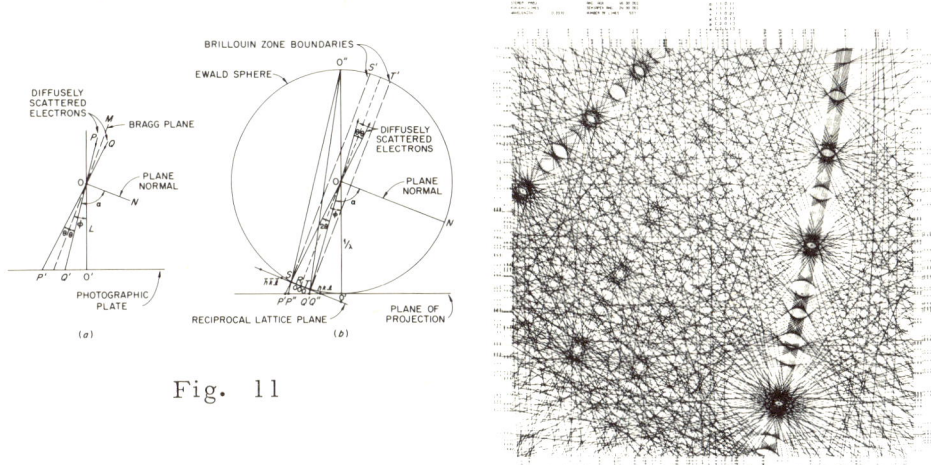

Fig. 11

Fig. 12

Brillouin zone boundaries. Distances RT and SR are each equal to 1/2d where d is the interplanar spacing, the Bragg angle is given by $\sin\theta = \lambda/2d$ and the semi-apex angle of the cone is $\alpha = \cos^{-1}(\lambda/2d)$. It is evident that the Kikuchi lines P' and Q' are the gnomic projections of the small circles SS' and TT'. φ is of the order of $1°$ so the more familiar stereographic projections, P" and Q", deviate very little from P' and Q'.

An example of a computed stereographic projection of the intersections of Brillouin zone boundaries with the Ewald sphere for topaz is reproduced in Figure 12. Other applications for these maps include the indexing of electron channelling and Kossel patterns, both of which are useful crystallographic tools in scanning electron microscopy.

"MOLECULAR" DESCRIPTION OF DEFECTS

The presence of a dislocation introduces local changes in both d and θ and evidently gives rise to important diffraction effects, see figure 13, the nature of which are essentially the same for all crystals and are discussed in a subsequent lecture by Dr. Hobbs. Very briefly, $\underline{g}\cdot\underline{u}_n$ gives the effect of the defect, where $\underline{u}_n$ is the displacement of

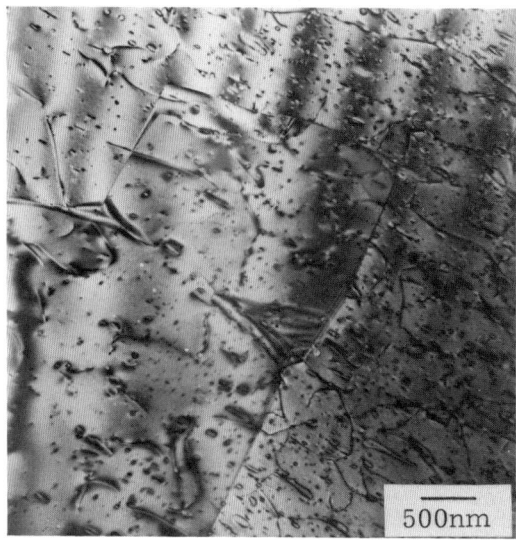

Fig. 13 Dislocations and Dauphiné twins in α-quartz. Notice that the diffraction conditions for the inner region are slightly deviated from those which give rise to the sharp dislocation contrast beyond the twin boundary. $\underline{g}\ 1\bar{1}01$

lattice site $\underline{r}'_n$ from its normal location $\underline{r}_n$

$$\underline{r}'_n = \underline{r}_n + \underline{u}_n.$$

Of particular interest, since it corresponds to zero contrast from the defect and hence provides a means for measuring its displacement field, is the condition $\underline{g} \cdot \underline{u}_n = 0$. In the following discussion, we shall confine our attention to details of geometrical interpretations, peculiar to non-metallic compounds, of displacement vectors measured by using this diffraction condition.

Dislocations are elastic defects and, in crystals whose bonding is ionic or partially ionic, they are expected to adopt core structures which avoid the large energies associated with electrostatic charges. Uniformly neutral cores are easy to visualise in some ionic crystals, for example rocksalt, but in general require careful consideration of ion positions. A convenient way of visualising neutral cores is to regard dislocations in ceramic crystals as dislocations in the arrangements of molecular units. As a matter of fact, molecular units often take the form of regular geometrical units which are so stable that the introduction of dislocations is expected to leave them more or less intact. To illustrate this, consider the geometry of dislocations in crystalline quartz.

The basic structural unit common to all forms of silica is the SiO_4 tetrahedron in which a silicon atom fits interstitially between four oxygen atoms. This unit is extremely stable ; it is only the spatial arrangement and precise geometry of the unit which varies from one form of silica to another. Consequently the nature of defects such as dislocations is expected to be such that the basic unit is not destroyed.

The Burgers vectors for the primary slip systems in α- and β-quartz are a and c, both of which are parallel to the axes of channels defined by helical chains of SiO_4 tetrahedra joined together at their corners, as illustrated in Figure 14 for morphologically right-handed β-quartz ($C6_42$). There are two kinds of channel parallel to both a and c, one of whose cross section is a hexagon and the other either a rectangle (parallel to a) or a triangle (parallel to c). The rectangular and triangular channels are defined each by a single helix with, respectively, four and three tetrahedra per pitch and of pitch equal in magnitude to the Burgers vector, so a screw dislocation along the axis of either channel changes the helix

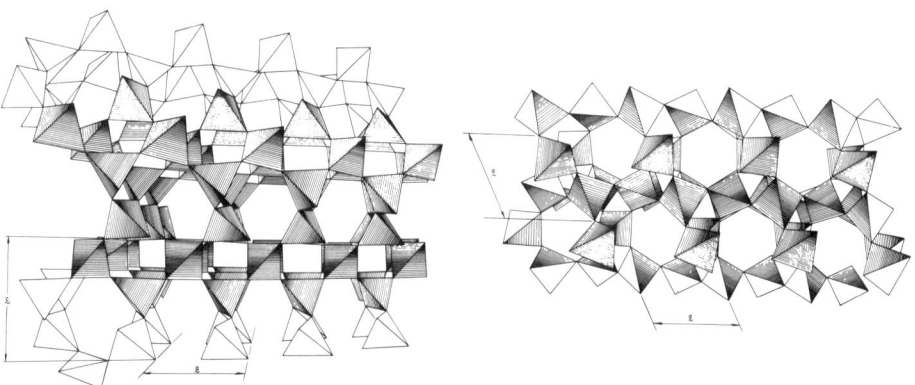

Fig. 14 Perspective views of a model of β-quartz. In α-quartz (rhombohedral), the SiO_4 tetrahedra are slightly rotated away from the β-quartz positions in such a way as to destroy one set of two-fold axes.

into rings of tetrahedra or into a double helix of twice pitch, depending on the sign of the Burgers vector.

The hexagonal channels parallel to $\underline{a}$ and $\underline{c}$ are each defined by a double helix with six tetrahedra per pitch and of pitch equal in magnitude to twice the modulus of the Burgers vector. Depending on the sign of the Burgers vector $\underline{b}$, a screw dislocation converts the double helix either into a single helix of half pitch or into a triple helix of three halves pitch. Both these core distortions are illustrated for a right handed double helix in Figure 15. If the screw dislocation does not thread the full length of the original double helix, it is terminated by an edge dislocation which, in physical terms, is simply the end or start of a single helix. Such a termination is shown at one end only of each of the screw dislocations illustrated in Figures 15(b) and (c). In a three-dimensional piece of quartz, the edge dislocation would reach out across the slip plane normal to the Burgers vector.

Similar interpretations of dislocation cores in other ceramic crystals have permitted the analysis of fault surfaces including translation twins. As an example, Figure 16 shows several nodes formed between extended and perfect dislocations in boron carbide, the occurrence of which is readily understood when dislocations in this material are described in terms of dislocations in the rhombohedral arrangement of B_{12}

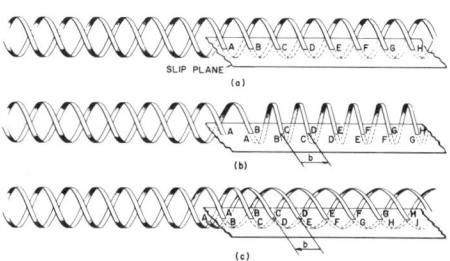

Fig. 15 (a) To introduce a screw dislocation along the axis of a β-quartz double helix, let the individual helices be scissioned at the points labelled A, B, C, etc. (b) With that part of the double helix above the slip plane held fixed in space, let the part below the slip plane be sheared to the right by a displacement equal to the Burgers vector $\underline{b}$. (c) The dislocation introduced into (a) by a shear that is equal and opposite to that shown in (b). Note: Some elastic relaxation from the configurations shown in (b) and (c) is expected, especially near the edge dislocations.

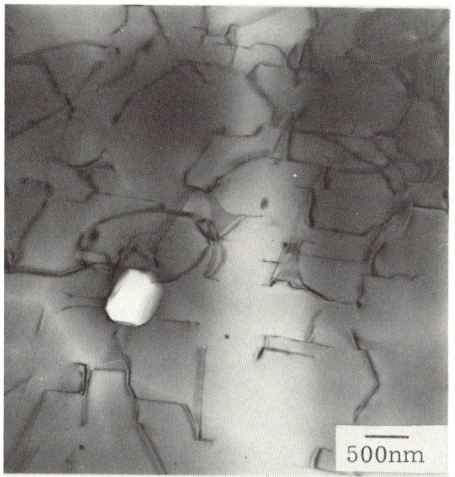 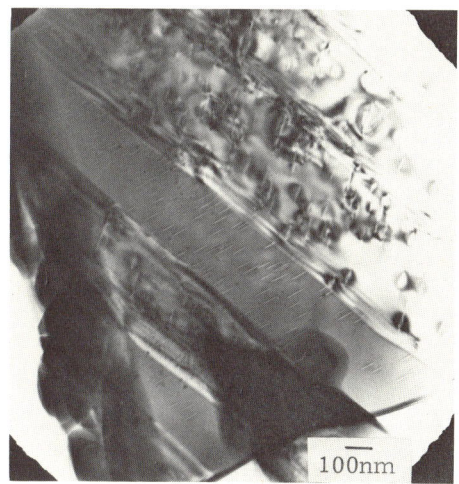

Fig. 16 Examples of nodes between perfect dislocations on the (111) plane and dislocations extended on rhombohedral planes in boron carbide

Fig. 17 Flat plate-shaped voids in a twinned crystal of boron carbide

icosahedra, the rhombohedral angle for which (65.5°) is close to that corresponding to face centred cubic. Perfect dislocations with Burgers vectors $\underline{a}_1$, $\underline{a}_2$ and $\underline{a}_1 - \underline{a}_2$, can dissociate into pairs of Shockley partial dislocations but only on rhombohedral planes. On the basal plane, dissociation is obstructed by the presence of C-B-C chains. As a consequence, threefold nodes can take the form of mixed perfect and imperfect dislocations. The small deviation from cubic symmetry is also responsible for the occurrence of prismatic loops on only one of what would be the four octahedral planes if the crystal were cubic, see Figure 17.

The defects seen in Figure 17 have been produced by neutron irradiation. The electron beam itself is also a source of radiation damage in some non-metals, the so-called beam sensitive materials. α-quartz is such a material. After two or three minutes observation at 200 keV, contrast appears which is characteristic of small perfect prismatic dislocation loops lying on rhombohedral planes, see Figure 18. Figure 19(a) shows a perspective view along an a-axis of the quartz structure. Since each oxygen is shared by two

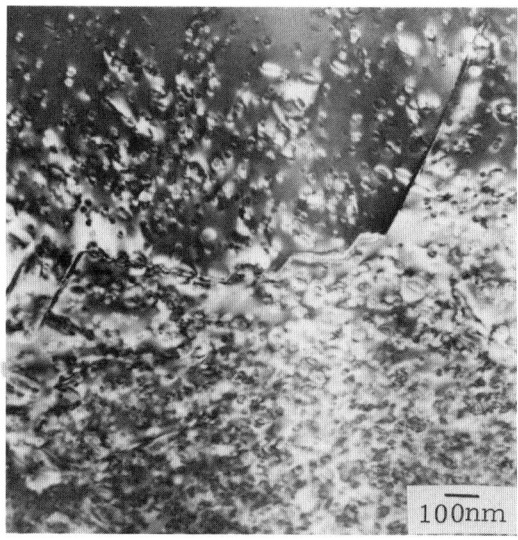

Fig. 18 Early stages of damage by the electron beam in α-quartz. Dark field micrograph, g 10$\bar{1}$1.

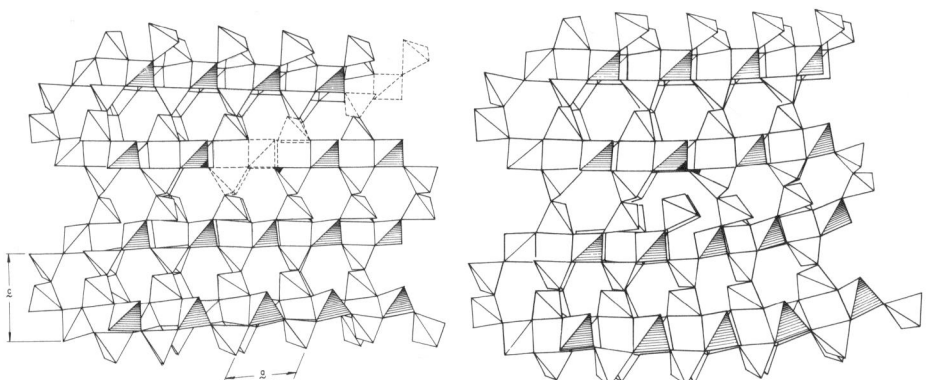

Fig. 19(a) and (b). Formation of perfect prismatic dislocation on α-quartz rhombohedral plane.

tetrahedra, the aggregation of stoichiometric numbers of point defects to form a disc on a rhombohedral plane is represented by the removal or addition of a rhombohedral layer of tetrahedra. Figure 19(b) is the same drawing with part of a single rhombohedral layer of tetrahedra removed and the

X-RAYS AND ELECTRON MICROSCOPY

adjacent layers brought together to form a perfect prismatic dislocation loop. During continued exposure to the electron beam, the lobes of strain contrast around each loop that are evident in Figure 18, progressively wane and eventually disappear. At this time it is found that the diffraction pattern has also vanished, indicating that the irradiated region has become metamict.

The dislocation images presented so far have been obtained by electron microscopy. Dislocations in crystals which, for one reason or another, cannot be thinned for for transmission electron microscopy, can be imaged by X-ray diffraction. Figure 20(a) shows a Lang topograph of

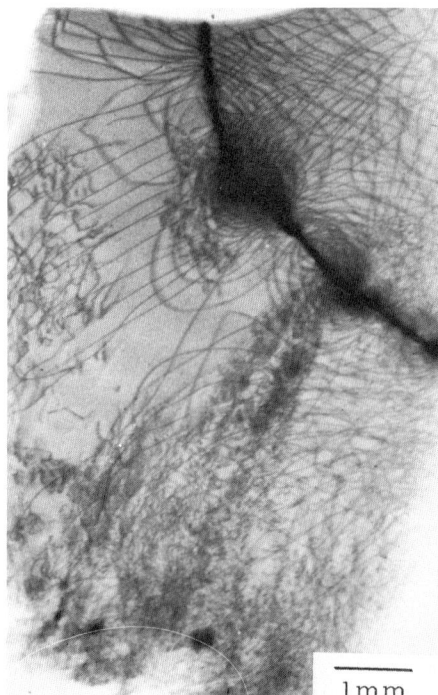

Fig. 20(a) X-ray topograph of basal dislocations associated with a fracture in ice (courtesy of C. A. English and J. S. Thompson) $\underline{g}$ $1\bar{1}00$

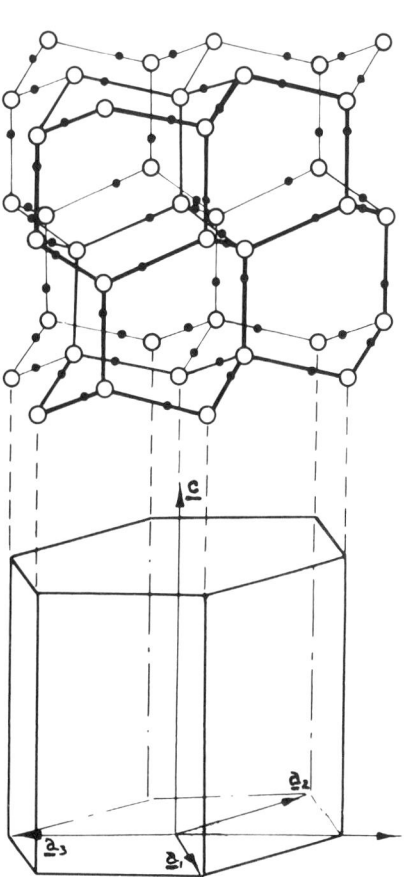

Fig. 20(b) Crystal structure of ice. Open circles denote oxygen sites, small solid circles denote hydrogen sites.

dislocations in ice. Ice is hexagonal and the primary slip system is $(0001) <11\bar{2}0>$. Examination of the crystal structure, Figure 20(b), reveals that basal glide disturbs the arrangement of hydrogen atoms; across the slip plane, neighbouring molecular units are obliged to adopt abnormal orientation relationships with respect to each other. Such orientational abnormalities in ice are known as Bjerrum defects. X-ray topography is also a powerful non-destructive method for investigating the growth histories of gemstones. Natural diamonds, for example, are often characterised by growth sectors which are distinguishable one from another because of differences in their populations of "diffraction contrast producing bodies" such as nitrogen platelets. The 'St. George Cross' within the stone shown in Figure 21 is an example of impurity variations arising from mixed habit growth. Ionic crystals are even more sensitive to damage by the beam of an electron microscope than are ionic-covalent crystals such as quartz and, for such materials, X-ray topography is the only method available for direct observation of lattice imperfections. Dislocation structures in LiF, for example, are most conveniently studied by X-ray topography, see figure 22(a). Dislocations in CaF_2, another beam sensitive material, have also been extensively studied by X-ray topography; the slip system is $\{001\} <110>$ and application of the principle of 'dislocations in the arrangement of molecular units' suggests

Fig. 21. X-ray topograph through a natural diamond. g 440. Vertical height of stone is 5 mm (courtesy of S. Suzuki)

Fig. 22(a) X-ray topograph showing dislocation images in LiF. Note that the scale of the defects is much larger than that which can be observed by electron microscopy. $\underline{g}$ 200 (courtesy of A. R. Lang)

Fig. 22(b) $(1\bar{1}0)$ projection of $(00\bar{1})$ $[110]$ edge dislocation in fluorite. Squares represent cation sites, circles anion sites and the separation between 'open' and 'full' sites is $\frac{1}{4}[1\bar{1}0]$.

the core structure for edge dislocations that is shown in Figure 22(b).

The defect contrast observed in the bright field electron microscope images shown in figures 13, 16, 17 and 18 is due to diffraction; an aperture is placed around the electron beam in the back focal plane of the objective so as to block off all the diffracted beams. In some compounds, a different kind of defect, arising from the need to accommodate non-stoichiometry, can be observed by using as many beams as possible so as to form a Fourier image of the diffraction pattern at the photographic plate. Figure 23 shows such an image of lattice fringes in β-alumina, $Na_2O.11 Al_2O_3$, the solid electrolyte used in the sodium-sulphur battery. Planar defects, believed to be due to migration of sodium ions out of the crystal, are resolved as abnormally wide white fringes each having a thin line of dark contrast along its centre. After removal of a monolayer of sodium ions, the two half crystals are brought together, with or without a displacement parallel to the missing layer, to create a planar defect geometrically similar to the intrinsic stacking fault

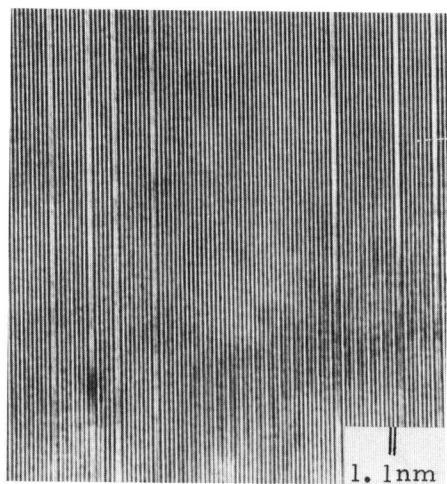

Fig. 23 Lattice image of β-alumina. The ideal structure (hexagonal) consists of monolayers of Na^+ sandwiched between 11Å thick blocks of Al^{3+} and O^{2-} ions in a spinel arrangement (courtesy of A. W. Holloway)

enclosed by a Frank dislocation loop in a close packed metal. Similar defects in mixed oxides of certain transition metals have received considerable investigation by inorganic chemists and are now known as 'crystallographic shear planes'.

LATTICE PARAMETER MAPPING

In some materials, doped tantalum pentoxide for example, local composition variation manifests itself as a smoothly varying lattice fringe spacing. The precise measurement of extremely small (1 in 10^7) variations in lattice parameter generated locally by impurity diffusion are of interest in the manufacture of semiconducting devices, and demonstrates a potential application of interferometry.

Figure 24 shows the layout of the Laue-case X-ray interferometer. (The distinction between Laue-case and Bragg-case diffraction is illustrated in Figure 25). The interferometer consists of three identically oriented single crystal wafers. Identical orientation is ensured by manufacturing the device from one single crystal; partial removal of two parallel-sided slabs leaves the wafers standing proud of a common base. The thickness of the

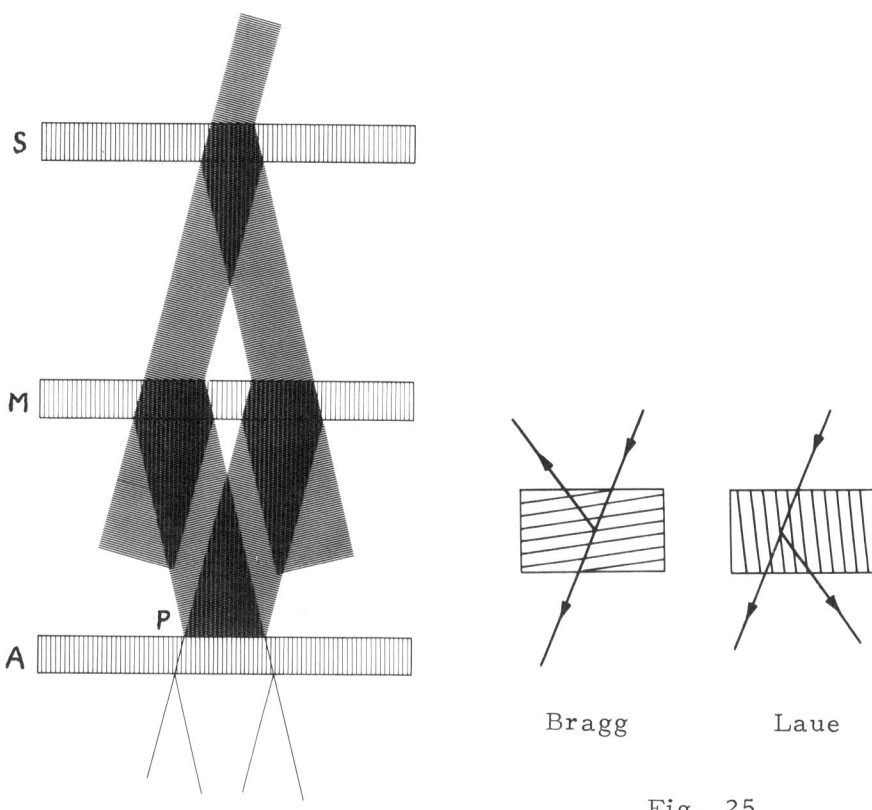

Fig. 24 The Bonse and Hart X-ray interferometer.

Fig. 25

wafers is such that only the wave with nodes at the atomic planes is excited (refer to Figure 6(a)). Phase-coherent beams are formed at the beam splitter S and at the transmission mirror M, and culminate in an atomic-scale standing wavefield at P in front of the analyzing crystal A. Superposition of this wavefield with the atomic planes of A produces a moiré pattern which can be observed on a piece of film inserted into one of the exit beams.

Figure 26 shows a moire pattern generated when an aluminium disc was evaporated on to the analyzer of a silicon interferometer, (a) before and (b) after annealing. The elastic strain is a maximum at the centre of the disc where the moire spacing in (b), 0.2 mm, corresponds to $\delta d/d = 10^{-6}$.

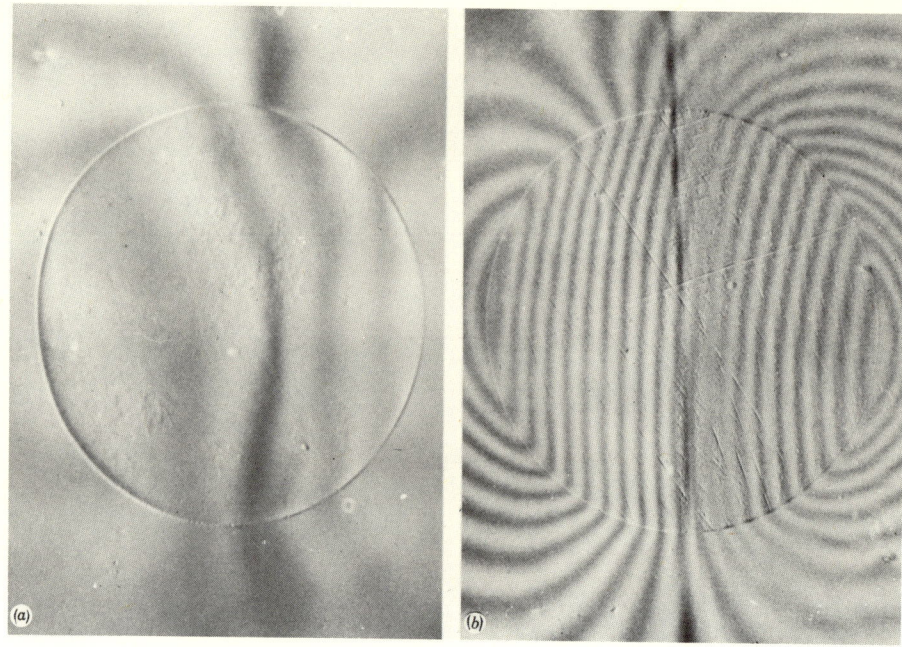

Fig. 26 (courtesy of J. Helliar)

"Wave propagation in periodic structures" by Leon Brillouin is a standard text for the subject of the lecture on waves in periodic structures. "Interaction of waves in crystals" by J. C. Slater, Rev. Mod. Phys(1958) 30 (1) 197-221 is also a recommended reference. Derivations of the dispersion relationship for the particular cases of X-rays and electrons, and applications to X-ray diffraction and electron microscopy respectively may be found in "The optical principles of the diffraction of X-rays" by R. W. James and "Electron microscopy of thin crystals" by P. B. Hirsch, A. Howie, R. B. Nicholson, D. W. Pashley and M. J. Whelan.

A FORTRAN programme for constructing maps of Bragg contours, Kikuchi lines and Kossel lines by B. T. M. Loh and K. H. G. Ashbee has been published as an Oak Ridge National Laboratory technical memorandum ORNL-TM-3557 (1971). An account of real space crystallography by J. A. Eades and J. W. Steeds is published in Physics Bulletin 26 (1975) 108.

Recent reviews of X-ray topography and X-ray interferometry respectively, have been published by A. R. Lang in "Modern Diffraction and Imaging Techniques in Material Science" edited by S. Amelinckx, R. Gevers, G. Remaut and I. Van Landuyt (North Holland 1970) and M. Hart, Proc. Roy. Soc. Lond. A346, 1-22 (1975).

Much of the content of these lectures is taken from the University of Bristol M. Sc. Course in the Physics of Materials and the author gratefully acknowledges the ideas and contributions of Professor F. C. Frank F. R. S., Dr. A. R. Lang F. R. S., Dr. M. Hart and Dr. J. W. Steeds.

TRANSMISSION ELECTRON MICROSCOPY OF DEFECT AGGREGATES IN NON-METALLIC CRYSTALLINE SOLIDS

L. W. Hobbs

Materials Development Division, Atomic Energy Research Establishment, Harwell, U.K.

Defect probes such as spectroscopy, magnetic resonance, electrical and thermal conductivity are <u>indirect</u> techniques since they rely on the <u>average</u> response of a large number of similar defects to the probe, whether it be a light photon, microwave quantum, conduction electron or lattice phonon. Microscopy is a <u>direct</u> technique for observing structure of <u>individual</u> defects.

ELECTRON MICROSCOPY AND FUNDAMENTAL RESOLUTION

Microscopy is based on the ability to retrieve spatial information about an object, usually in magnified form, from diffraction of electromagnetic radiation. Spatial information about an object observed in transmission is contained in the object <u>transmission function</u> $\phi(\underline{r})$ which represents diffracted amplitude and phase at the exit surface as a function of position $\underline{r}$. We can expand $\phi(\underline{r})$ in a two-dimensional Fourier series

$$\phi(\underline{r}) = \int_{\text{all } \underline{K}} \tilde{\phi}(\underline{K}) \exp(2\pi i \, \underline{K} \cdot \underline{r}) \, d^2\underline{K} . \qquad (1)$$

If we interpret vectors $\underline{r}$ in object space as characteristic separations ('wavelengths') in the object, for example lattice periodicities, then vectors $\underline{K}$ in Fourier space represent spatial 'frequencies' corresponding to $\underline{r}$. When the radiation source and point of observation are effectively at infinity (the case for <u>focussed</u> images in systems containing lenses), the result is <u>Fraunhofer</u> diffraction, and the diffracted amplitudes in the focal plane (or at infinity) are just given by the Fourier transform in (1) of the object transmission function

$$\tilde{\phi}(\underline{K}) = \int_{\text{all } \underline{r}} \phi(\underline{r}) \exp(-2\pi i \, \underline{K} \cdot \underline{r}) \, d^2\underline{r} \, . \tag{2}$$

If we assign to our incident radiation of wavelength λ a wave vector $\underline{\chi}$ in the direction of incidence with magnitude $\chi = 1/\lambda$, the vectors K in the focal or <u>diffraction</u> plane represent the change $\Delta\underline{\chi}$ in the incident wave vector after diffraction. If source and point of observation are at finite distances (the case for out-of-focus images in systems containing lenses, the result is <u>Fresnel</u> diffraction, for which the phases as well as amplitudes of the diffracted radiation are important. If the amplitudes $\tilde{\phi}(\underline{K})$ in the diffraction plane serve as the object for a second Fraunhofer diffraction, the amplitudes observed at the second diffraction plane are

$$\phi'(\underline{r}') = \int_{\text{all } \underline{K}} \tilde{\phi}(\underline{K}) \exp(-2\pi i \, \underline{K} \cdot \underline{r}') \, d^2\underline{K} = \phi(-\underline{r}) \tag{3}$$

where r' = Mr and M is the magnification of the lens system. The last result follows immediately from (1), and we obtain an inverted image of the object. Of course, what we ultimately measure is the intensity $\phi'(\underline{r}) \, \phi'^*(\underline{r})$; intensities may be substituted in (3) provided conditions are strictly those of Fraunhofer diffraction. In a microscope, the lenses perform the Fourier transforms for us; this makes it possible to define a lens <u>transfer function</u> $\tilde{T}(\underline{K})$ which transfers spatial frequencies associated with the object to the image

$$\tilde{\phi}'(\underline{K}) = \tilde{T}(\underline{K}) \, \tilde{\phi}(\underline{K}) \, . \tag{4}$$

The transfer function can be used to characterise the effects of lens aberration and defocus on the image.

Diffraction nevertheless fundamentally limits spatial resolution to object distances of order λ; this follows from the uncertainty principle. To probe directly the structures of defects at the atomic level, we must therefore employ radiation with wavelengths of order 1Å. X-rays, and fast electrons (kinetic energy U > 1 keV, Table 1) satisfy this requirement. X-rays, however, cannot be focussed, and although topographic imaging techniques exist, these are limited by the resolution of the recording media and, as will become apparent, by the large extinction distances for X-rays ($\xi \sim 25\mu m$). Fast electrons can be focussed by strong electromagnetic lenses, and we can thus construct a transmission electron microscope (Figure 1). The source provides a beam of roughly monochromatic electrons accelerated to energies between 40 keV and 3 MeV which is focussed on an area 1 - 10 μm across on the specimen object. Electrons passing through the specimen are brought to a focus in the diffraction plane of an electromagnetic objective lens from which a real image is formed and magnified in

U, eV	λ, Å
100	1.2
1000	0.39
10^4	0.12
10^5	0.037
10^6	0.009

Table 1. Wavelengths of fast electrons

subsequent lenses and ultimately recorded; overall magnification can exceed 10^6.

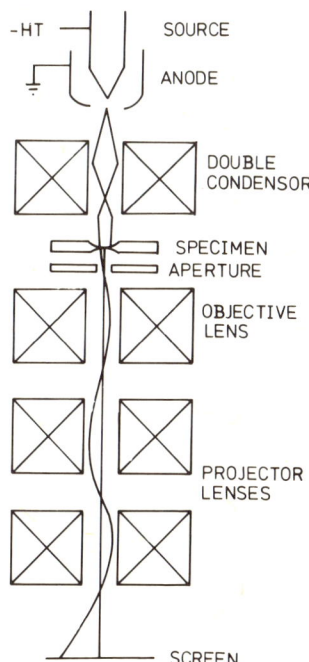

Figure 1. Schematic diagram of a transmission electron microscope.

Electron Scattering

We can assign to the electrons incident on the specimen a wave vector χ in the incident electron direction with magnitude $\chi = 1/\lambda = \sqrt{(2mU/h^2)}$. In passing through the specimen, incident electrons are scattered by atoms in the specimen both elastically (without energy loss) and inelastically (with energy loss). Elastic scattering results in scattering through an angle θ and a change $\Delta\chi$ in the incident wave vector direction, where $|\Delta\chi| = 2 \cdot \sin\theta/\lambda$. Elastic scattering from a single atom j in the specimen can be expressed by electron scattering factor f_j^{el} which represents the scattered amplitude from an incident electron wave of unit amplitude; incident electrons are scattered both by the atomic electron distribution $\rho_j(\underline{r})$ and by the nucleus (point charge $+Z_j$ at $\underline{r} = 0$) and provided incident electron velocity is very much greater than orbital electron velocities,

$$f_j^{el} = [2\,me^2/h^2|\Delta\chi|^2] \int [Z_j\,\delta(\underline{r}) - \rho(\underline{r})]\,d\tau$$

$$= [2\,me^2/h^2|\Delta\chi|^2]\,(Z_j - f_j^x) = (me^2/2h^2)(\lambda/\sin\theta)^2$$

$$\cdot (Z_j - f_j^x), \qquad (5)$$

where f_j^x is the X-ray scattering factor. To calculate scattering from <u>ions</u>, which are frequently encountered in non-metallic solids, we must add an additional Coulomb term to (5),

$$f_j^{el} = (me^2/2h^2)(\lambda/\sin\theta)^2 (Z_j - f_j^x + q_j/4\pi\,\varepsilon_o) \qquad (6)$$

where ε_o is the static dielectric constant and q_j the ion charge. f_j^{el} is appreciable for $|\Delta\chi| < 1\text{Å}^{-1}$ or $\theta < 1°$, so scattering is strongly peaked in the forward direction.

Inelastic scattering results in a change $\Delta\chi$ in the magnitude of the incident wave vector, and thus a change $\Delta U = h^2\,(\Delta\chi)^2/2m$ in the incident electron energy. The processes responsible are principally electron-phonon interactions ($\Delta U \sim 10^{-2}$ eV), exciton creation ($\Delta U \sim 10$ eV), collective oscillations (plasmons) of valence or conduction electrons ($\Delta U \sim 25$ eV) and core electron excitations ($\Delta U \sim 10^3$ eV). The average energy loss rate can be given by a Bethe-Bloch expression

$$dU/dz \simeq (B/U)\,\ln(U/\bar{I}) \qquad (7)$$

where $B \sim 10^{14}$ eV^2 m^{-1} and $\bar{I}$ is a sort of mean ionisation energy;

dU/dz initially amounts to $\sim 10^9$ eV m^{-1} for fast electrons, from which it can be seen that the range of 100 keV electrons is of order 100 μm. Specimen thickness must clearly be less than this, and for other reasons discussed below considerably less than this.

Practical Resolution

The wavelength of fast electrons is $\ll 1$Å, but in an electron microscope resolution is ultimately limited by aberrations in the electron lenses. These are of two sorts, spherical aberration in which electrons travelling considerably off-axis are brought to a closer focus than nearly axial electrons, and chromatic aberration in which electrons of different energies are focussed to different points. Spherical aberration may be reduced by imposing a small aperture after the specimen to limit the angular spread of scattered electrons entering the objective lens, but this can only be narrowed to the point where Fresnel diffraction from the aperture itself begins to limit resolution (typically $\theta \sim 5 \times 10^{-3}$ radians); in doing so we also limit the spatial frequencies we can resolve in the image. Appreciable intensity is still retained, since scattering is predominately within $\theta \sim 5 \times 10^{-2}$ radians. Chromatic aberration may only be reduced by keeping specimen thickness, and thus energy spread ΔU, small. For example, an average loss rate of 10^9 eV m^{-1} implies an energy spread of order 100 eV in a 1000Å thick specimen, which limits resolution to the order of 100Å. Plasmon oscillations, with a mean free excitation path about 1000Å, are of most importance for such thicknesses. The best point-to-point resolution (presently around 2Å) is therefore achieved only for very thin specimens <100Å thick.

Electrons are emitted comparatively incoherently in time and space from the electron source, consequently image resolution is governed by Poisson statistics and ultimate resolution is achieved only if a statistically sufficient number of electrons is collected to form the image. The minimum electron density ϕ_o at the object plane needed for resolution of object detail d_o is

$$\phi_o = (\kappa/d_o \Gamma)^2 \tag{8}$$

where Γ is the image contrast and $\kappa \sim 5\text{-}30$ is the physiologically determined signal-to-noise ratio required for recognition of detail. For example, to resolve $d_o \sim 50$Å in the object, with $\Gamma = 0.2$ (reasonable for diffraction contrast), requires $\phi_o \sim 10^{20}$C m^{-2}. The required fluence incident on the specimen must be considerably larger (typically $\phi_i > 20 \phi_o$) because the objective lens aperture subtends only a fraction of the angular scatter of electrons emerging from the foil; increasing foil thickness increases the width of this scatter and thus decreases intensity. We therefore require large incident electron current densities $\sim 10^{22}$e m^{-2} (1 kC m^{-2}).

We must subsequently focus and record the image, so response of the recording media is important. It turns out that photographic emulsions are practically noiseless electron detectors and record essentially every electron, though their speed for electrons bears only an approximate relation to their speed for light. Phosphor screens (for initial observation and focussing) are less efficient, and of course the eye only integrates over ~ 0.1 s; provided we collect emitted light over a somewhat wider solid angle than the eye's (by using subsequent optical lenses) we can nevertheless limit defocussing error to about the statistical resolution in recording.

CONTRAST FROM PERFECT CRYSTALS

The theory of contrast from perfect crystals considers the propagation of electrons of mass m and energy U incident along direction χ approximately normal to the top surface of the specimen foil, and evaluates intensities at the bottom surface. The electron wave function $\Psi(\underline{r})$ must satisfy the time-independent Schrödinger wave equation for electrons moving in a potential field $V(\underline{r})$

$$\nabla^2 \Psi(\underline{r}) + 8\pi m[U + eV(\underline{r})] \Psi(\underline{r}) = 0 . \tag{9}$$

In free space where $V(\underline{r}) = 0$, solutions to (9) are plane waves of unit amplitude

$$\Psi(\underline{r}) = \exp(2\pi i \underline{\chi} \cdot \underline{r}) . \tag{10}$$

Within the crystal, $V(\underline{r})$ is clearly periodic and can be expanded in a Fourier series

$$V(\underline{r}) = \sum_g V_g \exp(2\pi i \underline{g} \cdot \underline{r}) \tag{11}$$

with the same periodicity as the lattice, $\underline{g}$ being a reciprocal lattice vector. The Fourier components $\{V_g\}$ can be related to the electron scattering factors f_j^{el} for atoms (or ions) $j = 1, \ldots, \ell$ within a unit cell of volume Ω

$$V_g = \frac{m}{m_o} \frac{\exp(-M_g)}{\pi \Omega} \sum_{j=1}^{\ell} f_j \exp(-2\pi i \underline{g} \cdot \underline{r}_j) . \tag{12}$$

M_g is the Debye-Waller factor which takes into account thermal excursions from normal atom positions $\underline{r}_j$ and m_o is the electron rest mass. The sum in (12)

$$F_g = \sum_{j=1}^{\ell} f_j \exp(-2\pi i \underline{g} \cdot \underline{r}_j) \tag{13}$$

TRANSMISSION ELECTRON MICROSCOPY OF DEFECT AGGREGATES 437

is called the structure factor of the unit cell. In non-metallic solids, we often have more than one kind of atom (or ion) in the unit cell; this reflects in F_g. For example, in NaCl, F_g becomes

$$F_{hkl} = \begin{cases} (4 \ (f_{Na^+} + f_{Cl^-}) & \text{for h,k,l all even} \\ (4 \ (f_{Na^+} - f_{Cl^-}) & \text{for h,k,l all odd} \\ 0 & \text{otherwise} \end{cases} \quad (14)$$

for $g = (h,k,l)$. Table 2 lists F_g and V_g for several low order reciprocal lattice vectors g in NaCl. Since $eV_o \ll U$, we normally neglect the slight refraction of χ in the crystal due to V_o.

The effect of the potential is to change the wave vector from χ to $\chi'(\underline{r})$ locally, where

$$\chi'(\underline{r}) = \sqrt{(2m\{U + eV(\underline{r})\}/h^2)}. \quad (15)$$

Since $eV(\underline{r}) \ll U$ we can write

$$\chi'(\underline{r}) \simeq \chi\{1 + \tfrac{1}{2} eV(\underline{r})/U\} = \chi + (me/h^2\chi)V(\underline{r}) . \quad (16)$$

If the plane wave (10) is incident upon a thin enough slab of thickness δz, the amplitude is not much changed but the <u>phase</u> of the wave is changed,

$$\Psi'(\underline{r}) = \exp 2\pi i \ \chi'\cdot\underline{r}$$
$$\simeq \exp (2\pi i \ \chi \cdot \underline{r}) \exp\{2\pi i \ (me/h^2\chi) \ \delta z \ V(\underline{r})\}. \quad (17a)$$

In the limit as $V(\underline{r}) \to 0$ or $\delta z \to 0$,

$$\Psi'(\underline{r}) \simeq \exp(2\pi i \ \chi\cdot\underline{r}) [1 + 2\pi i \ (me/h^2\chi) \ \delta z \ V(\underline{r})]$$
$$= \Psi(\underline{r}) + (2\pi i \ me/h^2\chi) \ \delta z \ \sum_g V_g \exp 2\pi i (\chi+\underline{g}) \cdot \underline{r} . \quad (17b)$$

We therefore obtain plane wave solutions for (9) in the crystal in the form of Bragg diffracted plane waves

$$\Psi(\underline{r}) = \sum_g \phi_g \exp 2\pi i \ (\chi + \underline{g}) \cdot \underline{r} \quad (18)$$

travelling in directions

$$\theta_g = 2 \sin^{-1} (g/2\chi) \quad (19)$$

wherever the electrons scattered from the contents of each unit all scatter in phase. Condition (19) is an equivalent statement of Bragg's law and can be represented by a geometrical construction due to Ewald (Figure 2). A vector χ is drawn from an origin P in

g	V_g	V_g^i	F_g	ξ_g	ξ_g^i
000	8.42	0.46	31.6	404	7420
111	-1.46	-0.06	-5.5	2330	52100
200	5.27	0.12	19.7	645	27400
220	3.89	0.11	14.6	874	32100
311	-0.98	-0.05	-3.7	3480	72300
222	3.09	0.10	11.6	1100	34700

Table 2. Ionised atom electron diffraction parameters for 100 keV electrons in NaCl at 20 K. V_g in volts; F_g, ξ_g, ξ_g^i in Å. Origin taken at Na$^+$ ion site.

reciprocal space terminating at the h = 0 reciprocal lattice point along the direction of the incident beam; the Bragg condition $\underline{g}$ is then satisfied wherever a sphere of radius $\chi = 1/\lambda$ centred on P passes through reciprocal lattice point g. $s_g = g\Delta\theta$ represents the deviation from the Bragg condition. χ_g is typically of order 100; consequently, the Ewald sphere is practically a plane, and many Bragg beams can be appreciably excited at once. The Bragg angles θ_g are correspondingly very small ($\sim\frac{1}{2}°$). Electrons travelling in

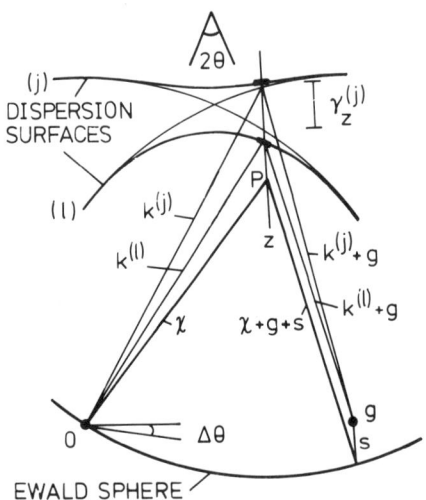

Figure 2. Ewald sphere construction showing dispersion surfaces.

TRANSMISSION ELECTRON MICROSCOPY OF DEFECT AGGREGATES

the same direction are brought to a focus at the same point in the focal (or diffraction) plane of the objective lens, forming the diffraction pattern which contains intensity maxima corresponding to the various Bragg beams $\underline{h}$ = 0, $\underline{g}$, ..., etc. The objective aperture is placed in this plane (Figure 1) and can be used to sample intensities in one or more Bragg beams.

The weak phase object approximation inherent in (17b) is typically valid only for $\delta z \simeq 10\text{Å}$. For thicker slabs we retain (17a) and, provided crystal thickness t remains small enough that $\phi_g << \phi_o \simeq 1$ (crystal remains a phase object), we can integrate (17a) to obtain

$$\Psi'(x,y,t) = \exp(2\pi i \underline{\chi} \cdot \underline{r}) \exp[(2\pi i m e/h^2 \chi) \int_0^t V(x,y,z) \, dz] \,. \quad (20)$$

This is the phase grating approximation and is valid so long as the diffracted beams are still sampling more or less the same region of the potential. This occurs only if $t\theta_g \simeq tg/\chi << g^{-1}$ or $g^2 t/\chi << 1$; since $g \sim 0.5\text{Å}^{-1}$ and $\chi \sim 100\text{Å}^{-1}$, this occurs for $t < 100\text{Å}$. If two or more Bragg beams are included in the objective aperture and used to form an image, the corresponding spatial frequencies will be represented in the images; for example, including two Bragg beams $\underline{h},\underline{g}$ allows us to recover lattice periodicity $\underline{h}-\underline{g}$ in the form of lattice fringes (Dr. Ashbee's Figure 23). If we include enough Bragg beams, we can reconstruct the projection of the lattice charge density $(1/4\pi) \nabla^2 \int V(x,y) \, dz$ together with certain artefacts - for example, we always introduce a centre of symmetry, lens aberrations and aperture truncation. (We cannot reconstruct the projected potential $\int V(x,y) \, dz$ because, in the approximation that the crystal is a phase object, the phase terms disappear in the intensity unless we defocus by an amount $\zeta \neq 0$ which introduces an interference term $\{1 - (me/h^2\chi^2) \zeta \nabla^2 \int V(x,y) \, dz\}$ into the image intensity.) By this powerful method, we can directly observe (projected) lattice structure on an atomic scale (e.g. so-called 'shear structures' in transition metal oxides, Figure 3, which accommodate deviations in oxide stoichiometry). Some caution must be exercised in making the direct correspondence between lattice images and actual atom positions, since image details are sensitive to specimen orientation and thickness, spherical aberration and defocus, but recent results indicate that under conditions where the approximation(20) is valid, the naive projected charge density interpretation provides a fairly unambiguous indication of structure. However, it is always safer to compare the reconstructed image with computer-simulated images of likely structures, particularly in the case of extended defects.

For $t > 100$ Å, the phase grating approximation breaks down at high resolution, but we can still represent the behaviour of electrons in the crystal in terms of Bragg-diffracted plane waves, provided that dynamic interactions between them are properly accounted for. We must also allow for the curvature of the Ewald

Figure 3. High resolution many-beam image of VNb_9O_{24}. The distance between adjacent white dots, which correspond to unfilled channels between metal-oxygen octahedra, is 3.8Å (courtesy Dr. J. L. Hutchison).

sphere (the approximations (17) assume a flat Ewald sphere) by seeking Bragg wave solutions of the form $\phi_g \exp 2\pi i(\underline{\chi} + \underline{g} + \underline{s}_g) \cdot \underline{r}$ in (18). Under these conditions, lattice images bear a much less apparent relation to lattice structure, and we usually revert to a less direct imaging mode by collecting intensity in only one of the Bragg diffracted beams. We choose the forward scattered beam ($h = 0$) to form the <u>bright field</u> image and any of the other beams ($\underline{h} = \underline{g}$) to form a <u>dark field</u> image.

The difficulty with plane wave solutions to (9) in thick crystals is that the Bragg plane waves are continually being scattered and rescattered by the perfect crystal lattice. As t increases, the amplitude of a Bragg beam ϕ_g builds up to a maximum and falls again to a minimum over a distance

$$\xi_g = [(2me/h^2) \, V_g/\chi]^{-1} \tag{21}$$

called the extinction distance, which is typically of order several hundred Å (Table 2). This makes it very difficult to assess the effects of varying crystal thickness, bent foils or presence of defects. An alternative solution to (9) is in the form of Bloch waves

$$b^{(j)}(\underline{k},\underline{r}) = \sum_g C_g^{(j)} \exp 2\pi i \, (\underline{k}^{(j)} + \underline{g}) \cdot \underline{r} \tag{22}$$

which are standing waves of electron density which propagate with constant phase through the perfect crystal wtihout scattering. The amplitude coefficients $\{C_g^{(j)}\}$ form an orthonormal set

$$\sum_g C_g^{(j)} C_g^{(1)*} = \delta_{j1}$$

$$\sum_j C_g^{(j)*} C_h^{(j)} = \delta_{gh} \quad . \tag{23}$$

The total electron wave function $\Psi(\underline{r})$ can be represented as a linear combination of these Bloch waves

$$\psi(\underline{r}) = \sum_j \psi^{(j)} \sum_g C_g^{(j)} \exp 2\pi i \, (\underline{k}^{(j)} + \underline{g}) \cdot \underline{r} \tag{24}$$

with coefficients $\{\psi^{(j)}\}$ representing the respective excitations of each Bloch wave.

If we substitute (24) into (9), we obtain a set of equations for the Bloch wave amplitudes $\{C_g\}$,

$$(h^2/2m) \{\chi^2 - (\underline{k}^{(j)} + \underline{g})^2\} C_g + \sum_{h \neq 0} e V_h C_{g-h} = 0 \, . \tag{25}$$

This is a set of dispersion equations relating energy $h^2\chi^2/2m$ to wave vector $\underline{k}$. Since $eV \ll U$, the dispersion surfaces are from (25) approximately spheres of radius χ centered on the reciprocal lattice points $\underline{g}$, except at the Brillouin zone boundary where $k = |\underline{k} + \underline{g}|$ (Figure 2). The Bloch wave vectors must have the same tangential components as the incident wave (i.e. $k_x^{(j)} = \chi_x$; $k_y^{(j)} = \chi_y$), so solutions to (25) are given by the intersection of the crystal normal (here assumed to be approximately parallel to z) and the dispersion surfaces. The terms $\{\chi^2 - (\underline{k}^{(j)} + \underline{g})^2\}$ in (25) can be written

$$\{\chi^2 - (\underline{k}^{(j)} + \underline{g})^2\} = \{\chi + |\underline{k}^{(j)} + \underline{g}|\}\{\chi - |\underline{k}^{(j)} + \underline{g}|\}$$

$$\approx 2\chi \{\chi - |\underline{k}^{(j)} + \underline{g}|\} \approx -2\chi[\gamma^{(j)} - s_g] \tag{26}$$

where $\gamma^{(j)} = k_z^{(j)} - \chi_z$. The last approximation can be made because the Bragg angles are small ($\sim \tfrac{1}{2}°$) and only the z-components of the wave vectors are important. We see that $\gamma^{(j)}$ represents the distance along the z direction between a sphere of radius χ centred on $h = 0$ and the jth branch of the dispersion surface (Figure 2), and that the ends of the Bloch wave vectors are constrained to lie on the dispersion surfaces. Equation (25) then becomes

$$C_g \gamma^{(j)} = C_g s_g + \sum_{h \neq 0} (me/h^2)(V_h/\chi) C_{g-h} \quad (27)$$

which is an eigenvalue equation of the form

$$\underset{\sim}{C} \{\gamma^{(j)}\} = \underset{\sim}{A} \underset{\sim}{C} \quad (28)$$

where C is the matrix of Bloch wave amplitudes $C_g^{(j)}$, $\{\gamma^{(j)}\}$ is a diagonal matrix, and the terms of $\underset{\sim}{A}$ are

$$A_{oo} = 0$$
$$A_{gg} = s_g$$
$$A_{gh} = (me/h^2)(V_{g-h}/\chi) = \tfrac{1}{2} \xi_{g-h}^{-1} . \quad (29)$$

At any interface, for example at the bottom crystal surface ($z = t$), the Bloch waves can be recombined to form a set of Bragg diffracted plane waves whose amplitudes ϕ_g are continuous through the interface.

Combining (18) and (24), we find that the amplitudes of the Bragg diffracted beams are

$$\phi_g = \sum_j \psi^{(j)} C_g^{(j)} \exp 2\pi i \, (\underline{k}^{(j)} - \underline{\chi}) \cdot \underline{r}$$
$$\simeq \sum_j \psi^{(j)} C_g^{(j)} \exp 2\pi i \, \gamma^{(j)} z . \quad (30)$$

Equation (30) can be expressed in matrix form as

$$\Phi(z) = \underset{\sim}{C} \{\exp 2\pi i \, \gamma^{(j)} z\} \underset{\sim}{\Psi}$$
$$= \underset{\sim}{C} \{\exp 2\pi i \, \gamma^{(j)} z\} \underset{\sim}{C}^{-1} \Phi(z=0) \quad (31)$$

where $\{\exp 2\pi i \, \gamma^{(j)} z\}$ is a diagonal matrix. For near normal incidence, boundary conditions require that $\psi^{(j)} \simeq C_o^{(j)}$, so that the intensities of the Bragg beams are

$$I_g(z) = \sum_{j,1} C_o^{(j)} C_g^{(j)} C_o^{(1)*} C_g^{(1)*} \exp 2\pi i \, z[\gamma^{(j)} - \gamma^{(1)}] . \quad (32)$$

The form of the A matrix (29) shows that diffracted intensities for perfect crystals depend only on ξ_g, foil orientation and foil thickness. For only two Bragg beams principally excited ($\underline{h} = 0, \underline{g}$), variation in thickness z gives thickness fringes (Figure 4) with depth periodicity ξ_g at $s = 0$ in bright field (I_o) or dark field (I_g) due to beating between the two Bloch waves $b^{(j)}$ and $b^{(1)}$ represented by the term

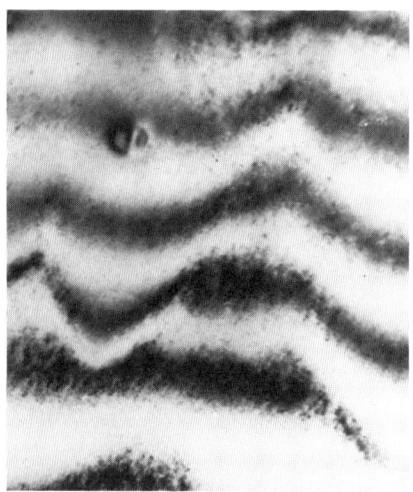

Figure 4. Thickness fringes in a wedge crystal at the Bragg position. Small inclusions are most visible at the fringe boundaries.

$\exp 2\pi i z [\gamma^{(j)} - \gamma^{(1)}] = \exp 2\pi i z [k^{(j)} - k^{(1)}]$ in (32).

Absorption can be included phenomenologically by making the lattice potential $V(\underline{r})$ in (11) complex,

$$V(\underline{r}) = \sum_g (V_g + i V_g^i) \exp 2\pi i \, \underline{g} \cdot \underline{r} , \qquad (33)$$

resulting in modifications to the $\underset{\sim}{A}$ matrix

$$\left. \begin{array}{l} A_{oo} = (ime/h^2)(V_o^i/\chi) = i/2 \, \xi_o^i \\ \\ A_{gg} = s_g + (ime/h^2)(V_o^i/\chi) = s_g + i/2 \, \xi_o^i \\ \\ A_{gh} = (me/h^2)(V_{g-h} + i \, V_{g-h}^i)/\chi = \tfrac{1}{2}(1/\xi_{g-h} - i/\xi_{g-h}^i) \end{array} \right\} \quad (34)$$

where $\xi_g^i = [2(ime/h^2)(V_g^i/\chi)]^{-1}$. Table 2 lists representative values of V_g^i and ξ_g^i. Since $V_g^i/V_g \sim 0.1$, first order perturbation theory may be applied to ascertain the change $\Delta k_z^{(j)} = i q^{(j)}$ in each Bloch wave vector component $k_z^{(j)}$ arising from V_g^i,

$$q^{(j)} = (me/h^2 \, \chi_z) \sum_{g,h} C_g^{(j)*} V_{g-h}^i C_h^{(j)} , \qquad (35)$$

which exponentially damps the Bloch wave amplitudes,

$$b^{(j)} = \exp(-2\pi q^{(j)} z) \sum_g C_g \exp 2\pi i \, (\underline{k}^{(j)} + \underline{g}) \cdot \underline{r} . \qquad (36)$$

The Bragg diffracted wave amplitudes become

$$\phi_g = \sum_j C_o^{(j)} C_g^{(j)} \exp 2\pi i z [\gamma^{(j)} + iq^{(j)}], \quad (37)$$

or in matrix form

$$\underset{\sim}{\Phi}(z) = \underset{\sim}{C} \{\exp 2\pi i z [\gamma^{(j)} + iq^{(j)}]\} \underset{\sim}{\Psi}$$

$$= \underset{\sim}{C} \{\exp 2\pi i z [\gamma^{(j)} + iq^{(j)}]\} \underset{\sim}{C}^{-1} \underset{\sim}{\Phi}(z=0). \quad (38)$$

The corresponding intensities are

$$I_g(z) = \sum_{j,1} C_o^{(j)} C_g^{(j)} C_o^{(1)*} C_g^{(1)*}$$

$$\cdot \exp(-2\pi z [q^{(j)} - q^{(1)}]) \exp 2\pi i z [\gamma^{(j)} - \gamma^{(1)}] . \quad (39)$$

We see that one effect of absorption is to damp out thickness fringes (Figure 4) in a wedge crystal. Another is to introduce an asymmetry in the variation of intensity across the Bragg position (rocking curve) in bright field (Figure 5). This explains the form of bend extinction contours in bent foils (Dr. Ashbee's Figure 7a). Maximum transmitted intensity occurs for $\xi_g s_g \sim 0.5 - 1$ in bright field and for $\xi_g s_g = 0$ in dark field. Absorption also explains an inelastic scattering effect in the diffraction pattern called Kikuchi lines (Dr. Ashbee's Figure 8). Electrons scattered inelastically or quasi-elastically (for example by phonons) can be scattered through large angles (several Bragg angles); they are thereafter subject to normal elastic (Bragg) scattering, so the diffuse background will have structure (Kikuchi lines) superimposed on it, arising from the same origin as bend extinction contours, since taken together the inelastically scattered electrons represent the whole range of incident electron orientations. The structure of the diffuse scattering is therefore equivalent to the complement of a superposition of all bright-field and dark field rocking curves, i.e. what is left after scattering into all possible Bragg beams. Kikuchi lines thus appear as if they were fixed rigidly to the crystal, unlike Bragg beams which remain fixed in relation to the incident beam direction, and can be used to accurately determine foil orientation.

CONTRAST FROM DEFECTS IN IMPERFECT CRYSTALS

Contrast from crystal imperfections such as lattice defects arises from two effects: (i) lattice atoms are displaced from their normal equilibrium sites, and (ii) lattice atoms are replaced by other atoms, not necessarily in the same positions, e.g. a precipitate. Often both effects are present for a single defect,

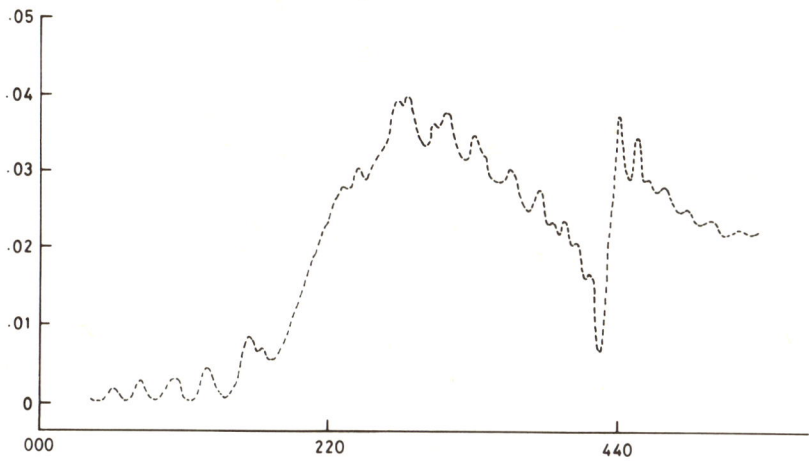

Figure 5. Bright field rocking curve, 220 systematics, 100 kV electrons in KI 4000 Å thick.

e.g. a misfitting coherent precipitate. In treating defect images, it is convenient to make two approximations. In the first, the <u>column approximation</u>, we divide the crystal foil along the x and y axes into columns (say 0.1 ξ_g wide) extending through the foil in the z direction, and assume that diffraction events in each column are independent; this is justified for the case of small Bragg angles in thin foils and introduces only small errors. In the second, we assume we have aligned the crystal so that only a single line of systematic Bragg beams ($\underline{h} = \underline{g}$, $2\underline{g}$ ..., $n\underline{g}$) is appreciably excited; this is done by rotating the crystal away from a pole of high symmetry containing $\underline{g}$ about $\underline{g}$ as an axis. The first approximation allows us to calculate the <u>projected</u> image of a defect from an assembly of independent parallelopipeds, in each of which the displacements or replacements due to the defect are constant over x,y and vary only with depth z. The second ensures that we are sampling the displacements or replacements along a single direction in the x,y planes, and so considerably simplifies image interpretation.

Displacement Contrast

Displacement contrast can usually be treated on the <u>deformable ion</u> model in which it is assumed that, for a displacement field $\underline{R}(\underline{r})$, the potential at point $\underline{r}$ in the imperfect crystal is the same as that at point $\underline{r} - \underline{R}(\underline{r})$ in the perfect crystal, i.e. the effect of $\underline{R}(\underline{r})$ is to shift the lattice potential $V(\underline{r})$ spatially. Thus

$$V^D(\underline{r}) = \sum_{\underline{g}} V_{\underline{g}} \exp 2\pi i \, \underline{g} \cdot (\underline{r} - \underline{R})$$

$$= \sum_{\underline{g}} V_{\underline{g}} \exp(-2\pi i \, \underline{g} \cdot \underline{R}) \exp 2\pi i \, \underline{g} \cdot \underline{r} \cdot \quad (40)$$

The displaced potential multiplies all the off-diagonal elements $A_{gh} = (me/h^2)V_{g-h}/\chi$ in the $\underset{\sim}{A}$ matrix (29) by the factor $\exp\{-2\pi i \, (\underline{g} - \underline{h}) \cdot \underline{R}\}$, producing a new matrix for the deformed crystal

$$\underset{\sim}{A}^D(z) = \underset{\sim}{Q}^{-1}(z) \underset{\sim}{A} \underset{\sim}{Q}(z) \quad (41)$$

where

$$\underset{\sim}{Q}(z) = \{\exp[-2\pi i \, (\underline{g} - \underline{h}) \cdot \underline{R}]\} \cdot \quad (42)$$

From the eigenvalue equation (28) we can write that in the deformed crystal

$$\underset{\sim}{A}^D(z) = \underset{\sim}{C}^D(z) \{\gamma^{(j)}\} [\underset{\sim}{C}^D(z)]^{-1} \quad (43)$$

and therefore that

$$\underset{\sim}{C}^D(z) = \underset{\sim}{Q}^{-1}(z) \underset{\sim}{C} \quad . \quad (44)$$

If the displacement $\underline{R}(\underline{r})$ introduces a sudden discontinuous shift in atom positions at some depth $z = z'$ in column (x',y'), thus dividing the crystal into two regions I and II displaced relative to each other, for example at a fault in the crystal (Figure 6), we can write using (31) that

$$\underset{\sim}{\Phi}(z = z') = \underset{\sim}{C}_I \{\exp 2\pi i \, \gamma^{(j)} z'\} \underset{\sim}{C}_I^{-1} \underset{\sim}{\Phi}(z = 0) \text{ in region I} \quad (45a)$$

$$\underset{\sim}{\Phi}(z = t) = \underset{\sim}{C}_{II} \{\exp 2\pi i \, \gamma^{(j)} (t - z')\} (\underset{\sim}{C}_{II})^{-1} \underset{\sim}{\Phi}(z = z')$$

$$= \underset{\sim}{Q}^{-1} \underset{\sim}{C}_I \{\exp 2\pi i \, \gamma^{(j)} (t - z')\} \underset{\sim}{C}_I^{-1} \underset{\sim}{Q}$$

$$\cdot \underset{\sim}{C}_I \{\exp 2\pi i \, \gamma^{(j)} z'\} \underset{\sim}{C}_I^{-1} \underset{\sim}{\Phi} \, (z = 0) \text{ in region II} \quad (45b)$$

because the diffracted beam amplitudes must be continuous across the fault. The term $\{\exp 2\pi i \, \gamma^{(j)}(t - z')\}$ means that inclined faults will exhibit sinusoidal fringes with a depth periodicity ξ_g (Figure 8) at the Bragg position just as the thickness fringes in (32); the effects of absorption can be similarly included, as in (39). If the fault is not inclined, there will nevertheless be a

TRANSMISSION ELECTRON MICROSCOPY OF DEFECT AGGREGATES 447

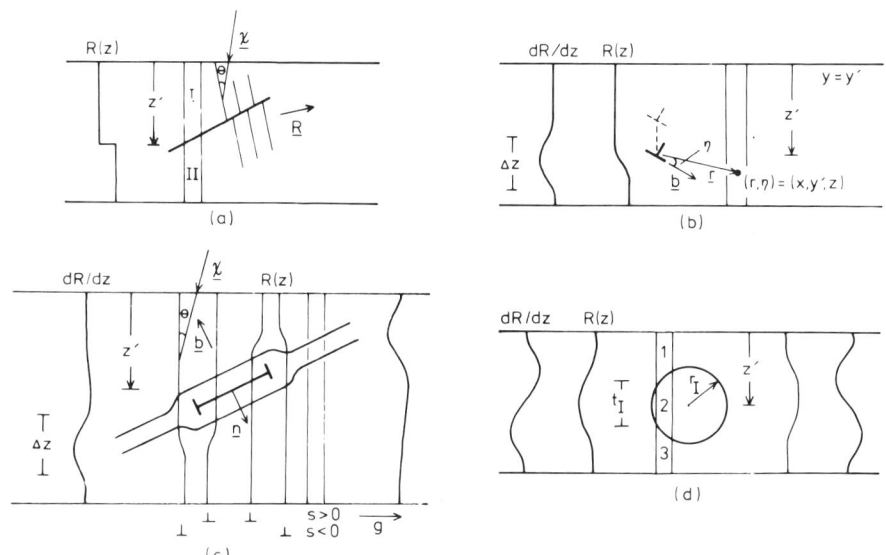

Figure 6. Defect strain and replacement fields. (a) fault, (b) dislocation, (c) large dislocation loop, (d) inclusion or small dislocation loop.

constant difference in intensity from the matrix crystal right across the fault. We note that contrast from the fault vanishes whenever $Q = \{\exp[-2\pi i\,(\underline{g} - \underline{h}) \cdot \underline{R}]\} = I$, either because $2\pi i\,(\underline{g} - \underline{h}) \cdot \underline{R} = 0, 2\pi, 4\pi, \ldots, 2n\pi$ for every $\underline{g}, \underline{h}$, which amounts to $\underline{R}$ being an integral lattice translation vector, or because having established systematic conditions $\underline{h} = \underline{g}, 2\underline{g}, \ldots, n\underline{g}$, $\underline{R}$ is normal to $\underline{g}$.

The effect of the fault on Bloch wave excitations $\psi^{(j)}$ is revealing, for while $\Phi(t')$ is continuous across the fault, $\Psi(z')$ is not. Using (45b) and inverting (31), we can write

$$\underset{\sim}{\Psi}_{II} = \{\exp(-2\pi i\,\gamma^{(j)}\,z')\}\,\underset{\sim}{C}_I^{-1}\,\underset{\sim}{Q}\,\underset{\sim}{C}_I\,\{\exp 2\pi i\,\gamma^{(j)}\,z'\}\,\underset{\sim}{\Psi}_I \quad (46)$$

from which we immediately see that the effect of the displacement $\underline{R}$, represented by Q, is to alter the Bloch wave excitations, i.e. the sudden displacement <u>scatters</u> the Bloch waves which propagate in perfect crystal without scattering. This is not altogether surprising, since for $2\pi\,\underline{g} \cdot \underline{R} = \pi$ (the case depicted in Figure 6), a Bloch wave (1) peaking between the atom positions will, upon encountering the fault, find itself transformed into another Bloch wave, say (j), which peaks <u>at</u> the atom positions.

If $\underline{R}$ varies slowly and continuously with z, for example, around a dislocation or misfitting inclusion (Figure 6) we can write

$$\underline{\Phi}(z + \delta z) = \underline{C}^D \{\exp 2\pi i \, \underline{\gamma}^{(j)} \, \delta z\} \, (\underline{C}^D)^{-1} \, \underline{\Phi}(z)$$

$$\simeq \underline{C}^D \{1 + 2\pi i \, \underline{\gamma}^{(j)} \, \delta z\} \, (\underline{C}_D)^{-1} \, \underline{\Phi}(z) \quad \text{for small } \delta z \quad (47)$$

$$\simeq \underline{\Phi}(z) + 2\pi i \, \underline{C}^D \{\underline{\gamma}^{(j)}\} \, (\underline{C}^D)^{-1} \, \underline{\Phi}(z) \, \delta z.$$

Utilising (43),

$$d\underline{\Phi}/dz = 2\pi i \, \underline{A}^D(z) \, \underline{\Phi}(z) = 2\pi i \, \underline{Q}^{-1}(z) \, \underline{A} \, \underline{Q}(z) \tag{48a}$$

which we can integrate through the crystal thickness t to obtain diffracted beam amplitudes $\Phi(t)$ at the bottom surface. Using (48a) and the definition of the $\tilde{A}$ matrix in (34), we can write for each Bragg beam amplitude ϕ_g

$$\partial \phi_g(\underline{r})/\partial z = \pi i \sum_h (1/\xi_h - i/\xi_h^i) \, \phi_{g-h}(\underline{r}) \, \exp[-2\pi i \, (s_h z + \underline{h} \cdot \underline{R})] \, . \tag{48b}$$

The form of (48a) is particularly convenient because it requires knowledge only of the displacement field $\underline{R}(z)$ and the A matrix for the perfect crystal which does not depend on z. In general, the integration must be performed numerically, and this does not tell us anything analytical about the nature of the contrast.

It is again more instructive to consider the change in Bloch wave excitations $\Delta\psi^{(j)}$, which we can obtain directly by defining a perturbation ΔV^D in the potential V due to the displacement field $\underline{R}$, which from (40) is

$$\Delta V^D = V^D - V = \{\exp(-2\pi i \, \underline{g} \cdot \underline{R}) - 1\} \exp 2\pi i \, \underline{g} \cdot \underline{r} \, . \tag{49}$$

Provided ΔV^D is small and localised over a small region $\alpha < z < \beta$ (Figure 6) – this is equivalent to requiring that the Bloch waves do not lose their identity – we can write that the perturbation $\Delta\psi^{(j)}$ is, to first order,

$$\Delta\psi^{(j)} = 2\pi i \, (me/h^2) \sum_1 \psi^{(1)}(z) \int_\alpha^\beta b^{(j)*} \, \Delta V^D \, b^{(1)} \, dz, \tag{50a}$$

or

$$\Delta\psi^{(j)} = \pi i \, (me/h^2) \sum_1 \psi^{(1)}(z) \sum_{g,h} C_h^{(j)*} \, C_{h-g}^{(1)} \, (V_g/k_z^{(j)})$$

$$\cdot \int_\alpha^\beta [\exp(-2\pi i \, \underline{g} \cdot \underline{R}) - 1] \exp 2\pi i \, z |k_z^{(1)} - k_z^{(j)}| \, dz \, . \tag{50b}$$

$\Delta\psi^{(j)}$ is a useful quantity because for strain centres in thick foils, it is in general possible to arrange that principally one Bloch wave only, say $b^{(j)}$, is well transmitted (for example, under two beam conditions $\underline{h} = 0$, $\underline{g}$ with $s_g \sim 0$). Under these conditions in bright field

$$\phi_o(z) \simeq \psi^{(j)} c_o^{(j)} \exp 2\pi i \gamma^{(j)} z = \psi^{(j)^2}(o) \exp 2\pi i \gamma^{(j)} z \; ;$$

$$\Delta\phi_o(z) \simeq \Delta\psi^{(j)} c_o^{(j)} \exp 2\pi i \gamma^{(j)} z = \Delta\psi^{(j)} \psi^{(j)}(o) \exp 2\pi i \gamma^{(j)} z. \tag{51}$$

The change in intensity due to Bloch wave scattering is then

$$\Delta|\phi_o(z)|^2 = |\phi_o(z) + \Delta\phi_o(z)|^2 - |\phi_o|^2$$

$$= \psi^{(j)^2}(o) \psi^{(j)*}(o) \Delta\psi^{(j)*} + \psi^{(j)^3}(o) \Delta\psi^{(j)} + \Delta\psi^{(j)^2}. \tag{52}$$

Neglecting the small term $\Delta\psi^{(j)^2}$, and since $\psi^{(j)}(o)$ is real for the perfect crystal,

$$\Delta|\phi_o(z)|^2 = \psi^{(j)^3}(o) (\Delta\psi^{(j)*} + \Delta\psi^{(j)}) = 2\psi^{(j)^3}(o) \text{Re} (\Delta\psi^{(j)}) \tag{53}$$

and $\text{Re}(\Delta\psi^{(j)})$ is a direct measure of the contrast.

It is often easier to envisage the strain field $d\underline{R}/dz$ than a displacement field $\underline{R}(z)$ because it represents bending of the lattice planes around the imperfection (Figure 6). Inverting and differentiating (31) and substituting (48), we obtain

$$d\underline{\Psi}/dz = 2\pi i \{\exp(-2\pi i \gamma^{(j)} z)\} \underline{C}^{-1} \{d(\underline{g} \cdot \underline{R})/dz\}$$

$$\cdot \underline{C} \{\exp 2\pi i \gamma^{(j)} z\} \underline{\Psi}. \tag{54}$$

Neglecting the small dilatation dg/dz, the term $d(\underline{g} \cdot \underline{R})/dz \simeq \underline{g} \cdot d\underline{R}/dz$ represents the projection upon $\underline{g}$ of the strain field due to the displacement $\underline{R}$. The change in each Bloch wave excitation may be written

$$d\psi^{(j)}/dz = 2\pi i \sum_1 \psi^{(1)} \sum_g c_g^{(j)*} c_g^{(1)} \{d(\underline{g} \cdot \underline{R})/dz\}$$

$$\cdot \exp 2\pi i |k_z^{(1)} - k_z^{(j)}| z \; . \tag{55}$$

Equations (54) and (55) illustrate the useful property that no contrast is produced if $\sum_g \underline{g} \cdot d\underline{R}/dz = 0$; in practice this condition can be achieved under systematic conditions $\underline{h} = \underline{g}, 2\underline{g}, \ldots, n\underline{g}$. All contrast thus vanishes wherever $\underline{R}$ is normal to $\underline{g}$. This property can be used to determine the direction of nearly uniaxial strain fields, e.g. a dislocation Burgers vector.

Equations (50) and (55) both represent the scattering into Bloch wave (j) arising from all other Bloch waves (1). The contribution to $\psi^{(j)}$ when $1 = j$ is called <u>intrabranch</u> scattering because it represents transitions from one dispersion surface to another point on the same dispersion surface (Figure 2); in this case the Bloch waves retain their identity essentially by following the curvature of the lattice plane bending. Intraband transitions give rise only to weak contrast because the change $\Delta k_z^{1',1} = |k_z^{(1')} - k_z^{(1)}|$ in the z component of the wave vector is in general small. Where the lattice bending becomes too severe, the Bloch waves cannot follow the lattice planes and <u>interbranch</u> scattering Δk_z^{1j} occurs with $1 \neq j$; such transitions are clearly more effective in producing contrast. Provided the scattering is again weak and localised, we can assume that the $\psi^{(1)}$ in (55) remain roughly constant ($\simeq \psi^{(1)}(o)$) and integrate (55) to obtain

$$\Delta\psi^{(j)}(t) = \psi^{(j)}(t) - \psi^{(j)}(o)$$
$$\simeq 2\pi i \sum_1 \psi^{(1)}(o) \sum_g C_g^{(j)*} C_g^{(1)} \int_0^t \underline{g} \cdot d\underline{R}/dz$$
$$\cdot \exp 2\pi i \Delta k_z^{1j} z \, dz. \qquad (56)$$

We see that the $\Delta\psi^{(j)}$ depend on the Fourier transform of the quantity $\underline{g} \cdot d\underline{R}/dz$, and for transitions to occur, $\underline{g} \cdot d\underline{R}/dz$ must have a Fourier component at the relevant wave vector Δk_z^{1j}. For interbranch transitions occurring under dynamical conditions ($s_g \sim 0$), $\Delta k^{1j} \simeq \xi_g^{-1}$, thus the part of the strain field sampled must have a a wavelength $\sim \xi_g$ and dynamical defect images will have dimensions of the same order.

To detect more rapidly varying strain fields, for example closer to dislocation cores or the details of very small defect aggregates, larger values of Δk_z^{1j} must be utilised. This is the basis of the <u>weak beam effect</u>. If a Bragg beam $\underline{g}$ is only weakly excited, i.e. $s_g \gg 0$, we can write

$$\phi_g(z) = \sum_j \psi^{(j)} C_g^{(j)} \exp 2\pi i \gamma^{(j)} z \ll 1 \qquad (57)$$

and therefore

$$\psi^{(j)} C_g^{(j)} \ll 1 \text{ for all } j.$$

A small change $\Delta\psi^{(j)}$ in Bloch wave excitations will then produce a change

$$|\Delta\phi_g| \simeq |\sum_j \Delta\psi^{(j)} c_g^{(j)} \exp 2\pi i \gamma^{(j)} z| \qquad (59)$$

in ϕ_g. In order to keep the image peak as narrow as possible, we must ensure that only one Bloch wave transition Δk_z^{1j} is effective in producing contrast; we can do so by ensuring that the transition is made from that Bloch wave, say (1), with the largest possible $\psi^{(1)}(o)$ to that Bloch wave, say (k), with the largest $c_g^{(k)}$; i.e. we require

$$|\psi^{(1)}(o)| \gg |\psi^{(j)}(o)| \quad \text{for all } j \neq 1$$

$$|c_g^{(k)}| \gg |c_g^{(j)}| \quad \text{for all } j \neq k \qquad (60)$$

If we number the branches of the dispersion surface in a special way (the numbering is arbitrary) such that at a given orientation the jth dispersion surface is closest to the sphere of radius $\chi = 1/\lambda$ centred on reciprocal lattice point j (Figure 7), then for $|s_j|$ large the C matrix (which from (28) is the matrix of eigenvectors of the A matrix) is close to diagonal, i.e. only the diagonal elements c_g^g are large. In such cases, perturbation theory gives the off-diagonal elements of $\underset{\sim}{C}$, to first order, as

$$c_n^m = A_n^m/(A_m^m - A_n^n) = 1/[2\xi_{m-n}(s_m - s_n)] \text{ for } m \neq n. \qquad (61)$$

We can thus rewrite the weak-beam image conditions (58) as

$$|\psi^{(1)}(o)| \gg |\psi^{(j)}(o)| = c_o^{(j)} = 1/[2\xi_j s_j] \text{ for all } j \neq 1 \qquad (62)$$

$$|c_g^{(g)}| \gg |c_g^{(j)}| = 1/[2\xi_{j-g}(s_j - s_g)] \text{ for all } j \neq g. \qquad (62)$$

These two conditions amount to the same thing, viz. establishing conditions such that all s_j are large and appreciably different,

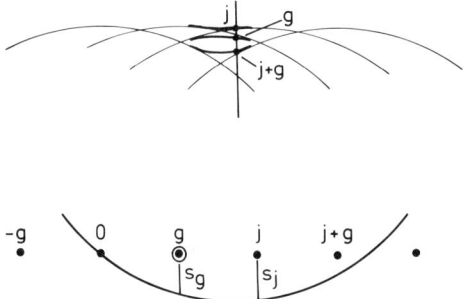

Figure 7. Ewald sphere orientation for weak beam condition g showing special numbering of the dispersion surfaces.

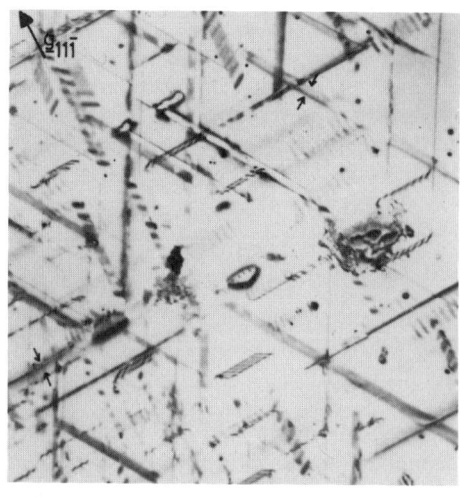 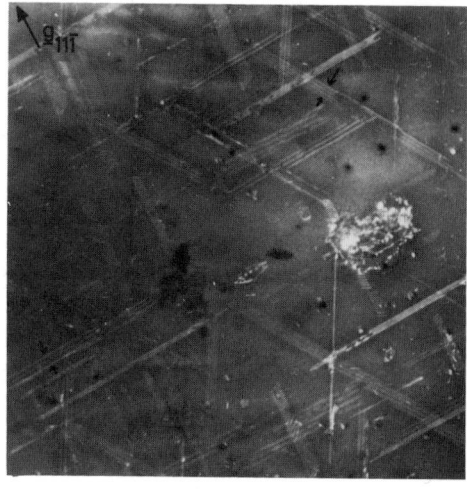

(a)   (b)

Figure 8. Planar faulted loops in irradiated silicon. (a) bright field, (b) weak beam dark field. Note closer fringe spacing in (b) corresponding to larger γ. (courtesy Dr. P. K. Madden).

since if $s_j = s_g$ for some j (Figure 7) violating (62b), then also $s_{j+g} = 0$, which violates (62a). We therefore carefully avoid satisfying any Bragg condition, while keeping s_g large and forming our image from Bragg beam $\underline{g}$.

Under conditions (62), we can rewrite (59) as

$$\Delta\phi_g = \left| \sum_j \Delta\psi^j \, c_g^{(j)} \, \exp 2\pi i \, \gamma^{(j)} z \right| \simeq \Delta\psi^{(g)} \tag{63}$$

for which we have derived an expression (50) already. Using (50b) and combining exponentials, we obtain

$$\Delta\phi_g \simeq \Delta\psi^{(g)} = \pi i \, (me/h^2) \, (V_g/k_z^{(g)}) \, \psi^{(1)}(o)^2 \, c_g^{(g)*}$$

$$\cdot \left[\int_o^t \exp\{-2\pi i \, (\underline{g} \cdot \underline{R} + \Delta k_z^{1g} \, z)\} dz - \int_o^t \exp 2\pi i \, \Delta k_z^{1g} \, z \, dz \right].$$

Maximum contrast occurs when the first integral is maximised, the condition for which is $(\underline{g} \cdot \underline{R} + \Delta k_z^{1g} \, z)$ = a constant integer. Therefore,

$$\frac{d}{dz}(\underline{g} \cdot \underline{R} + \Delta k_z^{1g} \, z) = \underline{g} \cdot d\underline{R}/dz + \Delta k_z^{1g} = 0 \tag{65}$$

and the weak beam peak occurs in that column for which $\underline{g} \cdot d\underline{R}/dz =$

TRANSMISSION ELECTRON MICROSCOPY OF DEFECT AGGREGATES 453

$-\Delta k_z^{-1}g$ at a turning point of $\underline{g} \cdot dR/dz$, since $\underline{g} \cdot d^2R/dz^2 = 0$ in (65). The width of the peak will be of order $0.3 \Delta k_z^{-1}j$ (in the above treatment, it is limited to a column width in the column approximation) and the position can be brought arbitrarily close to the actual defect with increasing values of s_g (Figure 8). However, as the intensity in ϕ_g decreases with increasing s_g, a compromise must be reached, which occurs for $\xi_g s_g \sim 5$.

Replacement Contrast

Replacing some part of the crystal with atoms of different scattering factor or different crystal structure, e.g. a precipitate, inclusion or cavity, results in regions of different potential characterised by new Fourier coefficients $\{V_g^R\}$. These in turn define a new matrix.

$$\underline{A}^R(z) = \underline{A} + \Delta \underline{A}(z) . \tag{66}$$

The Bloch wave excitations may therefore vary with z down a column, so

$$\underline{\Phi} = \underline{C} \{\exp 2\pi i \gamma^{(j)} z\} \underline{\Psi}(z) . \tag{67}$$

In exactly analogous fashion to the case of a deformed crystal, we can write

$$d\underline{\Phi}/dz = 2\pi i \underline{A}^R(z) \underline{\Phi} = 2\pi i \underline{A}^R(z) \underline{C} \{\exp 2\pi i \gamma^{(j)} z\} \underline{\Psi}(z) . \tag{68}$$

Differentiating (67) and setting the result equal to (68), we obtain

$$d\underline{\Psi}(z)/dz = 2\pi i \{\exp(-2\pi i \gamma^{(j)} z)\} \underline{C}^{-1} \Delta \underline{A} \, \underline{C} \{\exp 2\pi i \gamma^{(j)} z\} \underline{\Psi}(z) \tag{69}$$

and we see that $\Delta \underline{A}$ plays the same role as $\underline{g} \cdot dR/dz$ does in a deformed crystal. This is sometimes called <u>structure factor</u> contrast.

If the inclusion is <u>coherent</u> with the original lattice and small compared to the foil thickness, there will be variations only in the off-diagonal elements of $\underline{A}$, <u>viz</u>. the terms of $\Delta \underline{A}$ will be

$$\begin{aligned}\Delta A_{oo} &= \Delta A_{gg} = 0 \\ \Delta A_{gh} &= (me/h^2)[(V_{g-h}^R - V_{g-h})/\chi] \\ &= \tfrac{1}{2} (1/\xi_{g-h}^R - 1/\xi_{g-h})\end{aligned} \tag{70}$$

If the inclusion is sufficiently thick, absorption must be included, and the diagonal terms ΔA_{gg} will be non-zero,

$$\Delta A_{gg} = (i/2)(1/\xi_o^{Ri} - 1/\xi_o^i) \tag{71}$$

and there will be corresponding changes in the non-diagonal terms ΔA_{gh}.

In the presence of both replacement and displacement, we can write

$$\underset{\sim}{A}^{R,D}(z) = \underset{\sim}{Q}^{-1}[\underset{\sim}{A}^R(z)]\underset{\sim}{Q} = \underset{\sim}{Q}^{-1}[\underset{\sim}{A} + \underset{\sim}{\Delta A}(z)]\underset{\sim}{Q} \tag{72}$$

from which we derive

$$d\underset{\sim}{\Psi}(z)/dz = 2\pi i \{\exp(-2\pi i \underset{\sim}{\gamma}^{(j)}z)\}\underset{\sim}{C}^{-1}[\underset{\sim}{\Delta A} + \{\underset{\sim}{g}\cdot d\underset{\sim}{R}/dz\}]\underset{\sim}{C}$$
$$\cdot \{\exp 2\pi i \underset{\sim}{\gamma}^{(j)}z\}\underset{\sim}{\Psi}(z). \tag{73}$$

Fresnel Contrast

Equations (48) and (68) hold for the image, as well as the transmission function, strictly for the case of Fraunhofer diffraction (images in focus). If we defocus the objective lens slightly (either over- or under-focus, typically by an amount $\zeta \sim 1$ μm), the different phases of the diffracted waves at the bottom are allowed to interfere; we must therefore use the total wave function $\Psi(\underline{r})$, which includes phase, in place of just the amplitude $\phi(\underline{r})$ in (1), in which case (3) becomes

$$\Psi'(\underline{r}',\zeta) = \int_{\text{all } \underline{K}} \tilde{\Psi}(\underline{K}) \exp(-2\pi i \underline{K}\cdot \underline{r}')\exp[-2\pi i\zeta \sqrt{(\chi^2-K^2)}]d^2\underline{K}. \tag{74}$$

Since $K \ll \chi$, $\sqrt{(\chi^2 - K^2)} \simeq \chi - K^2/2\chi$, and (66) becomes

$$\Psi'(\underline{r}',\zeta) = \exp(-2\pi i \chi\zeta)\int_{\text{all }\underline{K}} \tilde{\Psi}(\underline{K})$$
$$\cdot \exp\{-2\pi i[\underline{K}\cdot\underline{r}' - (K^2/2\chi)\zeta]\}d^2\underline{K}. \tag{75}$$

The effect of the defocus ζ is thus to enhance the image with one or more Fresnel fringes.

IMAGES OF DEFECT AGGREGATES

Point defects such as interstitial atoms or ions, vacant lattice sites, etc. can, at temperatures where they are mobile or at high density, interact through their elastic or electrostatic fields to form large aggregates; aggregation in general lowers the total energy stored in lattice defects. Defect contrast theory enables us to predict contrast images from the planar and three-dimensional defect aggregates commonly encountered in non-metallic solids, viz. dislocation loops and inclusions (either strained or unstrained) arising from phase separation or voidage.

Dislocation Loops

For simplicity, we first consider the case of a monatomic solid; interstitials or vacancies are thus no different from normal lattice atoms. If they aggregate in a planar array (Figure 6) a section of new lattice plane is formed which terminates at the edges in a bounding dislocation loop with Burgers vector $\underline{b}$. If we define a unit vector $\underline{n}$ normal to the loop such that $\underline{n} \cdot \underline{z} \geq 0$, and a unit vector $\underline{\beta} = \underline{b}/b$ along the Burgers vector, we define the sense of $\underline{b}$ so that

$$\underline{n} \cdot \underline{\beta} > 0 \quad \text{for a vacancy loop}$$
$$\underline{n} \cdot \underline{\beta} < 0 \quad \text{for an interstitial loop.} \tag{76}$$

Then $\underline{n} \cdot \underline{\beta} = \pm 1$ defines a pure edge dislocation loop and $\underline{n} \cdot \underline{\beta} = 0$ a loop of pure shear. If $\underline{b}$ is not an integral unit cell translation vector, the loop is <u>faulted</u>; for example, the additional or missing atom plane may alter the normal stacking sequence of planes. Inclined faulted loops exhibit fault fringes (Figure 8) with depth periodicity ξ_g for $s_g \sim 0$ due to the term $\{\exp 2\pi i \gamma^{(j)} (t - z')\}$ in (45); uninclined faulted loops exhibit a more or less uniform intensity across the loop differing from background and depending on loop depth, in both cases provided that $\underline{g} \cdot \underline{R} \neq 2n\pi$ ($n = 0, 1, 2, \ldots$) at the fault.

A segment of <u>large</u> dislocation loop can be treated as an isolated straight dislocation lying along direction $\underline{u}$ in the $x = 0$ plane and inclined at an angle υ to the foil plane (Figure 6). A point $\underline{r}$ (x, y', z) in the column (x, y') can then be defined as point (r, η) in a plane normal to the dislocation line. The slip plane (containing $\underline{b}$ and $\underline{u}$) intersects this plane in a line making an angle ω with respect to the intersection of the foil plane. If z' is the depth of the dislocation in the $y = y'$ plane,

$$\eta = \tan^{-1} [\cos \upsilon \ (z-z')/x] - \omega . \tag{77}$$

The displacement field $\underline{R}$ $(x, y', z) = \underline{R}$ (r, η) in an isotropic medium is thus given by

$$\underline{R} = \frac{b}{2\pi} \{ (\underline{\beta} \cdot \underline{u})\underline{u}\eta + (\underline{u} \wedge \underline{\beta} \wedge \underline{u})[\eta + \frac{\sin 2\eta}{4(1-\nu)}]$$

$$+ (\underline{\beta} \wedge \underline{u})[\frac{(1-2\nu)}{2(1-\nu)} \ln |r| + \frac{\cos 2\eta}{4(1-\nu)}]\} \; . \quad (78)$$

The first term in (78) is along the direction of the screw component $\underline{b}_s$ of the Burgers vector, the second along the edge component $\underline{b}_e$, the third normal to the slip plane.

From the first term in (78), a screw dislocation lying in the foil plane will have

$$\underline{g} \cdot d\underline{R}/dz = (\underline{g} \cdot \underline{b}/2\pi) \, x/r \quad (79)$$

which at large distance r means $\underline{g} \cdot d\underline{R}/dz \propto x^{-1}$. Due to elastic interactions across the loop, the strain field from small loops falls off more quickly than this. For strong interbranch scattering, the 'width' of the $\underline{g} \cdot d\underline{R}/dz$ peak down a column (Figure 6) must be $\Delta k^{-1} \simeq \xi_g$ at s = 0. The width of $\underline{g} \cdot d\underline{R}/dz$ at, say, 20% maximum is, from (79), $\Delta z = 4x \simeq \xi_g$. Therefore the overall image width Δx at s = 0 is $\sim \xi_g/2$. The image intensity will be asymmetric across the dislocation ('black/white') because $\underline{g} \cdot d\underline{R}/dz$ is an odd function of x for all z.

From the second term of (78), we see that an edge dislocation image will have width larger by about a factor $(3-2\nu)/2(1-\nu)$ than a screw dislocation; since $\nu \sim 0.2$, this means a factor of 2. We

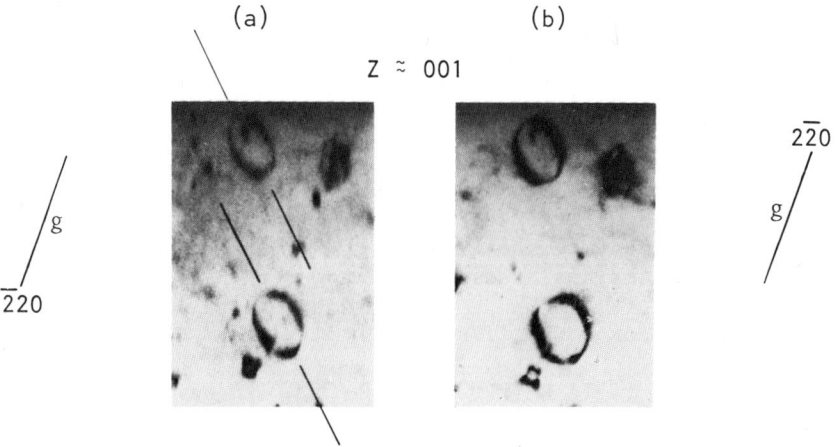

Figure 9. Loop normal determination for inclined dislocation loops produced by irradiation of KI. Note image behaviour in ± g.

similarly note that the image from a dislocation with $\underline{b}$ inclined to the foil will appear wider than one with $\underline{b}$ in the foil plane. We further note from (77) that the width of inclined segments of a loop will be narrowed by the factor cos $\cup$ (Figure 9). The loop normal n for an inclined loop may be determined either by tilting the foil until the loop is edge-on and determining the foil orientation (from Kikuchi lines) or, for a circular loop, by plotting the directions of the major axis of the projected ellipse for two different foil orientations (Figure 9); $\underline{n}$ must be normal to each of the major axis projections. For $|s| > 0$, the image will also be narrowed and, if an edge component is present, will lie according to (48b) on that side of the dislocation for which dR/dz decreases $|s|$. For a loop (Figure 6), provided $\underline{n} \cdot \underline{b}$ and $\underline{n} \cdot \underline{z}$ are not too small (i.e. the loop is substantially edge and not too steeply inclined), the image of an interstitial loop will from (76) lie inside the actual loop for $(\underline{g} \cdot \underline{b}) s_g > 0$, outside for $(\underline{g} \cdot \underline{b}) s_g < 0$ and conversely for a vacancy loop; confusion can arise only for small $|\underline{n} \cdot \underline{b}|$ and $\underline{n} \cdot \underline{z}$. By this means we can ascertain whether a loop arises from aggregation of interstitials or vacancies.

The direction of $\underline{b}$ may be determined from the property (55) that contrast vanishes where $\sum_g \underline{g} \cdot \underline{R} = 0$, provided that only systematic beams $n\underline{g}$ are excited and (from (79)) that a condition can be found where $\underline{g} \cdot \underline{b} = 0$ and $\underline{g} \cdot \underline{b} \wedge \underline{u} = 0$. Except for pure shear loops, the last two conditions cannot be satisfied simultaneously all around the loop; in general, one must decide whether $\underline{g} \cdot \underline{b}$ or $\underline{g} \cdot \underline{b} \wedge \underline{u}$ contrast is primarily observed. For loops with $\underline{b} \cdot \underline{z} = 1$, $\underline{g} \cdot \underline{b} = 0$ for all $\underline{g}$, but contrast from $\underline{g} \cdot \underline{b} \wedge \underline{u}$ can be quite strong (Figure 10) provided $\underline{g} \cdot \underline{b} \wedge \underline{u} > 0.5$; since the displacement $\underline{b} \wedge \underline{u}$ is radial for a loop, $\underline{g} \cdot \underline{b} \wedge \underline{u}$ contrast will vanish

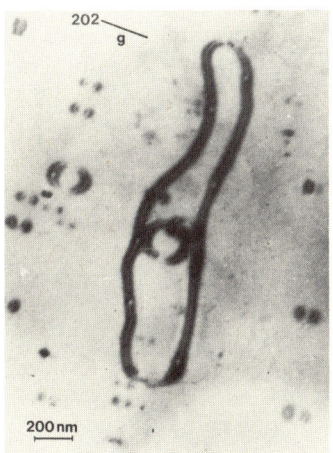

Figure 10. Dislocation loops in KI with Burgers vector $\underline{b}$ normal to the foil plane, exhibiting $\underline{g} \cdot \underline{b} \wedge \underline{u}$ contrast.

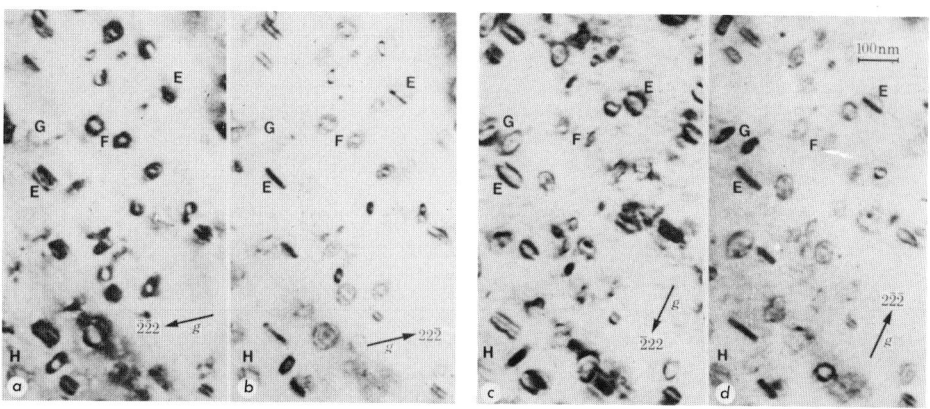

Figure 11. Burgers vector determination for loops in KI. Loops F have $\underline{g}\cdot\underline{b} = 2$ in (a) and (b), $\underline{g}\cdot\underline{b} = 0$ in (c) and (d).

wherever the projection of $\underline{u}$ is parallel to $\underline{g}$. For loops with a shear component, the term $\sin \eta/4(1-\nu)$ along $\underline{b}_e$ in (79) will also contribute contrast for inclined segments of the loop. These three sorts of contrast can be distinguished by imaging in $\pm \underline{g}$ (Figure 11), since only $\underline{g}\cdot\underline{b}$ contrast will show inside/outside image behaviour.

It is also possible to say something about the magnitude of b from the symmetry of the images across the dislocation. From ($\overline{48}$b), the image at the position of the dislocation core depends essentially on the value of $\exp 2\pi i (s_g z + \underline{g}\cdot\underline{R})$. At the dislocation core, $\eta = 0$ and $\underline{R} = 0$ (neglecting the $\underline{g}\cdot\underline{b} \wedge \underline{u}$ term) immediately above the dislocation, and $\eta = \pi$ and $\underline{R} = \underline{b}/2$ immediately below; therefore the dislocation introduces a fault with phase change $2\pi \underline{g} \cdot \underline{R} = (\underline{g}\cdot\underline{b})\pi$ in those columns passing through the core. For $s \sim 0$ and $\underline{g}\cdot\underline{b} = 1$, there will be fault contrast at the core, while for $\underline{g}\cdot\underline{b} = 2$ there is no fault contrast at the core, but there will be image intensity on either side (i.e. image doubling, Figure 11). Images deriving from $\underline{g}\cdot\underline{b} \wedge \underline{u}$ contrast when $\underline{b}\cdot\underline{z} = 1$ are similarly doubled, but for different reasons. In this case, $\omega = \pi/2$ and the strain field is given by

$$d\underline{R}/dz = [1/2(1-\nu)]\{\frac{(1-2\nu)(z-z')}{r^2} + \frac{2(z-z')^3}{r^4}\} \tag{80}$$

which is an odd function of $(z-z')$ but an even function of x; the image will therefore appear symmetric across the core. The displacements in the $(x = 0)$ column are

$$\underline{R}(z) = [\underline{b} \wedge \underline{u}/4\pi(1-\nu)][(1-2\nu) \ln |z - z'| + 1/2] \quad (81)$$

which vary too slowly with z to produce appreciable interbranch scattering, so there will be an intensity minimum at the core and the image will appear doubled (Figure 10).

Unstrained Inclusions

Voids or cavities formed from vacancy aggregation are examples of relatively unstrained inclusions. In a diatomic or multiatomic solid, there is the additional possibility that only one vacancy species may aggregate, and this can result in an inclusion of a second phase, which may be unstrained. Aggregation of well-fitting substitutional defects can produce a similar result.

In such cases we consider a matrix M of thickness t containing an inclusion I of thickness t_I characterised by matrices $\underset{\sim}{A}_M$ and $\underset{\sim}{A}_I$ which reflect the respective extinction distances ξ_g, absorption lengths ξ_g^i and orientations s_g of matrix and inclusion. We calculate the Bragg amplitudes $\phi_g(z = t)$ in a column passing through the inclusion (Figure 6); the foil is thus divided into regions i (i = 1,2,3). Since ϕ_g must be continuous across the interfaces, as in the case of a stacking fault,

$$\underset{\sim}{\Phi}_g (z = t) = \prod_{i=3}^{1} \underset{\sim}{C}_i \{\exp 2\pi i \, \gamma_i^{(j)} \, t_i\} \underset{\sim}{C}_i^{-1} \underset{\sim}{\Phi}(z = 0). \quad (82)$$

$\underset{\sim}{C}_i$ and $\{\gamma_i^{(j)}\}$ are derived from the eigenvalue relation (28) using $\underset{\sim}{\tilde{A}}_M$ for i = 1,3 and $\underset{\sim}{\tilde{A}}_I$ for i = 2. If the inclusion is <u>coherent</u> with the matrix (or void) and we restrict our consideration to the two-beam case $\underline{h} = 0$, $\underline{g}$ with $s_g \sim 0$, $\underset{\sim}{C}_i = \underset{\sim}{C}_n = \underset{\sim}{C}$ for all i, and (82) reduces to

$$\underset{\sim}{\Phi}_g (z = t) = \underset{\sim}{C} \{\exp 2\pi i \sum_{i=1}^{3} \gamma_i^{(j)} \, t_i\} \underset{\sim}{C}^{-1} \underset{\sim}{\Phi}(z = 0). \quad (83)$$

The corresponding intensities are

$$I_g (z = t) = \sum_{j,l} C_o^{(j)} C_g^{(j)} C_o^{(1)} C_g^{(1)*} \exp 2\pi i \, (\sum_{i=1}^{3} \Delta\gamma_i^{jl} \, t_i) \quad (84)$$

which from (32) are identical to $I_g (z = t')$ for a perfect crystal of thickness t' such that

$$\Delta\gamma_M^{jl} \, t' = \Delta\gamma_M^{jl} (t_1 + t_3) + \Delta\gamma_I^{jl} \, t_2 . \quad (85)$$

In the two-beam case at s = 0, $\Delta\gamma = \xi_g^{-1}$, so

$$t' = t + t_I [(\xi_{gM}/\xi_{gI}) - 1] . \quad (86)$$

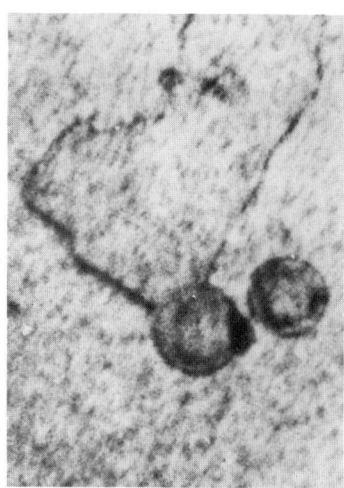

Figure 12. Spherical potassium metal inclusions in KCl precipitating on dislocations, exhibiting concentric thickness fringes and weak parallel Moiré fringes normal to $\underline{g}$.

The inclusion therefore acts to increase or decrease the effective foil thickness and thus the intensity I_g. A special case is a void or cavity, for which $\xi_{gI} \propto \chi/V_{gI} = \infty$, so $t' = t - t_I$. Thus an inclusion (even a void) may appear darker or lighter than background, depending on inclusion and foil thickness.

If the inclusion varies in thickness with (x,y), the result will be thickness fringes, just as in the case of a wedge crystal; for a sphere of radius r_I (Figure 6), $t_I = 2\sqrt{(r_I^2 - x^2)}$, and the fringes will be concentric circles (Figure 12) provided $t_I > \xi_{gM}\xi_{gI}/(\xi_{gM}/\xi_{gI})$. If ξ_{gM} and ξ_{gI} are not too different, the C_i will not be very different, and the result (79) will also hold for $s \neq 0$; the effect of increasing s is to increase $\Delta\gamma^{j1}$ and thus the number of fringes. Similarly if $\xi^i_{gM} \sim \xi^i_{gI}$, the result (39) holds with $z = t'$, and the fringes will be damped out with increasing inclusion thickness. For large $|s|$, diffraction contrast will wash out, but differences in mean absorption represented by $(i/\xi^i_{oM} - i/\xi^i_{oI})$ will produce absorption contrast.

Strained Inclusions

An inclusion such as a gas bubble or coherent precipitate may normally occupy a volume v_I^* more or less than the lattice volume v_M it replaces (the superscript * refers to the inclusion in an unstrained state). The volume fraction V^* by which the matrix must be increased to accommodate the inclusion, with the inclusion in a strain-free state is then $V^* = (v_I^* - v_M)/v_M$; for example for a coherent inclusion with a lattice parameter misfit $\delta^* = (a_I^* - a_M)/a_M$, $v^* = 3\delta^*$. We can replace inclusion I^* by another inclusion I having the same elastic properties as the matrix such that identical elastic strain fields are sustained when both inclusions are embedded in the matrix; the volume change associated with I is

$$V = \frac{B_I^* V^*}{\alpha B_I^* + 4 G_M \alpha/3} \tag{87}$$

where $\alpha = (1 + v_M)/3(1-v_M)$, v_M is Poisson's ratio, B_I the bulk modulus of the inclusion and G_M the shear modulus of the matrix. For a spherical inclusion of radius r_I, the displacements are purely radial

$$\underline{R}(\underline{r}) = -\alpha V \underline{r}/3 \quad \text{for } r \le r_I$$
$$\underline{R}(\underline{r}) = -\alpha V r_I^3 \underline{r}/3r^3 \quad \text{for } r > r_I. \tag{88}$$

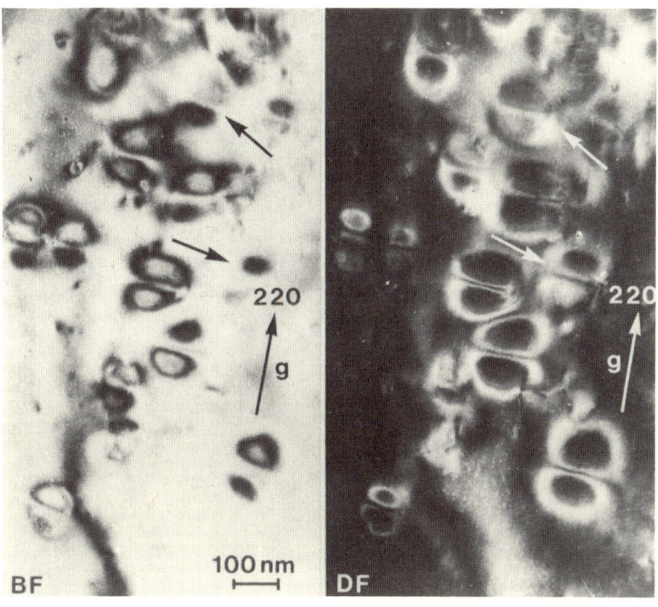

Figure 13. Highly strained inclusions in heavily irradiated KCl in bright and dark field. Anomalous contrast of arrowed inclusions allows the sign of their strain to be determined.

Inside the inclusion, the strain field is constant but outside falls off are r^{-3},

$$dR(r)/dr = \varepsilon_I = -\alpha V/3 \quad \text{for } r \leq r_I$$
$$dR(r)/dr = 2 \alpha V r_I^3 \underline{r}/3r^3 = 2 \varepsilon r_I^3 \underline{r}/r^4 \quad \text{for } r > r_I . \quad (89)$$

$\varepsilon = -\varepsilon_I$ is called the accommodation strain and is positive for the matrix in compression; from (87) and (89)

$$\varepsilon = \frac{B_I V^*}{3B_I + 4G_M} . \quad (90)$$

Since the strain is radial, there will always be a line of zero contrast normal to $\underline{g}$ for systematic $n\underline{g}$ excited, leading to a characteristic 'double loop' image (Figure 13).

The quantity $\underline{g} \cdot dR/dz$ for a column $(x, y = 0)$ outside the inclusion (x along $\underline{g}$, centre of the inclusion at ($x = 0$, $y = 0$, $z = z'$)) is

$$\underline{g} \cdot dR/dz = -2\varepsilon g \, r_I^3 \, x \, (z - z')/r^5 \quad (91)$$

where $r = \sqrt{[x^2 + (z - z')^2]}$. For the two-beam case $h_- = 0, \underline{g}$ and $\underline{g} \cdot \underline{R}$ small (weak scattering), (56) becomes

$$\Delta\psi^{(j)} = 2\pi i \sum_1 \psi^{(1)}(o) \, C_g^{(j)*} \, C_g^{(1)} \int_o^t (\underline{g} \cdot d\underline{R}/dz) \exp(2\pi i \, z/\xi_g) dz . \quad (92)$$

For defects far from the foil surfaces, the width of the image where scattering is weak depends on the term

$$\Delta = 2\pi i \int_{-\infty}^{\infty} (\underline{g} \cdot d\underline{R}/dz) \exp(2\pi i \, z'/\xi_g) \exp[2\pi i \, (z-z')/\xi_g] d(z-z'); \quad (93)$$

substituting (91) for columns $(x, y = 0)$ yields

$$\Delta = \exp(2\pi i \, z'/\xi_g)(32\pi^3/3)(\varepsilon g \, r_I^3/\xi_g^2)[\frac{x}{|x|} K_1(2\pi|x|/\xi_g)] \quad (94)$$

where K_1 is a modified Bessel function. For constant values of Δ (say 10% image contrast), (94) gives a relation between the overall image width $2x/\xi_g$ and the quantity $(\varepsilon g \, r_I^3/\xi_g^2)$; we can thus measure ε provided g, ξ_g and r_I are known. The size of the inclusion $2r_I$ can be estimated approximately from the fact that, since the strain in the inclusion is constant, the image contours are straight inside

TRANSMISSION ELECTRON MICROSCOPY OF DEFECT AGGREGATES 463

the inclusion and deviate abruptly at the perimeter.

If the lattice parameter of large inclusions is very close to that of the matrix, there may be insufficient misfit to produce a contrast width greater than inclusion diameter. There may, however, be Moiré interference fringes (Figure 12) due to the difference in spacing of the relevant diffracting lattice planes in the constrained inclusion, g_I^{-1} (not g_I^{-1*}), and unaffected distant matrix, g_M^{-1}. The fringe spacing will be $\lambda = g_M^{-1} g_I^{-1}/(g_M^{-1} - g_I^{-1})$, which using the fact that $V = (a_I - a_M)/a_M$ and $\varepsilon = \alpha V/3$ can be rewritten

$$\lambda = \alpha/\varepsilon g . \tag{95}$$

In e.g. the alkali halides, $\nu_M = 0.2$, so $\alpha = 0.5$ and $\lambda = 1/2\varepsilon g$.

Small Defect Aggregates

Small defects present special problems both of visibility and interpretation. Regardless of whether Bloch wave scattering arises from $\Delta \underline{A}$ or $g.d\underline{R}/dz$ terms in (73), the scattering will be small and we seek to maximise its effect. For example, (86) shows that a small unstrained inclusion of thickness t_I changes the effective foil thickness by an amount $\Delta t = t_I \xi_{gM} (1/\xi_{gI} - 1/\xi_{gM}) \ll \xi_{gM}$ if $t_I \ll \xi_{gM}$. The largest contrast will be obtained where ϕ_g changes most rapidly with thickness, viz. at the edges of thickness fringes (Figure 4) where $t = (2n + 1)\overline{\xi}_g/4$, $(n = 0,1,2,...)$, for $s_g = 0$ in bright field. As the foil gets thicker, absorption damps out the thickness fringes, consequently small unstrained inclusions are visible only in very thin crystal regions. Low index g are more effective in imaging than larger g, since ξ_g is smaller and $(1/\xi_{gI} - 1/\xi_{gM})$ larger; likewise lower energy electrons, since

$\xi_g = (me/h^2) \chi/V_g \propto \sqrt{U}$. If $s_g > 0$, inclusion visibility becomes depth dependent, alternately visible and invisible at depth multiples of $\xi_g/4$. The visibility of small unstrained inclusions can be considerably enhanced by out-of-focus imaging due to the Fresnel interference fringe produced (Figure 14).

Somewhat different criteria hold for very small strain-producing defects. For a very small pure edge loop, for example, with radius r_1 situated at depth z with $\underline{b}$ along the x direction, the quantity $\underline{g}.d\underline{R}/dz$ in the y = 0 plane is

$$\underline{g}.d\underline{R}/dz = \frac{3 \underline{g}.\underline{b} r_1^2}{8(1-\nu)} [\frac{(1-2\nu) x (z-z')}{r^5} + \frac{5 x^3 (z-z')}{r^7}] . \tag{96}$$

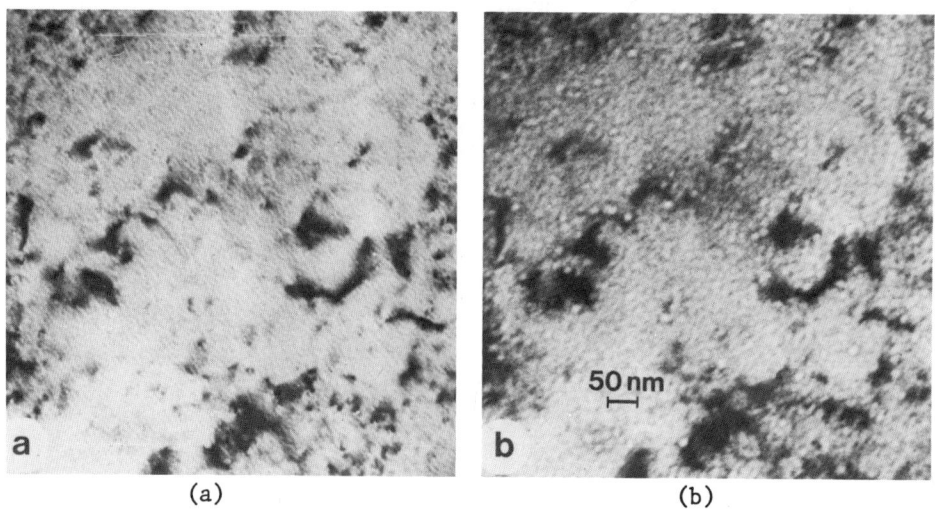

Figure 14. Unstrained inclusions (probably sodium metal) in heavily irradiated NaCl imaged (a) in focus, (b) under focussed.

The integral Δ in (93) becomes

$$\Delta = \exp(2\pi i\, z'/\xi_g)\, \{\pi^3/(1-\nu)\}\, (r_1^2\, \underline{g}\cdot\underline{b}/\xi_g^2)$$
$$\cdot\, [\{2(1-2\nu)\frac{x}{|x|} + 15\frac{x^3}{|x|^3}\}\, K_1(2\pi\,|x|/\xi_g)]. \quad (97)$$

This is qualitatively similar to Δ for a small inclusion derived in (94) with $r_1^2\, \underline{g}\cdot\underline{b}/\xi_g^2$ replacing $\varepsilon g\, r_I^3/\xi_g^2$. Since in either case the strain field diminishes rapidly with decreasing defect size, for small defects ($\sim 0.1\, \xi_g$) appreciable interbranch scattering only occurs well within a quarter period of $\exp(2\pi i\, z'/\xi_g)$, and we can use Δ to ascertain the image details right across the defect. The quantity in square brackets [ ] in (94) or (97) in an odd function of x; therefore the defect image is asymmetric ('black/white') across the defect (Figure 15). For a strained inclusion, (94), the direction of the black/white asymmetry depends only on $\underline{g}$ and the sense on the direction of $\underline{g}$ and the sign of ε; for a small loop, (97), the direction depends on both $\underline{g}$ and $\underline{b}$ and the sense on $\underline{g},\underline{b}$ and whether the loop is vacancy or interstitial.

Since, from (53), $\Delta I_o \propto \mathrm{Re}(\Delta\psi^{(j)}) \propto \mathrm{Re}(\Delta)$, we see that $\mathrm{Re}(\Delta)$ arises from the $\cos(2\pi i\, z'/\xi_g)$ part of the term $\exp(2\pi i\, z'/\xi_g)$ in

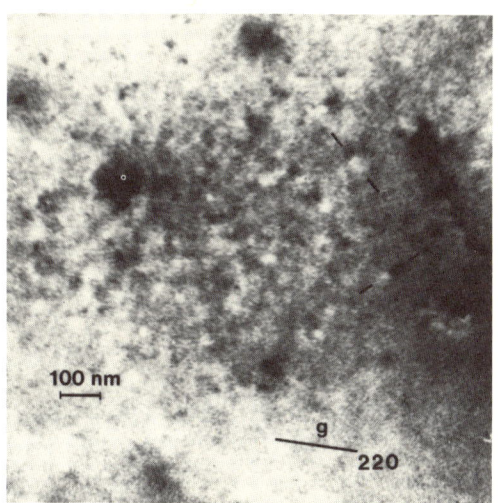

Figure 15. 'Black/white' contrast of small interstitial dislocation loops is irradiated KCl. The direction of black/white streaking is approximately along the projection of **b**.

(94) or (97). The sense of black/white asymmetry will thus be the same for all defects of the same kind in a layer $0 < z' < \xi_g/4$, but will reverse for $\xi_g/4 < z' < 3\xi_g/4$, etc. Hence the contrast can be used to determine the sign of the defect strain, provided the depth of the defect is known. This depth can be determined from stereoscopy measurements, maintaining the same diffraction conditions but rotating the specimen about **g** as an axis. Since the black/white asymmetry depends on interbranch Bloch wave scattering, absorption will damp out this feature for defect depths larger than about $3\xi_g/2$ from either foil surface.

For large inclusions ($\varepsilon_g\, r_I^3/\xi_g^2 > 0.2$) or large loops ($\underline{g}\cdot\underline{b}\, r_1^2/\xi_g^2 > 1$) lying within $\sim\xi_g$ of the foil surfaces, surface relaxations alter the sense of the black/white asymmetry in the second layer ($\xi_g/4 < z' < 3\xi_g/4$), so that for $0 < z' < 5\xi_g/4$ the sense of the black/white asymmetry in bright field will always be the same and depend only on the sign of the strain and the foil surface (top or bottom) which is closest to the defect; in dark field, images at top and bottom of the foil are always similar, so the asymmetry then depends uniquely on the sign of the strain. This property enables the nature of large inclusions and loops to be ascertained (Figure 13).

DEFECT AGGREGATE BEHAVIOUR AT HIGH DEFECT DENSITY

In the appropriate regimes of defect density and temperature, point defects interact with each other to form large aggregates, which in turn can interact with other aggregates. Electron microscopy reveals several basic features of defect stabilisation in these regimes.

Dislocation Loops

The driving force for condensation into dislocation loops is the point defect elastic energy which can be lowered by aggregation. The elastic energy of a loop of radius r_1 is

$$U_1 = \frac{G b^2 r_1}{2(1-\nu)} \ln\{\frac{r_1}{b} + \frac{5}{3}\} ; \qquad (98)$$

for loops of moderate size ($r_1 \sim 10b$), $U_1 \sim \pi G b^2 r_1$. A faulted loop has an additional contribution $U_f = \pi r_1^2 \gamma_f$ where γ_f is the fault energy per unit area. Since the number of defects contained in the loop is of order $N \simeq \pi r_1^2/b^2$, the elastic energy per defect

$$U_1/N \simeq G b^4 / r_1 \qquad (99)$$

decreases continuously as loop radius increases. Large loops or isolated dislocations are thus particularly good sinks for interstitials or vacancies. For a given number of condensed defects, a large dislocation loop has minimum perimeter and thus energy in an isotropic medium when it is pure edge, but interactions across smaller loops and anisotropy can alter the orientation for minimum energy substantially. A faulted loop has an additional contribution from the fault energy and is generally constrained by the fault to lie in the fault plane, unless it can unfault by shear, in which case it may rotate to reduce its line energy.

In diatomic or multiatomic solids, unfaulted loops must contain an excess or deficiency of atoms or ions from all sublattices in stoicheiometric proportions; yet in some systems, for example, alkali halides where irradiation produces only halogen interstitials (and vacancies), unfaulted loops are observed to arise from condensation of defects belonging to one sublattice alone. It is difficult to envisage a form of planar precipitation for defects from a single sublattice which does not contain a displacement fault. Due to long-range Coulomb fields, faults in strongly ionic crystals generally have high energies, and the lattice will try to avoid them. One possibility for interstitial condensation in a diatomic lattice is for interstitial defects on one sublattice to occupy

TRANSMISSION ELECTRON MICROSCOPY OF DEFECT AGGREGATES

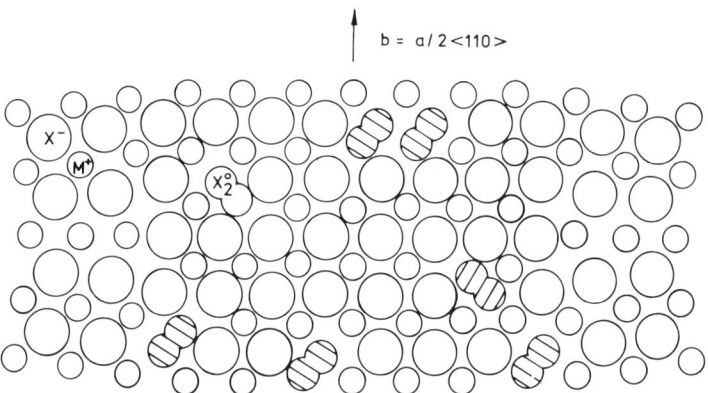

Figure 16. Model for interstitial anion stabilisation in a diatomic lattice with molecule-forming anions, showing dislocation loop and substitutional molecules.

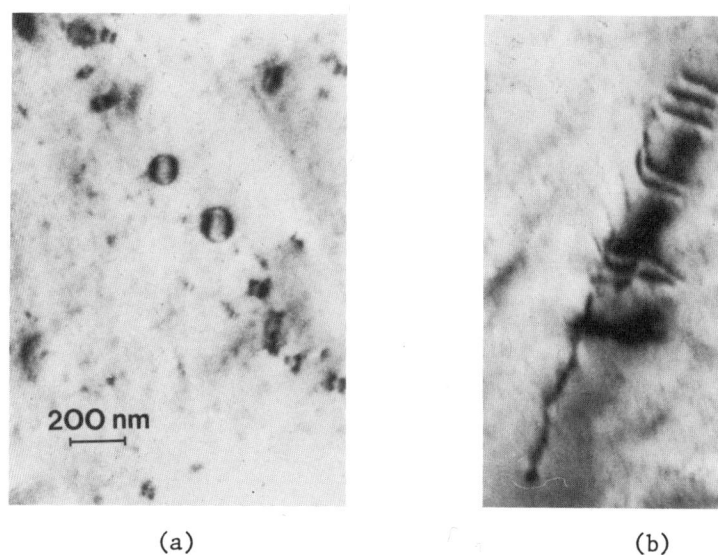

Figure 17. Incorporation of a chemical excess of anion species (iodine in KI) (a) dislocation loops probably formed on the model of Figure 16, (b) aggregation of substitutional molecules to form unstrained inclusion after annealing 30 m at 648 K.

substitutional sites on both sublattices, displacing ions on both sublattices in stoicheiometric proportions into perfect dislocation loops (Figure 16). Calculations (Table 3) show this process to be energetically feasible in systems where anion interstitials can form substitutional molecular defects, for example halogen molecules in halides; the molecular binding energy provides a large driving force for occupation of substitutional sites. We can actually produce such loops and molecular centres by chemical addition of excess halogen diffused into an alkali halide crystal (Figure 17). Provided the substitutional molecular defects remain relatively dispersed, replacement and displacement fields from the molecules will be short range and contrast negligible, so only the loops are observed. Spectroscopy, however, confirms the presence of the molecular defects.

Interaction between the strain fields of dislocation loops and point defects provides the mechanism for attraction of additional defects. Strain fields from interstitials are usually longer range than from vacancies, so loops of either sign will preferentially attract interstitials, and vacancy loops will tend to shrink in the presence of both interstitials and vacancies. Most forms of interstitial are considerably more mobile than vacancies, so interstitial loops will nucleate and grow at much lower temperatures and

Two separated H centres	1.68
Cl_2^o interstitial molecule along <100> (Molecular binding energy -2.51)	1.58
Cl_3^- along <111> in Cl^- site	1.08
Cl_2^o substitutional molecule (-0.57) + Dislocation loop	

N	r_1, Å	E_1/N	
10	6	1.85	1.28
10^2	19	0.91	0.34
10^3	60	0.39	0.18
10^4	189	0.16	-0.41
10^5	600	0.06	-0.51

Table 3. Energies (eV) for interstitial halogen stabilisation in NaCl relative to two isolated halogen atoms at infinity.

defect densities than vacancy loops. Loops grow most easily by the addition of new defects to existing jogs on the dislocation, and loop growth modes are often influenced by jog energy considerations. For example, in alkali halides and silicon, in certain temperature regions, long dipole loops can form instead of round loops, perhaps because pipe diffusion can supply defects to the lowest energy jog. Annealing these dipole loops causes them to break up into two or more round loops (Figure 18) again by the process of pipe diffusion which involves transport of material along the core of a dislocation, in this case to or from the ends of the dipole.

At high enough density and temperature, dislocation loops can themselves interact in two different ways. In the first, closely-spaced loops (e.g. those created by dipole break up) move together and coalesce by a process called glide and self-climb. A loop can glide conservatively on the glide cylinder which contains the loop core and the same Burgers vector to become coplanar (this reduces overall elastic energy). The loops can move towards each

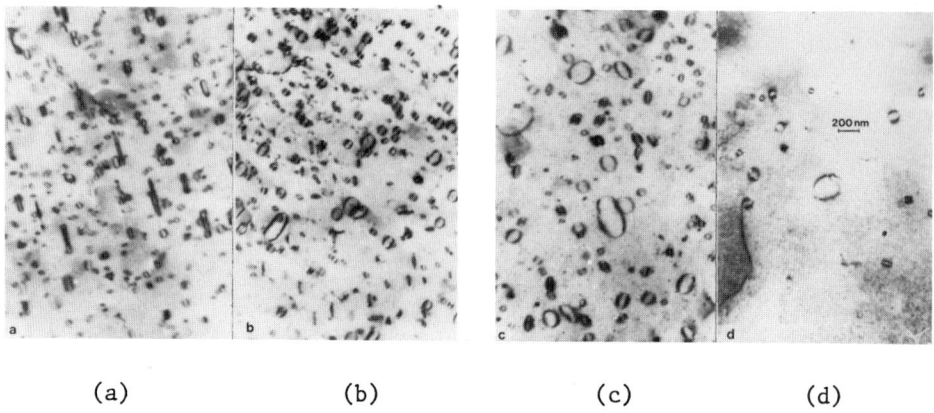

(a)　　　　　　(b)　　　　　　(c)　　　　　　(d)

Figure 18. Annealing of dipole loops in irradiated NaCl.
12 h at (a) 423 K, (b) 523 K, (c) 573 K, (d) 598 K.

other normal to their Burgers vectors by the transport of material by pipe diffusion around their dislocation cores (self-climb). The glide and self-climb process is the principal coarsening mechanism for loop distributions produced by irradiation, both during the course of irradiation (Figure 19) and during annealing afterwards (Figure 18). Loop distributions coarsened by this process are characterised by two distribution peaks due to persistance of small loops separated too far to coalesce. Loops with different (but crystallographically equivalent) Burgers vectors cannot coalesce, but can intersect to form dislocation junctions. When loops are supplied indefinitely with defects, combined coalescence and repeated loop intersections lead to formation of dense dislocation networks (Figure 19e).

The second process is redistribution of defects amongst loops by transport of re-emitted defects by bulk diffusion through the lattice, a process called Ostwald ripening. In this case, defects are re-emitted from the loop into the thermal equilibrium defect concentration, the loop being in local equilibrium with a defect concentration immediately surrounding the loop core which depends on loop curvature and is different from bulk crystal. Because of the high formation energy of interstitials, interstitial loops coarsen by this mechanism via creation, emission and absorption of vacancies. Large loops will grow at the expense of smaller loops, essentially because the energy required to expand a large loop by one defect is smaller than that gained by shrinking a smaller loop by one defect (cf. (99)). Since surfaces have a large radius of curvature, bulk diffusion is the principal mechanism for final loss of defects to the surface (Figure 18) during annealing.

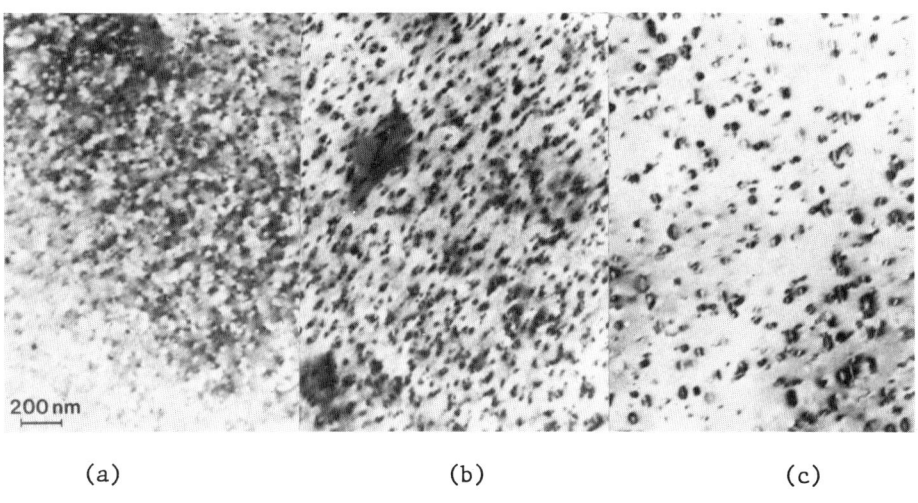

(a) (b) (c)

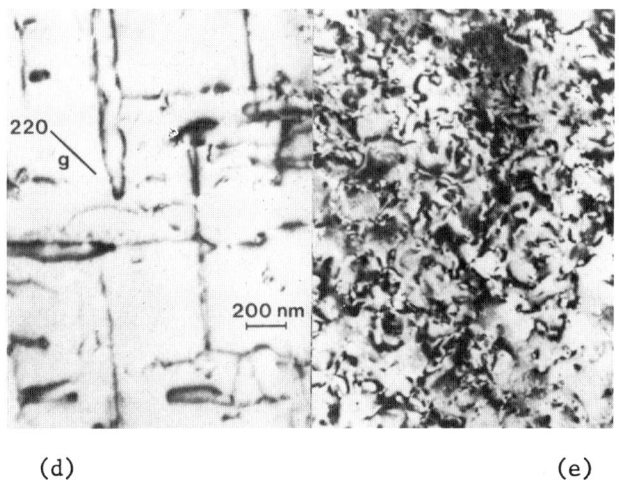

(d) (e)

Figure 19. Development of interstitial dislocation loop structure in irradiated NaCl. (a) 200 M rad at 298 K, (b) 500 M rad at 298K, (c) 2 G rad at 298 K, (d) 4 G rad at 423 K, (e) 40 G rad at 423 K.

Inclusions

The energy of an unstrained inclusion of radius r_I is largely contained in the interfacial (surface) energy $U_s = 4\pi r_I^2 \gamma$ where γ is the interface energy per unit area. Since the number of defects, each separately occupying volume v, contained in the inclusion is $N = 4\pi r_I^3/3v$, the energy per defect,

$$U_s/N = 3 v \gamma/r_I, \qquad (100)$$

again decreases as the inclusion radius increases in the same way as a loop. For the case of vacancies condensing into dislocation loops or voids, the ratio of the loop energy to inclusion energy for the same number of defects stabilised,

$$U_s(N)/U_l(N) \simeq (\gamma/5Gb^3) N^{\frac{3}{4}}, \qquad (101)$$

shows that energies are about comparable for very small loops or voids, but loops are more stable for larger aggregates. Voids may collapse into loops unless stabilised by internal pressure of trapped impurity gas atoms. Internal gas pressure can also stabilise voids against re-emission of vacancies.

In a diatomic or multiatomic solid, there is the possibility that a vacancy species on a single sublattice may alone aggregate. This results in precipitation of a new phase which stabilises the aggregate against the collapse possible in monatomic voids. If energy considerations were based solely on interfacial energy γ, then a three dimensional aggregate (say, spherical) will always be preferred to a planar aggregate, with the preference increasing as $N^{1/3}$. A good example is aggregation of F centres in alkali halides to form inclusions of alkali metal (Figure 12). In this case, there will be additionally a binding energy for F centres which must be subtracted from (100). Information in the diffraction pattern suggests that the alkali metal atoms retain their normal ion positions, the alkali metal inclusion remaining coherent with the matrix; that is, the alkali atoms retain the alkali ion f.c.c. structure rather than normal b.c.c. structure of alkali metals. An alkali metal atom occupies more space than an ion, but coincidentally only about as much space as an alkali-halogen ion pair, so there is little lattice misfit, and only small strains are sustained, due to the high compressibility of the alkali metal (Table 4). Figure 12 indicates that such inclusions can readily nucleate on existing lattice defects such as dislocations; this occurs either because the stress field of a dislocation enhances F centre diffusion, the charge acquired by dislocations in ionic crystals alters the electronic state (and thus mobility) of the F centre, or there is pipe diffusion of F centres down dislocation cores. Large inclusions can often lower their energy by forming facets, since the interface energy γ depends on the crystallography of the interface.

For highly strained inclusions, much of the defect energy is in the strain field. The strain energy depends on the relative elastic constants, inclusion shape, presence of nearby surfaces and whether the inclusion is coherent or incoherent, and competition between strain energy and interface energy can result in morphological changes during inclusion growth. An example is shown in Figure 20 in KCl heavily irradiated in the microscope; the radial strain is imaged as in $\underline{g}.\underline{b} \wedge \underline{u}$ dislocation contrast. These inclusions are probably halogen gas bubbles, and are similar to inclusions seen in irradiated heavy metal halides, mica and molybdenum disulphide. They may well arise from aggregation of substitutional halogen molecules, formed during interstitial stabilisation to form high pressure halogen gas bubbles. We can calculate the internal pressure from the observed lattice strain (Table 4),

$$P = 4 G \varepsilon, \qquad (102)$$

and a chlorine inclusion would be a highly compressed fluid. In the special case of $KI:I_2$, we expect the inclusion to be both unstrained and amorphous; in this case the only inclusion contrast mechanism

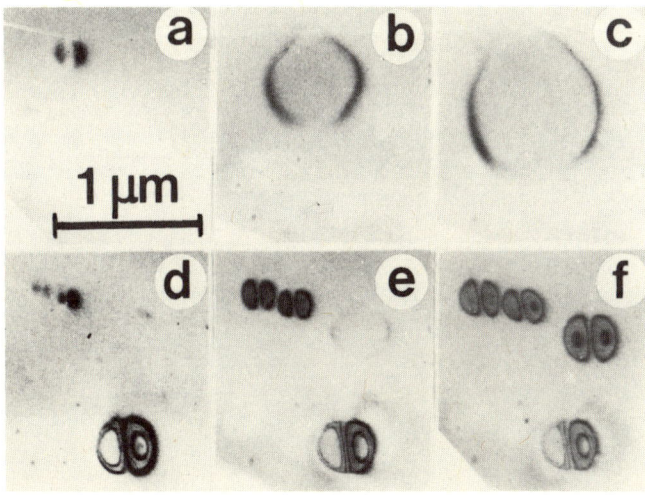

Figure 20. Development of strain inclusions from planar aggregates in irradiated KCl (courtesy Dr. Y. Kawamata).

observed to operate is mean absorption (Figure 17b). It is sometimes useful to detect such unstrained inclusions by artificial means, for example, by quenching in temperature and observing dislocations nucleated to relieve differential contraction stresses.

Inclusions can interact with each other in much the same ways as dislocation loops. Strained inclusions can move together and coalesce by diffusion of defects along the matrix-inclusion interface. If defects are aggregating in a dynamic radiation environment, it may be possible to effectively move even a large aggregate by preferential addition of defects to one side and annihilation on the other. Finally, a distribution of inclusions can undergo Ostwald ripening by re-emission of defects from small inclusions to large ones via the matrix. If there are even small elastic interactions between inclusions, it may be possible by one of these mechanisms to arrive at the lowest energy configuration of inclusions which may be an <u>ordered array</u> (Figure 21), analogous to the void lattice observed in metals. At sufficiently high temperatures, inclusions can lose stability in one of two ways: they can re-emit defects into an elevated thermal equilibrium defect concentration and thus redissolve, or during Ostwald ripening re-emitted defects can encounter antidefects and annihilate. The former is the

	ε, %
LiF : Li	1.1
NaCl : Na	-1.6
KCl : K	0.7
KBr : K	-0.14
KI : K	-1.7

	ε, %	P, GN m^{-2}
LiF : F	81.0	33.6
NaCl : Cl	22.8	13.8
KCl : Cl	7.0	3.2
KBr : Br	6.1	2.3
KI : I	-0.35	0

Table 4. Calculated accommodation strains and pressures for inclusions in alkali halides. Alkali metal colloids from aggregation of F centres, halogen inclusions from condensation of substitutional molecular centres. 10000 atmospheres $\simeq$ 1 GN m^{-2}.

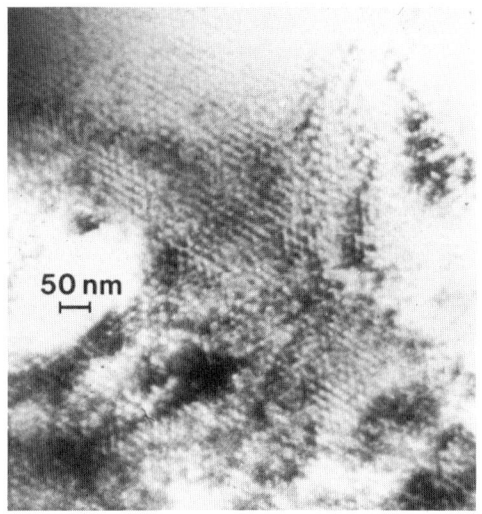

Figure 21. Ordered array of inclusions (probably calcium metal) in irradiated CaF_2.

mechanism responsible for eventual loss of alkali metal from alkali metal colloids in additively-coloured alkali and alkaline earth halides during annealing, the latter that probably responsible for loss of colloid stability in irradiated alkali or alkaline earth halides above a certain temperature. In the latter case, instead of coarsening the colloid distribution as in additively-coloured crystals, re-emitted F centres encounter dispersed molecular centres and annihilate.

Stabilisation of Radiation Defects

Radiation defects are almost always produced in complementary pairs, e.g. electrons and holes, vacancies and interstitials, and there is thus the possibility that recombination is an important mode of defect stabilisation. The presence of defect aggregates is critically important in determining the defect content of the lattice, because aggregation and recombination are competing processes. Three limiting simple cases illustrate expected behaviour.

First, suppose that a perfect aggregate sink exists for one defect (1) of a complementary defect pair (1,2), and that the other (2) remains isolated (it need not remain motionless). We can consider that around each isolated defect (2) exists a recombination volume v_2 within which a defect of the opposite sort (1) spontaneously annihilates. There is comparatively little defect

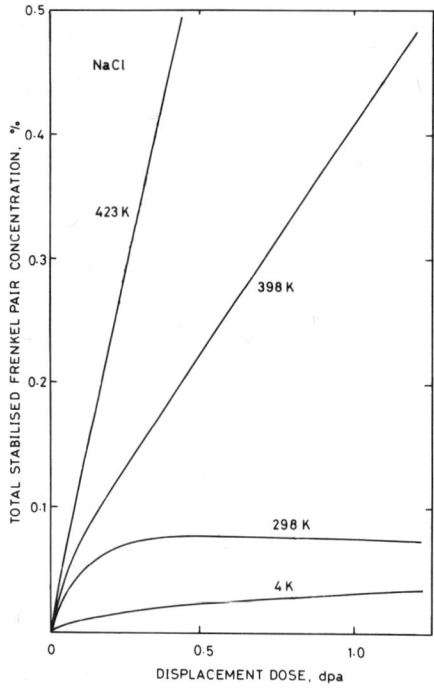

Figure 22. Three forms of Frenkel pair defect growth kinetics in NaCl irradiated up to 1 displacement per atom.

recombination associated with the sink for defects (1), since the sink density will be much less than the isolated defect (2) density. As the concentration of stabilised defects increases, the recombination volumes Σv_2 occupy increasingly more crystal and recombination becomes more efficient; the efficiency for defect stabilisation declines and stabilised defect concentration $c_1 = c_2 = c$ eventually saturates (Figure 22). The kinetics of saturation are described by the defect concentration at time t

$$c(t) = (1/v_2) \ln v_2 R t \qquad (103)$$

where R represents the basic rate of defect production.

If no aggregate sinks exist for either defect species, then around each defect (1) as well there is an additional recombination volume v_1, and recombination is more effective still. If $v_1 \simeq v_2 = v$, we can write

$$c(t) \simeq (1/4v) \ln 4 v R t \qquad (104)$$

which represents much slower defect accumulation (Figure 22). Finally, if aggregate sinks exist for both defects, then recombination is initially not very important, and defect density grows approximately linearly with dose (Figure 22)

$$c(t) \simeq R\,t. \qquad (105)$$

If aggregate density becomes high enough, this can actually increase R, since very close defect pairs may stabilise (thus survive) which would otherwise annihilate in perfect lattice. Eventual saturation occurs only through interactions between aggregates or from the finite possibility that a defect, on its way to its own sink, will encounter a defect or defect aggregate of the opposite kind.

Saturation kinetics are very important to microscopy of electron beam-sensitive materials, e.g. alkali halides, organic solids, because the electron microscope deposits ionising energy at enormous rates (typically $> 10^{30}$ eV m^{-3} s^{-1}) leading to displacement rates $R \sim 1\text{-}10$ displacements per atom (dpa) s^{-1}. In strongly ionic solids, such enormous displacement rates lead to formation of dense defect aggregates (Figure 19) which obscure existing information; in covalent or Van der Waals solids (e.g. quartz, organic molecular crystals), it can lead to eventual reduction to the amorphous state (Figure 23). One must be very careful that electron beam heating of the specimen in the microscope does not alter defect behaviour from kinetics (103) to kinetics (105); for example, onset of vacancy mobility leads to this transition for temperatures above 50°C in

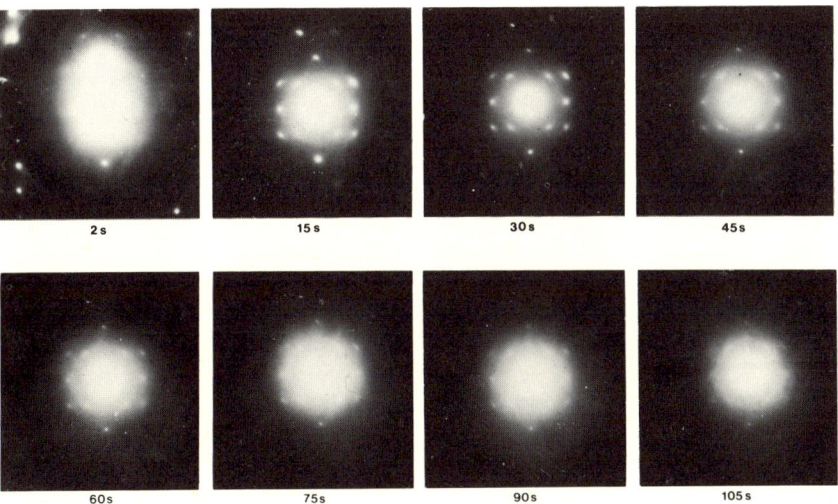

Figure 23. Reduction of crystalline anthracence to an amorphous hydrocarbon glass during irradiation in the electron microscope.

CaF$_2$ and 100°C in NaCl, temperatures which are easily achieved in non-metallic foils observed at room temperature under modest electron beam current density. There is a corresponding catastrophic alteration in aggregate density (cf. Figures 19 and 29).

The only way to moderate the effect of the electron beam is to operate in that regime where recombination is most efficient; this usually means at low temperatures (usually <30K) where all point defects are immobile and kinetics (104) apply. Such favourable kinetics cannot continue indefinitely, however, because defects of the same kind sufficiently close together can aggregate <u>athermally</u>, eventually altering kinetics (104) to (103). In the case of organic molecular or polymeric solids, creation of a defect may irreversibly alter the molecular building blocks of the crystal, and kinetics more like (105) may always apply, even at low temperature, and reduction to an amorphous solid (usually mostly carbon) inevitably ensues.

REFERENCES

Transfer theory of the microscope

K.-J. Hanszen, Advances in Optical and Electron Microscopy, 1971, 4, 1-84.

F. A. Lenz, in 'Electron Microscopy in Material Science', ed. U. Valdrè, Academic Press, New York and London, 1971, pp 540-69.

P. W. Hawkes, 'Electron Optics and Electron Microscopy', Taylor and Francis, London, 1972.

Lattice Images

J. M. Cowley and A. F. Moodie, Proc. Phys. Soc., 1960, 76, 378.

J. S. Anderson in 'Surface and Defect Properties of Solids', ed. M. W. Roberts and J. M. Thomas, The Chemical Society, London, 1972, vol. 1, pp. 1-53; J. S. Anderson and R. J. D. Tilley, ibid., 1974, vol. 3, pp. 1-56.

D. J. H. Cockayne, J. R. Parsons and C. W. Hoelke, Phil. Mag., 1971, 24, 139.

J. G. Allpress, E. A. Hewat, A. F. Moodie and J. V. Sanders, Acta Cryst., 1972, A28, 528.

J. G. Allpress and J. V. Sanders, J. Appl. Cryst., 1973, 6, 165.

S. Iijima, S. Kimura and M. Goto, Acta Cryst., 1973, A29, 632; 1974, A30, 251.

S. Iijima and J. G. Allpress, Acta Cryst., 1974, A30, 22; 29.

D. F. Lynch, A. F. Moodie and M. A. O'Keefe, Acta Cryst., 1975, A31, 300.

Dynamical theory of electron diffraction

P. B. Hirsch, A. Howie, R.B. Nicholson, D.W. Pashley and M. J. Whelan, 'Electron Microscopy of Thin Crystals', Butterworths, London, 1967.

M. J. Whelan, in 'Modern Diffraction and Imaging Techniques in Materials Science', ed S. Amelinckx et al., North Holland, Amsterdam, 1970, pp. 35-91.

A. Howie, in 'Electron Microscopy in Material Science', ed. U. Valdre, Academic Press, New York and London, 1971, pp. 274-300.

M. Wilkens, Phys. Stat. Sol., 1964, 6, 939.

B. Jouffrey and D. Taupin, Phil. Mag., 1967, 16, 703.

A. Howie and Z. S. Basinski, Phil. Mag., 1968, 17, 1039.

R. Serneels and R. Gevers, Phys. Stat. Sol. (b), 1972, 50, 99; 1973, 56, 681.

M. Wilkens, K. H. Katerbau and M. Ruhle, Z. Naturforschung, 1973, 28a, 681.

Weak beam technique

D. J. H. Cockayne, Z. Naturforschung, 1972, 27a, 452.

R. de Ridder and S. Amelinckx, Phys. Stat. Sol. (b), 1971, 43, 541.

D. J. H. Cockayne, J. Microscopy, 1973, 98, 116.

R. Sandstrom, Phys. Stat. Sol. (a), 1973, 18, 639.

R. C. Perrin and E.J. Savino, J. Microscopy, 1973, 98, 214.

Electron microscopy of ionic crystals

L. W. Hobbs, J. de Physique, 1973, 34 (C9), 227.

L. W. Hobbs, in 'Surface and Defect Properties of Solids', ed. M. W. Roberts and J. M. Thomas, The Chemical Society, London, 1975, Vol. 4, pp.152-250.

Electron microscopy of organic solids

D. T. Grubb, J. Materials Science, 1974, 9, 1715.

R. M. Glaeser, in 'Physical Aspects of Electron Microscopy and Microbeam Analysis', ed. B.M. Siegel and D.R. Beaman, John Wiley, New York, 1975, pp. 205-229.

W. Jones, J.M. Thomas, J. O. Williams and L. W. Hobbs, J. Chem. Soc. Faraday Trans. II, 1975, 71, 138.

Electron microscopy of semiconductors

H. Alexander and P. Haasen, Solid State Physics, 1968, 22, 28.

R. S. Nelson, in 'Radiation Damage and Defects in Semiconductors', The Institute of Physics, London, 1973, pp. 140-158.

Dislocation contrast

A. K. Head, M. H. Loretto and P. Humble, Phys. Stat. Sol., 1967, 20, 505.

P. Humble, Phys. Stat. Sol., 1967, 21, 733.

M. J. Norgett, R. C. Perrin and E. J. Savino, J. Phys. F., 1972, 2, L73.

A. K. Head, P. Humble, L. M. Clarebrough, A. J. Morton and C. T. Forwood, 'Computed Electron Micrographs and Defect Identification', Defects in Crystalline Solids, ed. S. Amelinckx, R. Gevers and J. Nihoul, Vol. 7, North Holland, Amsterdam, 1973.

Dislocation loop contrast

R. Bullough, D. H. Maher and R. C. Perrin, Phys. Stat. Sol. (a), 1970, 43, 689, 707.

D. M. Maher and B. L. Eyre, Phil. Mag., 1971, 23, 409.

D. M. Maher, M. H. Loretto and A. F. Bartlett, Phil. Mag., 1971, 24, 181.

M. Wilkens, in 'Modern Diffraction and Imaging Techniques in Material Science', ed S. Amelinckx et al., North Holland, Amsterdam, 1970, pp. 233-256.

M. Wilkens, in 'Vacancies and Interstitials in Metals', ed A. Seeger et al., North Holland, Amsterdam, 1970, pp. 485-529.

M. Wilkens and M. Ruhle, Phys. Stat. Sol.(b), 1972, 49, 749.

F. Haussermann, M. Ruhle and M. Wilkens, Phys. Stat. Sol.(b), 1972, 50, 445.

Inclusion contrast

J. van Landuyt, R. Gevers and S. Amelinckx, Phys. Stat. Sol., 1965, 10, 319.

H. Gleiter, Phil. Mag., 1968, 18, 847.

M. Ruhle, in 'Radiation Damage in Reactor Materials', IAEA, Vienna, 1969, IAEA-SM-120/B-2, pp. 113-159.

M. Ruhle, in 'Radiation Induced Voids in Metals', ed. J. W. Corbett and L. C. Ianiello, USAEC. Springfield, Va., 1972, pp. 255-291.

M. Ruhle and M. Wilkens, Proc. 5th Europ. Conf, on Electron Microscopy, Manchester, 1972, p. 416.

APPENDIX I

CONTRIBUTED PAPERS

K.W. BLAZEY   "Photochromic centers in $SrTiO_3$"

J.-J. PILLOUD   "EPR of X_2^- centers in Cs halides"

M. SAIDOH   "Interstitial trapping in alkali halides"

O.E. MOGENSEN   "Application of positron annihilation to defect studies."

K. ROSSLER and J. WINTER   "Investigation of damaged molecular units by ITC and d.c. conductivity"

F. FELIX   "Rare gas diffusion in ionic crystals"

K. ROSSLER   "Computer simulation of collision processes"

F. MODINE   "Magnetic circular dichroism"

E. LILLEY and Y. AL-JAMMAL   "Precipitate dissolution and the formation of vacancy pairs"

J. DAVENAS   "Formation of metallic aggregates by ion bombardment of alkali halides"

E. DYKES   "Defects in the apatite lattice"

C. KAGARAKIS   "X-ray and magnetic studies of iron oxides"

H. SPALT   "X-Ray Scattering from Defects in Alkali Halides"

R. BALZER   "X-Ray Measurements of Dimensional Changes at F and F_A centres in KCl"

APPENDIX II

REFERENCES - DEFECTS IN NON-METALLIC SOLIDS

<u>General</u> (The initials of the titles are used in subsequent references)

PCC : "Physics of Color Centers" ed. W. Beall Fowler (Academic Press 1968)

PDS : "Point Defects in Solids" Vols I & II ed. J.H. Crawford and L. Slifkin (Plenum Press 1972, 1975)

TDS : "Theory of Defects in Solids" A.M. Stoneham (Oxford University Press 1975)

DCS : "Defects in Crystalline Solids" B. Henderson (Arnold 1972)

CCIS : "Colour Centres and Imperfections in Solids" J.C. Kelly and P.D. Townsend (Chatto and Windus 1973)

DIS : "Diffusion in Solids" P.G. Sharman (McGraw Hill 1963)

RSR : "Real Solids and Radiation" A.E. Hughes and D. Pooley (Wykeham Publications, Taylor and Francis 1975)

CFS : "Crystals with the Fluorite Structure" ed. W. Hayes (Oxford Univ. Press 1974)

<u>A.B. LIDIARD - Structure and Properties of defects in non-metals</u>

1. DCS   2. DIS   3. PDS   4. CFS   5. TDS

6. "Theory of Imperfect Crystalline Solids" Trieste Lectures 1970 (Int. Atomic Energy Agency Vienna 1971)

7. A.B. Lidiard in "Orbital theories of molecules and solids" ed. N.H. March (Oxford Univ. Press 1974)

P. DOBSON - Defects and diffusion

1. PDS Vol I: chapter by R.G. Fuller.   2. DIS
3. "Diffusion in semiconductors" ed. D. Shaw (Plenum Press 1973)
4. "Point defects and diffusion" P. Flynn (Oxford Univ. Press 1970)

W.A. SIBLEY - The Photon as a Probe

1. PCC   2. PDS   3. DCS
4. J. Ferguson, Prog. Inorg. Chem. $\underline{12}$, 159 (1970)
5. "Light Scattering by small particles" H.C. van de Hulst (Wiley 1957)
6. "Light scattering in solids" ed. Balkanski (Paris 1973)
7. Bird et al. Phys.Rev. $\underline{B5}$, 1800 (1972)

A.E. HUGHES - Optical techniques and defect symmetry

1. PCC: chapters by W. Beall Fowler, C.H. Henry and D.B. Fitchen
2. "The Jahn Teller effect in Solids" M.D. Sturge in Solid State Physics ed. Seitz and Turnbull 1967.
3. "Group Theory and Quantum Mechanics" M Tinkham (McGraw Hill 1964)
4. "Chemical Applications of Group Theory" F.A. Cotton (Wiley 1971)

J.M. SPAETH - Magnetic resonance of vacancy centres in ionic crystals

1. PCC: chapter by H. Seidel and H.C. Wolf
2. CFS   3. DCS
4. PDS Vol.I: chapter by A.E. Hughes and B. Henderson
5. "Principles of magnetic resonance" C.P. Slichter (Harper & Row 1963)

APPENDIX II: REFERENCES

D. SCHOEMAKER - Interstitial centres

1. PDS Vol.I: chapter by M.N. Kabler

2. CFS

3. "Principles of magnetic resonance" C.P. Slichter (Harper and Row 1963) Chapter 7

4. N. Itch, Crystal Lattice Defects $\underline{3}$, 115 (1972)

G.D. WATKINS - Magnetic Resonance of Defects in Semiconductors

1. PDS Vol.II   2. DCS

3. "Electron irradiation damage in metals and semiconductors" J.W. Corbett, Suppl. to Solid State Physics ed. Seitz and Turnbull (Academic Press)

4. G.D. Watkins in:
   Si ("Radiation damage in semiconductors" (Dunod, Paris 1965) p. 97
   ("Lattice defects in semiconductors" (Inst. of Physics 1975) p. 1

   II-VI ("Radiation effects in semiconductors (Gordon & Breach 1971) p. 301
   ("Phys. Rev. Letters" $\underline{33}$, 223 (1974)

   Overreview: "Radiation damage and defects in semiconductors" (Inst. of Physics 1973) p.228

R.C. NEWMAN - Infrared Studies of Defects

1. "Infrared Studies of crystal defects" R.C. Newman (Taylor and Francis 1973)

2. L.J. Cheng et al., Phys. Rev. $\underline{152}$, 761 (1966)

3. "Radiation damage and defects in semiconductors" ed. J. Whitehouse (Inst. of Physics 1973)

4. "Lattice defects in semiconductors" (Prof. Freiburg Conf - Inst. of Physics 1975)

W. VON DER OSTEN - Vacancy aggregate centres in ionic crystals

1. PCC: chapter by D.B. Fitchen

2. A.A. Kaplyanskii, J. de Physique 28 Suppl 8/9 C4-39 (1967)

3. W. von der Osten, Z. Angew Physik 24, 365 (1968)

4. W.D. Compton and H. Rabin in Solid State Physics ed. Seitz and Turnbull (Academic Press) Vol. 16 p.121 (L964)

Y. MERLE D'AUBIGNE - Perturbation spectroscopy and excited states

1. PCC: chapters by C.H. Henry and C.P. Slichter, and D.B. Fitchen

2. C.H. Henry, S. Schnatterly and C.P. Slichter, Phys.Rev.137, A583 (1965)

3. "Electron paramagnetic resonance" ed. S. Geschwind (Plenum Press 1972) chapters by F.S. Ham (Jahn Teller effect) and S. Geschwind (Optical detection of resonance)

4. F.S. Ham, Phys. Rev. 138, A1727 (1965)

5. PDS Vol.I: Chapters by A.E. Hughes and B. Henderson, and M.N. Kabler.

G. SPINOLO - Excited State Properties

1. PCC: chapters by W. Beall Fowler, F. Luty, and H. Mahr; also chapter by C.H. Henry and C.P. Slichter.

2. TDS

3. R. Smoluchowski in "Proc. XV Colloque Ampere" ed. P. Averbuch (Grenoble 1968) p.120

R. DE BATIST - Internal friction and defects near dislocations

1. "Internal friction of defects in crystalline solids" R. de Batist (N. Holland)

2. "Anelastic relaxation in crystalline solids" A.S. Nowick and B.S. Berry (Academic Press)

APPENDIX II: REFERENCES

3. Most volumes of "Physical Acoustics" ed. W.P. Mason (Academic Press.

D.L. GRISCOM - Magnetic resonance techniques and defects in glasses

1. "Paramagnetic resonance" G.E. Pake (Benjamin 1962)

2. "Principles of Magnetic Resonance" C.P. Slichter (Harper and Row 1963)

3. "The structure of inorganic radicals" P.W. Atkins and M.C.R. Symons (Elsevier 1967)

4. "Transition metal ions" A. Abragam and B. Bleaney (Oxford Univ. Press 1972)

5. "Inorganic glass forming systems" H. Rawson (Academic Press 1967)

6. D.L. Griscom, J. Non-Crystalline Solids $\underline{13}$, 251 (1973/74)

A.M. STONEHAM - Defects at Surfaces

No references given, but for general surface physics and chemistry see articles in "Surface and defect properties of solids" ed. M.W. Roberts and J.M. Thomas, published annually by The Chemical Society.

H. PEISL - X-ray diffraction and scattering

1. "The theory of x-ray and thermal neutron scattering by real crystals" M.A. Krivoglaz (Plenum Press 1969)

2. P. Diederichs, J. Phys. F. (Metal Physics) $\underline{3}$, 471 (1973)

3. H. Peisl and H. Trinkhaus, Comments Sol.St.Phys. $\underline{5}$, 167 (1973)

4. Proc. Int. Disc. meeting on studies of lattice distortions and local atomic arrangements by x-ray, neutron and electron diffraction: J.Appl.Cryst. $\underline{8}$, 79 (1975).

K.H.G. ASHBEE - X-rays and electron microscopy

1. "Extended defects in non-metallic solids" ed. Leroy Eyring and M. O'Keefe (North Holland 1970)

2. "Recent applications of x-ray topography" A.R. Lang in "Modern diffraction and imaging techniques in materials science" ed. S. Amelinckx et al. (North Holland 1970)

3. "A decade of x-ray interferometry" M. Hart Proc.Roy.Soc. (1975)

   General references on dispersion surfaces listed by L.W. Hobbs

L.W. HOBBS - Electron microscopy of Defects in Insulators and Semi-conductors.

1. "Electron microscopy in materials science" ed. U. Valdre (Academic Press 1971)

2. "Electron microscopy of thin crystals" P.B. Hirsch et al. (Butterworths 1967)

3. "Modern diffraction and imaging techniques in materials science" ed. S. Amelinckx et al. (North Holland 1970).

4. L.W. Hobbs, Chapter 6 of "Surface and defect properties of solids" ed. M.W. Roberts and J.M. Thomas (The Chem. Soc. 1975)

5. "Computed electron micrographs and defect identification" A.K. Head et al. (North Holland 1973)

6. "Physical aspects of electron microscopy and microbeam analysis" ed. B.M. Siegel and D.R. Beaman (Wiley 1975)

APPENDIX III

PARTICIPANTS AT NATO ADVANCED STUDY INSTITUTE, EXETER 1975

Mr. A. Ayensu — H.H. Wills Physics Laboratory, University of Bristol, Tyndall Avenue, Bristol, England.

Dr. R. Balzer — Experimentalphysik I, 61 Darmstadt, Schlossgartenstr 7, W. Germany.

Dr. R.C. Barklie — Physical Laboratory, Trinity College, Dublin 2, Ireland.

Dr. K.W. Blazey — IBM Research Laboratory, 8803 Ruschlikon, Switzerland.

Dr. L.A. Boatner — Ecole Polytechnique Federale - Lausanne, Laboratoire de Physique Experimentale, 33 Av. de Cour, Lausanne, Switzerland.

Dr. M. Cetincelik — Nuclear Energy Forum of Turkey, P.O. Box 37 - Bakanliklar, Ankara, Turkey.

Mr. S. Datta — Department of Metallurgy Room 524, Imperial College, Prince Consort Road, London S.W.7.

Dr. J. Davenas — Department de Physique des Materiaux, Université Cl. Bernard Lyon, 43 Bd. du 11 Novembre, 69621 Villeurbanne, France.

Dr. G. Davies — Physics Department, Kings College, Strand, London W.C.2.

Dr. E. Dykes — The London Hospital Medical College, Department of Dental Anatomy, Turner Street, London E.1 2AD

Professor F.W. Felix — Hahn-Meitner-Institut für Kern-Forschung gmbH, Glienickerstr 100, D1 Berlin 39, Germany.

Dr. P. Gagliardelli	Istituto di Fisica del Politecnico, piazza L. da Vinci 32 - 20133 Milano, Italy.
Dr. R. Galloni	Lab. Lamel C.N.R., Via Castagnoli I - 40126, Bologna, Italy.
Dr. G. Gambarini	Istituto di Fisica del Politecnico, piazza L. da Vinci 32 - 20133 Milano, Italy.
Mr. M. Ghomi	Université Paris-Sud, Centre d'Orsay, Faculté des Sciences, Batîment 490, 91405 Orsay, France
Mr. W. Hagen	Max-Planck-Institut für Festkorperforschung, 7000 Stuttgart 1, Heilbronner Str. 69, W. Germany.
Mr. T. Hangleiter	Fachbereich 6 - Experimentalphysik, Gesamthochschule Paderborn, 479 Paderborn, Pohlweg 55, W. Germany.
Professor H. Herman	Department of Materials Science, State University of New York, Stoneybrook, New York, U.S.A.
Mr. H. Hubbard	Dept Inorganic and Structural Chemistry, University of Leeds, Leeds LS2 9JT, England.
Mr. H. Hübner	Physikalisches Institut, Teilinstitut 2, Universität Stuttgart, 7000 Stuttgart 80, W. Germany.
Mr. Y. Al-Jammal	Applied Sciences, University of Sussex, England.
Dr. C. Kagarakis	National Technical University of Athens, Laboratory of General Chemistry, 42 Patission Street, Athens (147) Greece.
Dr. S. W. Kennedy	Physical & Inorganic Chemistry Dept., University of Adelaide, Australia.
Mr. A. Kung	Physics Department, Manitoba University, Winnipeg, Manitoba, Canada.

APPENDIX III: PARTICIPANTS

Dr. A.D. Lagendijk	Physics Dept., University of Antwerp, (U.I.A.), B 2610 Wilrijk, Belgium.
Professor A.L. Laskar	Physics Dept., University of Clemson, Clemson, South Carolina 29631, U.S.A.
Dr. E. Lilley	School of Applied Sciences, University of Sussex, England.
Mr. L. Mamel	Dept. of Physics, Lehigh University, Bethlehem, Pennsylvania 18015, U.S.A.
Dr. F.A. Modine	Solid State Division, Oak Ridge National Laboratory, Oak Ridge, Tennessee 37830, U.S.A.
Dr. O.E. Mogensen	Chemistry Dept., Danish AEC, Risø, DK-4000 Roskilde, Denmark.
Mr. J. Niklas	Fachbereich 6 - Experimentalphysik, Gesamthochschule Paderborn, 479 Paderborn, Pohlweg 55, W. Germany.
Dr. Nouailhat	I.N.S.A. - Laboratoire de Physique de la Matiére - Batîment 502, 20 Avenue A. Einstein, 69621 Villeurbanne, France.
Mr. K. O'Donnell	Physical Laboratory, Trinity College, Dublin 2, Ireland.
Mr. J.P. Omaggio	Ohio State University, Physics Dept., Columbus, Ohio 43210, U.S.A.
Mr. J.M. Ortega	Laboratoire de Physique du Solide, Batîment 510, Faculté d'Orsay, Orsay 91, France.
Mr. I.B. Owen	Clarendon Laboratory, Parks Road, Oxford, England.
Dr. L. Passari	Lab. CNR Lamel, via de Castagnoli 1, Bologna, Italy.
Mr. J.J. Pilloud	Institut de Physique, Université de Neuchatel, 2000 Neuchatel, Switzerland.
Dr. K. Rössler	Institut für Chemie 1: Nuklearchemie der Kernforschungsanlage Jülich, D-5170 Jülich, W. Germany.

Dr. M. Saidoh	School of Mathematical and Physical Science, University of Sussex, Falmer, Brighton BN1 9QH, England.
Dr. A. Schoenberg	Physics Dept., University of Keele, Staffordshire ST5 5BG, England.
Professor J. Shackleford	Department of Mechanical Engineering, University of California, Davis, California 95616, U.S.A.
Dr. H. Spalt	Experimentalphysik I, Technische Hochschule, 61 Darmstadt, Schlossgartenstr. 7, W. Germany.
Mr. H. Stolz	Experimentalphysik I, Technische Hochschule, 61 Darmstadt, Schlossgartenstr. 7, W. Germany.
Mr. J. Troxell	Physics Department, Lehigh University, Bethlehem, Pennsylvania 18015, U.S.A.
Professor J. Vail	Physics Department, University of Manitoba, Winnipeg, Manitoba, Canada.
Dr. Ved-Mitra	Istituto di Physica Generale "A Volta", Universita di Pavia, 27100 Pavia, Italy.
Mr. J. Weber	Experimentalphysik I, Technische Hochschule, 61 Darmstadt, Schlossgartenstr. 7, W. Germany.
Mr. J. Winter	Institut für Chemie 1: Nuklearchemie der KFA Jülich, D-5170 Jülich, W. Germany.
Mr. G. Woysch	Physikalisches Institut, Teilinstitut 2, University of Stuttgart, 7000 Stuttgart 80, W. Germany.
Mr. Yuste	Laboratoire de Physique Cristalline, Batîment 490, Faculté des Sciences, Paris-Sud, 91405 Orsay, France.

INDEX

Adatoms, 363
Aggregates, 222
   dislocation loops, 453, 464
   electric field effects, 241
   electron microscopy, 452
   electronic degeneracies, 245
   in ionic crystals, 236
   orientational degeneracy, 241
   polarization effects in absorption and emission, 238
   small, 461
   strained inclusions, 458
   transmission electron microscopy, 429
   uniaxial stress in, 241
   unstrained inclusions, 457
   vibrational properties, 248
Alkali borate glasses, clusters in, 342
   ESR spectra, 343
Alkali halides, 173, 217
   aggregate centres, 237
   complex defects in, 14
   copper$^+$ substitutional ion, 293
   diffusion in, 33
   doped with divalent cations, 44
   electromechanical coupling in, 304
   electron deficient centres in, 81
   electronic degeneracies, 245

Akali halides (cont'd)
   electronic transition in, 108
   ENDOR spectra of F centres, 162
   ESR spectra of F centres, 162
   excess halogen diffused into, 466
   F-aggregate centres, 237
      ESR and ENDOR spectra, 169
   F centres in, 147, 393, 470, 472
      electronic structure of, 167
      optical spin memory, 277
   Frenkel defects in, 46, 62
   H_2O^- centre in, 169
   number and kind of defects, 173
   optical studies of, 133
   saturation kinetics of, 475
   Schottky defects in, 46, 61, 62
   structure and defect, 10, 13, 15
   surface diffusion, 365, 366
   surface geometry, 354
   V_k centre, ESR spectra, 175
   vibrational properties, 248
   Z centres in, magnetic resonance studies in, 169
Alkaline earth fluorides
   doped with higher valency cations, 45
   electron deficient centres, 81
   F centres, ENDOR spectra, 164
   Frenkel defects in, 61
   Frenkel formation energies in, 63
   M centres, multiplet states, 170
Alkaline earth halides, electronic transition in, 108

Alkaline earth oxides
  F centres
    ENDOR spectra, 164
    ESR spectra, 162
  intrinsic adatoms, 363
  structure and defect, 12
  surfaces, vacancy centres, 362
Alumina, 19
  lattice fringes in, 423
Amorphous peroxyborates, spectra of, 326
Anelastic effects from dislocations, 304, 305
Anisotropic defects, Huang scattering from, 386
Anthracene, 475
Asymptotic distortion X-ray scattering, 289
Atomic theory of defects, 46

Beryl, electron diffraction patterns, 413
Bonse interferometer, 425
Bordoni peak, 304
Boron E' centre, 339
Boron-oxygen hole centre, 338
Borosilicate glass
  bridging-oxygen hole centre in, 335
  paramagnetic defect concentration in, 337
Bragg contours, 426
Bridging-oxygen hole centre in glasses, 335
Brillioun zones, 112, 248, 406, 409
  maps of boundaries, 411
Bulk light scattering, 118

Calcium fluoride, Frenkel and Schottky defects in, 63
Calcium oxide, F centre in, 272
Carbides, surface diffusion, 367
Catalysis, mechanism of, 371
Centres
  electronic structure of, basic aspects, 282

Centres (cont'd)
  excited state properties of, 281
  in common oxide glasses, 325
Charged dislocations, 315
Chemical potential, 20
Clusters
  charged, 14
  Huang scattering and, 388
  in quartz, 341
  small, optical investigation, 120
  X-ray scattering from, 400
Complex defects, 8
  structure of, 14
Compound solids, defects in, 5
Concentrations of defects in thermal equilibrium, 2
Copper$^+$ substitutional ion, 293, 296
Correlation factor, 29
Coupling strengths, 140
Crystals
  growth from melt, 358
  imperfect
    electron microscopy, 442
    lattice defects, 442
    X-ray scattering from, 383
  lattice fringes, 337, 441
  macroscopic defects in, 118
  neutral cores in, 416
  perfect, 107
    electron microscopy, 434
  plasticity of, 303
  surface energies, 355
  with impurities
    ENDOR spectra, 168
    vacancy centres in, 168

Debye-Hückel screening, 369
Debye relaxation, 34
Debye-Waller factor, 383, 434
Defects
  as elastic dipoles, 380
  atomic theory of, 46
  classification, 222
  differences between, 2
  electronic theory of, 67, 83
  energies studied with X-rays, 379

INDEX

Defects (cont'd)
  formation energy, 392
  identification of, 277
  interactions among, 8, 10
  molecular description of, 415
  near dislocations, 303
  on surfaces, 361
Defect molecule, concept of, 207
Defect movements
  statistical theory of, 24
    relaxation processes, 33
Defect volumes and elastic dipole tensor, 395
Deformation-dipole models, 61
Diamond
  atomic theory of defect, 53
  formation of vacancies in, 61
  infrared absorption, 110
  role of structure in defect, 9
  surface energies, 357
  X-ray topography, 422
Diffusion, 24-33, 95-105
  coefficient D, 95, 100
  correlation factor, 29
  dislocation, 365
  electrical mobility and, 31
  enhancement of, 365
  experiments using marker layers, 103
  H centres, 186
  impurity, 28, 29
  in ionic solids, 97
  interstitial, 96
  ionic conductivity and, 98
  isotopic defects, 23, 31
  macroscopic, 24
  random walk theory of, 25
  self-, 28
    coefficient, 101
    enhancement of, 101
  significance of factors in D, 27
  surface, 365
  temperature and, 95
  V_k centres, 180
  vacancy mechanism, 96, 97

Di interstitial centres, 200
Dislocations
  anelastic effects from, 304, 305
  charged, 315
  defects near, 303
  diffusion, 365
  effect of line charge on damping constant, 317
  hysteresis, 308
  inducing piezoelectricity, 316
  relaxation effects, 304, 305
  resonance, 305
Dislocation loops, 453
Dislocation-point defect interaction, 303
  equilibrium effects, 312
  in non-metallic crystals, 310
  recovery kinetics, 310
Displacement rates, 20
Distortion scattering of X-rays, 385
Divacancies, 225, 226, 281
  structure of, 227
Donor and acceptor states, 70, 71

E' centres, 342
  mechanism of formation, 339
E' defect
  in silicon oxide, 325, 346
Elastic anisotropy of surface, 361
Elastic dipoles, defects as, 380
Elastic dipole tensor, 42, 395, 398
Electric fields, effect on zero-phonon lines, 245
Electrodes, 267
Electrolytes, 267
Electrons
  emission, 433
  wavelengths of, 431, 433
Electron deficient centres, 81
Electron excess centres, 156
Electron excess F-like centres, 72
Electron microscopy, 405
  displacement contrast, 443
  Fresnel contrast, 452
  fundamental resolution and, 429
  phase grating approximation, 437
  practical resolution, 433

Electron microscopy (cont'd)
  replacement contrast, 451
  structure factor contrast, 451
  transmission, 429
    diagram of instrument, 431
  weak beam effect, 448
Electron nuclear double resonance, 155
Electron-phonon interaction, 250, 251, 262, 283, 432
Electron scattering, 432
  inelastic, 432, 442
  intrabranch, 448
Electron spin resonance, 155
  anisotropy, 175
  of oxide glasses, 324
Electron theory of defects, 67
  limitations of, 83
Electronic states of defects, 133, 137, 141, 142, 145
  degenerate, 145
Electronic structure
  basic aspects of, 282
  F centres, 167
  H centres, 185
  of silicon oxide with defects, 345
  V_k centre, 176
Elementary movements of point defects, 19
Energies of defects studied with X-rays, 379
Energy level scheme, 285
EPR spectra, identification of defects with, 277
EPR studies in lattice defects, 203
Ewald sphere, 412, 436
Excited state
  energy level scheme, 285
  in localized centres, 281
  relaxed, 289
  spectroscopy, 295
  unrelaxed, 283
External field perturbation, 286

F_A centre, 281

F aggregate centre
  electronic state, 240
  ESR and ENDOR spectra, 169
  polarization effects in absorption and emission, 238
  relaxation times, 169
F centres, 236, 281
  allowed transitions of, 288
  dislocations and, 318
  electronic structure of, 167
  EPR spectra, 272
  ESR and ENDOR spectra of, 156, 158, 159, 161, 162
  excited states spectroscopy, 295
  external field perturbation, 286
  Hamiltonian, 263
  in alkali halides, 470, 472
  lifetime problem, 289
  optical spin memory, 277
  resonance of pairs, 279
  spin Hamiltonian, 156
  $^3T_{1u}$ state of, 275
  X-ray studies, 393, 398
$F_3(R)$ centre, Jahn-Teller distortion, 253
F_s^+ centres, 362
Fx^- centres, <111> orientated, 188
Faraday effect, 287
Ferromagnetic precipitates, 343
Fibre optics, 321, 345
Fluorite compounds, 19
  doped with lower valency cations, 43
Fluorite, X-ray topography, 423
Franck-Condon principle, 47
Fraünhofer diffraction, 429, 430
Frenkel disorder, 10, 12, 13, 17, 19, 46, 59, 61, 62, 63, 98, 394
Frenkel pairs, 392
Fresnel contrast, 452
Fresnel diffraction, 430

Gallium, interstitials, 232
Gallium arsenide, 221, 232
Gallium phosphide, 221, 232
Germanium
  atomic theory of defect, 53
  diffusion in, 100

INDEX

Germanium (cont'd)
  doped, vibrational
     absorption, 232
  E' centre, 342
  electron binding, 69
  electronic absorption, 229
  phonon absorption, 223, 224
  role of structure in
     defects, 9
  self diffusion in,
     enhancement of, 101
  shallow acceptors, 70, 71
  surface energeis, 357
Glasses, radiation-induced
     defects in, 335
Gouy-Chapman model, 269
Grain boundaries
  diffusion along, 365
  segregation and phase
     separation near, 369
Granato-Lücke description of
     hysteretic dislocation,
     309, 312, 313, 317
Group theory, 145, 241

H centres, 174
  diffusion, 186
  electronic structure of, 185
  ESR spectrum, 182
  <110> and <111> oriented,
     184
  optical absorption, 185
  thermal reorientation, 186
  thermal stability, 182
  uniaxial stress, 186
  X-ray studies, 398
H_a (Li) centre
  anisotropic optical
     absorption, 194
  ESR, 192
  uniaxial stress, 194
H_a(Na) centre
  thermal reorientation
     motions, 191
H_a(Na) centre (V_1 centre), 190
H_a(Na) centre in LiF, 196
H_A (Li) centre in NaF:Li, 197
H_{AA} centre, 198
$H_{A'A}$ centres, 198

HADES program, 50, 64, 81
Ham reduction factor, 143
Harper behaviour, 311
Hart X-ray interferometer, 425
Helmholz-Perrin model, 269
Huang diffuse X-ray scattering,
     385
  cluster formation and, 388
  from point defects, 397
Huang-Rhys factor, 138
Hysteresis, dislocation, 308

Ice, structure of, 421
Impurities
  adsorbed, 364
  defect, 112
Impurity adatoms, 363
Inclusions, 470
Infrared absorption in
     insulators, 110
Infrared studies, 221
Inhomogeneous broadening, 162
Insulators, infrared absorption
     in, 110
Insulator surfaces, 353
Interactions among defects, 8, 10
Interfaces, dynamic phenomena at,
     370
Internal friction, 303
Interstitial defects, 4, 21, 23,
     56
  in phosphorus diffused silicon,
     104
  in silicon, 212
  optical absorption and magnetic
     resonance, 173
  stabilising, 199
  strain fields, 466
  vibrational absorption, 230
Interstitial clusters, structure
     of, 61
Interstitialcy mechanisms, 28
Interstitial diffusion mechanism,
     96
Interstitial halogen atom centre
     See H centre
Ionic conductivity, diffusion
     and, 98

Ionic crystals
  F band energies in, 74
  interstitials in, 56
  recovery effects in, 312
  role of structure in
    defects, 10
  Schottky and Frenkel
    defects in, 59
  surfaces, 362
  surface energies, 356
  surface geometry, 354, 355
  surface states, 359
  vacancy aggregate centres
    in, 236
  vacancy centres, 167
Ionic defects, 31, 33
Ionic solids
  atomic theory of defects, 51
  diffusion in, 97
  Frenkel defects in, 98
  highly disordered, 17
  Schottky defects in, 98
Ionic thermocurrents, 36
Ion-size effects in F Band, 79
Irradiation, Mn^{2+} perturbed
  centres from, 126
Isotopic defects
  Huang scattering from, 385
  in diffusion, 23, 31

Jahn-Teller distortions, 84,
    141, 217, 237, 247, 252,
    259, 264, 282
  consequences for optical
    spectra, 144
  in II-VI compounds, 214
  in silicon, 206, 208, 209,
    210
  optical properties, 143
Jahn-Teller energy, 272
Jump frequency, 20, 21, 22,
    23, 24, 28, 32, 41, 55

Kanzaki method, 49, 61
Kikuchi lines, 412, 414, 426,
    442, 455
Kossel lines, 412, 414,
    415, 426

Lamb waves, 361

Lang topograph, 421
Lattice, 67
  absorption by, 224
  configurations, 22
  defects in, 379, 442
  distortion of, 47, 74, 78
  dynamics, 53
  fringes, 437, 441
  infrared active, 111
  interstitial anion
    stabilization, 465
  in V_k centres, 181
  nuclear spin resonance of
    nuclei, 159
  parameter changes, 391
  parameter mapping, 424
  parameter versus length change,
    391
  point-ion model, 74
  relaxations, 67, 84, 356
  static, general formalism
    for, 47
  symmetry of, 223
  vibrations, 22, 52
Lattice defects, 1
  EPR studies in semiconductors,
    203
  in F aggregate centres, 237
  magnetic resonance studies of,
    155
Laue-case X-ray interferometer,
    424
Laue scattering, 385
Lifetime, temperature and, 294
Lifetime problem in F centre, 289
Light, interaction with solids,
    107
Light scattering in defects, 117
Lithium fluoride
  absorption band, 141
  charged dislocations, 316
  clusters in, X-ray studies, 400
  Jahn-Teller distortion, 252,
    253, 254
  optical absorption in, 114
  vibrational properties, 250
  X-ray small angle scattering,
    401
  X-ray topography, 423

INDEX

Lithium impurity, ENDOR spectra, 168
Love waves, 361

M centres
  magnetic resonance, 170
  multiplet states, 170
  transition energies, 76
Macroscopic defects, 127
  introduction of, 118
Macroscopic diffusion, 24
Magnesium fluoride, electric and magnetic dipoles, 122
Magnesium oxide
  adsorption on microcrystals of, 369
  defects in, X-ray studies, 399
  internal friction, 313
  surface diffusion, 365
Magnetic circular dichroism, 117, 122, 259, 261
  measurements, 250
Magnetic fields, orientation of, 160
Magnetic resonance, 173
  of vacancy centres in ionic crystals, 155
  polarized luminescence in detection of, 268
Majority defect, 4
Metals
  atomic theory of defects, 54
  epitaxial, on surfaces, 370
  oxidation of, 370
Metal oxides, shear structures in, 437
Metal-oxide-semiconductors, 321
Methods of moments, 262
Microscopy, 107, 117
  phase-contrast, 118
Molecular description of defects, 415
Mott-Littleton method, 49, 61, 64, 82

Neutron irradiation, 419

Nonbridging-oxygen hole centre in silicate glasses, 330
Non-Crystalline oxides, defects in, 321
Non-stoichiometric oxides, role of structure in defect, 15
Norgett's HADES program, 50, 64, 81
Nuclear magnetic relaxation times, 38

Optical absorption, 173
Optical spectra
  form of, 137
  in Jahn-Teller effect, 144
Optical spin memory, 277
Optical techniques, 107, 133
  instruments, 134
Orbitally degenerate ground states, 265
Orthogonalised wave function, method of, 167
Orthorhombic defects, 40, 41, 44
Oscillator strength, 283
Ostwald ripening, 468, 472
Oxidation of metals, 370
Oxides
  surface diffusion, 365
  trapped holes in, 82
Oxide glasses, 321
  band structures of, 348
  boron E' centre in, 339
  bridging-oxygen hole centre in, 335
  cationic species, 322
  clusters in, 342
  defect centres in, 325
  ESR spectral analysis, 324
  ferromagnetic precipitates, 343
  lead-bearing, 342
  nonbridging-oxygen hole centres in, 330
  schematic view of, 323
  structure of, 322
  vacancies in, 323
Oxygen as impurity in silicon, 209

Paramagnetic defects, 155
  in borosilicate glasses, 337

Paramagnetic diinterstitial centres, 200
Paramagnetic resonance
  optical detection of, 259
  in excited states, 268
Pekarian bandshapes, 139
Pendellosüng effects, 411
Peroxyborates, spectra of, 326
Perturbation spectroscopy, 259
Phase grating approximation, 437
Phosphorus, diffusion in silicon, 102
Phonon assisted transitions, 139
Phonon modes near surfaces, 361
Photography, 368
Photons as probes, 107
Piezoelectricity
  dislocation induced, 316
  effects on surface, 361
Point ion approximation, M-centre transition energies in, 76
Point ion models, 72
Point-ion potential, 78
Polarized luminescence, 148, 238
  in detection of magnetic resonance, 268
Potassium bromide, 297
  cluster formation in, 401
  EPR spectra, 276
  Huang scattering from defects, 397
Potassium chloride
  H centre, 182
  inclusions in, 471
  lattice parameter of, 395
  <111> oriented FX^- centres, 188
  optical absorption and ESR studies, 174
  small interstitial dislocations, 463
  strained inclusions in, 459
  V_k centres, 175
Potassium chloride: Lithium, H_A (Li) centre in, 192

Potassium chloride: Sodium, H_a(Na) centre, 190
Potassium iodide, F centre lifetime, 295
Potassium silicate glasses, defect centres in, 333
Powder pattern, 325
Pseudo-Stark effect, 244

Quartz
  atomic clusters in, 341
  E' centre formation in, 339
  molecular structure, 417
  screw dislocation along axis, 418

R centres, magnetic resonance, 170
Radiation, semiconductors under, 20
Radiation defects, 112
  stabilization of, 473
Random walk theory of diffusion, 25
Rare earth ions
  absorption energies, 114
  spectra from, 115
Rare gas solids, 5
  atomic theory of defects, 51
Rayleigh waves, 361
Relaxation effects, 33, 65, 84
  anelastic, 33, 34
  dielectric, 33, 34, 45
  dislocation, 304, 305
  ionic thermocurrents, 36
  relation to defect properties, 38
  single applied frequency, 35
  sudden application of electric field, 36
Relaxation times
  in F aggregate centres in alkali halides, 169
  measurement of, 271
  nuclear magnetic, 38
Resonance, dislocation, 305
Reststrahl-band, 221
Ruby, light scattering from, 128

Schottky disorder, 10, 13, 17, 46, 59, 61, 62, 63, 98, 100, 392, 394
Shottky formation energy, 12
Self-diffusion, 28, 29
  coefficient, 101

INDEX

Self-diffusion (cont'd)
  enhancement of, 101
  in silicon, 367
Self-trapped exciton,
    resonance of, 275
Self trapped hole centre,
    See V_k centre
Semiconductors
  carrier concentrations near
    surface, 369
  covalent
    atomic theory of
      defect, 53
    defects in, 59
    shallow states in, 68
    vacancies in, 83
  defects
    patterns, 216
    role of structure, 9
  degenerate, 8
  diffusion in, 100
  impurities in, 364
  lattice defects in, EPR
    studies, 203
  nondegenerate conduction
    band minima in, 70
  shallow states in, 68
  surfaces of, 353, 359
  surface energies, 357
  surface geometry, 355
  under irradiation, 20
Shallow donors, 224
Shell models, 52, 61
Shockley defect, 419
Shockley states, on
    surface, 359
Silica
  absorbed impurities, 364
  E' centre, 342
  molecular structure, 417
  surface
    intrinsic defects, 363
  surface energies, 357
  surface state, 360
Silicon
  atomic theory of defect, 53
  damage mechanism, 212
  diffusion in, 100
  divacancies, 225, 226

Silicon (cont'd)
  electron binding, 69
  electronic absorption, 225
  impurities in, 230
  infra red studies, 221
  interstitial defects in, 212
  irradiation of, 203, 231
  lattice defects, EPR studies, 203
  lattice vacancies, 205
  oxygen as impurity, 209
  phonon absorption, 223, 224
  phosphorus diffusion in, 102
  planar faulter loops in, 450
  role of structure in defect, 9
  self-diffusion, 367
  shallow acceptors, 70, 71
  V_2 centre, 225, 228
  vacancy aggregates, 210
  vacancy-impurity pairs in, 208, 209
  vacancy mobility, 208
  vibrational absorption, 230
Silicon-germanium alloys,
    diffusion in, 102
Silicon oxide
  clusters in, 348
  defects in, electronic
    structure, 345
  E' centre, 346
  glass, 322
    E' centre, 325
    nonbridging-oxygen hole centre
      in, 330
Silver chalcogenides, 18
Silver halides
  photographic effect and, 368
  structure and defect, 10, 18
Single defect approximation, 384
Sodium bromide, 297
Sodium chloride
  charged dislocations in, 316
  diffusion in, 99
  electron diffraction, 436
  F centres, 318
  F centre lifetime, 295
  interstitial dislocation loops
    in, 469
  surface state, 359
Sodium fluoride, dislocation in, 316

Sodium silicate glasses,
    defect centres in, 333
Solids, interaction with
    light, 107
Solid rare gases
  defects in, 58
  entalpies of vacancy
      formation, 58
  vacancies in, 55
Spectra, form of, 137
Spectroscopy, 107
  high resolution, 136
  instruments, 134
  low resolution, 135
  of excited states, 295
  units and symbols, 134
Split-valency defects, 56
Stark effect, 151, 244, 245,
    259, 282, 287
Stoneley waves, 361
Strained inclusions, 458
Structure, 1
  investigation, 148
  X-ray studies, 379
Surfaces
  accumulation layer, 369
  band bending, 369
  defects on, 361
  depletion layer, 269
  diffusion, 365
  ease of cleavage in, 358
  elastic anisotropy, 361
  energies, 355
    importance of, 358
  epitaxial metal on, 370
  fracture mechanism, 358
  free from imperfections, 354
    energies, 355
    geometry, 354
  inversion layer, 369
  perfect, 359
  phenomena, 353
  phonon modes near, 361
  piezoelectric effects on,
      361
  vacancy centres, 362
Surface tension, 356
Symmetry, inversion, 151
Symmetry properties, 133

Tamm states, 359
Tantalum pentoxide, 424
Temperature
  DPDI processes and, 309
  defects and, 174, 181, 194, 344
  diffusion and, 95
  dislocation loops and, 467
  in defects, 188
  lifetime and, 294
  role of, 15, 18, 20
Tetragonal defects, 40
Tewary method, 49
Thermal equilibrium,
    concentrations of defects
    in, 2
Titanium oxide, intrinsic
    adatoms in, 363
Transition metal compounds, 284
Transition metal ions
  point defects and small
      clusters, 120
  spectra from, 115
Trigonal defects, 40, 41
Two wave approximation, 408

Ultramicroscopy, 118
Uniaxial stress, 149, 186, 282
Unstrained inclusions, 457

V_1 centre
    See $H_A(Na)$ centre
V_2 centres, 225, 228
V_k centres, 174
  diffusion, 180
  electronic structure, 176
  ESR spectrum, 175
  optical transitions, 178
  thermal reorientation, 180
Vacancies, 23
  concentration and doping, 10
  formation, enthalpies of, 58
  in crystals with impurities, 168
  in oxide glasses, 323
  in sub-lattices, 5
  mechanism, 96, 97
  mobility, 208
  structure of, 56
Vacancy aggregate centres, 210
  in ionic crystals, 236

Vacancy-impurity pairs, 208, 209
Vacancy-interstitial ion pairs, 395
Valence crystals, intrinsic defects, 363
Vibrational absorption, 230
Vibrational states of defects, 145, 248
Vibronic coupling, 140

Waves in periodic structures, 405

X-rays
  structural information and defect energies studied by, 379
  two wave approximation, 408
X-ray scattering, 379
  asymptotic distortion, 389
  diffuse, 383, 384
    far away from Bragg peaks, 389
    from defect clusters, 400
  distortion, 385
  from crystal lattices, 381

X-ray scattering (cont'd)
  from defect free crystal lattice, 381
  from defective crystal lattices, 382
  Huang, 385
    from point defects, 397
  Laue, 385
  small angle, 390, 401
X-ray studies, 405
  theory of, 380
X-ray topograhy, 422

Z centres, magnetic resonance studies of, 169
Zeeman splitting, 288
Zero-phonon lines, 149, 237, 243, 264
  effect of electric fields on, 245
Zero-phonon transition, 139
Zinc compounds
  EPR studies, 214
  interstitials, 215
  low temperature damage mechanism, 216
  metal vacancy, 214
Zinc oxide, intrinsic adatoms in, 363